THE CHARACTER AND ORIGINS OF SMOG AEROSOLS

EDITORIAL COMMITTEE

THE CHARACTER AND ORIGINS OF SMOG AEROSOLS

A Digest of Results from the California Aerosol Characterization Experiment (ACHEX)

Edited by

GEORGE M. HIDY and PETER K. MUELLER
Environmental Research & Technology, Inc.
Westlake, California

DANIEL GROSJEAN
Statewide Air Pollution Research Center
University of California, Riverside

BRUCE R. APPEL and JEROME J. WESOLOWSKI
Air and Industrial Hygiene Laboratory
California Department of Public Health, Berkeley

A Wiley-Interscience Publication

JOHN WILEY & SONS
New York / Chichester / Brisbane / Toronto

Library of Congress Cataloging in Publication Data

Main entry under title:
The character and origins of smog aerosols.

(Advances in environmental science and technology;
■ ■)
"A Wiley-Interscience publication."
Includes indexes.
1. Smog—Addresses, essays, lectures. 2. Aerosols—Addresses, essays, lectures. 3. Air—Pollution—Addresses, essays, lectures. I. Hidy, George M.
II. Series.
TD180.A38 ■. ■ [TD884.3] 628'.08s [628.5'3'2] Vol. 9
ISBN 0-471-04899-2 79-4585

Printed in the United States of America

10 9 8 7 6 5 4 3 2 1

Acknowledgment

The editors are deeply indebted to Mrs. Dolores Tanno and Mrs. Glenna Paschal for their painstaking effort in the final typing of this volume of the Series.

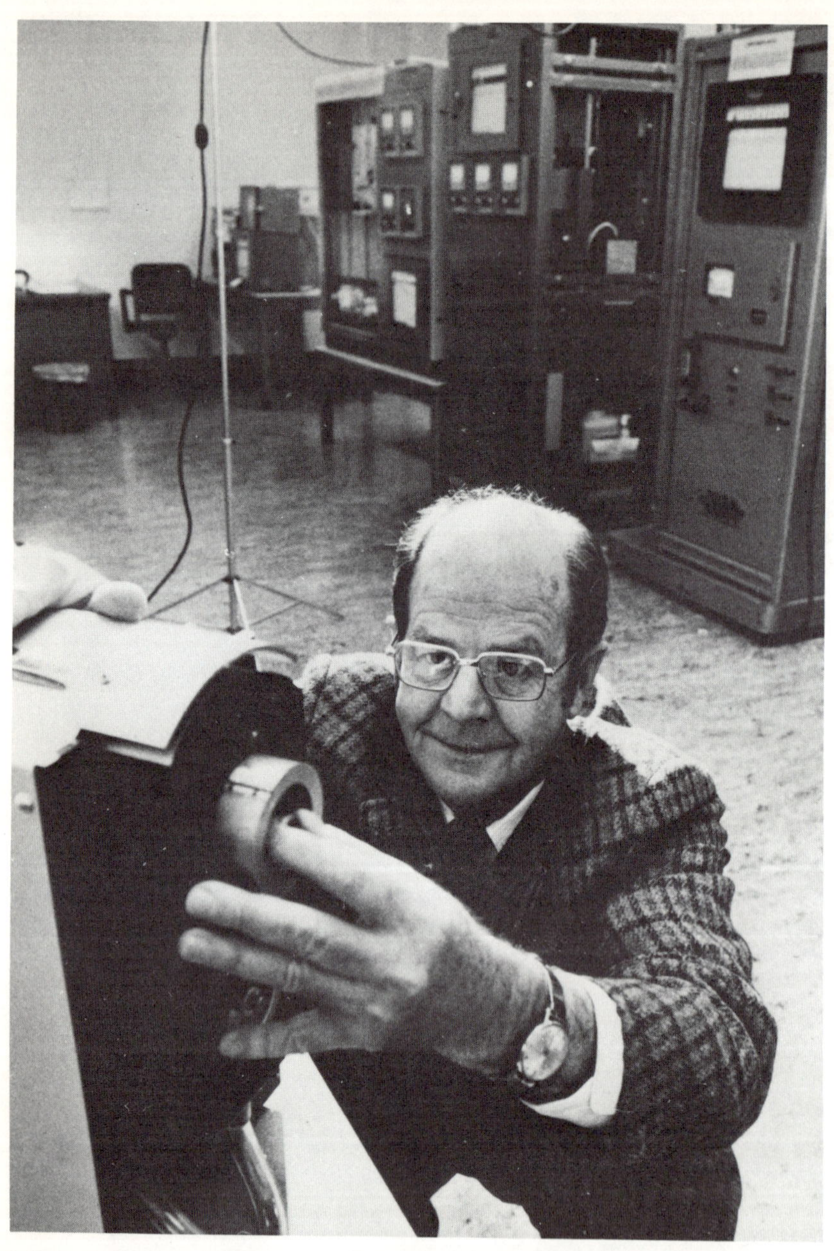

Dale Hutchinson 1900–1973

Dale Hutchinson

To initiate and maintain a multimillion dollar experiment like the Aerosol Characterization Experiment (ACHEX), a willing and dedicated sponsor is required. The ACHEX design was conceived by the coinvestigators and was implemented through funds of the California Air Resources Board (ARB) to take substantial steps in finding a scientific basis for lessening the visibility degradation accompanying photochemical smog. Several members of the ARB's research advisory committee, including Dr. Arie Haagen-Smit and Mr. Robert Brattain, experienced deep concern for the visibility problem in California and made a strong commitment to the ACHEX on behalf of the state's interests.

The ACHEX and its related studies were a major component of the ARB research program, which was greatly enhanced in 1970 as a result of funds committed by the California State Legislature. The ARB had the good fortune to retain Mr. Dale Hutchinson from Stanford Research Institute to manage its new research projects. As a part of his activities, he provided the ACHEX staff with substantial guidance to achieve the goals of the study. Dale Hutchinson's guidance of this program terminated with his sudden death in the late summer of 1973. This was a severe loss to the management of ARB research programs since Dale Hutchinson had spent much effort in organizing and coordinating many pioneering studies initiated by the Steering Committee. Despite early limitations of success in data collection resulting from anomalies in weather, the ACHEX program achieved essentially all of its research goals.

It is appropriate that Dale Hutchinson be recognized for his role in maintaining ACHEX support through an added season of investigators of the project. For those who knew Dale, the photograph on the facing page will serve as a reminder of his energies and dedication in guiding the ACHEX and other ARB-sponsored research to useful and timely results in support of the development of the California regulatory policy for air pollution abatement.

Introduction to the Series

Advances in Environmental Science and Technology is a series of multiauthored books devoted to the study of the quality of the environment and to the technology of its conservation. Environmental sciences relate, therefore, to the chemical, physical, and biological changes in the environment through contamination or modification; to the physical nature and biological behavior of air, water, soil, food, and waste as they are affected by man's agricultural, industrial, and social activities; and to the application of science and technology to the control and improvement of environmental quality.

The deterioration of environmental quality, which began when man first assembled into villages and utilized fire, has existed as a serious problem since the industrial revolution. In the second half of the twentieth century, under the ever-increasing impacts of exponentially growing population and of industrializing society, environmental contamination of air, water, soil, and food has become a threat to the continued existence of many plant and animal communities of the ecosystem and may ultimately threaten the very survival of the human race.

It seems clear that if we are to preserve for future generations some semblance of the existing biological order and if we hope to improve on the deteriorating standards of urban public health, environmental sciences and technology must quickly come to play a dominant role in designing our social and industrial structure for tomorrow. Scientifically rigorous criteria of environmental quality must be developed and, based in part on these, realistic standards must be established, so that our technological progress can be tailored to meet such standards. Civilization will continue to require increasing amounts of fuel, transportation, industrial chemicals, fertilizers, pesticides, and countless other products, as well as to produce waste products of all descriptions. What is urgently needed is a total systems approach to modern civilization through which the pooled talents of scientists and engineers, in cooperation with social scientists and the medical profession, can be focused on the development of order and equilibrium among the presently disparate segments of the human environment. Most of the skills and tools that are needed already exist. Surely a technology that has created manifold environmental problems is also capable of solving them. It is our hope that the series in Environmental

Science and Technology will not only serve to make this challenge more explicit to the established professional but will also help to stimulate the student toward the career opportunities in this vital area.

The chapters in this series of Advances are written by experts in their respective disciplines, who also are involved with the broad scope of environmental science. As editors, we asked the authors to give their "points of view" on key questions; we were not concerned simply with literature surveys. They have responded in a gratifying manner with thoughtful and challenging statements on critical environmental problems.

JAMES N. PITTS, JR., Editor
Statewide Air Pollution Research Center and Department of Chemistry
University of California
Riverside, CA 92521
Telephone: (714) 787-4584

ROBERT L. METCALF, Editor
Environmental Studies Institute
Departments of Biology and Entomology
University of Illinois
Urbana-Champaign, IL 61801
Telephone: (217) 333-3649

DANIEL GROSJEAN, Associate Editor
Environmental Research and Technology, Inc.
2625 Townsgate Road
Westlake Village, CA 91361
Telephone: (805) 497-0821

Preface

This volume contains a compendium of research papers associated with the California Aerosol Characterization Experiment (ACHEX). This study was a major field program sponsored by the California Air Resources Board. The objectives of the study focused on the description of the physical and chemical nature of smog aerosols, and their relationship to visibility reduction in major California air basins. In coordination with the ACHEX, the California Air Resources Board also sponsored a three-dimensional pollutant gradient study to document the three-dimensional distribution and transport of various pollutants during the ACHEX. Papers from the gradient study are included herein to provide a meteorological perspective for the ACHEX results.

The purpose of this volume is to provide a reasonably complete collection of important scientific papers derived from the ACHEX and the gradient study. From this compendium the investigators hope to provide a picture of the design strategy, the analytical methods development, and the results of this major program. This survey should be helpful to graduate students of the environmental sciences and engineering, especially air chemistry, and to the knowledgeable scientist and expert who wish to analyze the ACHEX as a basis for future studies. The contents of the book should be of assistance also in documenting a number of new and unique methods for detailed characterization of atmospheric aerosols and their sources. The limitations and advantages of the study approach using pollution episode experiments can be identified by careful reading of this work and other contract reports derived from the project.

For readers who wish to obtain more details, the results of the ACHEX were reported in 1975 in four major volumes:

1. Hidy, G. M. et al. "Characterization of Aerosols in California," Vol. I: "Summary," Final Report, Contract No. 358, Science Center, Rockwell International, Thousand Oaks, Calif. (NTIS No. PB-271733).

2. Hidy, G. M. et al. "Characterization of Aerosols in California," Vol. II: "Experimental Methods and Analytical Techniques," Final Report, Contract No.

358, Science Center, Rockwell International, Thousand Oaks, Calif. (NTIS No. PB-248799).

3. Hidy, G. M. et al., "Characterization of Aerosols in California," Vol. III: (Parts I and II) Observational Data," Final Report, Contract No. 358, Science Center, Rockwell International, Thousand Oaks, Calif. (NTIS Nos. PB-265714 and PB-266613).

4. Hidy, G. M. et al., "Characterization of Aerosols in California," Vol. IV: "Analysis and Interpretation of Data," Final Report, Contract No. 358, Science Center, Rockwell International, Thousand Oaks, Calif. (NTIS No. PB-247947).

The results of the gradient study were published in three volumes:

5. Blumenthal, D. L. et al. 1974. "Three-Dimensional Pollutant Gradient Study: 1972-1973 Program," MRI Final Report to the California Air Resources Board; Report No. MRI 74 FR-1262. Agreement Nos. ARB-631 and ARB2-1245 (NTIS PB 241982/LL).

6. Ensor, D. S. et al. 1973. "Three-Dimensional Pollutant Gradient Study: 1972 Program, Data Summary," Report No. MRI 73 FR-1082, prepared for the California Air Resources Board under Agreement No. ARB-631.

7. Blumenthal, D. L. et al. 1974. "Three-Dimensional Pollutant Gradient Study: 1973 Program, Data Summary," MRI unnumbered report, prepared for the California Air Resources Board under Agreement No. ARB-2-1245.

The data from the ACHEX are also available on computer magnetic tape from the Air Resources Board archives.

The contents of this book include:

(1) a brief description of the experiment in the light of previous work;

(2) a survey of sampling and analysis methods;

(3) selected reports of the findings on the characterization of particulate physical and chemical behavior;

(4) selected analyses and interpretations of the observations as related to the study objectives; and

(5) an epilogue discussing in retrospect the limitations and impact of the experiment on subsequent work.

The ACHEX involved more than 100 scientists from industry, government, and the university community. Initially the co-principal investigators included the following:

R. J. Charlson
University of Washington
Seattle, Washington

S. K. Friedlander
California Institute of Technology
Pasadena, California

G. M. Hidy* (Principal Investigator)
Science Center, Rockwell International
Thousand Oaks, California

P. K. Mueller*
Air and Industrial Hygiene Laboratory
California State Department of Public Health
Berkeley, California

T. B. Smith
Meteorology Research, Inc.
Altadena, California

J. J. Wesolowski
Air and Industrial Hygiene Laboratory
California State Department of Public Health
Berkeley, California

K. T. Whitby
University of Minnesota
Minneapolis, Minnesota

---

*Presently with Environmental Research and Technology, Inc. Westlake Village, CA.

Later in the program, the following persons participated as co-investigators:

B. Appel
Air and Industrial Hygiene Laboratory
California State Department of Public Health
Berkeley, California

W. Richards
Science Center, Rockwell International
Thousand Oaks, California

In addition, there were these co-investigators of the three-dimensional gradient study:

D. Blumenthal (Principal Investigator)
T. B. Smith
J. R. Stinson
Meteorology Research, Inc.
Altadena, California

The key scientific staff, by participating institution, were the following:

SCIENCE CENTER, ROCKWELL INTERNATIONAL

Technical	Administrative
C. A. Battle	K. Bader
C. S. Burton	J. L. Balderston
R. S. Carpenter	C. Coffindaffer
J. H. Davis	J. F. Davis
D. Dockweiler	W. Dinkin
P. Friedman	N. R. Johnson
J. R. Gardner	A. J. Lewin
R. M. Govan	T. Loucks
D. H. Hern	M. E. Lyon
J. R. Hribar	M. McGuire
K. Kubler	R. A. Mekaelian
C. Landon	K. L. Nolte
G. Lauer	P. K. Ocorr
R. A. Meyer	D. E. Palermo
R. L. Myers	R. L. Seale
E. P. Parry	H. Shannon
H. H. Wang	D. B. Stephens
	B. B. White

AIR AND INDUSTRIAL HYGIENE LABORATORY

A. E. Alcocer
T. Belsky
C. Bunce
P. Bussard
T. Cahill (University of California, Davis)
F. Jung
R. Kaifer (Lawrence Livermore Laboratory)
S. Kerns
D. Levaggi (BAAPCD)
J. Medaxian
H. Moore
D. Mosley
K. Noll (University of Tennessee)
C. Ong (BAAPCD)
L. Parton
L. Pierce
M. Fracchia
M. Hoffer
S. Hope
M. Imada
E. Jeung
A. Ragaini (Lawrence Livermore Laboratory)
A. del Rosario
W. Siu (BAAPCD)
K. Smith
R. Stanley
J. Tang
S. Twiss
S. Wall
B. Wright
F. Yoon

CALIFORNIA INSTITUTE OF TECHNOLOGY

P. Andriola
C. Davidson
G. Gartrell
D. Grosjean
S. L. Heisler
R. Husar
R. Lamb
D. Landis
Pamela McMurry
Peter McMurry
P. Roberts
J. Seinfeld
W. H. White

METEOROLOGY RESEARCH, INC.

D. Ensor
P. S. McMurry
S. Muller
A. Ostrye
S. Marsh
V. Mirabella
A. Ollivares
T. Sperling

UNIVERSITY OF CALIFORNIA, RIVERSIDE

T. Cree
G. Kats
L. M. Kienitz
M. Price
E. R. Stephens
R. Tenorio
C. R. Thompson

LAWRENCE LIVERMORE LABORATORY, BERKELEY

R. Garrett
R. Giauque
L. Goda
T. Novakov

UNIVERSITY OF MINNESOTA

B. Cantrell
B. Y. H. Liu
V. A. Marple
D. Y. H. Pui
W. Clark
G. Sem (Thermo-Systems, Inc.
G. M. Sverdrup
K. Willeke

FORD MOTOR COMPANY RESEARCH LABORATORIES

W. Pierson
H. Hammerle

GENERAL MOTORS RESEARCH LABORATORIES

P. Groblicki

U.S. ENVIRONMENTAL PROTECTION AGENCY

P. Hanst
W. Wilson
R. Patterson

U.S. NAVAL WEAPONS CENTER, CHINA LAKE

P. Davis
P. Owens
R. Kelso

UNIVERSITY OF WASHINGTON

N. Ahlquist
D. Covert
D. Cronn
P. Solem
A. Vanderpol
A. Waggoner

We are indebted to J. Suder of the California Air Resources Board (ARB) for his continued interest in this program as the ARB project officer. After Dale Hutchinson's death, Dr. Suder provided the stimulation and encouragement from the State of California necessary to assure the successful completion of the ACHEX. We are also grateful to H. Samuels, the first project officer of the ARB, who coordinated the ACHEX with the other ARB projects and acted, on many occasions, for Dale Hutchinson as representative of the ARB.

Finally, we express our appreciation to A. Bockian and J. Holmes of the ARB, who have worked closely with the ACHEX investigators to bring the achievements of the project into the regulatory practice of the State of California.

George M. Hidy
Peter K. Mueller

Westlake, California
April 1979

Contents

PART II. CHEMICAL CHARACTERIZATION

PART III. PHYSICAL PROPERTIES

PART IV. AEROMETRIC AND SOURCE FACTORS

THE CHARACTER AND ORIGINS OF SMOG AEROSOLS

Executive Summary

GEORGE M. HIDY AND PETER K. MUELLER
Environmental Research and Technology, Inc.
Westlake Village, California

I. INTRODUCTION

Among the more pressing questions of air pollution requiring improved knowledge are those concerning clouds of airborne particles, often called aerosols. Although these clouds make up only a small fraction of the trace constituent loading in air, they are a potential hazard to health. They also are the main contributor to impairment of visibility and a possible agent for weather modification. Despite a long history of investigation the origins and evolution of atmospheric aerosols remain poorly understood quantitatively compared with the behavior of many common trace gases. Recognizing this fact, the California Air Resources Board (ARB) sponsored a major experiment in the chemistry of aerosols in urban and nonurban sites of California.* The program,

*The work, called the Aerosol Characterization Experiment (ACHEX), was sponsored from funding allocated under Legislative Bill No. 848, State of California.

initiated in the fall of 1971, was extended in the spring of 1973. The principal results and details of methods were reported in 1975 via a series of four reports to the ARB. The project was also summarized in a 1975 article by Hidy et al.*

The experiment required the expertise of scientists and engineers from a number of institutions, including industry, the university community, and government agencies. The program was managed by the Science Center, Rockwell International; it involved directly the Air and Industrial Hygiene Laboratory of the California State Department of Health (AIHL), the California Institute of Technology, Meteorology Research, Inc., the Statewide Air Pollution Research Center (SAPRC) of the University of California, Riverside, the University of Minnesota, Thermo-Systems, Inc., the Lawrence Berkeley Laboratory, the Lawrence Livermore Laboratory, the University of Washington, the ARB El Monte Laboratories, as well as several other organizations.

The first phase of activity was devoted to the development of observational systems for the project and to the acquisition of a variety of data characterizing aerosols sampled primarily in late summer and early fall of 1972. The second phase, initiated in 1973, was a continuation of observations and analysis, with primary emphasis on photochemical smog aerosols in the South Coast Air Basin, which includes the Los Angeles area.

A. Objectives

The objectives of the Aerosol Characterization Experiment (ACHEX) are summarized as follows:

(1) To characterize aerosols in urban and nonurban air, primarily in the South Coast Air Basin and surroundings, in terms of their physical and chemical properties and their interactions with atmospheric gases.

(2) To investigate the origins and evolution of aerosols, mainly under conditions of photochemical smog formation, to improve knowledge about the secondary aerosol formation processes.

(3) To estimate the primary and secondary aerosol contributions to the total airborne particulate concentration in the South Coast Air Basin and in other California urban and nonurban locations.

*Hidy, G. M., et al. 1975. Summary of the California Aerosol Characterization Experiment, J. Air Pollut. Control Assoc. 25, 1106-1114.

(4) To evaluate the significance of anthropogenic sources and uncontrollables, including meteorological factors, natural sources, and water, in the evolution of the respirable* and visibility-degrading aerosol.

(5) To recommend from the characterization of aerosols and their sources the key ingredients for a control strategy in the South Coast Air Basin.

B. Structure of the ACHEX Volume

The ACHEX yielded a variety of contributions to aspects of aerosol chemistry. The results of the investigations can be identified with the program objectives, and can be categorized under four headings: (a) Analytical Methods and Sampling Technology, (b) Chemical Characterization of Airborne Particles, and (c) Physical Properties of Smog Aerosol Particles, and (d) Aerometric and Source Factors. This volume contains a collection of scientific papers written by the investigators concerning the four categories listed.

The reports cover a number of areas of measurement technology and data analysis which have provided incentives for continuing work in California, as well as other locations. Historically, it is difficult to identify specifically the impact of the ACHEX on later research. However, the investigators involved have made more recent contributions to the elucidation of visibility and atmospheric aerosol behavior in the United States which can be traced directly to this pioneering experiment. The final section of the volume includes a brief summary of the impact of the ACHEX on investigations that followed it.

To obtain a picture of pertinent activities surrounding the ACHEX, some papers have been included in this volume that describe aspects of work done at the same time in California, which enhanced our general knowledge of air pollution, especially in the South Coast Air Basin. Included in these related investigations were related studies of the ARB, such as the three-dimensional (3-D) gradient experiment, as well as experiments of the U.S. Environmental Protection Agency (EPA) and the Ford Motor Company's research laboratories.

---

*The respirable fraction consists of the airborne particles less than approximately 3 μm in diameter.

II. EXPERIMENTAL DESIGN AND OPERATIONAL METHODS

A. Approach to Objectives

The ACHEX was designed to achieve its objectives through intensive case studies involving detailed episodic measurements of aerosol behavior in the presence of trace gases and meteorological changes.

Generally, each observational interval in the field experiment consisted of a 24-hr period. The periods during which measurements would be made were selected on the basis of an aerometric forecast 1 day ahead of start-up, the principal criteria being conditions conducive to high levels of air pollution. In the South Coast Air Basin, runs were planned for oxidant levels exceeding 0.20 ppm, except for source-dominated cases.

The various measurables for the experiment are listed in Table 1. The ACHEX represents a first attempt to obtain observations of all these parameters simultaneously with a time resolution ranging from 10 min to 2 hr. Therefore, a significant amount of instrument and chemical method development was required in the ACHEX to achieve the needed measurement capability. Several important operational achievements are noted in Table 1.

B. Structure of Field Program

The intensive acquisition of data took place during a 2-year period from 1972 to 1973, with most of the activity concentrated in the late summer and early fall to cover times of highest photochemical smog intensity as measured by elevated oxidant concentrations. The focal point of the measurement program was a mobile laboratory equipped with a variety of instrumentation and designed and built especially for the ACHEX. The mobile unit moved to several different sites during the course of the field studies. These locations are listed in Table 2, along with a short description of each site. Sites were selected to provide key information on (a) source-dominated air pollutants, (b) receptor cases where pollutants were mixed and had reacted for a period of time, or (c) nonurban locations potentially characteristic of marine, desert, or photochemical background contamination.

The mobile laboratory operations were supplemented by fixed stations. In the first phase (1972) these stations were at the Keck Laboratories at Caltech, the SAPRC monitoring station at Riverside, and the Bay Area Air Pollution Control District (BAAPCD) station in San Jose. These sites were equipped with an array of instrumentation similar to that of the mobile laboratory, with the principal exceptions of the aerosol mobility analyzer and the large-particle optical counter. During the 1973 expedition the fixed stations were replaced in the South Coast Air Basin by a modification of four air monitoring stations: the Los Angeles Air Pollution Control District (LAAPCD) stations at downtown Los Angeles and Pomona, the ARB station in Riverside, and the Orange County APCD station in Anaheim. The gas monitoring data at these sites were complemented by sequential filter sampling every 2 hr for inorganic and organic aerosol analysis and by integrating nephelometer measurements.

During the two observational seasons, 1972 and 1973, more than 5 million items of data were collected and reviewed as part of the experiment.

The "core" ACHEX effort was complemented by a number of special studies in 1972 and 1973, including (a) ARB/MRI-Navy aircraft sampling experiments over the San Francisco Bay area and southern California, (b) a series of tracer experiments using fluorescent particles, (c) long-path infrared spectrometer measurements by the EPA, (d) gas and particle sampling by General Motors and Ford research groups, (e) lidar sensing by NASA Langley Center staff and investigations of airborne nitrogen oxides by the Battelle Columbus Laboratories. The ACHEX experiments in late summer of 1973 overlapped the period of a second major experiment, the Los Angeles Reactive Pollutant Program (LARPP), sponsored by the Coordinating Research Council, the EPA, and the ARB.

C. Aerosol Chemical Analysis

A key element of the ACHEX was the extensive chemical analysis undertaken on selected aerosol samples during the course of the experiment. The analyses attempted are listed in Table 3. Samples for study were chosen on the basis of the intensity of air pollution observed or on the basis of information required to meet the program objectives.

Many of the chemical methods used in the ACHEX required levels of sensitivity that exceeded those of procedures routine in the past. Therefore considerable development of methods was undertaken, including neutron activation analysis (NAA), x-ray fluorescence analysis (XRFA), photoelectron spectroscopy (ESCA), high-resolution mass spectroscopy (HRMS), and atomic absorption (AA), as well as more classical techniques covering water-soluble sulfate and nitrate, liquid water content, and noncarbonate carbon. Extensions of existing methods were also required for gas-phase chromatographic determination of organic and sulfur compounds, and for ammonia and oxygenated vapors.

To ensure the quality of the aerosol chemistry results, extensive method intercomparison studies and interlaboratory comparisons were made. Such investigations aided in improving some of the analytical procedures and verified the limitations of aerosol sampling and analytical procedures.

As a result of such studies, XRFA can be considered to be a useful operational method for analysis of sulfur and heavier elements in aerosol samples taken over 2-hr periods in urban air. These studies also reconfirmed previous efforts which had determined that NAA is a powerful method for obtaining concentrations of many elements on 2-hr samples. On the other hand, limitations remain in the application of ESCA, HRMS, and all organic analyses attempted for quantitative studies of aerosol chemistry.

D. Data Management and Processing

With the relatively large number of data accumulated, systematic procedures for their management had to be implemented. Many of the data were obtained on the computer acquisition systems and used only after review and validation. The routine chemical analyses were cataloged and listed to form a complementary bank. The continuous data are available as 10-min averaged information for selected cases and as 2-hr averages that are directly comparable with aerosol chemical analyses. Various specialized data obtained from the experiments were not amenable to cataloging for computerized output. Such information is accessible in the data bank, which is available from the California ARB. The general contents of the ACHEX data bank are listed in Table 4.

Even with advanced planning the processing and review of the data recorded in the ACHEX proved to be a major effort during the course of data bank development. For economy in processing, one of the minicomputers was expanded in capability with a tape recorder, a processing disk, and a line printer. The use of the machine required unique software development, and its operation for processing the data proved to be too slow for efficiency of operation. The development of the computerized data bank proved to be the greatest single limitation on completion of the analysis and interpretation of results. In future programs of this type, every effort should be made to develop systematically the data processing software in the earliest stages. Ample resources should be set aside for data manipulation on large, high-capacity computers.

The ACHEX data have been reviewed at several levels by the investigators of the project. The computerized data have been examined on a priority basis, treating most carefully these episodes that were selected for detailed chemical analysis. In general, it is believed that the results from the mobile laboratory are of the highest quality for quantitative analysis, and those from the fixed satellite stations, somewhat lower in reliability. The reason for this distinction is the priority on the mobile laboratory observations for data validation and the lower degree of quality control achieved in the fixed station opertions.

III. SUMMARY OF IMPORTANT OPERATIONAL RESULTS

At the time of this study investigations were designed to use the newest techniques of measurement to characterize the aerosols in considerable detail. Several methods involving the determination of physical and chemical properties were developed and implemented during the course of the ACHEX. Some important operational "firsts" included the following:

(1) A complete, transportable aerometric laboratory was designed, built, and operated for simultaneous observation of pollutant gases, aerosols, and meteorological variables.

(2) A minicomputer system was designed and built for continuous data acquisition and on-line surveillance of both gas and aerosol particle analyzers.

(3) Detailed data were taken over 24-hr periods for aerosols on a time resolution of 2-hr averages, which are compatible with significant gas chemistry changes.

(4) Aerosol chemical composition was characterized by using combined wet chemical and physical methods, including photoelectron spectroscopy, x-ray fluorescence analysis, neutron activation analysis, and high-resolution mass spectroscopy.

(5) More than 5 million bits of data were recorded and examined on the computerized acquisition system, and more than 30,000 chemical determinations were made on the aerosol samples.

IV. MAJOR CONCLUSIONS AND RECOMMENDATIONS

(1) The mass concentration of aerosol particles in California air varies widely during the day and is heavily influenced by the presence of sulfate, nitrate, and noncarbonate or organic carbon. These constituents are identified as products of atmospheric chemical reactions of pollutant gases SO_2, NO_x, and nonmethane hydrocarbon vapors.

(2) The population in eastern parts of the Los Angeles area is exposed in the extreme case to 2-hr average total mass concentrations of more than 450 $\mu g/m^3$, with (water-soluble) sulfate concentrations of 70 $\mu g/m^3$ or more, (water-soluble) nitrate concentrations of 240 $\mu g/m^3$ or more, and noncarbonate carbon concentrations of more than 50 $\mu g/m^3$. These maximal concentrations are substantially higher than observed 24-hr averaged concentrations.

(3) Measurements confirmed the (multimodal nature of the aerosol volume) size distribution. Typically, the aerosol was characterized by a bimodal distribution, with the fraction less than approximately 3 μm in diameter dominated by anthropogenic emissions and particles from atmospheric chemical reactions. The supermicron fraction greater than 3 μm was dominated by natural or quasi-natural sources such as sea salt or windblown or reentrained soil dust. The chemical composition of the modes was consistent with the source identification.

(4) Although sulfates, nitrates, and organics differ somewhat in their distribution in particle size, they fall within the submicron range, which is most significant for visibility degradation and hazards related to respiratory disease. On a unit mass basis, sulfate influences light scattering more strongly than do nitrates and organics. However, the influence of nitrates varies strongly with relative humidity.

(5) The aerosol growth accompanying the evolution of photochemical smog was found consistently to be concentrated in the diameter range of 0.1 to 1.0 μm, which is the particle size range most effective in reducing visibility.

(6) Evidence was found suggesting the contribution of photochemically related reactions to the production of particulate sulfate, nitrate, and organics that signficantly enhances the formation of these constituents. However, interpretation of the measurements also can involve nonphotochemical processes, including heterogeneous reactions in suspended particles.

(7) The natural or quasi-natural trace gases, ammonia and water vapor, have an important influence on the evolution of aerosol in the presence of photochemical processes. The sulfate and nitrate anions can be accounted

for as ammonium salts, with a minor fraction as sodium salts. Direct measurements of aerosol liquid water content ranged from less than 10% to more than 50% by weight, depending on humidity.

(8) The aerosol particles in Los Angeles air are identified with different source categories, including secondary material, primary emissions, and natural or quasi-natural background. Arguments were given to link sulfate mainly with stationary sources using fuel oil. Organics were linked with transportation sources using gasoline, while nitrates probably come from both stationary and transportation sources.

(9) More than half the aerosol sampled in Los Angeles can be accounted for by atmospheric chemical reactions. The remaining half can be divided roughly equally between material of primary origin from stationary or transportation sources, and background material such as soil dust or sea salt.

(10) Analysis of the 1973 results indicated that visibility degradation from secondary aerosol production in the western part of the Los Angeles area stems from SO_2 pollution from stationary sources. The visibility reduction over the central area is identified with roughly equal contributions from transportation (NO_x and organics) and stationary sources (SO_2). Evaluation of existing emission inventories and the ACHEX results indicates that the marked degradation of visibility on the eastern side of the basin is heavily influenced by transportation sources. However, particulate pollution from other sources, including primary emissions, cannot be disregarded as a signficant factor in visibility degradation throughout the basin.

(11) As a recommended control strategy, consideration must be given to a continuing program to reduce aerosol precursor gases, including SO_2, NO_x, and nonmethane hydrocarbons. Control of the high-molecular-weight reactive hydrocarbon vapors, including olefins, diolefins, and cyclic olefins with more than six carbon atoms per molecule, is particularly important. Control of SO_2 emissions is a high-priority consideration from the view of its projected influence on visibility degradation. Since primary particulate matter influences ambient aerosol levels in California air, strategy must continue to include control of such emissions, especially in the submicron range, 0.1- to 1.0-μm diameter.

(12) Three types of monitoring and surveillance for aerosol behavior were proposed for a second-generation system referencing the present high-volume filter sampling network: (a) continuation of the high-volume filter sampling program, with determination of water-soluble sulfate and nitrate, and total carbon concentrations; (b) surveillance of sensitive receptor and source-dominated areas for particle size-composition distributions of aerosols; and (c) limited special field-episodal studies aimed at extending the data derived from the ACHEX base.

(13) Further studies of the nitrogen oxide chemistry and the hydrocarbon vapor precursors are required to provide a fully defensible control strategy for nitrate and organic aerosols.

(14) Studies on the health effects of aerosols should be guided in the choice of materials by the extensive particle size and chemical composition data obtained in experiments like the ACHEX and other programs. To identify potential health hazards from specific components, considerable effort is required to refine present techniques for the chemical analysis of particulate material.

TABLE 1. Observables and Instrumentation

Instrument	Parameter Measured	Manufacturer	Model No.	Remarks
Aerosols				
Condensation Nuclei counter	Total aerosol number concentration	Environment One	Rich 100	
Optical counter	Aerosol size distribution (0.3 to 5-μm-diameter range)	Royco	220	First attempt to continuously measure aerosol distribution in range from 7.5-Å to 30-μm diameter.
Optical counter	Aerosol size distribution (5- to 30-μm-diameter range)	Royco	245	
Electrical mobility analyzer	Aerosol size distribution (0.0075- to 0.4-μm-diameter range)	Thermo-Systems, Inc.		
Beta-gauge mass monitor	Total mass concentration	Frieseke-Hoepfner	FH62A	First operational test of, and attenuation measurements for, ambient aerosol.
Rotating drum impactor	Mass distribution and chemical sampling			
High-volume sampler	Mass distribution and chemical sampling			
Integrating nephelmometer	Total aerosol light scattering	Meteorology Research, Inc.	1550	
Filter system	Liquid water content	Rockwell		First field observations of water in filter-collected aerosols
Waterometer	Dielectric constant - liquid water content	Rockwell		First semicontinuous observation of liquid water content in aerosols.
Surface film collector		Rockwell		
Dustfall collector	Particulate fallout	Weather Measure Corp.		
Two-stage high-volume sampler	Organics	Pacific Environmental		

TABLE 1. Observables and Instrumentation (continued)

Instrument	Parameter Measured	Manufacturer	Model No.	Remarks
Meteorology				
Meteorological measurement system	Wind speed and direction	Meteorology Research, Inc.	1074	First extensive measurements of atmospheric aerosols made simultaneously with trace gas observations and meteorological variables.
Thermometers	Temperature	Rosemount	4128	
Spectral pyranometer	Total radiation	Eppley Laboratory, Inc.	2(1154 F3)	
Ultraviolet radiometer	Ultraviolet radiation	Eppley Laboratory, Inc.	2(11692)	
Dew point hygrometer	Relative humidity	Cambridge	880	
Rain gauge	Rainfall	Weather Measure Corp.	P561	
Gaseous Pollutants				
Chromatograph	Total HC, CH_4, NMHC, C_2H_4, C_2H_2, and Carbon Monoxide	Beckman	6800	First use of automated chromatographic air monitoring in California
$NO-NO_2$ analyzer (Chemiluminescence)	Nitric oxide, nitrogen dioxide	Bendix	B101B	First use of chemiluminescent analyzers in surveillance in California
SO_2 analyzer (GC-flame photometric)	Sulfur dioxide, hydrogen sulfide	Varian	1440	
Sequential NH_3 analyzer	Ammonia	Gelman	Sequential bubbler	First attempt to relate changes in NH_3 to aerosol behavior
Ozone analyzer (Chemiluminescent)	Ozone	REM	612	
Data Acquisition				
Computer		Digital Equipment Corporation interfaced and modified by Rockwell	PDP-8/E	First use of minicomputer data acquisition for air characterization with aerosol instrumentation; first on-line computerized capability.

TABLE 2. Sites Selected for the Mobile Laboratory

Occupancy date	Site	Location	Environment
A. 1972			
July 15-30	Berkeley	College of Agriculture tract 2175 Hearst Avenue Berkeley, Calif.	Acceptance tests and preliminary checkout
Aug. 1-13	Richmond	San Pablo Water Pollution Control Plant 2377 Garden Tract Road San Pablo, Calif.	Urban-Industrial (downwind from chemical complex)
Aug. 15-27	San Francisco Airport	On the airport property at Bayshore Drive and end of north-south runway	Aircraft-enriched aerosol; receptor of aerosol from west San Francisco Bay area
Aug. 28-Sept. 11	Fresno	Fresno County Fairgrounds 1121 Chance Avenue Fresno, Calif.	Photochemical-agricultural
Sept. 12-18	Hunter-Liggett Military Reservation	Meadow areas, 1 mile south of the public road connecting Hunter-Liggett with Route 1	Vegetation-enriched (expected photochemical aerosol production from vegetation organic emissions)
Sept. 19-Oct. 2	Freeway Loop	May Company parking lot, 33rd Street and Hope Street, Los Angeles, Calif.	Auto-enriched (100 ft. east of Harbor Freeway)
Oct. 3-30	Pomona	Los Angeles County Fairgrounds Pomona, Calif.	Los Angeles photochemical (receptor)
Oct. 30-Nov. 4	Goldstone	Goldstone Tracking Station Barstow, Calif.	Desert background
Nov. 6-10	Point Arguello	U.S. Coast Guard LORAN Station	Marine-enriched
B. 1973			
July 11-Aug. 9	West Covina	Hospital on Sunset Avenue West Covina, Calif.	Los Angeles photochemical (receptor)
Aug. 11-25	Pomona	Los Angeles County Fairgrounds Pomona, Calif.	Los Angeles photochemical (receptor)
Aug. 27-Sept. 26	Rubidoux (Riverside)	Rubidoux Water Treatment Plant Rubidoux, Calif.	Los Angeles photochemical (receptor)
Sept. 29-Oct. 11	Dominguez Hills	California State College Dominguez Hills, Calif.	Source-dominated (refineries, chemical industry)

TABLE 3. Projected Key Chemical Constituents of Collected Particulate Matter.

Element or molecular constituent	Analytical Method	Remarks and Possible Origins
Carbon (total and organic)	Selective solvent extraction, combustion to CO_2, thermal conductivity detection. HRMS, column chromatography, volatilization, flame ionization detection	Primary: natural and anthropogenic
Nitrogen (NO_3^-, NH_4^+, N^+)	ESCA, wet chemical	Sources - oxidation of NO_x, NH_3, fuel additives
Sodium	NAA, flame photometry	Mainly sea salt
Aluminum	NAA, XRFA, AA	Mainly soil - some possible anthropogenic
Silicon	XRFA, AA	Mainly soil
Sulfur (SO_4^{2-}, SO_3^{2-}, S...)	XRFA, wet chemical, ESCA	Mainly secondary production from SO_2 oxidation
Chlorine	NAA, XRFA, wet chemical	Mainly sea salt, but some man-made
Potassium	XRFA, NAA	Mainly natural (?)
Calcium	XRFA, NAA	Soil and cement
Titanium	XRFA, NAA	Anthropogenic and natural
Vanadium	NAA, XRFA	Power plant - fuel oil
Chromium	XRFA, NAA	Anthropogenic
Manganese	NAA, XRFA	Anthropogenic, smoke suppressant, soil
Iron	XRFA, NAA	Anthropogenic and natural
Nickel	XRFA, NAA	Anthropogenic
Copper	XRFA, NAA	Anthropogenic
Zinc	XRFA, NAA	Tire dust, smelting, fuel additives
Bromine	NAA, XRFA	Auto exhaust
Cadmium	NAA	Metal production and processing
Iodine	NAA, XRFA	Sea salt and others (?)
Barium	NAA	Diesel exhaust, lubricating oil automization
Lead	XRFA	Auto exhaust, industrial processes
Water	Gas chromatography, microwave waterometer, humidified nephelometer	Liquid water content is a potentially important inert ingredient leading to visibility reduction

TABLE 4. Summary Catalog of the ACHEX Data Bank

Element	1972	1973
A. Continuous Data Bank		
1. Mobile	x	x
2. Fixed		
a. Pasadena	x	
b. Riverside	x	
c. San Jose	x	
B. Chemical Analysis (AIHL/LODAMS)*	x	x
C. Meteorological Background	x	x
D. Special Experiment Data		
1. Ammonia	x	x
2. Hydrocarbon Vapor	x	x
3. Peroxyacetylnitrate (Riverside)	x	
4. Oxygenate Hydrocarbon Vapors		x
5. β-Attenuation Gauge	x	x
6. Water Filter	x	
7. Microwave Liquid Water	x	x
8. Aerosol Infrared Spectra	x	
9. High Resolution Mass Spectra	x	x
10. Aerosol Organic Solvent Extraction and Analysis	x	x
11. Aerosol Surface Flux	x	
12. Aerosol Dustfall	x	
13. Humidified Nephelometry	x	
14. Large Particle Impactor	x	
E. Blimp Experiment		x

*LODAMS = Logistics and Data Management System.

Prologue

GEORGE M. HIDY and PETER K. MUELLER
Environmental Research and Technology, Inc.
Westlake Village, California

I. INTRODUCTION

The coastal regions of California, especially southern California, have been known historically to suffer from frequent cases of reduced visibility. Although the severe impairment in visibility has been identified with air pollution in recent years, the susceptibility of southern California to such events has been documented since the earliest days of the state's history. The Spanish explorers in the eighteenth and nineteenth centuries, for example, named the Los Angeles area the "Valley of the Smokes" because of the widespread, persistent haze over the region. It is believed that the meteorological conditions in southern California are especially conducive to haze accumulation. The "background" sources of airborne particulate matter making up such hazes may be particularly intense in this area.

Investigators observed that the frequency of occurrence of severe visibility reduction increased in California with the increase of air pollution through the mid-1940s to 1960s. Keith's (15) analysis indicates a general degradation in visibility in the Los Angeles area from 1950 to 1962, which followed the increases in estimated primary and secondary sources of airborne particulate matter in the city. During the course of early studies of photochemical smog, workers identified the atmospheric chemical reactions in smog as an important contributor to the aerosol particle concentrations. In the 1950s Haagen-Smit and others noted that haze frequently accompanied the intensification of photochemical oxidant formation. Haagen-Smit demonstrated in one of his dramatic experiments that intense haze could be generated from reactions between ozone (the main contributor to oxidant) and certain organic vapors such as olefins.* The results of such investigations suggested that the haze in California could be identified with particles of primary origin, emitted directly from sources, and secondary material from a tropospheric chemical reaction.

Studies of the chemical composition of aerosol samples in California through the 1950s [e.g., Mader et al. (20)] suggested that airborne particulate matter was made up of material, including sulfate, nitrate, and organics, which could not be attributed completely to primary sources. However, the extent of the contributions of secondary particles to atmospheric aerosols in California was not fully appreciated until the late 1960s, with analyses like those of Hidy and Friedlander (8).

By the mid-1960s the measurement technology became available to characterize the physical and chemical changes in atmospheric aerosols on a time scale approaching that of trace gas measurement. In 1969 an experiment was conducted in Pasadena [Whitby et al. (27)] which provided aerosol data with pollutant gas observations showing the close dynamic correspondence in both components of Los Angeles smog. This experiment in combination with other information provided the technological basis for the Aerosol Characterization Experiment (ACHEX). The social incentive emerged from the increasing concern by the California Air Resources Board (ARB) that inadequate progress in improvement of visibility had

*Haagen-Smit's experiment actually used lemon oil vapors or gasoline vapors.

been achieved by the late 1960s despite major efforts to control both emissions of primary particles and reactive gases. The ARB was unquestionably motivated by the continuing popular concern that visibility impairment as a major symptom of pollution. As part of the ARB's research program, stimulated by the University of California's Project Clean Air, a major effort was initiated to investigate the nature and origins of smog aerosol, and to determine alternatives for a control strategy to improve visibility in California air basins.

This introductory paper provides some highlights of the features of the aerosol visibility problem in California on which the ACHEX was based. The following sections include a brief summary of an approach to control strategy, a discussion of the meteorological setting in southern California, and a summary of aerosol properties in California as known by 1971. With this information in mind the scientific reports contained in this volume can be read with perspective.

II. AIR QUALITY CONTROL STRATEGY

On an urban scale, metropolitan areas such as the Los Angeles basin-San Francisco Bay area represent "hot spots" of aerosol generation. Emissions from sources are high, and chemical reactions producing particulate matter in the atmosphere are accelerated by relatively high concentrations of reactants. Other parts of California, such as the San Joaquin Valley, are experiencing a degradation in air quality that is associated partly with the transport of airborne material from the neighboring urban complexes, as well as from material emitted from local sources. Over the past two decades much effort has been devoted to the study of the chemistry of polluted atmospheres such as those of Los Angeles and San Francisco. Early work established that aerosols were produced in photochemical reactions involving pollutant gases. Yet through the 1960s the mainstream of smog research tended to disregard particulate matter as an "active" constituent of smog. As a result much more is known about gas-phase photochemistry than about smog aerosol physics and chemistry.

Although air pollution in Los Angeles and San Francisco has received considerable attention, there is more recent concern for problems in the San Joaquin Valley of California. For example, Bakersfield in the southern part of the valley has one of the highest airborne particulate concentrates in California, according to the U.S. National Air Surveillance Network (NASN). Yet the pollution associated with trace gases has been considered less serious than for the other two basins. Because local sources of particulate matter in the San Joaquin Valley are small, except for blowing soil dust, this region is believed to be a receptor for aged aerosol transported south from the San Francisco area.

With the continued persistence and widespread degradation of visibility in major air basins in California, a control strategy beyond regulating primary particulate emissions was required in the early 1970s. The technical basis for such a step needed to be established through the ACHEX and other investigations.

Ideally, aerosol control strategy in a given region should be based on a consideration of

(1) a material balance relating measurement of the amount and chemical composition of the aerosol to the amounts and chemical compositions of the sources and to the natural background aerosol;

(2) a simulation model describing the concentration of particulate matter from sources and atmospheric processes as a composition as the material is diffused and advected through a geographical region;

(3) the optical properties as they relate visibility reduction to the size spectrum of the aerial suspension;

(4) the health effects of the aerosol in relationship to aerosol composition, particle size distribution, and humidity.

An essential ingredient for the development of a useful strategy for improving the control of aerosols from human activities is a quantitative understanding of the sources of primary particles, the sources of precursor gases, the transport of pollution, and the chemistry of interactions between trace gases and suspended particles. Thus extended knowledge of the combined influences of source intensity, meteorology, and the physical and chemical nature of the aerosols is a fundamental requirement for rational regulatory decisions, the adverse health and welfare impact of aerosols having been established. The ACHEX was designed to derive a material balance for airborne particulate matter and to identify key contributors to visibility reduction.

The intensity of the visibility reduction problem in California is greatest in the Los Angeles area. Therefore the focus of the ACHEX was directed strongly to this part of the state. The observations obtained during the study have been reviewed and interpreted largely as the basis of information concerning the distribution of pollution sources and the meteorology of southern California.

III. THE METEOROLOGICAL SETTING

Southern California is well known for its mild, Mediterranean-like climate. However, many of the features that make the area so attractive are crucial ingredients of weather in air pollution episodes. The sunlight, of course, is vital to the production of photochemical smog. The high concentrations of pollutants in the Los Angeles Air Basin result from the persistent stable atmospheric conditions over the city.

Several meteorological studies have been made of the climatological features of the southern California region. The relationship to the larger regional (or synoptic) scale* atmospheric processes over the northeastern Pacific Ocean was explored by Neiburger et al. (23), who noted the importance to California weather of the persistent anticyclonic air motion around the Pacific high-pressure zone. On a smaller scale, studies of the winds and density structure over Los Angeles have been available for several years. Summaries of measurements made before the late 1950s are given by the Air Pollution Foundation [e.g., Neiburger and Edinger (22)]. More recent data are available from the U.S. National Weather Service.

In general, the meteorology near the ground in the Los Angeles basin is dominated by the influence of the neighboring ocean and the local topography. The broad-scale features that characterize Los Angeles weather are (a) the Pacific high-pressure zone, which dominates the synoptic scale (anticyclonic) atmospheric motion from early spring until early fall; (b) the continental high-pressure region over the deserts and high plains to the east and north, which is present much of the time from fall through winter; and (c) the winter passage of cyclonic storms originating to the north, south, and west over the Pacific.

The conditions that seriously restrict the dispersion of material in the planetary boundary layer develop most frequently in the summer months and in the fall when the Pacific high induces a subsidence temperature inversion† over the basin. Conditions frequently occur in which the inversion is elevated a few hundred meters from the ground, and the air near the ground is stably stratified with little or no wind. This creates conditions where transport of pollutants away from the ground is restricted by the

---

* Synoptic scale refers to large-scale atmospheric processes occurring over horizontal extents of 100 to 1000 km.

† A temperature inversion is defined as a temperature increase with height. Normally there is a temperature decrease with height in the atmosphere.

inversion, and there is little or no removal of contaminants by horizontal transport away from the city. The resulting elevated inversion, together with the blocking by the surrounding mountains, forms a boxlike trap for pollutants discharged into the atmosphere. Periods when a subsidence inversion exists are accompanied by a strong onshore sea breeze during the day and diminished winds at night. Sometimes the wind reverses direction at night to blow offshore as a result of nocturnal radiation cooling and downslope drainage. In late fall and winter the passage of cyclones creates conditions of deep vertical mixing near the ground, giving rise to occasional favorable dispersion conditions. However, many of the winter months also show persistent shallow inversions, creating conditions for the accumulation of pollutants.

Outbreaks of airflow from intense high-pressure areas over the deserts to the east are observed often during the fall, resulting in easterly Santa Ana winds. This condition is marked by strong dry winds sweeping across the basin from the northeast, pushing the contaminants out to sea. As the Santa Ana winds diminish, however, shallow inversions sometimes form by radiative cooling during the night, which can hinder the dispersion of pollutants emitted at the Earth's surface. This class of inversion tends to dissipate quickly during the day, in contrast to the elevated subsidence inversions associated with the Pacific high.

Generally, the diffusion meteorology of the Los Angeles basin is identified with urban or mesoscale (10 to 100 km) horizontal ebbing and flowing of air, limited to mixing in a rather shallow surface layer [e.g., Neiburger and Edinger (22)]. The depth of this mixing layer below the inversion base at any particular time defines roughly the volume of air available for accumulating the pollutant emissions. The altitude of the inversion base varies, but averages about 300 to 800 m over Los Angeles during most of the year.

The period of late summer and early fall was chosen for the ACHEX because this is historically the period in which days of intensive air pollution are most frequent in the Los Angeles area. The weather is associated largely with the Pacific high-pressure zone, and at this time typical streamlines for surface airflow during the day look like those in Figure 1a. In such circumstances the westerly sea breeze penetrates the Los Angeles area, pushing pollutants emitted in the industrial and heavily populated portions of the city eastward and northeastward. At night the average wind patterns, as indicated in Figure 1b, show a tendency toward reversal in flow, returning air toward the sea.

Since there is often limited flushing of air in the marine layer as it ebbs out to sea, pollution may accumulate from day to

FIGURE 1a. Typical wind trajectories over southern California in summer: sea-breeze-dominated flow during midday. [(Diagram from DeMarrais et al. (3).]

day in the "tidal" flowing air mass. Some evidence of this sort of buildup over Los Angeles has been reported by Pack and Angell (24). Ordinarily the westerly inflow in the day is stronger than the easterly return at night, so that the ebb and flow over the city are unequal. Such conditions are believed to make the normal daily accumulation of pollutants in the marine layer worse, and they may be involved in severe pollution episodes over a sequence of 2 to 3 days. In any case it is expected that the changes in pollution over Los Angeles will constitute a complicated pattern involving emissions from many local and distant sources, but probably also heavily contributed to by motor vehicle emissions.

The weak mixing of air in surface layers over the Los Angeles area was exemplified in a visibility study a few years

FIGURE 1b. Typical wind trajectories over southern California in summer: reversal of flow inland and weakening of the sea breeze at night. [(Diagram from DeMarrais et al. (3).]

ago. Rosenthal (25) found that he could trace the movements of highly polluted air masses in terms of the displacements of zones of low visibility. The observed motion of such identified "clouds" of aerosols was qualitatively similar to the known wind patterns over Los Angeles. Evidently these masses of polluted air failed to disperse or mix extensively with their surroundings and could be traced at times out over the Pacific Ocean and into Ventura County or into Orange County as they followed the general sea-land breeze pattern.

IV. PROPERTIES OF AEROSOLS

A. Mass Concentration and Chemical Composition

Before the initiation of the ACHEX, the results of a variety of investigations of aerosol behavior in California were available. These included information on both the physical and the chemical properties of airborne particles in the Los Angeles urban setting, as well as surrounding rural areas. They provided a useful framework for design of the ACHEX.

Pre-1968 data for the total mass concentration and selected chemical constituents of airborne particles are listed in Table 1. Data from Los Angeles and Riverside data [Lundgren (18)] are compared with observations from California island sites, and the average data from urban or nonurban sites from the National Air Surveillance Network. Data are listed for total mass concentration and for primary particles, lead and iron, as well as for material from secondary processes. Sulfate, nitrate, and benzene-soluble organic material are in this category; the last-named may include organics from either primary or secondary sources. Riverside is downwind climatologically from Los Angeles at a distance of approximately 70 km. San Nicolas Island is an ocean site approximately 90 km west of Los Angeles. The data listed for downtown Los Angeles for 2 different years suggest a reduction in total mass concentration and benzene-soluble organic material, but little or no change in sulfate and nitrate components. Comparison with Riverside data suggests that the Riverside area, at least in 1968, experienced somewhat lower concentrations of particulate material, with significantly less organic component. The observations of Holzworth (11), taken on San Nicolas Island and southeast Farallon Island ($\sim$40 km west of San Francisco) in the mid-1950s, show high total mass concentration with substantial amounts of organic material, sulfate, and nitrate present. The early results from San Nicolas Island are somewhat puzzling because one would expect such a remote site to be similar to remote sites in the NASN set.

Generally the data from the Los Angeles stations were similar in mass concentrations to other urban areas characterized by NASN data. However, the content of benzene-soluble organics and nitrate are seen to be significantly larger than other urban data.

The results in Table 1 suggest that only a small fraction of the total mass of particulate material could be accounted for, even though it was suspected that the largest components by mass are sulfate, nitrate, and organic material. An additional ubiquitous component was thought to be liquid water. Atmospheric particulate matter is normally hygroscopic and can absorb substantial

TABLE 1. Comparison Between Samples of Aerosol in Photochemical Smog and Other Urban and Nonurban Aerosols.

Sample	Total Mass ($\mu g/m^3$)	Benzene-soluble fraction		Sulfate		Nitrate		Lead		Iron	
		($\mu g/m^3$)	(%)	($\mu g/m^3$)	(%)	($\mu g/m^3$)	(%)	($\mu g/m^3$)	(%)	($\mu g/m^3$)	(%)
Downtown Los Angeles[a] (1958)	213	30.4	14.2	16.0	7.5	9.4	4.4	1.7	0.1-0.9	-	-
Downtown Los Angeles[a] (1966)	119	15.2	12.8	14	11.8	13	10.9	1.6 (Burbank)	1.3	1.4 (Burbank)	1.2
Riverside, Calif. (summer-fall 1968)	82	-	-	8.3	10.1	11.7	14.3	0.59	0.72	0.69	0.84
San Nicolas Island, coastal and ocean (1956-1957)	69	3.6	5.2	8.4	12.2	2.8	2.9	-	-	-	-
San Nicolas Island (1956-1957)	110	10.1	9.2	9.3	8.4	5.4	4.9	-	-	-	-
S.E. Farallon Island (1956)	142	4.6	3.2	12.6	8.9	1.8	1.3	-	-	-	-
Urban (1966-1967)[a] 25 stations	102	6.7	6.6	10.1	9.9	2.4	2.4	1.11	1.07	1.43	1.38
Proximate (1966-1967)[a] 5 stations	45.0	2.5	5.6	10.0	22.2	1.4	3.1	0.21	0.47	0.56	1.24
Nonurban (1966-1967)[a] 15 stations	40.0	2.2	5.4	5.29	13.1	0.85	2.1	0.096	0.24	0.27	0.67
Remote (1966-1967)[a] 10 stations	21.0	1.1	5.1	2.51	11.8	0.46	2.2	0.22	0.10	0.15	0.71

[a]Annual average values from NASN data [U.S. Public Health Service (26), Ludwig et al. (17)].
[Source: Hidy and Friedlander (8).]

quantities of moisture. The change in apparent mass of particles with humidity was studied by Covert et al. (2).

As of the late 1960s detailed knowledge of the chemical constituents in the Los Angeles aerosols was quite limited. The weight fractions of several ions and elements in Los Angeles and San Francisco particulate samples are shown in Table 2. These results again suggest the dominance of (water-soluble) sulfate and nitrate, with organic material, in the particulate mass concentration.

Major differences between Los Angeles and San Francisco appear to be in the sea salt contribution (as exhibited by sodium and chlorine), potassium, and bromine (a halogen identified with motor vehicle emissions). Table 2 shows selected chemical constituents that are interpreted as indicators or tracers for certain sources. The benzene-soluble fraction has been identified mainly with primary emissions and the secondary production of particles from organic vapors. Lead and bromine are linked with auto emissions, and silicon and aluminum are largely contributed by soil dust. Potentially significant differences appear between the areas of interest in organic content, nitrate content, and possibly in automobile contributions.

The historical observations through the 1960s and the 1969 Pasadena experiment reinforced the need to improve the definition of the background or baseline aerosol in parts of California, including the coastal and desert areas. Because of the anomalously large mass concentrations observed earlier offshore of Los Angeles, further study of the marine background aerosol was necessary to (a) resolve the makeup of possible origins of this regional background, and (b) confirm that substantial amounts of particulate material could drift into southern California from offshore. With these goals in mind an observational program was undertaken on San Nicolas Island during the summer and fall of 1970 [Hidy et al. (10)]. The physical and chemical characteristics of airborne particles on the island during that period are given in Tables 3 and 4. The physical characteristics listed in Table 3 show Aitken nuclei concentrations and large-particle concentrations typical of marine air measured elsewhere. The total mass concentrations reported were somewhat less than those observed in the 1950s by Holzworth. The fraction of particles less than 3.5 μm in diameter was found to be about half the total mass concentration. The average chemical composition showed a similarity to other samplings in marine-coastal conditions, with heavy contributions from sulfate and nitrate, but only a relatively small fraction of organic noncarbonate carbon. The average Cl/Na ratio in these collected samples was anomously high compared with that of seawater. The water-extracted chloride present

TABLE 2. Chemical Analysis of Aerosols

Element	Amount (wt %)	
	Los Angeles[a]	San Francisco[b]
Aℓ	1.05	1.14
Ba	0.016	0.0191
Br	1.24	0.322
C	24.67	-
Ca[c]	1.76	-
Cℓ	0.29	3.95
Cu	0.06	0.0623
Fe	5.23	2.21
1[c]	0.0043	0.0086
ln	0.000044	-
K	0.51	0.0101
Mg	0.94	1.73
Mn	0.031	0.023
Na	1.49	4.43
Pb	4.57	-
Si[c]	0.90	-
Ti[d]	0.44	0.185
V	0.0084	0.0071
Zn	0.31	0.147
Cr[d]	0.079	0.0112
Sb[d]	0.000	0.0023
Ag	-	0.0002
Se	-	0.0013
Hg	-	-
Sulfate[e]	12.3	10.6
Nitrate[e]	10.9	5.0
Ammonium[e]	0.6	0.7
Organics (benzene-soluble extract)	12.8	9.8
Total	80.2	40.3

[a]Average of nine samples of Pasadena 1969 study from unpublished data of the Air and Industrial Hygiene Laboratory.
[b]Average of nine APCD Stations in the San Francisco Bay area from results of Johns et al. (13).
[c]Limits of detectability.
[d]NASN data for Burbank or downtown Los Angeles or San Francisco, 1965-1966 observations [U.S. Public Health Service (26)].
[e]Water-soluble component.

TABLE 3. Overall Averaged Properties of Aerosols Measured on San Nicolas Island in 1970

Property	July-August	October
Aitken nuclei concentration (number/cm^3)	2400	2600
Particles ≥0.5-μm diameter (number/cm^3)	30	93
$b_{scat} \times (10^4\ m^{-1})^a$	0.5	--
Total mass concentration ($\mu g/m^3$)[b]	29.8	
Refined mass concentration ($\mu g/m^3$)[c]	12.1	

[a]Inlet air at ≲70% relative humidity.

[b]Average of 22 glass fiber filter samples, excluding one taken at the spray zone which was 79.8 $\mu g/m^3$.

[c]Average of 22 glass fiber filter samples, excluding one taken at the spray zone which was 16.6 $\mu g/m^3$. Refined mass covers particles ≲3.5 μm in diameter at 50% removal efficiency.

could not be explained in terms of lead halides from gasoline, and was identified with aerosol aging and condensation of gaseous chlorine compounds in the air [see also Hidy et al. (10)].

Comparison of the 1970 sampling results with the reports of Holzworth suggested generally higher concentrations of carbon, sulfate, and nitrate in 1970 than in 1956-1957, but lower total mass concentrations. Comparison of the 1970 observations on San Nicolas Island with aircraft samples over the same area in 1968 showed similarities for trace metals. However, Si, Na, Cl, and S were significantly greater in the island samples than in those taken 250 km west of Santa Barbara at an altitude of 0.015 km [Gillette and Blifford (6)].

With lead and silicon as tracers for particles from automobile emissions and soil dust, respectively, and sodium as a tracer for sea salt and soil dust, a crude breakdown of particle origins at San Nicolas Island was undertaken. The results are shown in Table 5. Here the secondary constituents or those produced by atmospheric processes appear to dominate this "background" aerosol, like the urban aerosol.

TABLE 4. Overall Averaged Analysis for Chemical Constituents of 13 SNI Aerosol Samples in 1970

Constituent	Amount[a] (wt %)
Al	2.8 (±50%)
Si	5.0 (±50%)
Na	3.6 (±5%)
P	ND[b]
Cl	9.9 (±10%)
K	1.3 (±25%)
Ca	1.9 (±25%)
Ti	0.15 (±20%)
V	ND
Cr	0.02 (±20%)
Mn	0.04 (±20%)
Fe	1.6 (±20%)
Cu	0.6 (±20%)
Zn	0.2 (±20%)
Br	0.2 (±20%)
Pb	0.4 (±20%)
I	ND
$SO_4^{=}$	17.5 (±10%)
NO_3^-	5.7 (±10%)
NH_4^+	2.2 (±10%)
Total inorganics account for	53.1
Cyclohexane-soluble (organics)	2.0 (±10%)
Noncarbonate carbon	3.0
Sum of constituents	56.1
Total ash content (TAC)	74.6
Loss on ignition (100 - TAC)	25.4

Source: Hidy et al. (10).

[a]Percentage in parentheses in error estimated by analyst; x-ray fluorescence for α-particle excitation only.
[b]ND = less than minimum detectable concentration.

TABLE 5. Estimated Contributions by Source to SNI Aerosols

Source	Constituent	Amount (wt %)
Primary	Soil (based on 25% Si in soil)	20.0
	Sea salt (based on 30.6% Na in sea salt)	11.1
	Automobile lead (based on 40% lead in auto-emitted aerosol)	1.0
	Carbon[a]	3
Secondary	Sulfate	17.5
	Ammonium	2.2
	Nitrate	5.7
		61
Total volatiles other than carbon: [100 - (Total ash carbon)]		22
Total accounted for		83

[a]Some of this may be produced in the atmosphere by chemical reaction; however, it is identified with primary origin for this estimate. Source: Hidy et al. (10).

In late fall of 1971 a few exploratory samples were obtained by ACHEX investigators in selected urban and remote sites in California away from the Los Angeles area. These were as follows:

Site	Location	Comments
Big Sur	30 miles south of Monterey	Very remote coastal, heavily conifer forested
Big Basin	40 miles south of San Francisco	Remote, heavy redwood forest, 10 miles inland from ocean
Point Reyes	30 miles northwest of San Francisco	Very exposed ocean beach site
Goldstone	60 miles north of Barstow	Very remote desert site

Site	Location	Comments
Fresno	Central San Joaquin Valley	Urban-agricultural
Bakersfield	South San Joaquin Valley	Urban-local petroleum production and agriculture

This set of sites was selected to supply additional information about the chemical nature of the background and urban aerosols in central or southern California. The results are shown for high-volume filter-collected material in Table 6. They are interesting in comparison with earlier work in Table 1, and with the range of composition found for Los Angeles and San Francosco. The urban sites in the San Joaquin Valley show substantial amounts of suspended particulate material, similar to the data for Los Angeles. Samples from both Fresno and Bakersfield contained large quantities of nitrate and organics, but showed less sulfate than Los Angeles or San Francisco samples. The large quantities of silicon may reflect a high level of soil dust on the sampling days. Unlike the Los Angeles area, only half of the particulate concentration was in the range ≤ 2 μm diameter.

Generally, the remote, rural sites displayed considerably smaller suspended particle concentrations. The notable exception is Point Reyes, which evidently was influenced significantly by sea salt aerosol containing very large particles. The Big Basin sample contained much greater concentrations of small particles than the samples from other sites, possibly because of the influence of open campfires in the area. Even though Big Sur, Big Basin, and Point Reyes are all near the ocean, the Cl/Na ratios vary widely, instead of being near 1.5, the value for sea salt. The Cl/Na ratio at Point Reyes was very high, over 3.6. This is strikingly similar to the 1970 sampling on San Nicolas Island. Significant amounts of airborne sulfate were recorded at this site and at Goldstone.

Concentrations of the heavy metals were very low at the remote sites, but silicon appeared at Point Reyes, presumably from blowing sand grains. The Point Reyes sample was not unlike material sampled earlier on San Nicolas Island with significant amounts of sulfate and with organics at high total mass concentrations.

The results characterizing the chemical nature of aerosols in urban air and nearby locations, as sparse as they were in 1971, still showed a wide diversity of materials, which involve the universal presence of sulfate, nitrate, organic carbon, and metals.

TABLE 6. Selected Chemical Analyses of Filter-Collected Aerosols in California

Constituent or Parameter	Concentration[a] ($\mu g/m^3$)							
	Basin							
	Big Sur[b]	Big Basin[b]	Point Reyes[b]	Fresno[b]	Bakersfield[b]	Goldstone[b]	Los Angeles	San Francisco
Total Mass	35.1 (3.5)	27.8 (2.8)	129 (12.9)	203 (20.3)	170 (17.0)	48.9 (4.9)	119	68
Refined mass[c] (% of total)	42.4	82.4	19.3	48.9	49.7	38.4	66-84[d]	-
Total organics (cyclohexane extract)	8.92 (0.89)	4.48 (0.45)	4.46 (0.47)	23.2 (2.32)	14.2 (1.42)	4.04 (0.50)	15.2 (benzene-soluble extract)	6.0
Na	0.48 (0.03)	0.56 (0.03)	20.9 (1.05)	0.42 (0.02)	0.83 (0.04)	0.26 (0.02)	3.0	3.8
Cl	0.65 (0.07)	0.39 (0.04)	75.3 (7.5)	1.70 (0.17)	1.15 (0.12)	0.22 (0.02)	3.0	3.3
Ti	ND[e]	ND	ND	0.103 (0.031)	0.126 (0.038)	ND	-	-
Fe	0.128 (0.22)	0.219 (0.37)	0.337 (0.51)	1.84 (0.184)	2.35 (0.235)	0.4 (0.07)	-	-
Cu	ND	ND	ND	0.049 (0.015)	0.17 (0.005)	ND	-	-
Zn	ND	ND	ND	0.046 (0.014)	0.015 (0.005)	ND	-	-
Pb	ND	ND	ND	2.30 (0.230)	1.35 (0.135)	0.3 (0.06)	4.4	1.2
Br	ND	ND	0.076 (0.011)	0.910 (0.091)	0.446 (0.045)	0.05 (0.25)	1.4	0.27
$SO_4^{=}$	0.87 (0.15)	0.63 (0.09)	6.24 (0.60)	5.64 (0.57)	7.68 (0.78)	2.19 (0.21)	14.6	7.2
NO_3^-	0.44 (0.04)	0.84 (0.08)	0.40 (0.04)	19.4 (1.94)	21.8 (2.17)	1.59 (0.16)	12.4	3.4
NH_4^+	0.48 (0.04)	0.42 (0.03)	0.83 (0.08)	8.87 (0.88)	4.79 (0.18)	1.63 (0.17)	7.1	4.8
Si	0.50 (0.05)	0.13 (0.01)	2.52 (0.25)	14.77 (0.74)	13.42 (1.34)	3.72 (0.37)	2.5 - 6.1	2.5
Al	0.053 (0.011)	0.043 (0.002)	0.67 (0.13)	6.47 (0.65)	4.48 (0.22)	1.33 (0.07)	1.2	0.98
Total Ash	10.3 (5.2)	1.4 (1.4)	25.3 (3.0)	72.0 (2.9)	83.0 (4.2)	22.3 (3.3)	-	-

[a] Numbers in parentheses are standard deviations for analytical error.
[b] Samples collected during November-December 1971 with ICN constant flow on TFA 810 filter at ground level, one filter taken at each site. Analysis done by AIHL.
[c] Percent mass $\leq$2-μm diameter.
[d] Based on results reported by Lundgren (19).
[e] ND = not detected.

Characterizing the chemical composition requires the identification and the development of new methods of chemical analysis. Natural sources, particularly of marine origin, represent an important but variable part of the western California aerosol. In combination with fluctuations in humidity and temperature in the mixed surface layers of the atmosphere, the marine aerosol is believed to interact significantly with anthropogenic material on occasion to reduce visibility in southern California coastal areas.

B. Size Distributions

Workers have recognized for sometime that aerosol particles in the atmosphere occur over a very wide range a size and number concentration, N. Some number distributions are shown for illustration in Figure 2. Measurements in Pasadena during the 1969 experiment (Whitby et al., 28) show distributions similar to those typical of European continental conditions. In Los Angeles, however, significantly larger concentrations of very small particles were present in smog than would be expected from the European observations. The concentration of particles smaller than approximately 1 μm in diameter (D_p) found in urban air like that in Los Angeles is very much greater than observations in remote areas of California would suggest.

The details of the particle size distribution are sometimes lost in number distribution plots like those in Figure 2. By looking at the higher moments of the number distribution, corresponding to the total surface, S, distribution or the volume, V, distribution, one can find useful information. Whitby and coworkers (28) were the first to emphasize the multimodal nature of the surface and volume-size distributions. In particular, these workers found that continuous measurements of the particle size distribution revealed a submicron fraction of less than about 1 μm diameter.

An example of the growth of particles in the submicron range with the evolution of smog was found in the 1969 Pasadena experiment. The change during the day is shown in Figure 3. The growth of particulate matter in the submicron range is accompanied by an intense change in light scattering by particles, and by significant changes in particulate sulfur and nitrogen during the day. The growth was explained by Husar et al. (12) in terms of condensation of material on the particles present early in the day.

An important result from comparing the nephelometer observations and the particle size distributions emerged from the Pasadena study. The incremental light-scattering coefficients at several wavelengths are shown as a function of particle diameter

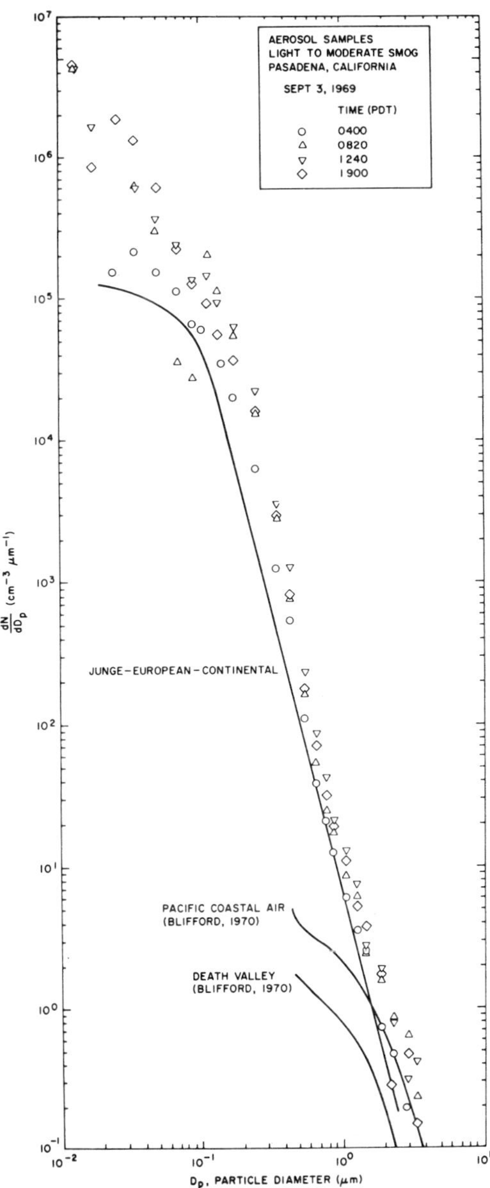

FIGURE 2. Comparison between aerosol size spectra in different geographical locations and spectra taken for light to moderate smog in Pasadena, California. [From Hidy and Friedlander (8).]

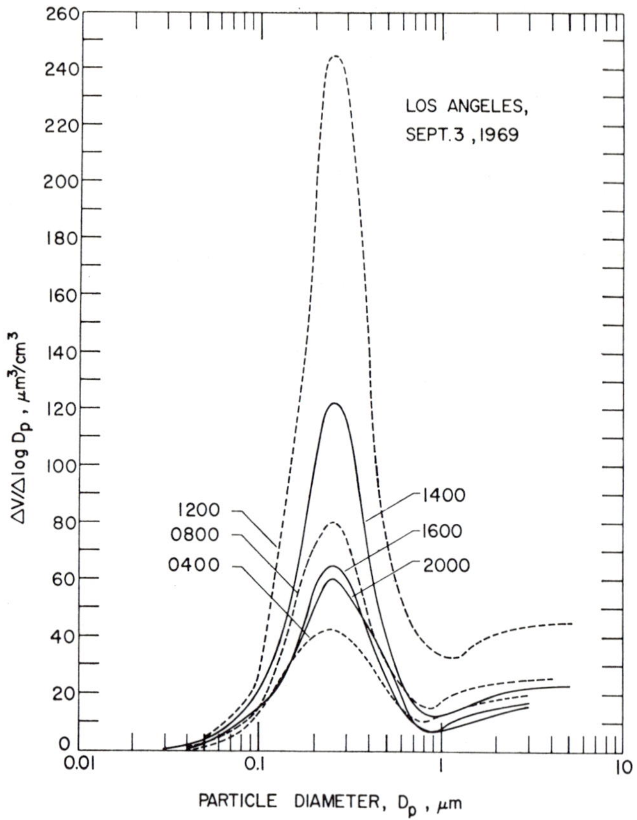

FIGURE 3. Development of the volume distribution with time in Los Angeles on September 3, 1969: Note the narrow log-normal distribution of the volume in the submicron range and the constancy of the size at which the mode is located. Also note that the maximum volume is reached at noon with the solar radiation. [From Whitby et al. (22)].

in Figure 4. This graph demonstrates that the strong intensity of light scattering comes from particles in the 0.1 to 1.0 μm diameter range.

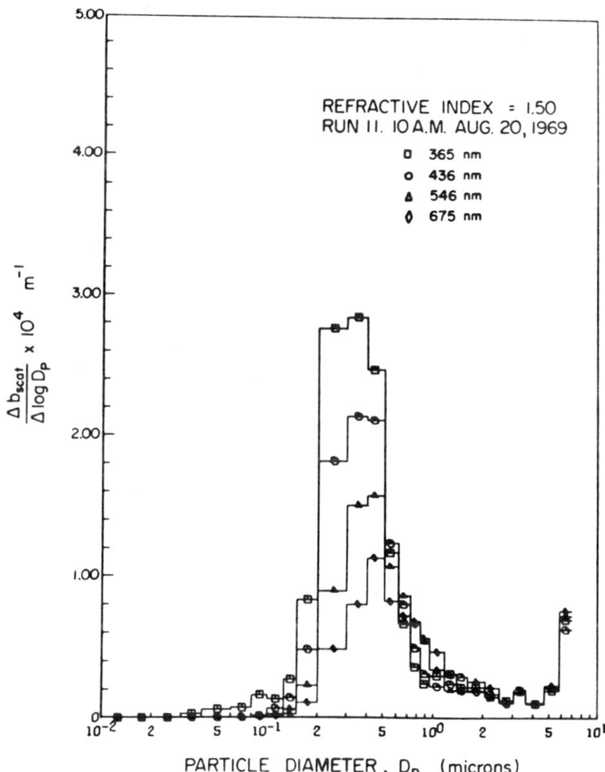

FIGURE 4. $\Delta b_{scat}/\Delta \log D_p$ as a function of particle diameter for a particle refractive index of 1.50. Linear -log plot. [From Ensor et al. (5)].

The observations from the 1969 Pasadena experiment led to the hypothesis that two major modes exist in the atmospheric aerosol mass distribution (Figure 5). The submicron part is largely dominated by anthropogenic sources of small particles generated by combustion or condensation of materials derived from atmospheric chemical reactions. The supermicron fraction is characterized by major natural sources such as wind blown soil dust or sea salt aerosol. Thus the two modes change more or less independently, with the larger particle fraction contributing strongly to the total mass concentration.

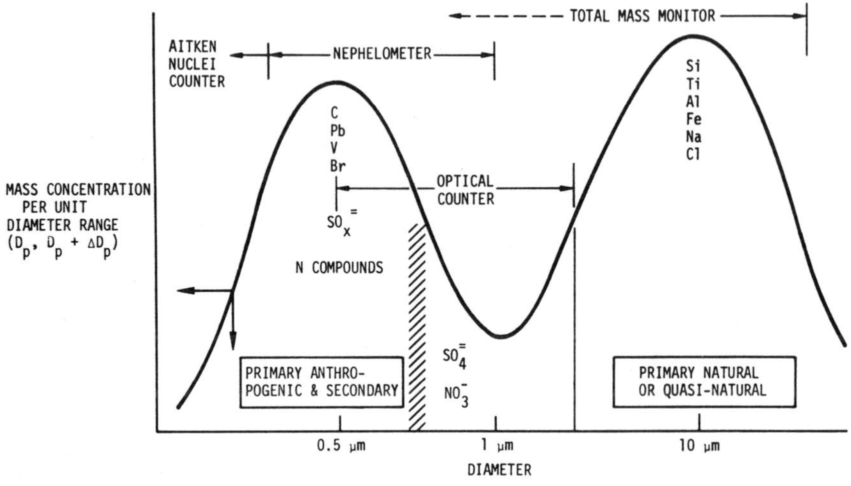

FIGURE 5. Hypothetical presentation of bimodal mass distribution.

Such behavior requires for its surveillance monitoring instrumentation for parts of the distribution. Existing instruments were identified; these included the Aitken nuclei counter for particles mainly less than 0.1 μm in diameter, the nephelometer for the fraction 0.1 to 1.0 μm in diameter, and the total mass monitor (high-volume filter) for the large-particle fraction.*

C. Dynamics of Aerosols

The strong changes in the gaseous reactants and products of photochemical smog have been recorded by many investigators. The picture that is normally given for the evolution of oxidant and its hydrocarbon and nitrogen oxide precursors is shown in Figure 6. These data were obtained in Pasadena during the 1969 aerosol experiment. Here the NO is shown to decrease in the morning,

*This distinction is not always unique because particle mass concentration in many cities is dominated by submicron particles, with greatly improved controls on sources of large particles.

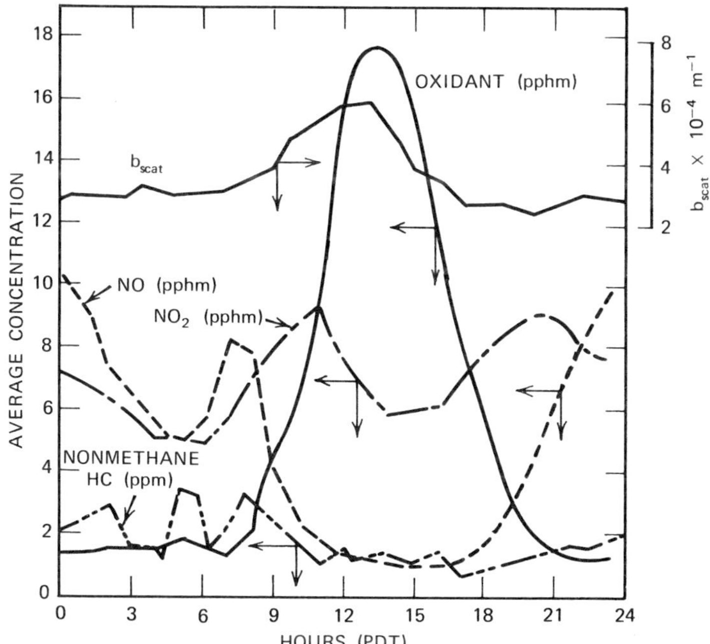

FIGURE 6. Composite data from Pasadena averaged over August to October 1969 for the daily changes in "classical" photochemical smog components; the light-scattering coefficient measured by an integrating nephelometer is added for comparison. [From Hidy et al. (9).]

followed by an increase in NO_2 to a maximum concentration by midmorning. Later, from midday to midafternoon, the maximum ozone is observed. Generally, the nonmethane hydrocarbons are found to decrease to a minimum by midday as they react to form oxidant precursors, and mix vertically.

Shown with the diurnal variation in gases is the change in the light-scattering coefficient, b_{scat}. This coefficient is a rough measure of visibility change; a high b_{scat} is a low visual range. According to the 1969 results, b_{scat} reaches a maximum, approximately corresponding to the oxidant peak, at midday. The composite b_{scat} behavior in Pasadena followed closely the changes in SO_2 concentration, as indicated by comparison of the data in Figures 6 and 7.

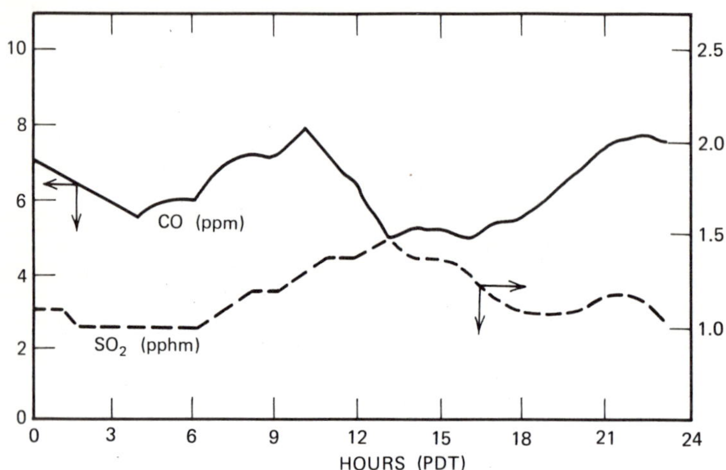

FIGURE 7. Composite Pasadena data for the period August-October 1969, showing the daily variation in carbon monoxide and sulfur dioxide. [From Hidy et al. (9)].

Examination of the observations taken in Pasadena suggested differences in the behavior of different parts of the particle size distribution during the day. Because of the shape of the aerosol size distribution, the total number concentration, N, is characteristic mainly of the smallest particles, while the surface area per unit volume, S, and the volume fraction, V, are weighted toward the submicron fraction. The composite results for several days are shown for Pasadena in 1969 in Figure 8, where a dramatic, broad maximum for N is seen through most of the day and night, reflecting general human activities. A secondary peak is seen in the morning with the peak in auto traffic. The behavior of S and V are contrasted to that of N with a midday maximum corresponding to the observed maximum in ozone and SO_2 concentration measured at the receptor site.

Chemical interpretation of the observed diurnal changes in both reactive gas and aerosol behavior at a fixed site is very complicated because of the air movement in the basin. However, it appeared qualitatively that the "coupling" between the gaeous photochemical processes and the particle behavior was present,

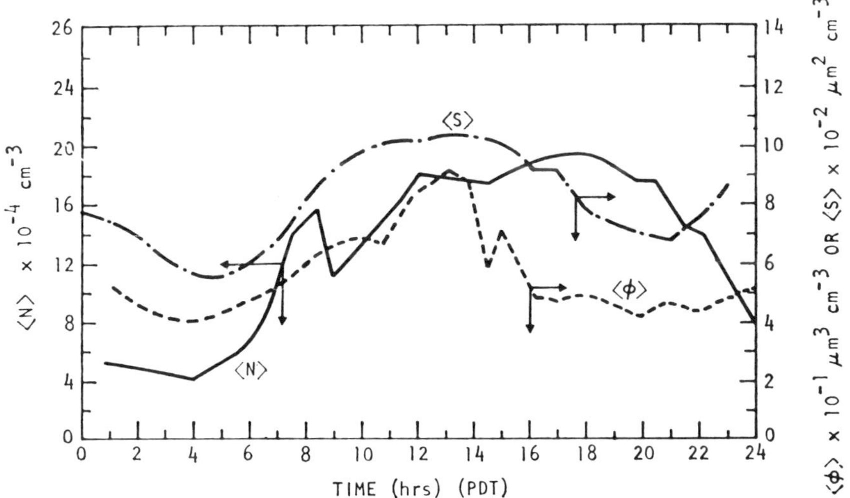

FIGURE 8. Composite Pasadena data for the total number, surface area, and volume fraction of aerosols during the period of the Pasadena experiment. [Data from Whitby et al. (28)].

and the circumstantial evidence for such an assumption was strong. In any case the basis was firm for using such a hypothesis as a framework for design of the ACHEX.

During the investigation of reactive pollutant behavior in the Los Angeles basin, studies such as those of Edinger (4) and of Hall and Ageno (7) suggested that the vertical heterogeneity in contaminant distribution was of importance in the interpretation of surface level data. In particular, a significant vertical layering of pollution and haze was observed. This indicated a complicated folding and vertical transport of material, increasing the ambiguity of data interpretation in terms of ground level observations, which were assumed to be characteristic of a well-mixed surface layer of air.

The nature of the vertical structure of pollution over the Los Angeles area as related to aerosol behavior was explored in late 1971 by aircraft profiling. Some results reported by Blumenthal and Smith are shown in Figure 9. The differences between a clear day and a smoggy day are illustrated for comparison. The parameter TURB is a relative measure of atmospheric

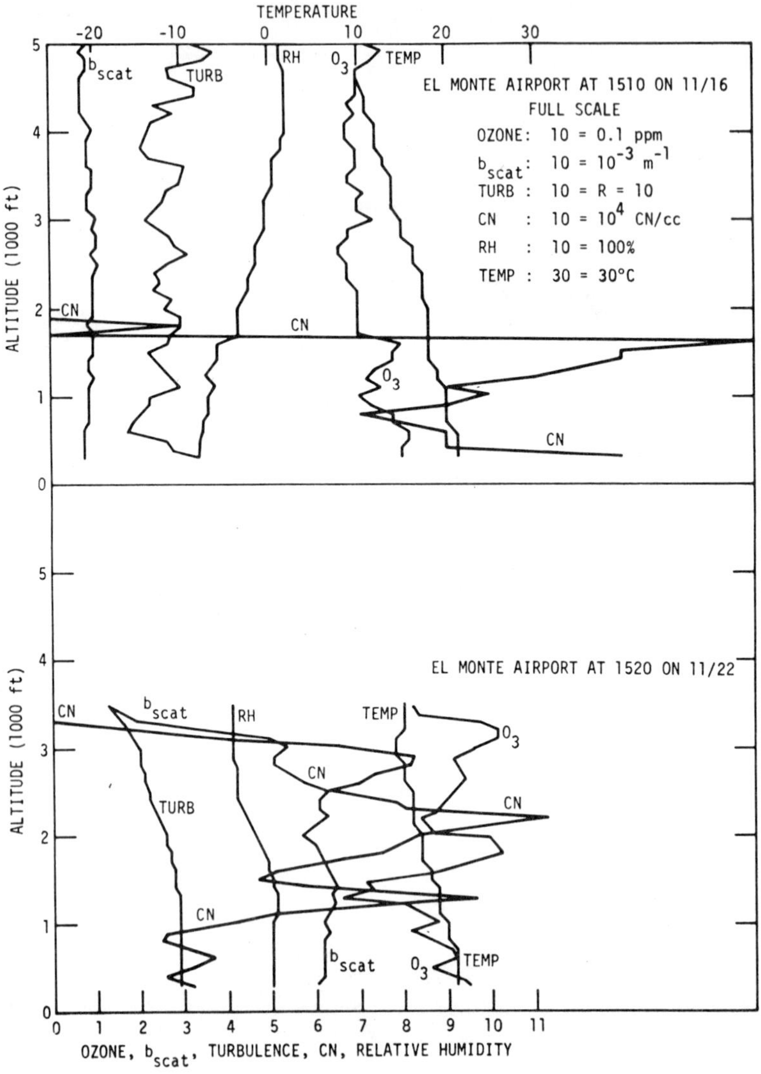

FIGURE 9. Vertical traverse comparison between a "clear" day (11/16/71) and a hazy day (11/22/71) in the Los Angeles basin (San Gabriel Valley). (Unpublished data from Blumenthal and Smith).

turbulence, RH is relative humidity, and CN is equivalent to the total number of suspended particles. There are minor differences in the profiles of most parameters measured measured except for aerosol behavior. On the clear day the CN has an intense maximum at the surface and at 1800 ft altitude over El Monte, while the value of the light-scattering coefficient, b_{scat}, is rather small. On the hazy day, the number concentration is minimal at the surface, but b_{scat} is much larger in the entire region below 3000 ft altitude. On the hazy day a zone of ozone maximum is found aloft as well as at the surface. However, the ozone concentrations at the surface were similar on both days. On comparing the two days, the humidity appears to correspond roughly with b_{scat} behavior. In particular, the dryness at the surface on November 22 was accompanied with "low" values of the light-scattering coefficient.

The results describing the vertical layering of pollutants, as shown in Figure 9, add to the complexity of data interpretation. However, this work demonstrated that information on aerosols and gases above the ground would be required as part of the ACHEX concept if the study objectives were to be achieved.

V. RELATION BETWEEN SOURCES AND THE ATMOSPHERIC AEROSOL

To construct a useful strategy for control of haze in California, the origins of the particulate material which contributed strongly to the mass concentration of particles and to the particulate light scattering had to be identified. The anthropogenic emissions inventories reported by the Los Angeles Air Pollution Control District included annual averaged data for the trace gases and for primary particles. Such an inventory is listed by major source category in Table 7, where the contributions to particulate emissions are seen to come both from stationary and from transportation sources. The contributions to sulfur oxides are identified mainly with fuel combustion and chemical processing, while nitrogen oxides and hydrocarbon emissions are heavily influenced by transportation.

Identification of the sources of airborne particles required taking into account the natural sources from marine processes and processes of natural and man-made soil disturbances, primary particle sources, as well as the products of chemical transformations of gaseous pollutants. In the late 1960s Friedlander and colleagues developed a chemical tracer method for estimating the source contributions to particulate material of primary emissions [e.g., Hidy and Friedlander (8), Miller et al. (21)]. This method was applied initially to the estimate of contributions from natural sources and from key sources of man-made primary particles.

Using the arguments of Hidy and Friedlander (8), one can identify certain tracers with emissions (see, e.g. Table 8) and estimate the relative contributions of sources to the aerosol particles over a given city. Essentially all of the atmospheric lead and bromine is believed to have its origin in automobile emissions. From the data of different investigations the lead concentraion in Los Angeles is found to average ∿3.7%; it is difficult to determine the value most representative of the city, but an average value of 4.4 μg/m^3 appears reasonable. If the lead content of automobile particulate is taken as 30%, this concentration corresponds to 15 μg/m^3 of automobile particulate. The other figures for the primary anthropogenic emissions shown in Table 8 were obtained by scaling the particulate emissions of Table 7 in proportion to the motor vehicle emissions. The estimate is rough since the other sources are located at sites quite different from the motor vehicles.

In Table 8, values are also given for the contribution of primary natural sources to the Los Angeles aerosol. The mineral dust concentration was obtained by assuming that all of the silicon originated in mineral dust, where it composed 30% of the total (its composition in igneous rock). All of the chloride and sodium in Table 3 were assumed to be of marine origin; this undoubtedly overestimates the strength of the marine source since there must be some anthropogenic and soil sources for these substances as well. Ammonium is evidently a minor contributor, identified mostly with production in soil; a small amount may be produced by combustion.

All of the nitrate and sulfate were assumed to be generated by chemical reactions involving NO_2 and SO_2, produced originally by man-made sources. The estimate of the organic aerosol formed naturally is due to Junge (14). The figure for the amount generated from anthropogenic sources is placed in parentheses, since it was calculated from an assumption of 1.5% conversion of reactive organic vapors to aerosol, based on speculations of Renzetti and Doyle [see Hidy and Friedlander (8)]. This value should be somewhat less than that for the benzene-soluble component (Table 6), partly because of the contribution of primary anthropogenic sources. There remains an uncertainty about this fraction because the material in the benzene-soluble portion should include the anthropogenic contribution plus the material from vegetation, and the total for all of these sources exceeds the measured value. This may not be serious, however, because the collection methods currently use loose material <0.3 μm in diameter; some of the organic material may be concentrated in very tiny aerosol.

TABLE 7. Illustrative Contaminant Emissions (tons/day) from Major Sources Within Los Angeles County, Average During 1968

Major Source	Organic gases Reactivity: High	Low	Total	Particulate	NO_x	SO_2	CO	Total
Motor vehicles	1255	475	1730	45	645	30	9470	11,920
Organic solvent usage	100	400	500	17	-	1	-	520
Petroleum	55	165	220	4	45	55	30	355
Aircraft	45	45	90	12	15	3	190	310
Combustion of fuels	-	9	9	15	235	40	1	300
Chemical	-	-	-	-	-	90	-	90
Other	-	3	3	16	11	3	4	35
Total (rounded)	1,455	1,095	2,550	110	950	225	9,695	13,530

Source: Lemke et al. (16).

TABLE 8. Aerosol Inventory for Los Angeles[a]. (Basis: 119 $\mu g/m^3$ total mass)

Source	Amount (wt %)
Natural background	
Primary	
Dust rise by wind (Si)	7-17
Sea salt (Na, Cℓ)	5
Spores, pollen, etc.	Unknown
	12-22
Secondary	
Vegetation (organic vapors)	2-5
Biological (soil bacterial action, decay of organics) - NH_3, NO_x, S, etc.	($NH_4^+ \approx 0.5$)
	2-5
Anthropogenic	
Primary	
Motor vehicles (Pb)	13
Aircraft	3
Chemical	3
Metallurgical	3
Mineral	1
Petroleum	3
Other industry	1
Combustion of fuels	5
	32
Secondary	
Reactive hydrocarbon vapors	9
NO_3^-	12
SO_4^{2-}	11
	32
Total accounted for	78-91

[a]Based on relative humidity less than 40%; chemical composition data based on Table 2.
[Source: After Hidy and Friedlander (8)].

The total aerosol, estimated in this way, was 78 to 92% of the total mass. About 30% of the Los Angeles aerosol originates from primary anthropogenic sources and perhaps a third from secondary sources; the remainder comes from natural sources.

A similar inventory can be made from the San Francisco Bay area. Using bromine to estimate lead concentration (motor vehicles) and aluminum to estimate silicon or windblown dust, one can make a calculation from historical data and results such as those in Table 6; the results are shown in Table 9. In this case only about two thirds of the aerosol mass can be accounted for. Of this amount about 40% comes from primary anthropogenic sources, and 25% from secondary anthropogenic sources; the remainder is from natural emissions. This crude breakdown suggests that secondary reactions tentatively identified with photochemical processes, may be somewhat less important than primary emissions in the San Francisco aerosol. However, the main differences between Los Angeles and San Francisco appear to be linked with lower nitrate content and lower total mass loading.

These surveys on aerosols in Los Angeles disclosed that as much as one-third of the suspended particles over this urban area may be of natural or quasi-natural origins, either from the sea, from vegetation, or from wind or traffic dust rise. Available evidence suggests that the remaining anthropogenic portion appears to be split evenly between material from primary origins and material produced secondarily by chemical or physical processes in the atmosphere. This preliminary evaluation indicates the importance of aerosols from natural origins and from secondary production mechanisms. Before the ACHEX, comparatively little was known about either of these factors in southern California, although it appeared that at least in Los Angeles and the San Francisco Bay area the automobile is the main source of both primary and secondary particles [e.g., Los Angeles APCD, Lemke et al. (16); San Francisco Bay area APCD (1)]. When one takes a broader view of California, outside of metropolitan Los Angeles and the San Francisco Bay area, however, it is not clear that the automobile is the main anthropogenic source of aerosols.

VI. SUMMARY

On the basis of research before the 1970s a considerable effort would be required to measure aerosol behavior in sufficient detail to understand the interactions between primary sources, atmospheric chemistry, and air transport. The work prior to ACHEX, especially the 1969 Pasadena experiment, and the effort during its earliest phases showed the following:

TABLE 9. Aerosol Particle Inventory for San Francisco[a]. (Basis: 68 μg/m³ total mass)

Source	Amount (wt %)
Natural Background	
Primary	
Dust rise by wind	12
Sea salt	7
Spores, pollen, etc.	Unknown
	19
Secondary	
Vegetation (organic vapors)	2-5
Biological	Unknown
	($NH_4^+ \approx 0.5$)
	2-5
Anthropogenic	
Primary	
Motor Vehicles	5
Aircraft	2
Other transport	3
Chemical	3
Metallurgical	4.5
Petroleum	0.5
Organic solvent use	0.7
Combustion of fuels	2
Incineration	0.2
Agricultural burning[b]	7
	38
Secondary	
Reactive organic (from 0.5x total hydrocarbon vapors)	~5
NO_3^-	3
SO_4^{2-}	7
	15
Total accounted for	64 - 72

[a]Based on relative humidity less than 40%; chemical composition data based on Table 2.
[b]Inventory for 1972 and later should exclude this particulate source.

(1) Visibility is closely related to changes in humidity and in particle concentration over the 0.1 to 1.0 μm diameter size range, and showed a maximum at midday which corresponded with a maximum in ozone concentration.

(2) The daytime interaction between aerosols and the photochemically reactive atmosphere was manifested mainly in the particle size range exceeding 0.1-μm diameter.

(3) There is a persistent bimodality in the mass distribution with respect to particle size. The pollution contribution by combustion is hypothesized to be primarily in the small particles, while the mechanically generated material often was identified principally with the large-particle fractions. The distribution of certain chemical elements reflected the bimodal pattern.

(4) There are significant diurnal changes in sulfur- and nitrogen-bearing compounds in the aerosol particles, and the suspended material was unusually rich in noncarbonate carbon ($\gtrsim$30% by mass).

(5) The vertical structure of pollutants in the atmosphere is a complicated fingerprint of aerosol chemistry and air transport which must be taken into account in elucidating the contribution to visibility changes.

(6) A method was tested for estimating the contributions for major primary sources to the Pasadena aerosol through its chemical composition. This procedure was found to be useful for characterizing the Los Angeles aerosol in relation to the aerosols of other cities, and identified the great importance of aerosol formation by atmospheric chemical transformations.

These considerations, combined with the urgent need of the ARB to obtain new information on the nature of haze, provided the foundations for the ACHEX.

Because of the success of the case study approach adopted in the 1969 Pasadena experiment, this method was used as the basis for designing the ACHEX rather than using the more conventional alternative of a monitoring station array.

In the subsequent papers the results of the experiment are reported, as interpreted by a number of investigators who participated in the project.

REFERENCES

1. Bay Area Air Pollution Control District (APCD). 1970. Air Pollution and the San Francisco Bay Area. Hansen Lithograph Co., Sausalito, Calif.

2. Covert, D., et al. 1972. J. Appl. Meteorol. 11, 968-976.

3. DeMarrais, G. A., Holzworth, G. C., and Hosler, C. R. 1965. "Meteorological Summaries Pertinent to Atmospheric Transport of Dispersion Over Southern California", Tech. Paper No. 54, Weather Bureau, U. S. Dept. of Commerce, U. S. Gov't Printing Office, Washington, D.C.

4. Edinger, J. 1973. Environ. Sci. Technol. 1, 247-252.

5. Ensor, D., et al. 1972. Multiwavelength Nephelometer Measurements in Los Angeles Aerosol, in Aerosol and Atmospheric Chemistry (G. M. Hidy, Ed.). Academic Press, New York, p. 323.

6. Gillette, D. A., and Blifford, I. H., Jr. 1971. J. Atmos. Sci. 28, 1199-1210.

7. Hall, F. F., Jr., and Ageno, H. Y. 1970. Laser Applications in the Geosciences (J. Gauger and F. F. Hall, Jr., Eds.). Western Periodicals Co., North Hollywood, Calif., pp. 17-32.

8. Hidy, G. M., and Friedlander, S. K. 1971. The Nature of the Los Angeles Aerosol, in Proceedings of the 2nd International Clean Air Congress (H. Englund and W. Burg, Eds.). Academic Press, New York, pp. 390-404.

9. Hidy, G. M., Mueller, P. K., Tokiwa, Y., and Twiss, S. 1972. Aerometric Factors Affecting the Evolution of the Pasadena Aerosol, in Aerosols and Atmospheric Chemistry (G. M. Hidy, Ed.). Academic Press, New York, pp. 219-236.

10. Hidy, G. M., et al. 1974. J. Appl. Meteorol. 13, 96-107.

11. Holzworth, G. C. 1959. J. Meteorol. 16, 68.

12. Husar, R. B., Whitby, K. T., and Liu, B. Y. H. 1972. Physical Mechanisms Governing the Dynamics of Los Angeles Smog Aerosol, in Aerosols and Atmospheric Chemistry (G. M. Hidy, Ed.), Academic Press, New York, pp. 271-284.

13. Johns, W., et al. 1973. Atmos. Environ. 1, 107-118.

14. Junge, C. E. 1962. Air Chemistry and Radioactivity. Academic Press, New York.

15. Keith, R. W. 1964. "A Study of Low Visibilities in the Los Angeles Basin, 1950-1961". Air Quality Rep. No. 53, Los Angeles County Air Pollution Control District, Los Angeles, Calif.

16. Lemke, E. E., Thomas, G., and Zwiecker, W. E. 1969. "Profile of Air Pollution Control in Los Angeles County", Los Angeles County.

17. Ludwig, J. H., Morgan, C. B., and McMullen, T. B. 1969. Trans. Amer. Geophys. Union 51, 468.

18. Lundgren, D. A. 1969. "Atmospheric Aerosol Composition and Concentration as a Function of Particle Size and of Time", presented at the 62nd Annual Meeting, Air Pollution Control Association, New York, June 1969.

19. Lundgren, D. A. 1972. Mass Distribution Data from the 1969 Pasadena Smog Experiment, in Aerosols and Atmospheric Chemistry Symposium (G. M. Hidy, Ed.). Academic Press, New York, pp. 265-270.

20. Mader, P., MacPhee, R. D., Lafberg, R. T., and Larson, C. 1952. Ind. Eng. Chem. 44, 1352-1355.

21. Miller, M., Friedlander, S. K., and Hidy, G. M. 1972. A Chemical Element Balance for the Pasadena Aerosol, in Aerosols and Atmospheric Chemistry (G. M. Hidy, Ed.). Academic Press, New York, pp. 301-311.

22. Neiburger, M., and Edinger, J. 1954. "Summary Report on the Meteorology of the Los Angeles Basin with Particular Respect to the Smog Problem", Tech. Rep. No. 1, Air Pollution Foundation, Los Angeles.

23. Neiburger, M., Johnson, D. S., and Chien, C. W. 1961. Univ. Calif. Berkeley Publ. Meteorol. 1, 1-94.

24. Pack, D. H., and Angell, J. K. 1963. Mon. Weather Rev. 91, 583-604.

25. Rosenthal, J. 1967. Presented to the 1967 AMS Conference on Air Weather Meteorology, Santa Barbara, Calif.

26. U.S. Public Health Service. 1968. "Air Quality Data from the National Air Sampling Networks, 1966", APTI 68-9, National Air Pollution Control Administration, Raleigh, N.C.

27. Whitby, K. T. (Ed.). 1971. "Aerosol Measurements in Los Angeles Smog", Vol. 1, Particle Technology Laboratory Publ. No. 141, University of Minnesota, Minneapolis; Air Pollution Control Office Publ. No. APTD-0630, Environmental Protection Agency, Research Triangle Park, N.C.

28. Whitby, K. T., Husar, R. B., and Liu, B. Y. H. 1972. The Aerosol Size Distribution of Los Angeles Smog, in _Aerosols and Atmospheric Chemistry_ (G. M. Hidy, Ed.). Academic Press, New York, pp. 237-264.

PART I

ANALYTICAL METHODS AND SAMPLING TECHNOLOGY

Design, Instrumentation, and Operation of a Large Mobile Air Pollution Laboratory for ACHEX

GILMORE J. SEM
TSI Incorporated
St. Paul, Minnesota

and

KENNETH T. WHITBY and GEORGE M. SVERDRUP*
Particle Technology Laboratory
Mechanical Engineering Department
University of Minnesota
Minneapolis, Minnesota

Abstract

The history, philosophy, design, construction, instrumentation, and operation of the 40-ft (12.2-m), semitrailer, mobile air pollution laboratory is described. The uniquely designed laboratory contained advanced meteorological, aerosol, and gas analyzers used for simultaneous, continuous measurements at 13 locations in the San Francisco Bay area, the Los Angeles Basin, Fresno, and three remote sites in California. A computerized data acquisition system provided some on-line data reduction and

*Present address: Battelle Columbus Laboratories, 505 King Avenue, Columbus, OH 43201.

recorded data on magnetic tape for subsequent detailed analysis. The laboratory was assembled and the instrumentation was calibrated and operated by a team of specialists from the University of Minnesota, TSI Incorporated, Rockwell International Science Center, and California State Department of Health.

A typical 1-week site visit consisted of 2 to 3 days of setup and calibration, 1 to 2 days of continuous data gathering, 1 day of packing, and 1 day of moving to the next site. Instrumentation routinely operated included the Minnesota Aerosol Analyzing System, SO_2, O_3, NO, NO_x, and hydrocarbon analyzers, wind speed and direction detectors, outdoor and indoor temperature-recording devices, solar radiation sensors, a nephelometer, two Lundgren impactors, several filter samplers, and a high-volume sampler.

I. INTRODUCTION

This paper serves as an index of relevant information concerning the design and construction of the vehicle and sampling system, the modification and calibration of instrumentation, the design and development of the data acquisition system, and the operation of the mobile air pollution laboratory for the Aerosol Characterization Experiment (ACHEX).

II. HISTORY AND PHILOSOPHY OF THE MOBILE LABORATORY

The ACHEX program evolved from the 1969 Los Angeles Smog Project described by Whitby et al. (1972). The objective of the 1969 program was to gain further insight into the mechanisms of formation and behavior of smog, particularly aerosols, by using a sufficient variety of meteorological, chemical, and aerosol measurement techniques on the same aerosol at the same place at the same time. The comprehensive 1969 collaborative research study was carried out by a group of investigators, each equipped with well-tested apparatus and each competent in his or her own area.

Although the 1969 program was a highly successful beginning, it was limited geographically to the air mass that passed the laboratory at the California Institute of Technology in Pasadena. The investigators recognized the need for simultaneous, continuous measurements with the most advanced meteorological, aerosol, and gas analyzers at several urban and rural sites to characterize the atmospheric aerosols of these areas, to attempt to identify relative contributions of natural and anthropogenic aerosol sources, and to learn more about the dynamics of atmospheric aerosols as a function of time and space. A computerized data acquisition system was needed to efficiently process the large quantity of data, both for the on-line analysis required to assure data quality and guide the activities of the crew, and for the detailed analysis to be performed later. Because the high cost of equipment prevented the use of many fixed stations, it became necessary to develop a movable laboratory capable of being moved to a location, set up, calibrated, operated for a continuous 24 to 48 hr period, and repacked for highway travel within 5 to 6 days by an experienced crew. The mobile laboratory described below was a functional departure from previous designs which met the many operational requirements.

The calibration and operation of the advanced instrumentation and the analysis of the data required a team of experienced specialists. Because no single organization possessed all the necessary disciplines, a team of organizations and people was needed. The success of the field program, that is, the development of the mobile laboratory and the gathering of high-quality data on schedule, required a truly cooperative team spirit rather than a strict contractual relationship.

Organizational responsibility for the mobile laboratory was as follows. Rockwell International Science Center, the prime contractor, coordinated the efforts of the other participants. It was also specifically responsible for the development of the hardware and software of the computerized data acquisition system, the site selection and preparation, the scheduling and transportation of the laboratory and personnel, the personal needs of the crew, the supply of a full-time crew member, and the performance of several experiments reported separately. Fruehauf Corporation built the semitrailer and performed many of its modifications. TSI Incorporated designed much of the laboratory, including the floor and roof layout, the air conditioning and electrical systems, and the aerosol- and gas-sampling systems. It also supplied one full-time crew member. The University of Minnesota was responsible for the total laboratory design, including the design and operational coordination of the trailer, instrumentation, and data acquisition systems. It was also responsible for the installation, calibration, and operation of continuous aerosol and meteorological instruments and supplied three or four full-time crew members. The Air and Industrial Hygiene Laboratory (AIHL) of the California State Department of Health was responsible for the installation, calibration, and operation of the gas analyzers and aerosol chemistry samplers and supplied a crew member. The particulate chemical analysis was performed by the AIHL and the Lawrence Berkeley and Lawrence Livermore Laboratories.

III. DESCRIPTION OF THE LABORATORY

The basic vehicle was a standard, commercial 40-ft (12.2-m) long Beaded Panel Van manufactured by Fruehauf Corporation. A semitrailer is the most maneuverable vehicle of its size, and the space inside the trailer is larger and more usable than that of any other commercial vehicle. Semitrailers are ruggedly made to withstand the difficulties of highway travel. Moreover, they are no more expensive than other, similarly sized vehicles and are routinely modified by the supplier to suit customer needs. Although the exterior dimensions are at or near the maximum legal limits, the trailer meets all highway regulations. Also, it readily connects to most standard semitractors.

The exterior of the trailer, shown with most equipment in place in Figure 1, was extensively modified as follows:

FIGURE 1. Photograph of the ACHEX mobile air pollution laboratory.

(1) A 2-in. high catwalk covered the entire roof, making the roof a working platform. Most of the equipment that operated on the roof attached directly to the catwalk by means of J-bolts. The entire trailer was sufficiently reinforced to carry the additional load of the catwalk and equipment.

(2) A single, air-suspended rear axle with dual wheels on each end was mounted near the rear of the trailer. A single axle was sufficient since the total load on the rear axle was well under the legal limit of 18,000 lb. Although double axles would have resulted in a smoother ride, the single axle allowed about 3 ft of additional space in the undercompartment described below. Airsuspension on the rear axle reduced the shock on equipment during transport.

(3) A special compartment below the floor of the trailer was used for transport and storage of miscellaneous equipment such as vacuum pumps and gas cylinders. This undercompartment, installed between the front supports and the rear axle, was about 2 ft high by 8 ft wide by 20 ft long with twin doors on both sides.

(4) A personnel door was located above the supports on the curb side. The door was 32 in. wide and 80 in. high, that is, large enough to allow equipment to mass through, but small enough to reduce interior temperature fluctuations when operating in cold or hot locations. This door plus the rear door provided emergency exists near each end of the trailer.

(5) A smaller door was located within the large curbside rear door. The large rear freight doors were locked shut except when loading or unloading major equipment.

(6) Both personnel doors had suitable stairs, landings, and handrails capable of disassembly and storage in the undercompartment during transport.

(7) Four access ports extended through the roof. The ports handled sampling tubes, electrical extension cords, and electrical signal lines.

(8) Six combination storm-screen windows were installed in the walls, two on each side and one on each end. The windows were about 11 in. high by 18 in. long, that is, large enough to allow the operators to see the surroundings, but small enough to discourage unauthorized personnel from entering.

The trailer was insulated by 2 in. of foam in all four walls and the ceiling. Twin air conditioner-heater units installed near the ceiling in the front exterior of the trailer had a combined cooling capacity of 5 tons and a heating capacity of 28 kw, large enough to maintain 25°C with outdoor temperatures between +37 and -10°C and with reasonable personnel and instrument loads and in-and-out traffic. The use of two identical units rather than a single larger unit increased reliability and resulted in less wear and operating cost since only one unit operated when two were not needed. The conditioned air passed through separate ceiling ducts running the full length of the trailer. The return duct was in the front interior wall of the trailer.

The interior walls and ceiling were finished with wood paneling which met strict fire code requirements. Logistic rails installed along both interior walls allowed equipment to be strapped down during transport. Electrical wiring was mounted on the surface of the interior paneling so that modifications could be made quickly and neatly. Twenty-four incandescent bulbs lit the trailer. Fluorescent lights would have given better lighting but could have interfered with some electronic circuits.

The floor layout, shown in Figure 2, was a departure from that of conventional mobile laboratories, having an aisle down the curbside with work stations branching toward the roadside. This effectively separated the traffic areas from the workplaces. The aerosol instruments were located on the rear island arrangement with the gas instruments on the forward island, as labeled in Figure 3. Instruments faced either forward or rearward on either island with some at desk-top height and others on a higher shelf. The rear of all instruments was readily accessible, and few instruments rested on top of others, so that repairs could be made without interfering with other equipment. Traffic flow was smoother than in conventional mobile laboratories because operators did not need to stand or sit in the main aisle. The primary sampling tubes were directly above the aerosol and gas islands. Some equipment was mounted on the roadside wall.

The data analayasis center with the minicomputer data acquisition system, teletype, plotter, and work tables was located in the center of the laboratory between the aerosol island and the gas island. Both groups of operators had ready access to the system.

The chemistry workbench with a sink and hose-fed water supply was located in the front of the laboratory near the gas

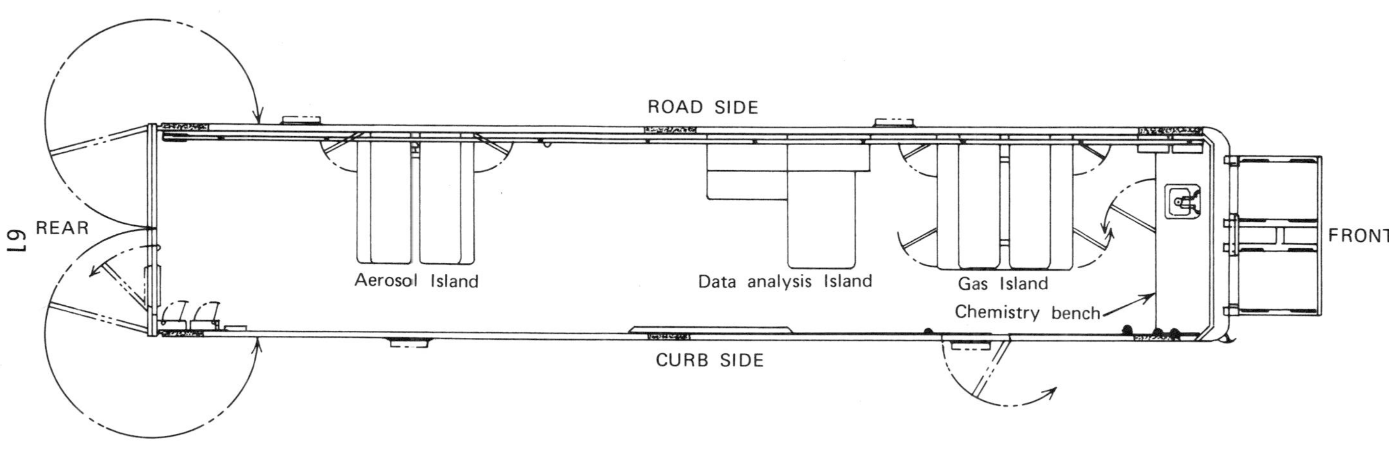

FIGURE 2. Floor plan of the ACHEX mobile laboratory.

island. A pencil board and a bulletin board were mounted opposite the data analysis center. The electrical service center was on the curbside wall near the rear door. Fire extinguishers were mounted on the wall near each door. The aerosol island, gas island, chemistry workbench, and data analysis center had storage drawers below the desk-top work level. Wall cabinets were mounted above the data analysis desk for storage of instructions manuals, data equipment, and the like.

Most interior equipment was strapped, with seat-belt-like straps, to the wall or to the island shelves during transport. Some of the gas instruments were bolted to the gas island. Small, loose items were placed in the drawers.

The layout of instruments on the roof was very flexible. J-bolts, which fastened most equipment to the roof, hooked into any of the thousands of slots in the perforated catwalk. Depending on the type of equipment on the roof and the location and orientation of the trailer, equipment could be shifted around to almost any roof location. Signal cables passed to the interior of the trailer through a modified electrical service entrance cap. Roof-top equipment was carried in either the undercompartment or the interior of the trailer during transport.

Electric power reached the laboratory through a large cable-and-plug assembly, which laid on the ground. The plug mated with a receptacle on a service entrance pole that was set up before the trailer reached a new site. The maximum power rating of the entrance cable was 285 A at 230 V, singlephase. Inside the trailer.the power fed to two service panels: one for air conditioners and the other for instruments and equipment. With both room heaters drawing maximum power, 160 A maximum at 230 V remained for instruments and equipment. Most interior instruments received power through a suspended busway system that allowed 115- or 230-V receptacles to be quickly moved to any point along the full length of the trailer. Additional conventional weatherproof outlets supplied the pumps and blowers in the undercompartment and the equipment on the ground and roof.

IV. SAMPLING SYSTEM

The two primary goals of the aerosol sampling system design were:

(1) To remove a representative sample of atmospheric aerosol, independent of wind speed and direction, from either of two altitudes.

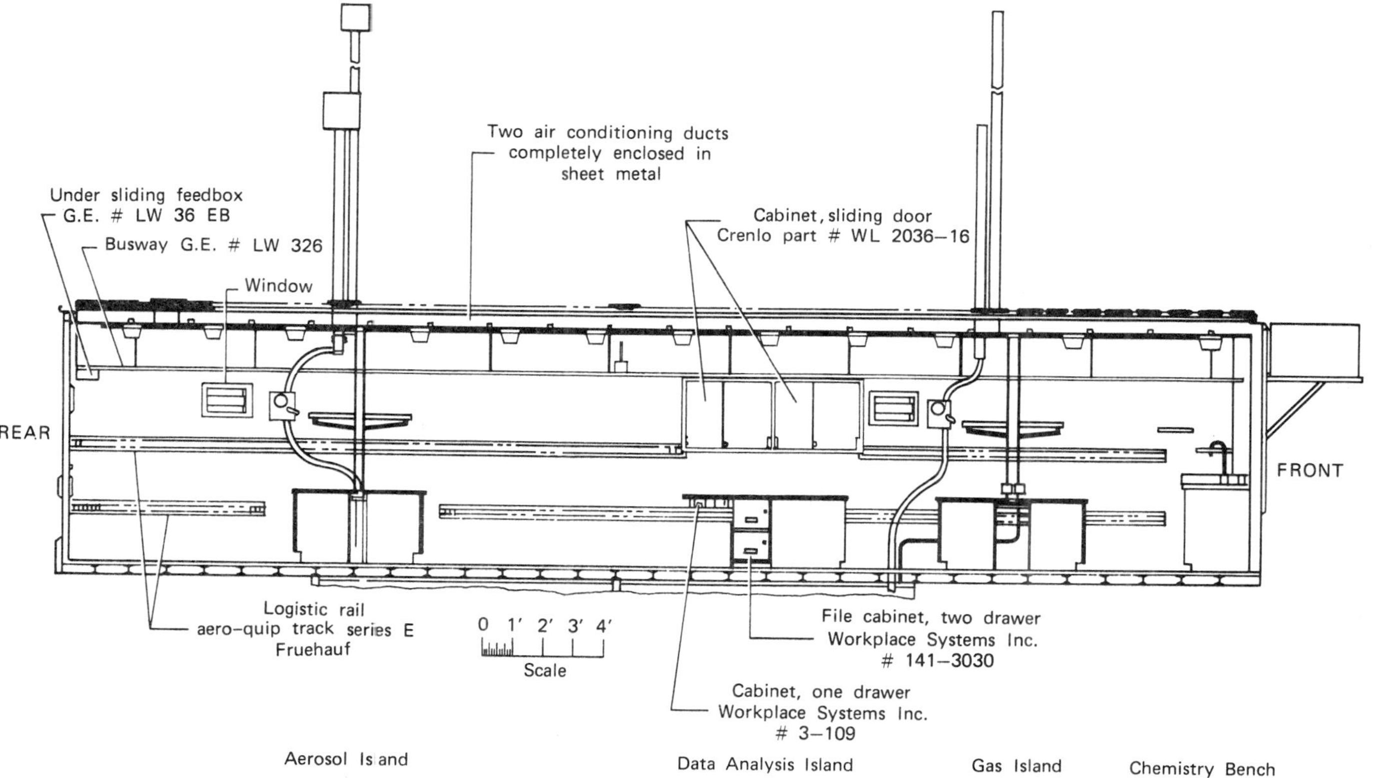

FIGURE 3. Interior roadside scale view of the ACHEX mobile laboratory, including aerosol and gas sampling tubes.

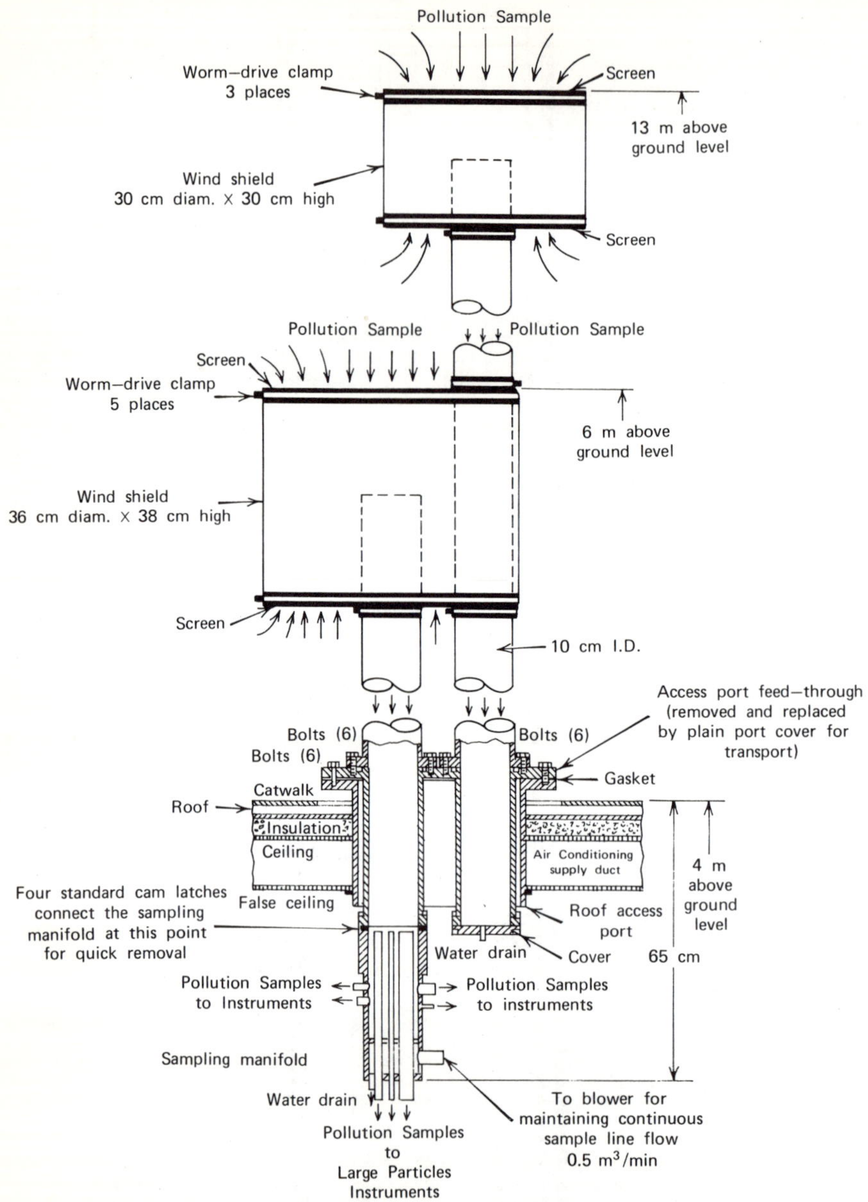

FIGURE 4. Detailed cross section of aerosol sampling tubes and manifold.

(2) To deliver that sample to the instruments inside the laboratory quickly with no significant particle loss.

The aerosol sampling system, shown partially in Figures 1 and 3 and schematically in Figure 4, consisted of two 10-cm-diameter aluminum tubes extending 2 and 9-m above the roof over the aerosol island. The 9-m tube required guy wires. Cylindrical wind shields at the inlets of the tubes reduced the horizontal velocity of particles before they entered the sampling tubes, resulting in less impaction of large particles on the inside of the tube inlet. Screens with 16 wires per centimeter protected the tube inlet from birds, twigs, leaves, bugs, and other foreign matter. A separate blower maintained a flow rate of about 0.5 m^3/min of air through the sampling tube to reduce particle losses, to lower the response time of the system (residence time through the manifold was 8.4 and 2.1 sec for the 9- and 2-m tubes, respectively), and to allow any of the instruments to operate separately or in combination with any other instruments without strongly affecting response time or particle loss. The entire sampling system was vertical, essentially eliminating gravitation losses of particles.

An aerosol manifold was attached to the lower end of the roof port feedthrough. The aerosol instruments, particularly the large-particle monitors and samplers, were arranged to sample as nearly vertically downward as possible. The inlet tubes of all large-particle samplers protruded upward about 20 cm into the vertical sampling tube and sampled approximately isokinetically at that point. Other samplers, including small-particle instruments, drew their samples through any of the many side tubes of the manifold. The manifold could be quickly removed so that the operator could switch from the 2-m to the 9-m sampling tube with little delay. A separate tube allowed rainwater to drain to the ground below the laboratory.

Table 1 shows the penetration efficiency of the sampling tubes for several particle diameters and two densities, using the calculation procedure of Liu and Agarwal (2) and the rough wall data of Agarwal (1). On the assumption of an average roughness height of 50 μm and a typical flow rate of 0.566 cm^3/min, Table 1 shows that the penetration efficiency of the long tube exceeded 94% and that of the short tube exceeded 98% for particles smaller than 20 μm with a density less than 2 g/cm^3. Larger, denser particles had lower penetration efficiencies.

Agarwal (1) has also analyzed the inlet sampling efficiency in calm air as a function of particle diameter, tube size, and average sampling velocity in the tube. For the typical operating condition of 1.16-m/sec sampling velocity, 10-cm tube diameter,

TABLE 1. Particle Penetration for Each Aerosol Sampling Tube as a Function of Particle Diameter and Density, Calculated from Liu and Agarwal (2) and Agarwal (1), Assuming 50-μm Roughness Height and 0.57-m^3/min Flow Rate.

Particle density, ρ (g/cm)	Tube Length (m)	Penetration Particle Diameter (μm)			
		10	20	30	35
1.0	9.8	0.999	0.988	0.928	0.901
	2.4	0.999	0.997	0.981	0.974
2.0	9.8	0.997	0.945	0.724	0.606
	2.4	0.999	0.986	0.922	0.882

and particle diameter smaller than 20 μm, Agarwal showed 100% sampling efficiency. For 30-μm particles Agarwal found about 103% sampling efficiency. The impaction losses onto the inside of the tube entrance, neglected by Agarwal, were considered negligible for wind velocities below 3 m/sec with the wind shields in place.

Figure 5 shows the distribution of aerosol instruments around the aerosol manifold. Instruments that were sensitive primarily to particles smaller than 1 μm sampled from side ports, while the supermicron instruments sampled directly downward. The two Lundgren samplers had to be tilted 20° from the vertical.

The sampling system above the gas analyzer island was similar to the aerosol system except that both 2- and 9-m tubes were 5 cm in diameter and no wind shields were necessary. Some instruments, such as ozone sensors, sampled from the center of the manifold.

V. INSTRUMENTATION

Whitby et al. (5) have listed the various instruments in the mobile laboratory and the groups responsible for their calibration and operation. The methods used for the calibration of the aerosol instrumentation have been described elsewhere [Liu et al. (3), Whitby et al. (5), Sverdrup et al. (4)].

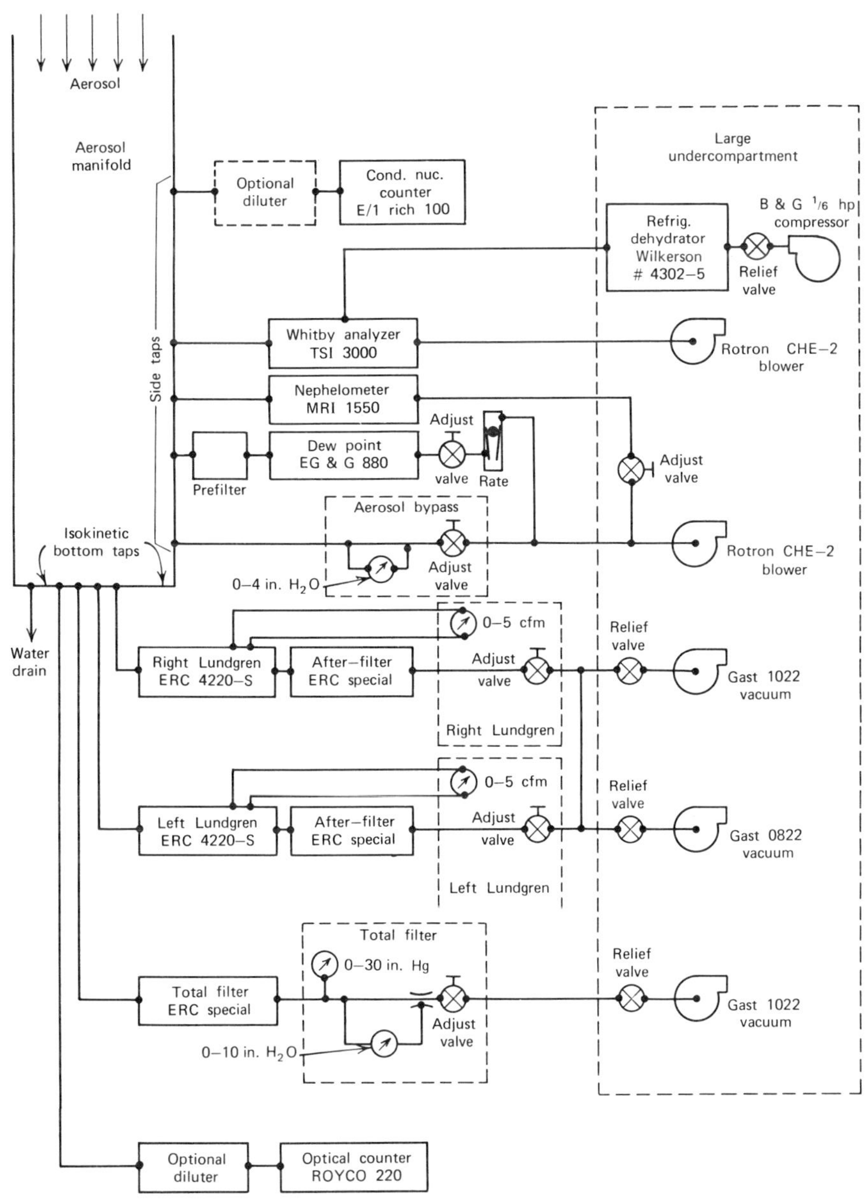

FIGURE 5. Schematic diagram of aerosol instrumentation flow sytem.

The aerosol instrumentation was calibrated at the University of Minnesota. The electrical aerosol analyzer (EAA) used for the particle size range 0.01 to 0.422 µm was calibrated based upon measurements of the flow rates and electrical parameters. The condensation nuceli counter was adjusted to give the same aerosol concentration as the EAA for aged aerosols of about 0.02-µm diameter.

Two optical particle counters (OPC) were used. A Royco Model 220 OPC served for the particle size range 0.422 to 5.62 µm, and Royco Model 245 for the size range 5.62 to 38 µm. The calibration of the OPCs has been described by Liu et al. (3).

Sverdrup et al. (4) have discussed the operational procedures used at the background sites to make size distribution measurements of dilute background aerosols.

REFERENCES

1. Agarwal, J. K. 1975. "Aerosol Sampling and Transport", Ph.D. Thesis, University of Minnesota, Minneapolis.

2. Liu, B. Y. H., and Agarwal, J. K. 1974. Experimental Observation of Aerosol Deposition in Turbulent Flow, J. Aerosol Sci. 5, 145-155.

3. Liu, B. Y. H., Berglund, R. N., and Agarwal, J. K. 1974. Experimental Studies of Optical Particle Counters, Atmos. Environ. 8, 717-732.

4. Sverdrup, G. M., Whitby, K. T., and Clark, W. E. 1975. Characterization of California Aerosols - II. Aerosol Size Distribution Measurements in the Mojave Desert, Atmos. Environ. 9, 483-494.

5. Whitby, K. T., Clark, W. E., Marple, V. A., Sverdrup, G. M., Sem, G. J., Willeke, K., Liu, B. Y. H., and Pui, D. Y. H. 1975. Characterization of California Aerosols - I. Size Distributions of Freeway Aerosol, Atmos. Environ. 9, 463-482.

6. Whitby, K. T., Liu, B. Y. H., Husar, R. B., and Barsic, N. J. (1972). The Minnesota Aerosol-Analyzing System Used in the Los Angeles Smog Project, J. Colloid Interface Sci. 39, 136-164.

Quality Assurance for the Chemistry of the Aerosol Characterization Experiment

BRUCE R. APPEL, JEROME J. WESOLOWSKI, ARTHUR ALCOCER, STEPHEN WALL, and SUZANNE TWISS
Air and Industrial Hygiene Laboratory
Laboratory Services Branch
California State Department of Health
Berkeley, California

ROBERT GIAUQUE
Lawrence Berkeley Laboratory
University of California
Berkeley, California

and

RICHARD RAGAINI and ROBERT RALSTON
Lawrence Livermore Laboratory
University of California
Livermore, California

I. INTRODUCTION

The concept of "quality assurance" in the chemistry for a program such as the Aerosol Characterization Experiment (ACHEX) is clearly multifaceted. To correctly characterize atmospheric aerosols, efforts were made at every step during sampling and analysis to minimize the likelihood of error.

This report discusses intermethod and intercollection media comparisons and some specific experiments undertaken to provide quality assurance as related to particulate sampling, handling, and analysis for the ACHEX. Users of the ACHEX data bank should find this report useful in assessing the accuracy of the data on individual chemical species.

Sampling for ACHEX during the summer and fall of 1972 is referred to as "Phase I", while that during 1973 is termed "Phase II". This report emphasizes the latter.

A number of topics relating to quality assurance for the ACHEX have been included in other papers, including an intermethod comparison of electron spectroscopy chemical analysis results (1) and descriptions and validation of methods for sulfate and nitrate analysis (2,3). These will not be discussed here.

II. GUIDE TO DATA BANK CHEMICALS RESULTS

Since time did not permit carrying out all the necessary validation work before the field experiment, a number of validation procedures were incorporated into the experiment itself. For example, the Lundgren impactors were employed to sample in pairs, with either the same or different collection media. Similarly, analyses for elements such as Na, Cl, Al, and S were obtained by more than one technique on the same sample.

Users of the ACHEX data banks may, therefore, find it necessary to make a choice of which analytical results would be best to use. Table 1 provides the Air and Industrial Hygiene Laboratory's recommendations for this choice. Users will find, in many cases, close agreement between the results of two methods of analysis on the same sample. Thus this table is not to be construed as suggesting that any of the methods are unsuitable for the determinations listed.

With regard to Phase I Lundgren impactor results, those employing sticky polyethylene (SPE) as impaction media are considered more reliable (both for impaction stages and after-filters) because of the diminished likelihood of wall loss and bounce-off problems.

In Phase II, Lundgren impactors sampled side by side with SPE on both; however, the systems differed in the after-filter medium. In a limited number of cases, stages from both impactors were analyzed by X-ray fluorescence analysis (XRFA) to confirm the equivalency of the sampling scheme. In these cases duplicate results were entered in the data bank. The XRFA results averaged over both impactors should be more reliable than those for either impactor individually.

III. WALL LOSSES IN THE LUNDGREN IMPACTOR

Previous work has indicated that SPE, as the impactor collection medium, has substantially greater efficiency than bare plastic strips (e.g., Mylar or Teflon) (4). Nevertheless, there remained substantial concern that SPE was far from ideal.

Figure 1, prepared by K. Rao, University of Minnesota Particle Technology Laboratory (5), compares a greased surface, SPE as used in the ACHEX, and bare foils as impaction surfaces for polystyrene lattices. Although better than the bare plastic foils, SPE was substantially less efficient than the greased surfaces, at least for particles with physical characteristics similar to those of polystyrene latex spheres.

TABLE 1. AIHL Choice of Analytical Results in Cases Where Values Were Obtained by Two or More Methods[a]

Species	Analytical Methods[b]	Choice	Comments
Elements heavier than Mg	DA,G,Q1[c], Q2,L1,L2,L3	Q1,Q2, L1,L2,L3	Exceptions are Br, because of volatility, and the soil-related elements Ca and Fe, because of loss of large-particle-related elements. In these cases we suggest using the G values.
Na	Q2,A	A	
Cl	G,Q2,A[d]	A	
Sulfate	G,A,DA[e]	A	A provides SO_4^{2-}; G and DA provide total S.
Al	Q2,A	A	
C	E,A	A	

[a]The choice refers only to the ACHEX data bank and does not necessarily imply lack of suitability of any method for the determination listed.

[b]
DA = α-XRFA, University of California at Davis.
G = XRFA, Lawrence Berkeley Laboratory.
Q1,Q2, L1,L2 L3 = NAA, Lawrence Livermore Laboratory (short and long irradiations and first, second, or third counts).
A = Wet chemical and/or instrumental, AIHL.
E = ESCA, Lawrence Berkeley Laboratory.

[c]Where a choice is required between DA and G, G is suggested.

[d]Where a choice is required betewen Q2 and G, Q2 is suggested.

[e]Where a choice is required between G and DA, G is suggested.

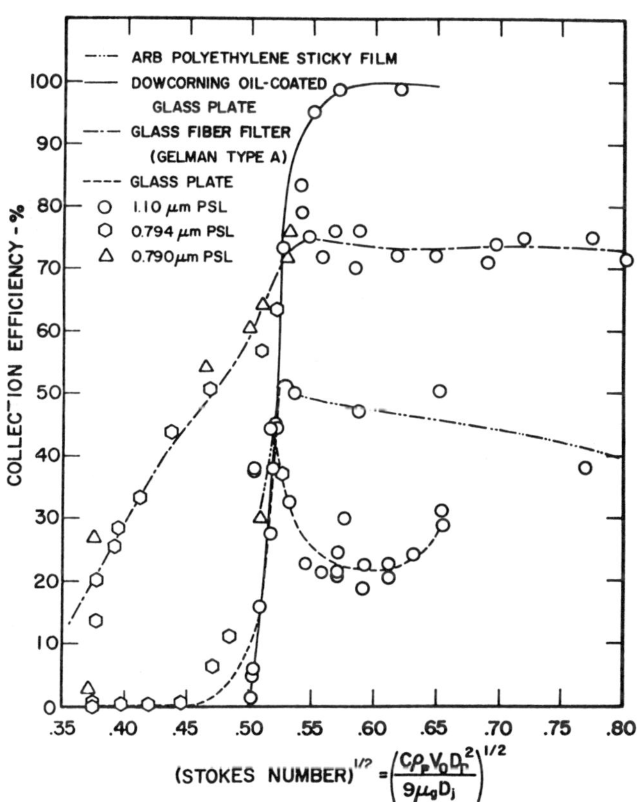

FIGURE 1. Collection efficiency of inertial impaction media (Minnesota Particle Technology Laboratory).

Wall loss was evaluated with atmospheric aerosols by comparing the quantity of a particular element summed over all stages of the Lundgren impactor, including the after-filter, to the amount collected on a total filter employing XRFA.* Both the Lundgren sampler and the total filter were designed to sample

*Wall losses for sulfate and nitrate are discussed separately in Section VI.

the same manifold air stream isokinetically, making quantitative agreement possible, in the absence of wall loss. Table 2 details such XRFA results for eight episodes obtained on 2-hr strips cut from the impaction medium, SPE, followed by a Gelman GA-1 membrane after-filter.

Agreement within 20% was obtained between the integrated Lundgren sampler stages and the total filter data for the elements S, Ni, Br, and Pb, which are present principally in smaller particles, that is, <4 μm. There were often losses of 40 to 50% of the larger particles, as indicated by the K, Ca, Ti, Mn, and Fe data. Losses were especially pronounced at Rubidoux, where a sizable fraction of the aerosols was probably composed of windblown soil dust.

IV. EQUIVALENCY OF THE TWO LUNDGREN IMPACTORS AND OF THE TWO TOTAL FILTER SAMPLERS

The equivalency of the two Lundgren impactors employed on the California Air Resources Board (ARB) mobile laboratory was tested by XRFA analysis of the stages from side by side sampling. Results for impactors A and B are shown in Figures 2 and 3 by plotting size distributions for four elements, Pb, Br, Fe, and Ca, chosen to represent small- and large-particle-related elements. The data shown are for samples collected at West Covina, each impactor using SPE impaction media. The available data indicate agreement within experimental error for Pb and Br size distributions. The size distributions for Fe and Ca reveal somewhat poorer agreement, compared to results for the elements principally associated with small particle sizes, although the results are not sufficiently different to cause differences in interpretation.

Two total-filter samplers were employed, sampling isokinetically from the same manifold used by the Lundgren impactors. One unit used Gelman GA-1 membrane filters; the other, Gelman A glass fiber. The comparison of the two total filters is based upon chemical analysis for large- and small-particle-related elements. Table 3 includes a comparison of Fe, Br, and Pb loadings obtained from the glass fiber (TV) and Gelman GA-1 membrane (TF) total filters during the July 23-24, 1973 (West Covina) episode. The ratio of results, TF/TV, was between 0.94 and 1.0, indicating quite good agreement. Previous work established, on a particle number basis, 98% efficiency for the GA-1 and 95% for the Gelman A toward atmospheric aerosol at the flow rate used here (3 cfm) as measured with a condensation nuclei counter (6). Accordingly,

TABLE 2. Comparison of Impactor Stages Plus After-Filter and Total-Filter Data: XRFA Analysis ($\mu g/m^3$)

Episode	Stage[a]	Time	S	K	Ca	Ti	Mn	Fe	Ni	Br	Pb
7/23-24/73	A	1200-1400	<400[b]	290±29	395±60	73±7	5±2	760±30	<3	16±1	42±2
	B		<400	39±25	<60	6±5	<6	156±6	<3	11±1	10±2
	2		<400	107±27	204±57	17±6	3±2	334±13	<3	21±1	42±3
	3		460±160	171±27	325±59	29±6	4±2	388±16	<4	53±2	184±7
	4		3,200±300	<78	<60	<15	<7	104±4	5±1	107±4	396±16
	AF		10,500±1,100	303±59	166±30	<46	9±5	306±12	41±3	488±20	2,930±120
	Σ(A-AF)[c]		14,200	910	1,090	125	21	2,048	46	696	3,604
	TF[d]		15,200±1,500	1,690±80	2,090±80	392±98	93±7	3,580±140	53±4	735±29	3,890±150
	Σ(A-AF)/TF		0.93	0.54	0.52	0.32	0.23	0.57	0.87	0.95	0.93
7/24-25/73	A	1200-1400	<420	389±39	626±31	118±7	21±3	1,030±40	6±1	21±2	88±4
	B		<420	70±29	113±16	36±6	3±2	186±9	2±1	13±2	17±2
	2		350±150	201±30	322±19	72±6	13±3	532±21	5±1	26±2	92±4
	3		340±160	154±30	306±19	49±6	7±2	425±17	4±1	52±2	195±8
	4		2,900±300	98±29	40±16	21±12	6±2	111±4	10±1	94±4	436±17
	AF		10,200±1,000	307±59	161±30	<60	22±5	330±13	52±3	490±20	3,220±130
	Σ(A-AF)		13,790	1,219	1,568	296	72	2,614	79	696	4,048
	TF		13,500±1,400	1,470±80	2,430±100	473±98	96±7	4,050±160	67±4	787±31	4,380±180
	Σ(A-AF)/TF		1.02	0.83	0.65	0.63	0.75	0.65	1.18	0.88	0.92

TABLE 2 (Continued)

Episode	Stage	Time	S	K	Ca	Ti	Mn	Fe	Ni	Br	Pb
7/26/73	A	1200-1400	<450	340±31	619±71	91±7	17±3	1,020±40	2±1	15±1	69±3
	B		<420	<78	<80	<15	<7	161±6	<3	8±1	15±2
	2		<420	81±27	179±57	18±6	6±2	300±12	<3	20±1	51±3
	3		490±160	139±27	206±58	18±6	<8	284±11	<3	38±2	129±5
	4		4,600±500	67±28	<60	<15	<8	100±4	2±1	98±4	375±15
	AF		3,400±500	192±52	111±27	18±10	<13	95±5	10±2	202±8	1,360±50
Σ (A-AF)			8,490	819	1,115	145	23	1,960	14	381	1,999
	TF		7,600±800	940±73	1,310±50	233+53	42±6	2,000±80	18±3	393±16	2,160±90
Σ(A-AF)/TF			1.12	0.87	0.85	0.62	0.55	0.98	0.78	0.97	0.93
8/16-17/73	A	1200-1400	<400	442±44	589±30	149±24	18±3	1,220±50	3±1	16±2	69±3
	B		<400	70±26	93±15	28±5	5±2	138±5	2±1	8±2	18±2
	2		<400	100±28	172±17	26±5	7±3	242±10	3±1	20±2	53±2
	3		820±160	153±28	214±17	40±17	10±3	374±15	4±1	51±2	187±8
	4		5,500±600	113±28	58±15	<18	5±3	136±5	8±1	140±6	593±24
	AF		7,500±800	<156	35±29	<35	20±5	167±7	47±3	272±11	2,020±80
Σ(A-AF)			13,820	878	1,161	243	65	2,277	67	507	2,940
	TF		12,300±1,200	1,150±120	2,110±80	395±94	84±7	3,420±140	67±4	507±20	3,020±120
Σ(A-AF)/TF			1.12	0.76	0.55	0.62	0.77	0.67	1.00	1.00	0.97

TABLE 2 (Continued)

Episode	Stage	Time	S	K	Ca	Ti	Mn	Fe	Ni	Br	Pb
9/5-6/73	A	1200-1400	<400	330±33	584±29	73±6	14±3	694±28	<3	8±2	18±2
	B		<400	164±27	241±17	30±12	5±2	257±10	2±1	4±1	14±2
	2		<400	207±28	400±20	44±17	7±2	386±16	<3	12±2	36±2
	3		<400	179±29	453±23	44±6	8±3	349±14	2±1	24±2	93±4
	4		1,200±200	98±27	63±15	<15	<7	56±3	4±1	59±2	244±10
	AF		2,600±400	<156	107±29	<24	<12	94±5	6±2	145±6	967±35
	Σ(A-AF)		3,800	978	1,848	191	34	1,836	14	252	1,272
	TF		5,300±500	1,370±140	3,280±130	340±39	69±6	3,150±130	10±3	294±12	1,500±60
	Σ(A-AF)/TF		0.72	0.71	0.56	0.56	0.49	0.58	1.4	0.86	0.85
9/18-19/73	A	1200-1400	400±150	540±54	1,410±70	148±17	28±3	1,430±60	< 8	5±1	12±2
	B		<800	225±30	687±34	68±17	9±3	604±24	< 8	5±2	7±2
	2		<800	297±30	584±29	69±20	13±2	523±21	2±1	7±1	34±2
	3		500±150	200±29	474±20	48±6	8±2	379±15	< 3	19±2	72±3
	4		4,100±400	43±29	76±16	< 21	6±3	81±3	10±1	107±4	509±20
	AF		3,900±400	< 169	193±29	< 30	< 14	151±6	13±2	129±5	754±30
	Σ(A-AF)		8,900	1,305	3,429	333	64	3,168	21	272	1,388
	TF		11,300±1,000	2,570±130	8,830±440	815±129	141±8	7,200±290	26±3	294±5	1,650±70
	Σ(A-AF)/TF		0.79	0.51	0.39	0.41	0.45	0.44	0.81	0.93	0.84

TABLE 2 (Continued)

Episode	Stage	Time	S	K	Ca	Ti	Mn	Fe	Ni	Br	Pb
10/4-5/73	A	1000-1200	250±150	102±27	188±16	35±11	6±2	293±12	3±1	7±1	23±2
	B		<400	39±27	28±15	15±5	<6	39±3	<3	5±1	5±2
	2		<400	<83	110±16	<15	<7	104±4	4±1	11±2	18±2
	3		950±150	96±27	174±16	20±8	7±2	160±6	7±1	26±2	72±3
	4		9,600±900	81±29	65±16	27±8	<8	166±6	13±1	61±2	510±20
	AF		4,400±400	<178	97±30	27±11	<14	59±5	28±3	89±4	637±25
	Σ(A-AF)		15,200	318	662	124	13	821	55	199	1,265
	TF		14,300±1,400	355±63	693±37	123±30	33±6	1,010±40	47±3	214±9	1,280±50
	Σ(A-AF)/TF		1.06	0.90	0.96	1.01	0.33	0.81	1.17	0.93	0.99
10/10-11/73	A	1200-1400	<400	50±27	102±15	15±11	3±2	134±5	3±1	9±1	26±2
	B		<400	37±27	35±15	14±5	<7	58±3	<3	8±1	8±2
	2		<400	76±27	101±16	14±10	<7	93±4	2±1	12±2	28±2
	3		<400	112±28	114±16	21±5	<7	91±4	4±1	30±2	57±3
	4		350±150	<83	41±15	<15	4±2	58±3	4±1	39±2	102±4
	AF		2,900±400	<178	84±29	<31	13±5	81±5	51±3	160±6	782±31
	Σ(A-AF)		3,250	275	477	64	20	515	64	258	1,003
	TF		4,400±400	220±68	424±37	44.5	26±6	561±22	80±4	220±9	1,130±50
	Σ(A-AF)/TF		0.74	1.25	1.13	1.44	0.77	0.92	0.80	1.17	0.89

[a]Stage designation: the cut points for these stages are discussed in reference 4.

[b]Values shown to be below the detection limit were assumed equal to zero in summing the stages.

[c]Sum of stages A through after-filter.

[d]Total filter.

TABLE 3. Comparison of Glass Fiber (TV) and Gelman GA-1 (TF) Total-Filter Loadings (ng/m^3) for West Covina, 7/23 - 24/73

Element	Time (PST)										Ratio of means, TF/TV[a]
	2100-0102		0806-1004		1007-1200		1600-1804		1805-2100		
	TF	TV	TF	TV	TF	TV	TF	TV	TF	TV	
Fe	2180	2510	3010	3400	4530	4250	2970	3040	2130	2570	0.94 ± 0.05
	(90)[b]	(130)	(120)	(170)	(180)	(210)	(120)	(150)	(90)	(130)	
Br	928	720	1460	1560	698	700	633	610	612	550	1.00 ± 0.07
	(37)	(50)	(60)	(80)	(28)	(40)	(25)	(30)	(25)	(30)	
Pb	4400	4460	6650	7470	3810	3760	3330	3220	3670	3690	0.97 ± 0.03
	(180)	(220)	(270)	(380)	(150)	(190)	(130)	(160)	(150)	(190)	

[a]The errors shown are 1σ values calculated to include the sample-to-sample variability, including both analytical and other sources of error (e.g., sampling error). The σ values listed were calculated treating TF and TV determinations as dependent variables. If they are treated as independent variables, the 1σ values increased by a factor of 3 to 6.

[b]Parenthetical values are analytical errors for the value immediately above.

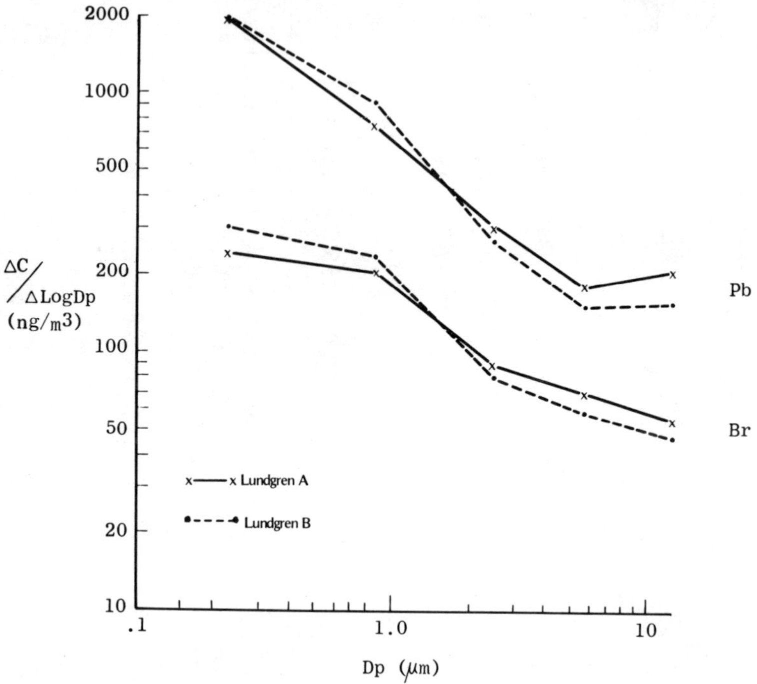

FIGURE 2. Size distribution of lead and bromine obtained with the Lundgren impactor: West Covina, July 26, 1973.

good agreement would be expected, assuming equivalence of sampling geometry and flow rates and high reproducibility in the analytical scheme.

The data presented in Sections III and IV demonstrate the symmetry of the manifold and sampling units and the equivalence of the flow calibration, as well as the good precision of the analytical methods used.

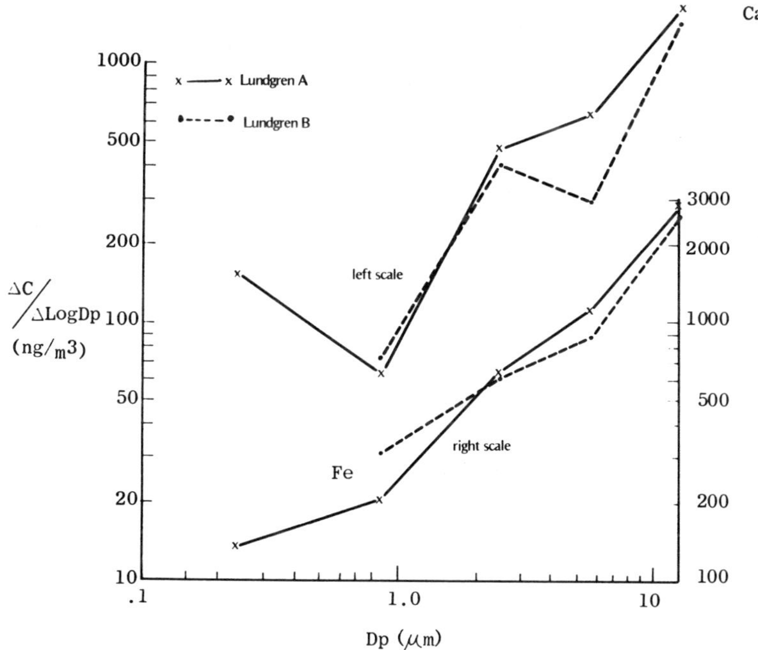

FIGURE 3. Size distribution of iron and calcium obtained with the Lundgren impactor: West Covina, July 26, 1973.

V. COMPARISON OF ELEMENTAL CONCENTRATIONS ON TOTAL VERSUS HIGH-VOLUME FILTER SAMPLES

Four elements were chosen for comparison of analyses of Gelman GA-1 membrane total-filter samples with analyses of 8 x 10 in. Whatman 41 high-volume filter samples: Pb, Br, Fe, and Ca as determined by XRFA. The first two represented the small-particle elements, while Fe and Ca were chosen to represent large-particle-related elements.

Figures 4 to 7 plot the calculated 24-hr average values obtained with the membrane total filters against the 24-hr (Whatman 41), high-volume filter value. The results for Pb and Br indicate, on the average, 18 to 19% higher results with the membrane

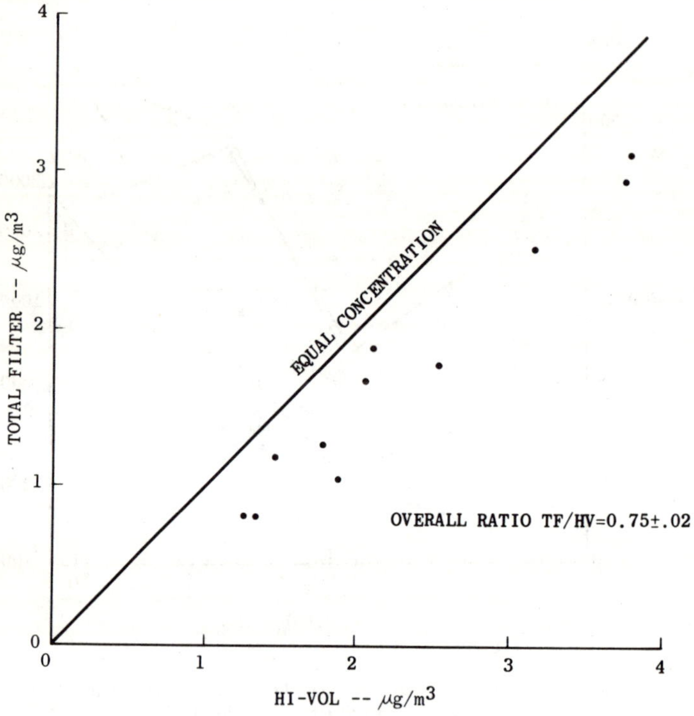

FIGURE 4. Twenty four hour calcium concentration determined on high-volume and total filters: South Coast Air Basin, July-October, 1973.

filters. These results are similar to those found on comparing Pb analyses on high-volume samples collected with glass filter versus Whatman 41. In studies conducted jointly by the Orange County APCD, the ARB, and the AIHL, the Pb values on cellulose (Whatman 41) were about 30% lower than those on glass fiber filters. This was shown to result from the incomplete collection of the small-particle-related Pb on the cellulose filter. These results are also consistent with laboratory studies comparing the collection efficiencies of Whatman 41, glass fiber, and cellulose ester membrane filters (6).

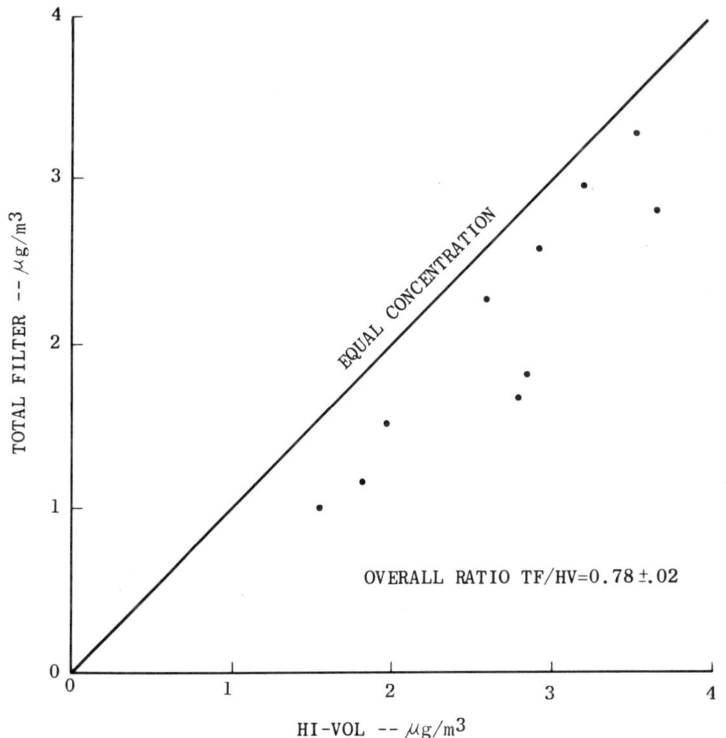

FIGURE 5. Twenty-four hour iron concentration determined on high-volume and total filters: South Coast Air Basin, July-October 1973.

The results for Ca and Fe are the reverse of those for Pb and Br; since the upper size cutoff for the total filter is about 20 μm and that for the high-volume sampler is substantially higher,* the higher results with the latter sampler are reasonable for these soil-related elements.

---

*A recent study demonstrated up to 34% efficiency for collection of 50-μm particles with a wind speed of 4.6 m/sec (8).

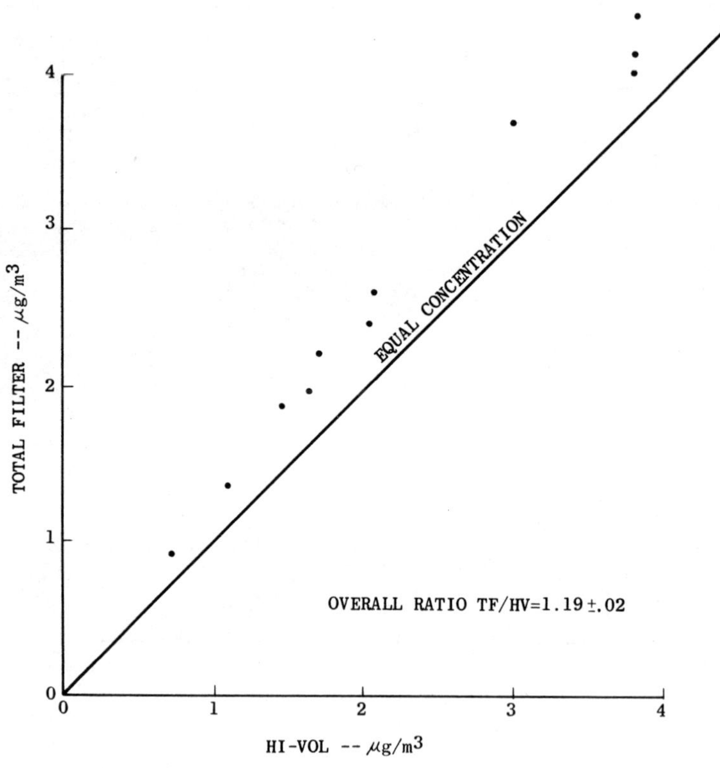

FIGURE 6: Twenty-four hour lead concentration determined on high-volume and total filters: South Coast Air Basin, July-October, 1973.

VI. SAMPLING AND ANALYTICAL ERRORS IN SULFATE AND NITRATE DETERMINATIONS

Sulfate and nitrate analyses were carried out on Gelman GA-1 membrane after-filters and total filters, on cellulose (Whatman 41) high-volume filters, and on SPE used with the Lundgren impactor. The analytical methods used have been detailed elsewhere (2,3).

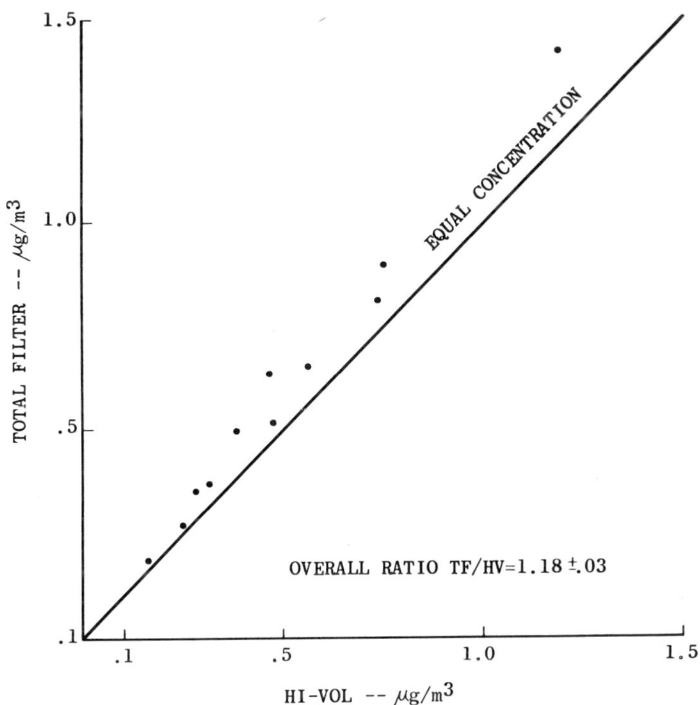

FIGURE 7. Twenty-four hour bromine concentration determined on high-volume and total filters: South Coast Air Basin, July-October, 1973.

Table 4 compares the sulfate and nitrate analyses obtained on high-volume filters to the calculated 24-hr average from the eight Gelman GA-1 total filters run during the same period. With the exception of episodes beginning 8/16, 9/5 and 9/24/73, the sulfate results show little scatter and support the equivalency of sulfate analyses obtained by the two sampling techniques. The results also imply minimal artifact sulfate formation on these filter media. In marked contrast to sulfate, the values for

TABLE 4. Comparison of Nitrate and Sulfate on Whatman 41 24-hr High-Volume Filters (HV) and on Gelman GA-1 Membrane Total Filters (TF).

		TF/HV[a]	
Site	Date	NO_3^-	SO_4^{2-}
West Covina	7/23-24/73	2.32 ± 0.34	0.88 ± 0.12
West Covina	7/24-25/73	2.02 ± 0.28	0.97 ± 0.13
West Covina	7/26/73	1.76 ± 0.26	1.01 ± 0.14
West Covina	8/8-9/73	1.76 ± 0.24	1.14 ± 0.16
Pomona	8/16-17/73	4.20 ± 0.62	1.67 ± 0.23
Rubidoux	9/5-6/73	1.21 ± 0.17	2.03 ± 0.29
Rubidoux	9/18-19/73	1.55 ± 0.23	0.99 ± 0.14
Rubidoux	9/24-25/73	1.47 ± 0.22	3.59 ± 0.53
Dominguez Hills	10/3-4/73	7.68 ± 1.28	1.01 ± 0.14
Dominguez Hills	10/10-11/73	1.53 ± 0.22	1.15 ± 0.16
Overall		1.67 ± 0.18[b]	1.14 ± 0.11[b]

[a]TF values are 24-hr averages calculated from eight filters collected during this period.

[b]The values were calculated as in footnote a, Table 3. If means are assumed to be independent, σ values are 0.66 and 0.30 for NO_3^- and SO_4^{2-}, respectively.

nitrate obtained on the Gelman total filters are systematically higher than those for the high-volume filters. Possible sources of this discrepancy include:

(1) Greater artifact particulate nitrate formation on the cellulose acetate membrane filter relative to that on the cellulose high volume filter, and

(2) Lower particle collection efficiency for cellulose filters.

The latter is probably not a viable argument, however, because of the good agreement between sulfate values on cellulose and membrane filter samples.

TABLE 5. Apparent Losses of Nitrate and Sulfate in the Lundgren Impactor.

Site	Date	Σ Lundgren Stages/TF[a]	
		NO_3^-	SO_4^{2-}
West Covina	7/23-24/73	0.71 ± 0.08	1.14 ± 0.14
West Covina	7/26/73	0.56 ± 0.07	0.66 ± 0.09
Pomona	8/16-17/73	0.52 ± 0.06	0.94 ± 0.11
Rubidoux	9/5-6/73	0.87 ± 0.11	1.05 ± 0.13
Rubidoux	9/18-19/73	0.96 ± 0.12	1.12 ± 0.13
Dominguez Hills	10/3-4/73	0.79 ± 0.11	1.07 ± 0.13
Dominguez Hills	10/10-11/73	0.66 ± 0.08	1.27 ± 0.15
Overall		0.79 ± 0.08[b]	1.02 ± 0.06[b]

[a]The sum of the concentrations for the stages of the Lundgren impactor divided by the calculated 24-hr average concentration for the Gelman GA-1 membrane total filters.

[b]Calculated as in footnote a, Table 3. If variables are assumed to be independent, σ values are 0.38 and 0.25 for overall NO_3^- and SO_4^{2-}, respectively.

The analysis of Lundgren impactor foils for sulfate and nitrate was done on sections representing 22 hr and 2 hr of sampling, the latter section being removed from an early afternoon period. The sum of these two (i.e., the 24-hr average sulfate and nitrate levels) from the foils plus after-filter was compared to the calculated 24-hr averages for the total filters run simultaneously, with the results shown in Table 5.*

The results for sulfate indicate, with two exceptions, recoveries close to the calculated values for the total filter. In contrast to sulfate, the nitrate values display systematically

*The filters were extracted with a micropercolation technique (2) into 5 ml H_2O, while the foils were extracted for 30 min in 20 ml H_2O at 80°C. Any differences in extraction efficiency between these techniques would contribute to any discrepancies observed for both nitrate and sulfate.

lower recoveries in the Lundgren impactor. Possible sources of this discrepancy include:

(1) Enhanced volatilization of nitrate from particles caught on impaction surfaces, perhaps because of the high air jet velocity.

(2) Volatilization of nitrate (as HNO_3) during neutron activation analysis (NAA). All foils, as well as selected midday after-filters and total filters, were first analyzed by NAA. The samples are subjected to moderate heating during NAA (measurements indicated a maximum value of 57°C). In addition, nitrates may be decomposed by the gamma radiation and/or neutron bombardment; it is known that nitrates are subject to decomposition by ultraviolet light (7).

(3) Collection of nitric acid, an artifact particulate nitrate precursor, on the inner walls of the impactor.

In an effort to elucidate the source of error, diurnal plots for NO_3^- were constructed for after-filter and total-filter data, seeking obvious loss of nitrate for the samples analyzed first by NAA. Figures 8 and 9 for Rubidoux and Dominguez Hills reveal that the samples analyzed first by NAA (indicated by the arrow) appear to be low compared to neighboring samples. Furthermore, this marked negative excursion in concentration is not paralleled by the nonvolatile species, lead. Similar plots for other episodes failed, however, to reveal a clear suggestion of such loss.

To assess directly the importance of nitrate loss under NAA conditions, three filter samples, collected during the 1972 sampling period (Phase I) in Riverside, were split in half with one piece subjected to NAA. The pieces were then analyzed for nitrate with the following results.

Sample	NO_3^-/NO_3^- (after NAA)
1	1.02
2	0.89
3	0.78

These results fail to reveal a significant loss of nitrate by exposure ty NAA conditions. However, the samples available for this study had been maintained (in closed Millipore filter boxes) at room temperature for nearly 2 years, conditions that may have made them inappropriate for this study.

The collection of gaseous nitric acid on the SPE impaction surfaces, cellulose acetate after filter, and the interior walls

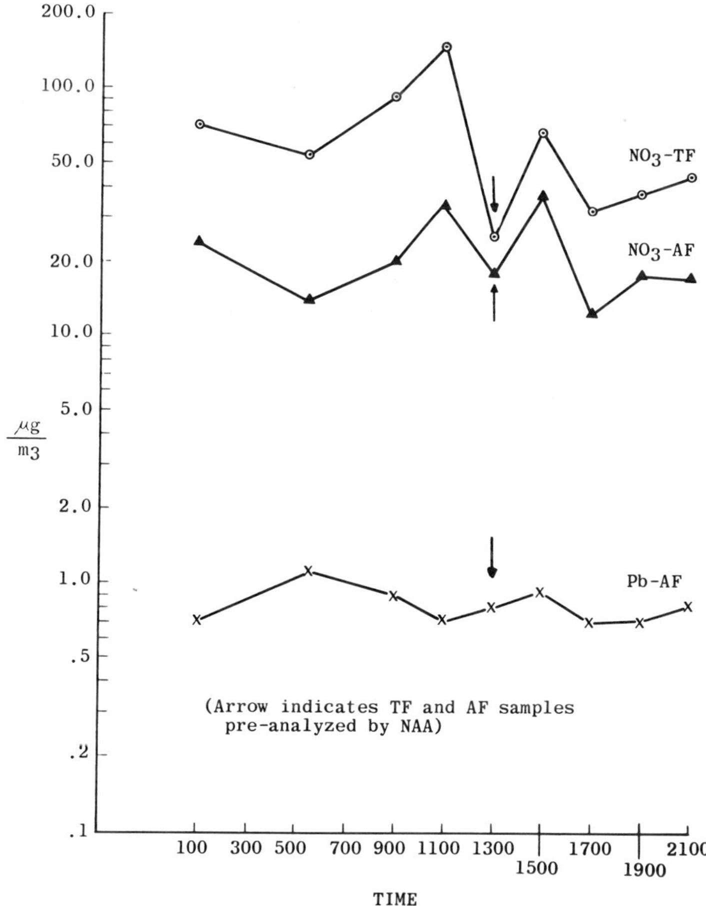

FIGURE 8. Diurnal variation of nitrate and lead; Rubidoux, September 18-19, 1973.

of a four stage Lundgren impactor was evaluated by sampling clean air containing 53 ± 1 ppb (138 $\mu g/m^3$) HNO_3 at 50% relative humidity at 20°C. After two hours sampling the nitrate levels for stages 1, 2, 3, 4, after-filter and interior walls were 4.3, 3.8,

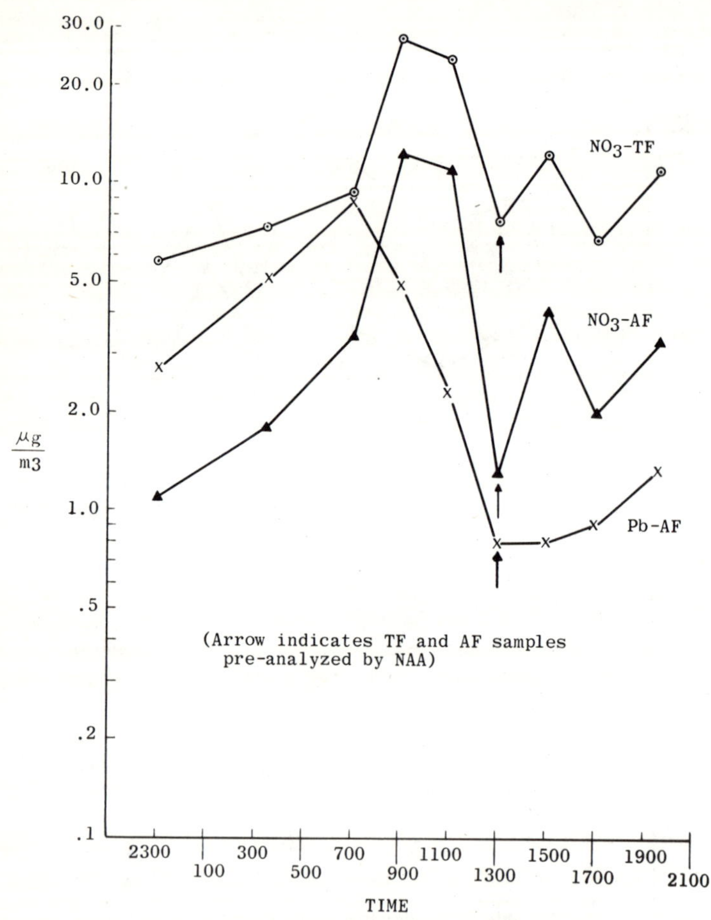

FIGURE 9. Diurnal variation of nitrate and lead; Dominguez Hills, October 10-11, 1973.

2,3, 1.1, 1.6, and 12.9 μg/m^3, respectively. Thus about 50% of the nitric acid retained and recovered by water extraction was collected on the inner walls of the impactor. This probably results in lower artifact particulate nitrate on the impaction surfaces plus after-filter compared to a cellulose acetate total-filter sampling in parallel. Cellulose acetate filters were shown to have about 12% efficiency for collection of nitric acid in sampling a similar amount of the acid (9).

VII. THE APPARENT LOSS OF PARTICULATE MATTER IN TRANSIT AND HANDLING

Preliminary findings during Phase I by R. Giauque and R. Ragaini suggested that large-particle-related elements were lost from filter samples before or during NAA. This led to lower results by NAA than when samples were initially analyzed by XRFA.

As part of the Phase II analysis scheme, all high-volume (Whatman 41) filter samples were first analyzed by XRFA, then by NAA (short irradiation only), and finally by XRFA again, using a single 1-in. disk. This permitted limited comparison of XRFA and NAA results, as well as direct measure of apparent loss of materials. These results are summarized in Table 6.

The elements Br and Mn were analyzed by both XRFA and NAA, while the other seven elements were analyzed by XRFA only. When the first XRFA is compared with the NAA results, both Br and Mn are seen to display about 15% (on the average) higher values by XRFA. The difference in Br values might be ascribed to volatile loss, while that for Mn might relate to other causes (e.g., mechanical loss). The reanalysis by XRFA yielded values for Mn and Br that generally agreed better with the NAA results.

The first XRFA values for the soil-related elements K, Ca, Ti, Mn and Fe were typically 20 to 30% higher than those for the second analysis. Combustion-related elements (Ni, Pb, Br), present in small particles, typically showed lesser changes. The somewhat greater loss of Br than of Pb suggests the importance of an effect other than mechanical loss (e.g., volatile loss).

We conclude that, in addition to possible losses of volatile species during NAA, loss of soil-related elements can be significant because of mechanical losses of the high-volume samples in shipment and/or under NAA instrumental conditions (e.g., rapid deceleration in the "rabbit" used for sample transport).

VIII. THE ANALYSIS OF TOTAL SULFUR COMPARED TO SULFATE DETERMINATIONS ON FILTER AND FOIL SAMPLES

Particulate matter collected on Gelman GA-1 membrane filters and on SPE foils were analyzed by XRFA for total sulfur and by the AIHL microsulfate method for water-soluble sulfate (2). Table 7 compares the results of these determination on the filter samples by episode, distinguishing between after-filters (<0.5-μm

TABLE 6. Analysis and Reanalysis of High-Volume Specimens Collected During Phase II of ACHEX Program.

Filter Number	Analysis Method	Analysis Date		Element (ng/m^3)								
				K	Ca	Ti	Mn	Fe	Ni	Zn	Br	Pb
TA0151 HV	XRF-1	12/28/73		721±72	1470±70	210±42	46±4	2090±80	23±2	152±6	302±12	1720±70
(West	NAA	3/12/74			1329±50	125±50	34±1				259±9	
Covina)	XRF-2	5/1/74		536±54	1120±60	124±36	34±3	1310±50	19±2	122±5	252±10	1470±60
			XRF(2/1)	0.74	0.76	0.59	0.74	0.63	0.83	0.80	0.83	0.85
TB0152 HV	XRF-1	12/28/73		957±96	2060±100	308±85	76±5	2920±120	50±2	462±18	749±30	3750±150
	NAA	3/12/74			2526±660	326±81	58±2				555±16	
	XRF-2	5/1/74		804±80	1610±80	226±67	59±4	2100±80	43±2	407±16	634±25	3290±130
			XRF(2/1)	0.84	0.78	0.73	0.78	0.72	0.86	0.88	0.85	0.88
TC0153 HV	XRF-1	12/28/73		1160±120	2530±130	328±106	83±6	3660±150	68±3	500±20	733±29	3750±150
	NAA	3/12/74			2278±328	301±93	64±2				639±23	
	XRF-2	5/1/74		809±81	1930±100	240±73	59±5	2490±120	63±3	423±17	642±26	3320±130
			XRF(2/1)	0.70	0.76	0.73	0.71	0.68	0.93	0.85	0.88	0.89
TD0154 HV	XRF-1	12/28/73		800±80	2110±110	270±77	64±6	2590±100	38±3	304±12	568±23	2940±120
	NAA	3/12/74			1843±754	251±92	46±2				521±15	
	XRF-2	5/1/74		631±63	1700±90	183±52	42±6	1900±80	39±3	285±11	534±21	2840±110
			XRF(2/1)	0.79	0.81	0.68	0.66	0.73	1.03	0.94	0.94	0.97
TE0155 HV	XRF-1	10/24/73		714±71	1340±70	209±28	48±3	1820±70	22±2	221±9	470±19	2050±80
	NAA	3/12/74			1400±484	290±128	34±1				420±15	
	XRF-2	5/1/74		416±42	984±49	123±32	33±3	1140±50	22±2	194±3	407±16	1900±80
			XRF(2/1)	0.58	0.73	0.59	0.69	0.63	1.00	0.88	0.87	0.93
UF0156 HV	XRF-1	10/24/73		933±93	1890±90	275±61	71±4	2770±110	42±2	169±7	473±19	2010±80
(Pomona)	NAA	3/12/74			2249±247	297±67	59±2				433±16	
	XRF-2	5/1/74		712±71	1560±80	193±53	58±4	2050±80	43±2	149±6	432±17	1920±80
			XRF(2/1)	0.76	0.83	0.70	0.82	0.74	1.02	0.88	0.91	0.96
UG0157 HV	XRF-1	10/24/73		886±89	1780±90	241±32	71±4	2830±110	27±2	221±9	387±15	1690±70
	NAA	3/13/74					47±2				354±13	
	XRF-2	5/1/74		667±67	1430±70	171±32	48±4	2010±80	28±2	207±8	359±14	1690±70
			XRF(2/1)	0.75	0.80	0.71	0.68	0.71	1.04	0.94	0.93	1.00

TABLE 6. (continued)

VH0158 HV (Rubidoux)	XRF-1	10/24/73		1280±130	3780±190	289±43	92±4	3520±140	34±2	172±7	289±12	1450±60
	NAA	3/13/74			4764±369	219±63	80±3				217±6	
	XRF-2	5/1/74		995±100	3240±160	192±44	80±4	2430±100	36±2	168±7	257±10	1460±60
			XRF(2/1)	0.78	0.86	0.66	0.87	0.69	1.06	0.98	0.89	1.01
V10159 HV	XRF-1	10/24/73		1200±120	3730±190	300±42	84±4	3200±130	36±2	220±9	312±12	1630±70
	NAA	3/13/74			4705±376	355±65	74±2				260±9	
	XRF-2	5/1/74		843±84	2980±150	194±50	72±4	2160±90	36±2	213±9	269±11	1610±70
			XRF(2/1)	0.70	0.80	0.65	0.86	0.68	1.00	0.97	0.86	0.99
VJ0160 HV	XRF-1	10/24/73		1450±150	3150±160	384±46	86±5	4150±170	11±2	116±5	162±6	706±28
	NAA	3/13/74			2706±321	265±64					131±4	
	XRF-2	5/1/74		770±77	1810±90	185±28	42±4	2090±80	10±2	81±2	130±5	640±26
			XRF(2/1)	0.53	0.57	0.48	0.49	0.50	0.91	0.70	0.80	0.91
WK0161 HV (Dominguez Hills)	XRF-1	10/24/73		568±57	1280±60	152±85	42±3	1540±60	65±3	353±14	251±10	1090±40
	NAA	3/13/74			1308±70	181±56	40±1				203±6	
	XRF-2	5/1/74		436±44	1120±60	114±81	37±3	1230±50	61±2	344±14	223±9	1050±40
			XRF(2/1)	0.77	0.88	0.75	0.88	0.80	0.94	0.97	0.89	0.96
WL0163 HV	XRF-1	10/24/73		602±60	1450±70	199±44	54±4	1960±80	41±2	290±12	1190±50	3750±150
	NAA	3/13/74			1936±60	<148	44±2				1212±34	
	XRF-2	5/1/74		441±44	1100±60	125±51	40±3	1360±60	38±2	268±11	1140±50	3720±150
			XRF(2/1)	0.73	0.76	0.63	0.74	0.69	0.93	0.92	0.96	0.99

TABLE 7. Comparison of X-Ray Fluorescence Total Sulfur and Wet Chemical Sulfate Determinations

Site	Date	SO_4^{2-} as S/XRFA S	
		After-Filters (AF)	Total Filters (TF)
West Covina	7/23-24/73	0.73 ± 0.04	0.78 ± 0.04
West Covina	7/24-25/73	0.89 ± 0.06	0.87 ± 0.05
West Covina	7/26/73	0.82 ± 0.06	0.79 ± 0.05
West Covina	8/8-9/73	1.44 ± 0.36	1.28 ± 0.11
Pomona	8/16-17/73	1.19 ± 0.08	1.14 ± 0.06
Rubidoux	9/5-6/73	1.13 ± 0.07	1.06 ± 0.06
Rubidoux	9/18-19/73	1.05 ± 0.07	0.99 ± 0.05
Rubidoux	9/24-25/73	1.09 ± 0.15	0.87 ± 0.06
Dominguez Hills	10/4-5/73	1.00 ± 0.05	0.99 ± 0.05
Dominguez Hills	10/10-11/73	1.00 ± 0.17	0.91 ± 0.14
Riverside	8/8-9/73		1.40 ± 0.10
Riverside	8/16-17/73		1.88 ± 0.05
Riverside	9/18-19/73		1.19 ± 0.06
Riverside	10/4-5/73		1.24 ± 0.08
Anaheim	8/8-9/73		3.54 ± 0.73
Anaheim	8/16-17/73		1.02 ± 0.15
Anaheim	10/4-5/73		1.09 ± 0.07
Los Angeles	8/8-9/73		1.20 ± 0.14
Los Angeles	8/16-17/73		0.99 ± 0.05
Los Angeles	10/4-5/73		1.11 ± 0.07
Pomona	8/8-9/73		0.98 ± 0.06
Pomona	8/16-17/73		0.99 ± 0.05
Pomona	9/18-19/73		1.32 ± 0.18
Pomona	10/4-5/73		1.10 ± 0.06
Pasadena	8/8-9/73		0.85 ± 0.05
Pasadena	8/16-17/73		1.04 ± 0.07
Overall		0.96 ± 0.04[a)]	1.02 ± 0.02
TF and AF		1.01 ± 0.01	

[a]Calculated as in footnote a, Table 3. If variables are assumed to be independent, σ values are 0.14, 0.07, and 0.06 for after-filter, total filter, and TF and AF.

particles) and total filters (0 to 20-μm particles). Results for each episode are typically pooled values for nine filters. In addition, the overall pooled results are shown. The ratios of sulfate to total sulfur (as sulfate) for the first three episodes are consistently <1, suggesting that nonsulfate sulfur and/or insoluble sulfate (at least as extracted by the microper-colation method) is significant here. This finding could also imply that the sulfate analyses for these episodes were subject to negative interference. From subsequent method evaluation studies, the latter is considered more likely (2). The overall results, however, for both after-filters and total filters demonstrate that most of the sulfur at these locations was present as water-soluble sulfate.

A similar comparison for foil samples is shown in Table 8. In this case, foils were extracted by soaking in H_2O at 80°C for 2 hr. Again, in nearly all cases where comparison was possible, results were equal within analytical error.

IX. REPRODUCIBILITY OF X-RAY FLUORESCENCE ANALYSIS

Five polyethylene foil sections (2-hr strips) were re-analyzed by XRFA about 1 month after their initial analyses. Variability in these determinations reflects possible losses in handling and differences in positioning of the foil sections for analysis, in addition to the uncertainties due to counting statistics.

The results are shown in Table 9. In general, there is good agreement between the two sets of data except for the sulfur results for specimen UF0304S4. After reexamining the x-ray spectra, it appears that this specimen was inadvertently turned upside down for the first analysis. This error would make the data for the lower Z elements (S to Ti) too low, with the most pronounced effect for sulfur.

X. THE ANALYSIS OF ALUMINUM BY NEUTRON ACTIVATION ANALYSIS AND ATOMIC ABSORPTION ANALYSIS

The analysis of cellulose high-volume filters for aluminum by atomic absorption and NAA (as performed at the Lawrence Liver-more Laboratory (LLL) is compared in Figures 10 and 11 for Phases I and II. In spite of the use of freshly prepared standards for NAA, the discrepancy is greater in Phase II.

Previous interlaboratory comparisons between the LLL and other laboratories using NAA gave similar results for aluminum, but the samples analyzed did not have NBS-certified values for

TABLE 8. Comparison of Total Sulfur (XRFA) and Wet Chemical Sulfate Analysis on Sticky Polyethylene Foil Samples from Lundgren Impactors[a] ($\mu g/m^3$)

	Episode													
	7/23-24/73		7/26/73		8/16-17/73		9/5-6/73		9/18-19/73		10/4-5/73		10/10-11/73	
Stage	SO_4^{2-} as S	Total S	SO_4^{2-} as S	Total S	SO_4^{2-} as S	Total S	SO_4^{2-} as S	Total S	SO_4^{2-} as S	Total S	SO_4^{2-} as S	Total S	SO_4^{2-} as S	
A[b]		<0.40		<0.45		<0.4		<0.4		0.40±0.14		0.27±0.14		<0.4
	0.13±0.01		0.12±0.01		0.10±0.01		0.027±0.003		<0.013		0.63±0.006		0.107±0.011	
B		<0.40		<0.42		<0.4		0.17±0.13		<0.4		0.20±0.14		<0.4
2	0.13±0.01	<0.40	0.11±0.01	<0.42	0.12±0.01	<0.4	<0.015	0.16±0.14	<0.015	0.22±0.13	0.049±0.005	0.19±0.14	0.526±0.053	<0.4
3	0.66±0.07	0.46±0.16	0.56±0.06	0.49±0.16	0.66±0.07	0.82±0.16	<0.015	<0.4	0.40±0.04	0.47±0.15	0.82±0.08	0.96±0.15	0.125±0.013	0.23±0.15
4	2.69±0.27	3.20±0.30	4.83±0.05	4.60±0.50	4.73±0.47	5.5±0.04	1.45±0.15	1.3±0.2	4.19±0.42	4.1±0.4	9.53±0.95	9.6±1.0	0.41±0.041	0.35±0.15

[a]All samples are from sections corresponding to 1200-1400 hr.

[b]Stages A and B both have the same 50% cut-point, and were combined for wet chemical analysis.

TABLE 9. Reproducibility of XRFA as Measured by Reanalysis of Foil Samples.

		Element (ng/m^3)								
Specimen	Analysis[a]	S	K	Ca	Ti	Mn	Fe	Ni	Br	Pb
UF 0300SA	1	< 500	410 ± 40	366 ± 28	127 ± 20	14 ± 3	1110 ± 40	< 4	16 ± 2	61 ± 3
	2	< 400	442 ± 44	589 ± 30	149 ± 24	18 ± 3	1220 ± 50	3 ± 1	16 ± 2	69 ± 3
UF 0301SB	1	< 400	133 ± 28	85 ± 16	< 12	6 ± 2	126 ± 5	< 4	7 ± 1	13 ± 3
	2	< 400	70 ± 26	93 ± 15	28 ± 5	5 ± 2	138 ± 5	2 ± 1	8 ± 2	18 ± 2
UF 0302S2	1	< 500	121 ± 29	159 ± 18	31 ± 6	< 7	233 ± 5	< 4	20 ± 2	43 ± 2
	2	< 400	100 ± 28	172 ± 17	26 ± 5	7 ± 3	242 ± 10	3 ± 1	20 ± 2	53 ± 2
UF 0303S3	1	740 ± 180	127 ± 30	266 ± 19	44 ± 14	9 ± 3	391 ± 16	< 4	58 ± 2	203 ± 8
	2	820 ± 160	153 ± 28	214 ± 17	40 ± 17	10 ± 3	374 ± 15	4 ± 1	51 ± 2	187 ± 8
UF 0304S4	1	< 600	46 ± 29	27 ± 16	< 13	7 ± 3	141 ± 6	5 ± 1	139 ± 6	586 ± 23
	2	5500 ± 600	113 ± 28	58 ± 15	< 18	5 ± 3	136 ± 5	8 ± 1	140 ± 6	593 ± 24

[a]Analyses 1 were performed on 11/30/73; analyses 2, on 1/4/74.

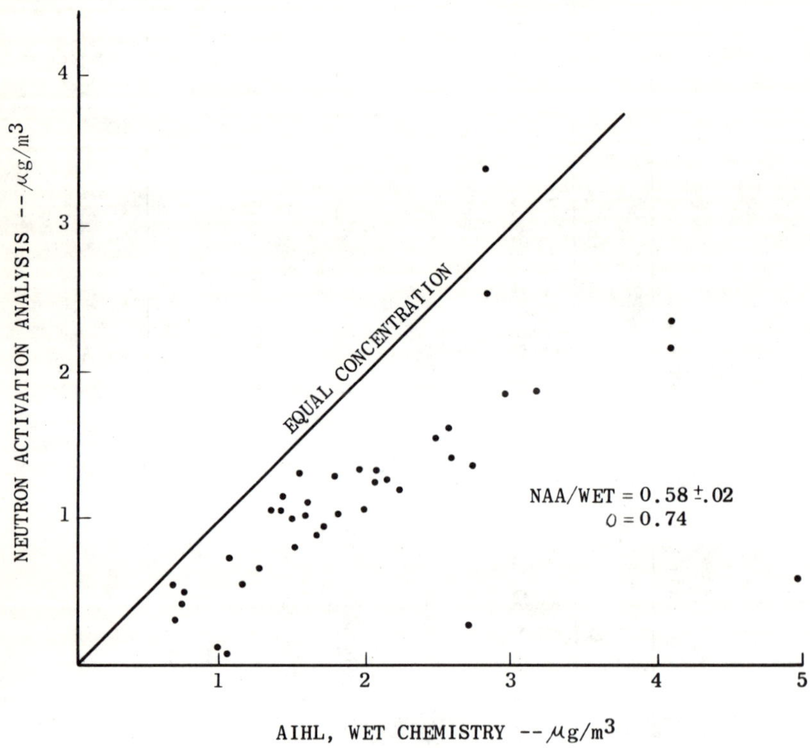

FIGURE 10. Aluminum determined by two methods, Phase I.

this element. As a further check on NAA results for aluminum, a sample of NBS Standard 104, burned magnesite, was analyzed by LLL for analysis with results as follows:

Element	LLL-NAA(%)	NBS (%)	Error (%)
Mg	40.5	51.5	-19.5
Cu	1.8	2.40	-25.0
Al	0.29	0.445	-35.0

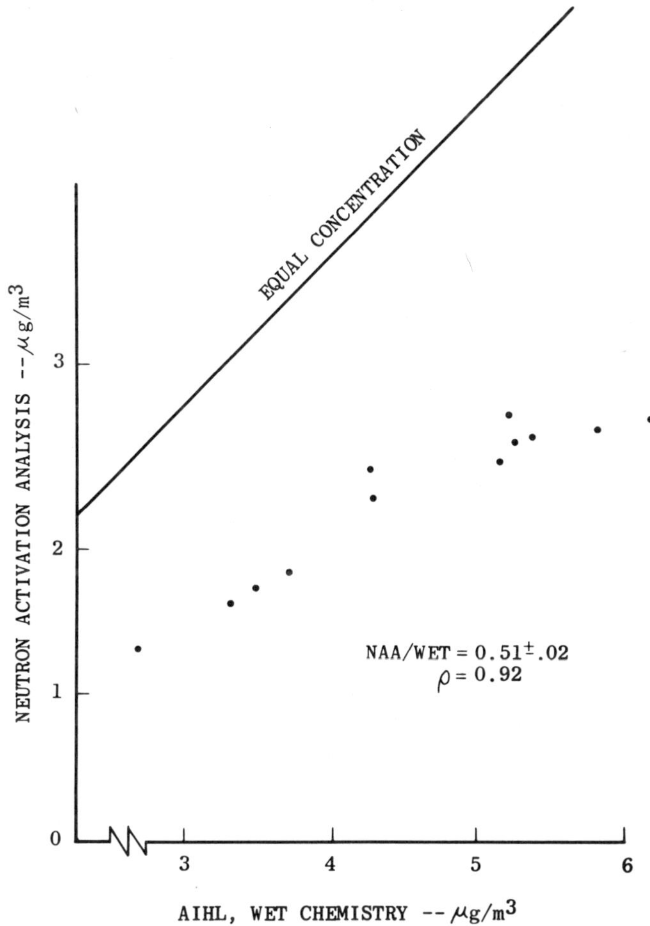

FIGURE 11. Aluminum determined by two methods, Phase II.

In contrast to these results, the analysis by atomic absorption agreed within 1% of the NBS value for this sample.

XI. THE ANALYSIS OF SODIUM AND CHLORINE BY NEUTRON ACTIVATION AND WET CHEMICAL ANALYSIS

The analyses of sodium by NAA and by AIHL's flame photometric procedure are compared in Figures 12 and 13 for Phases I and II, respectively. As previously found for aluminum, the

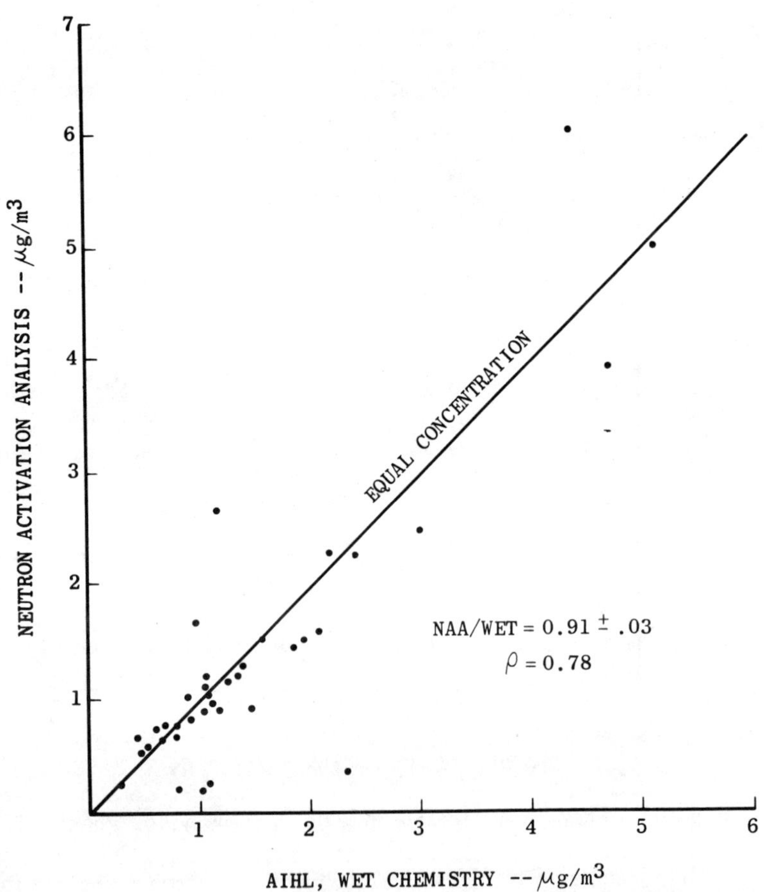

FIGURE 12. Sodium determined by two methods, Phase I.

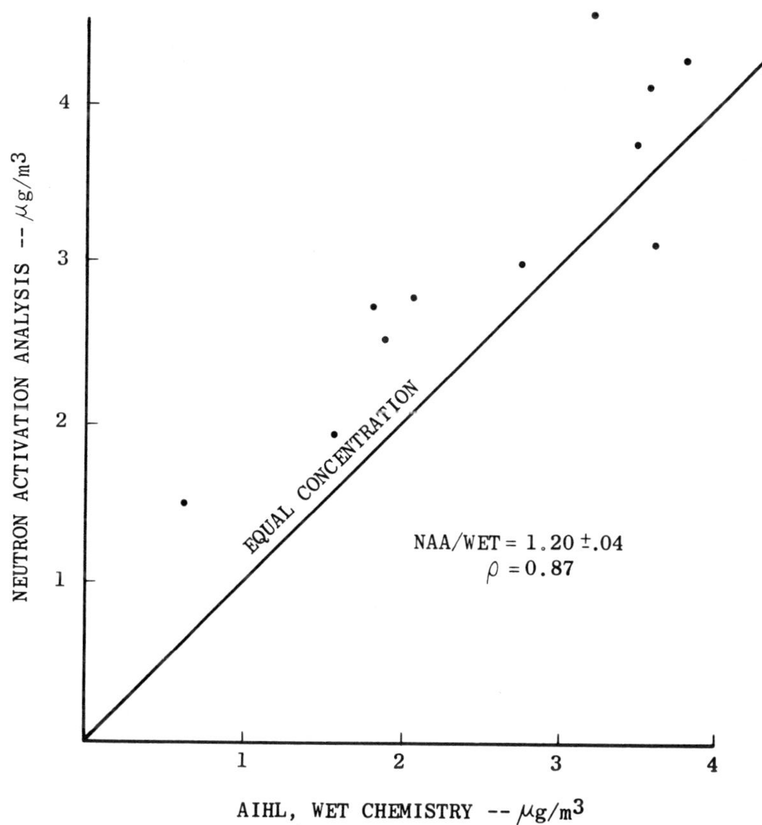

FIGURE 13. Sodium determined by two methods, Phase II.

Phase II results showed somewhat poorer agreement. However, considering that some of the sodium may be found in water-insoluble matrices such as clay and, therefore, lead to higher total Na values, the Phase II results, with the overall NAA/wet chemical ratio for Na = 1.2, appear reasonable.

The comparison between NAA and wet chemical results for chlorine (the latter as chloride) is shown in Figures 14 and 15 for Phases I and II. The Phase II results show substantial improvement in agreement.

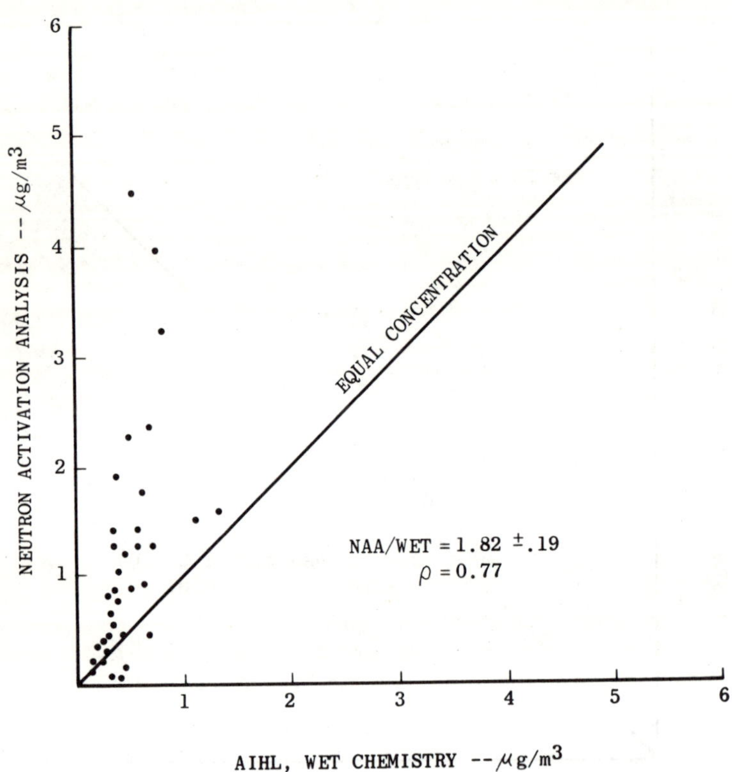

FIGURE 14. Chlorine determined by two methods, Phase I.

A comparison of Cl/Na ratios obtained by the two techniques is shown in Table 10 for Phase II results. The values often agreed within experimental error, with the 9/24/73 episode results showing the largest discrepancy. The NAA ratio for this episode seems anomalously low because of an unusually low value for chlorine; the AIHL results for this episode appear to be more reasonable. On the basis of these findings, the chlorine result for this episode by NAA was omitted from Figure 15.

XII. THE INTERNAL CONSISTENCY OF CARBON ANALYSES BY THE DU PONT THERMAL ANALYSIS METHOD

Carbon analyses were conducted on Gelman A glass fiber after-filters (AV), total filters (TV), and filters capturing particles not removed by a cyclone (RV). The approximate size fractions obtained on these filters are as follows:

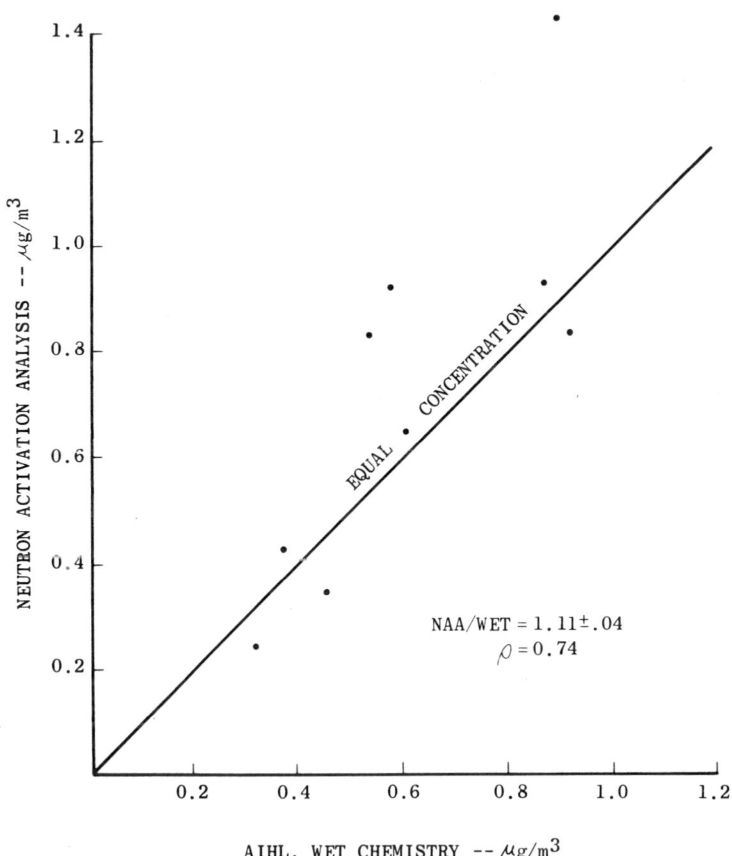

FIGURE 15. Chlorine determined by two methods, Phase II.

TABLE 10. Chlorine/Sodium Ratio by NAA and Wet Chemical Techniques for Phase II Analyses of High-Volume Filter Samples

Site	Date	Cl/Na (w/w) NAA	Cl/Na (w/w) Wet Chemical
West Covina	7/23-24/73	0.10 ± 0.01	0.11 ± 0.02
West Covina	7/24-25/73	0.14 ± 0.01	0.19 ± 0.03
West Covina	7/26/73	0.31 ± 0.02	0.31 ± 0.04
West Covina	8/8-9/73	0.30 ± 0.01	0.26 ± 0.04
Pomona	8/16-17/73	0.13 ± 0.01	0.25 ± 0.04
Pomona	8/22-23/73	0.24 ± 0.01	0.17 ± 0.02
Rubidoux	9/5-6/73	0.27 ± 0.01	0.31 ± 0.04
Rubidoux	9/18-19/73	0.43 ± 0.02	0.58 ± 0.08
Rubidoux	9/24-25/73	0.006± 0.006	1.14 ± 0.16
Dominguez Hills	10/4-5/73	0.057± 0.004	0.083± 0.01
Dominguez Hills	10/10-11/73	0.56 ± 0.03	0.47 ± 0.07

AV(μm)	RV(μm)	TV(μm)
0-0.5	0-3.5	0-20

To evaluate the internal consistency of these results the total carbon values obtained by analyzing 74 AV and RV filters were compared to that for the corresponding TV filters. The mean ratio for AV/TV was 0.65 ± 0.06, and for RV/TV, 1.04 ± 0.09. Since the quantity of carbon on particles >3.5 μm is expected to be small, a mean ratio of 1.0 was consistent with expectations.

REFERENCES

1. Appel, B. R., Wesolowski, J. J., Hoffer, E., Twiss, S., Wall, S., Chang, S. G., and Novakov, T. An Intermethod Comparison of X-Ray Photoelectron Spectroscopic (ESCA) Analysis of Atmospheric Particulate Matter, this volume.

2. Hoffer, E., Kothny, E. L., and Appel, B. R. 1979. Atmos. Environ. 13, 303.

3. Hoffer, E. 1974. Nitrate Method for Low-Volume Filters, AIHL Method No. 50.

4. Wesolowski, J. J., A. Alcocer, and B. R. Appel. The Validation of the Lundgren Impactor, this volume.

5. Rao, K. 1975. Ph.D. Thesis, University of Minnesota.

6. Appel, B. R. and Wesolowski, J. J. Selection of Filter Media for Particulate Sampling with a Lundgren Impactor, this volume.

7. Petriconi, G. L. and Papee, H. M. 1972. Water, Air Soil Pollut. 1, 117.

8. Wedding, J. B. et al. 1977. Environ. Sci. Technol. 11, 387.

9. Appel, B. R., Wall, S. M., Tokiwa, Y., and Haik, M. 1979. Atmos. Environ. 13, 319.

Selection of Filter Media for Particulate Sampling with a Lundgren Impactor

BRUCE R. APPEL and JEROME J. WESOLOWSKI
Air and Industrial Hygiene Laboratory
California State Department of Health
Berkeley, California 94704

I. INTRODUCTION

Selected filter media were examined to determine their suitability as the after-filters of the Lundgren impactors for the Aerosol Characterization Experiment (ACHEX). Since the particles to be collected on this filter were to be analyzed for elemental composition by x-ray fluorescence analysis (XRFA) and neutron activation analysis (NAA), and for information on S, N, P, and Pb contents and oxidation states by electron spectroscopy for chemical analysis (ESCA), severe restrictions were placed upon the filter media.

In the 1969 Pasadena experiment (1) Mitef (Teflon) filters 5 or 10 μ in pore size were used for this purpose. However, the fibrous nature of these filters, with consequent deep penetration of particulate matter below the surface, resulted in inaccurate particulate analysis by surface characterization techniques such as ESCA and XRFA for lighter elements.

The present study was therefore directed toward finding a filter medium that permits all desired analyses while maintaining desirable sampling properties. The requirements of the medium being sought included:

<u>For Sampling</u>:

(1) High efficiency ($\geq$ 95%) for removal of particles in the <0.5 μ size range.

(2) An airflow rate $\geq$ 3.0 cfm at a reasonable pressure drop across the Lundgren impactor plus filter, with minimum possible filter diameter. A ΔP = 25 cm Hg was considered tolerable in these applications. A small diameter was not necessary for NAA but was important for ESCA and XRFA, where the area analyzed is usually small.

<u>For Analysis</u>:

(1) Reproducibility in tare weight.

(2) Surface collection of the particles (not important for NAA).

(3) Low blanks for interfering elements. In the case of ESCA, only the impurities in the first 30 Å are important.

Since membrane filters have been shown to provide the desired surface collection of particles, various such materials were examined. Data on selected fiber filters are also shown for comparison, but these filters were not considered candidates for the present application.

This paper will summarize studies of the following:

(1) the degree to which membrane filters collect particles on their outer surface (from published literature);

(2) initial filtration efficiencies and flow characteristics of filter media;

(3) trace element contents of filter media with and without water washing;

(4) the hygroscopicity of membrane filters;

(5) the suitability of membrane filters for use in analysis by ESCA.

II. SURFACE DEPOSITION ON CELLULOSE ESTER MEMBRANE FILTERS

Goetz and Tsuneishi (2) have shown that particle penetration and deposition on cellulose ester membrane filters are confined to a layer 10 μ and 50 μ from the surface for pore sizes 0.2 ± 0.1 μ and 0.6 ± 0.1 μ, respectively. Lindeken et al. (3) investigated the properties of cellulose ester filters of large pore size, seeking a filter that combined high flow rates with retention of radioactive particles at the surface. By measuring the energy attenuation of α-particles emitted by the radioactive particles, the relative degrees of surface shielding by the filters were obtained. Some of Lindeken's data are shown in Table 1.

These data reveal, first, only a 10% increase in surface shielding on changing from an 0.8- to a 5-μ MF Millipore filter if the "front" face of the filter is used. Second, the front and back faces of the MF Millipore (a mixture of cellulose nitrate and acetate) differ greatly in shielding. Third, the 5-μ Gelman filter, although compositionally similar to the Millipore, is substantially poorer than the front face of the 5-μ Millipore filter, but exhibits no significant change between its front and back face.

The above observations suggest that the 5-μ MF Millipore, if used in the current study, might permit particulate analysis by ESCA, ignoring for the moment the question of analytical interference by nitrogen in the filter. The Gelman filter used by Lindeken is no longer available. The present studies included

TABLE 1. Degree of Filter Penetration of Airborne Radon C' as Measured by Energy Loss of Emitted α-Particles

Filter	Resolution, (KeV)[a]
0.8 μ Millipore (AA)	404
5 μ MF Millipore (SM)	
(Front)	450
(Back)	826
5 μ Gelman AM-1	
(Front)	985
(Back)	1080

[a]The attenuation of α-particle energy results in a broadening of the energy peak. These values measure relative degrees of peak broadening. The higher the number, the greater is the surface shielding.

a 5-μ Gelman cellulose ester filter which may not resemble its predecessor as regards surface deposition.

III. INITIAL FILTRATION EFFICIENCIES AND FLOW CHARACTERISTICS OF FILTER MEDIA

A. Experimental

Room aerosol was used for trials of filter efficiencies, using the experimental setup shown in Figure 1. The filters were tested downstream from the Lundgren impactor since this represents the actual setup used in the ACHEX. Flow rate was measured with the built-in flowmeter of the Lundgren and regulated with a screw clamp on the tubing near the pump. For trials with a 47-mm filter, a holder was added in series between the 90-mm holder and the glass "Y".

To determine the filter efficiency for particles in the size range 0.003 to 0.1 μ, the particle count was determined by a GE PC-1 portable condensation nuclei counter (CNC) with and without the filter in place. The measurements with and without filters were made at the same flow rate and vacuum by adjusting a screw clamp adjacent to the holder. The flow rates used were between 1 and 4.5 cfm.

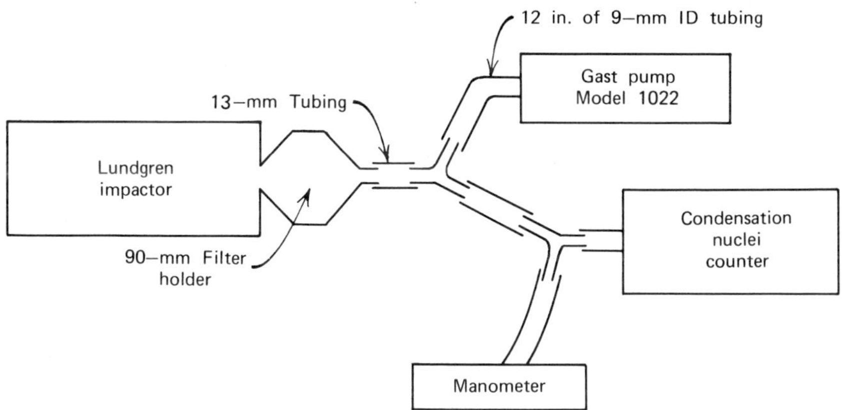

FIGURE 1. Experimental setup.

The CNC readings obtained were a function of the pressure drop of the system, reflecting both the real decrease in particles per cubic centimeter and the influence of reduced sampling pressure on the nuclei counter. For example, with an inlet pressure 23 cm Hg below atmospheric, at 20°C, the instrumental reading is about 40% below the correct value (4). At atmospheric pressure, room air contained 50,000 to 100,000 particles per cubic centimeter.

To determine efficiencies as a function of particle size, a Royco optical particle counter Model PC 200B was used with flow of 300 ml/min. The filter holder was attached directly over the sample entry tube. Flow through the filter was verified with a flowmeter attached momentarily on the filter. The OPC was used to provide particle counts for room aerosol of equivalent spherical diameters 0.32, 0.4, 0.5, 0.64, 0.8, 1.0, 1.3, and 1.6 μ. Operation with subambient sampling pressure (such as exists downstream of a filter sampling air at $\geq$1 cfm) was impractical with this instrument. Therefore the total flow through the filters was the sampling rate of the Royco (300 ml/min). Face velocities, accordingly, decreased by about a factor of 300 from those with a 3-cfm flow. The particle sizes are considered reliable to $\pm$50%.

B. Results

Table 2 details the flow behavior and filtration efficiency of filters in place on a Lundgren impactor. The observed pressure drop consists of that across the four Lundgren stages and filter holders(s) (about 10 cm Hg at 3.0 cfm) plus that across the filter media. In studies with 47-mm filters the system contained an unnecessary (90-mm) filter holder that contributed about 2 cm Hg to the total ΔP.

During the 1969 Pasadena study the Lundgren impactor employed (5- or 10-μ Teflon Millipore (Mitef) filters 90 mm in diameter. At a typical flow rate of 3.0 cfm, ΔP = 30 and 11 cm Hg for the 5- and 10-μ filters, respectively. With 5 μ filters of only 47-mm diameter the ΔP was approximately double. Increasing the pore size to 8 μ reduced the ΔP for the 90-mm filter by a factor of 3.

The flow behavior of Nuclepore filters varied dramatically with pore size; with an 0.8 μ pore ΔP reached >50 cm Hg at only 1.5 cfm, while with a 5 μ pore ΔP was nearly unchanged from its value without the filter in place. The 1.2 μ silver membrane filter exhibited initial airflow behavior comparable to that of the 5 μ MF Millipore.

All Millipore filters, as well as the Gelman cellulose ester filters tested, exhibited $\geq$95% removal of room particulates $\geq$0.003 μ. The silver membrane and Nuclepore filters exhibited efficiencies in the range of 50 to 72%.

The influence of face velocity on filtration efficiency, although not studied systematically, was small at high flow rates with Millipore filters; comparison of data obtained at 1 and 3 cfm with the same filter revealed no significant difference.

An additional indication of the lack of importance of threefold changes in face velocity is provided by comparing filtration efficiencies for the same filter in 47- and 90-mm sizes. Such comparisons made with both 3- and 5-μ MF Millipore filters revealed no change beyond the $\pm$2% precision generally encountered in this range.

Stern et al. have demonstrated the enhancement in filter efficiency produced by lowering the inlet pressure below atmospheric (5). The ΔP across the Lundgren provides an inlet pressure of about 67 cm Hg to the after-filter, making such improvements possible. However, Stern observed significant effects only at inlet pressures <0.5 atm. Thus this effect is probably negligible with the Lundgren impactor.

TABLE 2. Initial Filtration Efficiency for Room Air Condensation Nuclei and Flow Behavior of the Lundgren Impactor as a Function of the After-Filter Medium[a]

Filter	Size (mm)[b]	Number of Trials	Flow Rate (cfm)	Overall Pressure Drop (cm,Hg)	Efficiency by CNC(%)
None	90 (holders only)	1	4.3 (max.)	25.0	-
			3.0	10.3	-
			2.5	8.0	-
None	90 + 47 (holders only)	1	4.2 (max.)	26.6	-
			3.0	9.0	-
			2.5	6.9	-
Membrane Filters					
0.8 μ Nucleopore	47	2	1.5 (max.)	50.8	72 ± 1[c]
5 μ Nucleopore	47	1	3.0	11.5	≤50
3.0 μ MF Millipore (SS)	47	1	3.4 (max.)	32.2	
			3.0	25.6	
			2.5	19.9	98
3.0 μ MF Millipore (SS)	90	2	4.1 ± 2(max.)	26.3 ± 2.1	
			3.0	14.2 ± 2.4	96 ± 2
			2.5	10.8 ± 1.8	98
5.0 μ MF Millipore (SM)	47	1	3.7 (max.)	31.7	
			3.0	22.6	96
			2.5	17.6	
5.0 μ MF Millipore (SM)	90	1	4.4 (max.)	23.2	
			3.0	11.4	98
			2.5	8.6	98
8.0 μ MF Millipore (SC)	47	2	4.0 (max.)	27.1	
			3.0	16.5	97 ± 1
			2.5	13.1	97 ± 1
Gelman GA-1 5 μ Cellulose acetate	47	2	3.8 ± 0.1(max.)	30.2 ± 0.7	
			3.0	20.2 ± 0.2	98
			2.5	15.6 ± 0.2	97 ± 2
1.2 μ Silver membrane[d]	47	2	3.6 ± 0.2	29.3 ± 0.5	
			3.0	20.8 ± 2.7	
			2.5	16.4 ± 1.4	66
S & S 0.45 μ (B6A) cellulose acetate	47	1	1.5 (max.)	53.1	ND
			1.0	41	ND

TABLE 2 (continued)

Filter	Size (mm)[b]	Number of Trials	Flow Rate (cfm)	Overall Pressure Drop (cm,Hg)	Efficiency by CNC(%)
		Teflon Fiber Filters			
5.0 μ Teflon Millipore (LS)	90	2	3.2 ± 0.1(max.)	36.6 ± 0.6	
			3.0	30	
			2.5	24.8 ± 1.6	
			2.0	16.0	93
5 μ Teflon Millipore (LS)	47	1	1.7(max.)	55	-
10 μ Teflon Millipore (LC)	90	1	4.7(max.)	26	
			3.0	10.8	73
			2.5	8.6	
		Glass Fiber Filters			
Gelman A	47	2	3.9 ± 0.3(max.)	27.9 ± 1.1	
			3.0	17.6 ± 2.5	95
			2.5	13.1 ± 1.5	98
Reeves Angel No. 900AF	47	1	4.1(max.)	23.2	
			3.0	11.4	98
			2.5	8.6	98
MSA 1106 BH	47	1	4.2(max.)	26.4	
			3.0	14.8	99
			2.5	11.2	98
S & S No. 25 acid washed	47	2	3.6 ± 0.2(max.)	29.9 ± 0.7	
			3.0	21.6 ± 2.6	59
			2.5	16.0 ± 2.0	50 ± 12
		Cellulose Fiber Filters			
S & S Green Ribbon No. 599	47	1	3.2(max.)	32.9	
			3.0	28.1	
			2.5	20.5	60
TFA	47	2	3.7 ± 0.3(max.)	29.6 ± 1.0	
			3.0	20.1 ± 2.7	66
			2.5	14.8 ± 1.9	52 ± 14
Whatman 41	47	1	3.7(max.)	31.0	
			3.0	21.2	70
			2.5	16.0	81

[a]Face velocity at 3.0 cfm ≅ 89 cm/sec for 47-mm filter and ≅ 30 cm/sec for 90-mm filter.

[b]The effective sizes for the filters were 45 and 78 mm for 47- and 90-mm filters, respectivel

[c]Measured at 1.2 liters/min with the filter holder attached directly to the CNC.

[d]Selas Flotronics.

Although not considered candidates for the Lundgren afterfilter, glass and cellulose fiber filters were also examined, as summarized in Table 2. The glass fiber filters, while varying in flow resistance, yielded efficiencies $\geq$95% except for the S & S No. 25 (acid washed). The latter, of glass fiber with a cellulose binder, closely approximated in filtration behavior the all-cellulose fiber filters tested. These varied in initial efficiencies from 60 to 80%. However, as discussed below, particle buildup sharply increases their efficiency.

The influence on filtration efficiency of acid washing of glass fiber filters was examined using 25-mm Gelman A filters at a flow rate of 1.0 cfm. Their efficiency for removal of condensation nuclei, 99%, remained unchanged upon acid washing. The flow resistance, however, decreased from 14.1 $\pm$ 0.5 to 12.4 $\pm$ 0.2 cm Hg, suggesting removal of some binder material by the washing procedures.

Table 3 details particle filtration efficiencies as a function of particle size in the range 0.3 to 1.6 μ diameter. These data closely parallel those obtained by the nuclei counter, although the face velocities are now about 300 times smaller than at 3 cfm. Again, all Millipore filters gave very high efficiencies, the poorest being the 8- and 10 μ filters, which permitted 4% passage of the particles counted as having 0.3 μ diameters. The Gelman GA-1 5 μ cellulose ester filter was indistinguishable from the comparable Millipore. Again, the Nucleopore filters yielded the lowest efficiency of any membrane filter. With one exception the glass filters displayed $\geq$99.8% filtration efficiency for particles 0.3 μ and larger. The exception, S & S No. 25, again resembled the pure cellulose filters. The cellulose filters yielded low efficiencies. It is notable, however, that the filtration efficiency of TFA sharply increased after collecting room aerosol from 500 ft^3 of air (see Table 3, footnote b).

Table 4 compares filtration efficiency data from the present study with those obtained by other workers. Data are shown for membrane, glass, and cellulose fiber filters at high and low face velocities. The membrane filter listed, Gelman AM-1, was not available for the current work, but results for it were compared with the GA-1 data since both have a 5 μ nominal pore size.

The filtration efficiency of Nucleopore filters has been evaluated by Spurny et al. (6) and, more recently, by Liu and Lee (7). The latter authors reported for a Nucleopore filter, 0.6 μ pore size, at a face velocity of 5 cm/sec, about 90% efficiency for the collection of 0.3 μ particles. This compares to the value of 82% found here with filters of 0.8 μ pore size at 0.3 cm/sec.

TABLE 3. Initial Efficiency for Filtration of Room Air Particles $\geq$0.3 μ in Diameter[a]

Filter	Size (mm)	Number of Trials	Efficiency (%) for removal of particles of equivalent spherical diameters (μ) 0.32	0.4	0.5	0.64	0.8	1.0	1.3	1.6
			Membrane and Teflon Fiber Filters							
0.8 μ Nucleopore	47	1	82	96	99	99	100	100	100	100
5.0 μ Nucleopore	90	1	83	95	98	97	98	100	100	100
3.0 μ MF Millipore (SS)	90	3	99.93	99.98	100	100	100	100	100	100
5.0 μ MF Millipore (SM)	47	1	100	100	100	100	100	100	100	100
5.0 μ MF Millipore (SM)	90	1	100	100	100	100	100	100	100	100
5.0 μ Teflon Millipore (LS)	90	1	99	99.6	100	100	100	100	100	100
8.0 μ MF Millipore (SC)	47	1	96	99	99	99	96	99	100	-
10 μ Teflon Millipore (LC)	90	1	96	99	99	99	100	100	100	100
1.2 μ Silver Membrane	47	2	99.9	100	100	100	100	100	100	100
S & S 0.45 μ (B6A) cellulose acetate	47	1	100	100	100	100	100	100	100	100
Gelman GA-1 cellulose acetate	47	1	99	100	100	100	100	100	100	100
			Glass Fiber Filters							
Gelman A	47	2	99.8	99.8	99.6	99.7	99.1	99	99	97
Gelman A	25	1	100	100	100	100	100	100	100	100
Gelman A acid washed	25	1	99.9	100	100	100	100	100	100	100
Reeves Angel No. 900AF	47	1	99.9	100	99.9	100	100	100	100	100
MSA 1106BH	47	1	100	100	100	100	100	100	100	100
S & S No. 25 acid washed	47	3	26	53	65	64	69	71	79	88
			Cellulose Fiber Filters							
TFA[b]	47	2	38	67	80	81	83	85	84	87
Whatman 41	47	1	63	83	90	84	89	81	94	100
S & S Green Ribbon No. 599	47	1	69	84	88	87	93	98	97	100

[a]Face velocity (cm/sec) for 25-, 47-, and 90-mm filters is 0.82, 0.31, and 0.11, respectively.

[b]After sampling 500 ft^3 of room air, the filtration efficiency of TFA increased to 90% for 0.3 μ particles, with other values in the range 80 to 90% for sizes up to 1.6 μ. This efficiency was not significantly altered by discharging the filter over a radioactive source.

TABLE 4. Comparison of Filtration Efficiencies.

Filter	Particle						
	0.025 μ Ur[a]	0.27 μ Ur	0.30 μ DOP	Atm Aerosol[b]	Atm. Aerosol[c] (0.3 μ)	0.3 μ PS[d]	0.30 μ DOP[e]
Gelman AM-1	87	98	-	(98)[f]	-	-	
MSA 1106 BH	>99.9	>99.9	-	99	100	66	>99.99
Gelman A	-	-	99.99[e]	95	99.8	-	>99.99
Whatman 41	75	66	91[d]	70	63	-	56
Face Velocity (cm/sec)	← 95 ± 6 →				← 0.3 →		

[a]Uranine dye studies by S. Posner, quoted by M. Lippman in "Air Sampling Instruments - 1967", American Conference of Government Hygienists.
[b]Present study as measured by CNC.
[c]Present study as measured by Royco particle counter.
[d]Data of D. Rimberg, Am. Ind. Hyg. Assoc. J., July-August, 1969, p. 994.
[e]Data of L. B. Lockhart, et al., Naval Research Laboratory (Washington, D.C.) Rep. No. 6054.
[f]Measured with Gelman GA-1 filter. Both the AM-1 and GA-1 are 5 μ cellulose nitrate-acetate membrane filters.

TABLE 5. Comparison of Gelman and Millipore 5 μ Membrane Filters Sampling Ambient Air

Trial	Sampling time (min)	Filter	Flow Rate (cfm) Initial	Flow Rate (cfm) Final	Flow Rate (cfm) Δ	Pressure Drop (cmHg) Initial	Pressure Drop (cmHg) Final	TSP (μg/m3)[a]
I	320	MF Millipore (5 μ)	2.34	2.27	0.07	8.0	8.4	61.7
I	320	Gelman GA-1	2.34	2.10	0.24	8.4	10.2	66.3
II	123	MF Millipore (5 μ)	2.34	2.30	0.04	7.8	8.6	135
II	123	Gelman GA-1	2.34	2.24	0.10	11.8	12.4	145

[a] Corrected for relative humidity effects on filter media.

Our principal conclusion from this comparison is that the procedures employed provided values similar to those obtained by other workers. Quantitative agreement is not expected because of the gross differences in the nature of the particles used. The large difference in filtration efficiency for DOP (liquid) and uranine dye aerosols of nearly equal mass median diameters is striking.

On the basis of the preceding, both 5.0 μ MF Millipore and 5 μ Gelman GA-1 filters exhibit excellent filtration efficiencies with total pressure drops <25 cm Hg at 3 cfm. The behavior of both was therefore examined in greater detail.

IV. COMPARISON OF 5 μ MF MILLIPORE AND GELMAN GA-1 MEMBRANE FOR 2-HR SAMPLING

To compare with laboratory evaluations of filter efficiencies, as well as to observe the influence of particulate buildup on flow behavior, 5 μ, 47-mm MF Millipore and Gelman GA-1 filters were run in ambient air for varying periods side by side. Separate pumps were used, and flow rates and pressure drops evaluated initially and upon completion. As summarized in Table 5, the GA-1 exhibited a somewhat greater plugging tendency than the Millipore filter, but both appeared to provide excellent flow behavior. The Gelman filter loadings were higher by about 7.5% than the Millipore. These findings are in qualitative agreement with condensation nuclei results discussed earlier.

V. HYGROSCOPICITY OF FILTER MEDIA

The influence of relative humidity on the tare weights of the Gelman and Millipore filters was experimentally measured. Filters stored under ambient conditions were weighed at various relative humidities over the range 40 to 55%. The Gelman showed an average weight change of 62 μg/% RH change; the Millipore, 24 μg/% RH change. Thus it was clear that for the ACHEX, correction for RH changes would have to be made if these filters were used. Figure 2 shows the results of measurements on 47-mm filters equilibrated (for ≥7 hr) at 44% and then weighed in less than 5 min in ambient air of higher RH. After a period of rapid increase in weight (about 30 μg) further change was slight. The average of the first three weighings differing by <10 μg was employed. The changes in weight shown in Figure 2 are relative to the filter weights at 44% ambient humidity.

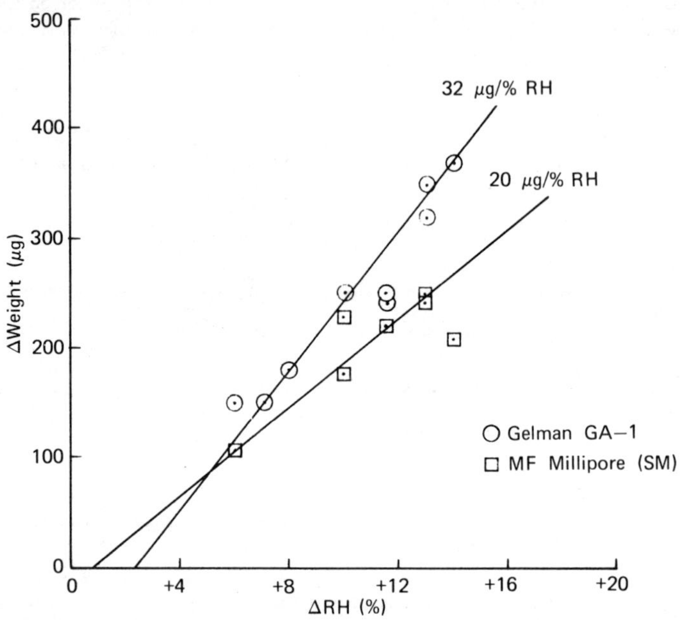

FIGURE 2. ($RH_{ambient}$ - 44%) versus change in weight for membrane filters stored at 44% RH and weighed under ambient conditions.

VI. FILTER IMPURITY LEVELS

Table 6 includes filter impurity data for a variety of filter media. Although data for GA-1 filters are not shown, these may be similar to those for the Gelman GA-6 filter, both of which are membrane types of predominantly cellulose acetate composition, differing only in pore size. If so, the Gelman filter is dirtier than MF Millipore filters to a degree that is intolerable for elemental analysis of particles collected in short-term samples. Accordingly, samples of both GA-1 and 5 μ MF Millipore filters were washed with distilled H_2O and dried under vacuum over silica gel. Samples of the water-washed and untreated filters were analyzed by neutron activation. The results obtained are shown in Table 7 and confirm Rahn's findings

TABLE 6. Filter Impurity Levels (ng/cm^2)[a].

Element	Filter			
	Whatman 41	MF Millipore, 0.8 μ (AA) mixed cellulose esters	Millipore, 0.5 μ (EH) cellulose acetate	Gelman Metricel (GA-6) 0.45 μ, cellulose acetate
Cl	100	1,000	1,800	600
Br	5	<2	6	4
S	-	-	-	-
Na	150	400	1,800	2,200
K	15	100	-	-
Mg	<80	200	-	-
Ca	140	370	570	1,250
Ba	<100	<100	-	-
Al	12	10	60	740
Sc	<0.05	<0.05	-	-
Ce	<0.5	<0.3	-	-
La	<0.2	<0.2	-	-
Ti	10	<10	-	-
Fe	40	40	-	-
Mn	0.5	2	6	2
Co	0.1	0.1	-	-
Ni	<10	<20	-	-
Ag	2	<1	-	-
Cu	<4	60	25	30
Zn	<25	7	-	-
Sb	0.15	1	-	-
Cr	3	15	-	-
Hg	0.5	0.5	-	-
V	<0.03	<0.05	<0.05	<0.05

[a]Source: K. A. Rahn, "Sources of Trace Elements in Aerosols, an Approach to Clean Air", University of Michigan Technical Rep. No. ORA 08930, May 1971. Also see R. Dams, K. A. Rahn, and J. W. Winchester, Environ. Sci. Technol. 6, 441 (1972).

TABLE 7. Comparison of Impurity Levels of Washed and Unwashed Membrane Filters with Minimum Levels Expected from Ambient Air Sampling (ng/cm^2)[a]

	Filter				
Element	Gelman GA-1	Gelman GA-1 (H_2O washed)	MF Millipore 5 μ (SM)	MF Millipore, 5 μ (SM) (H_2O washed)	Approximate Expected Minimum[b]
Cl	1480	92.7	1123	732	460
Br	–[c]	0.17	3.04	3.18	27
Na	983	23.9	497	272	725
I	–	0.34	0.72	0.32	c
Mg	552	55.9	150	24.6	280
Ca	911	115	637	105	c
Al	39.0	23.0	17.4	22.5	70
Mn	1.55	0.50	4.80	1.34	2.0
Cu	29.7	19.7	18.0	24.6	7.5

[a] Data from R. Kaifer, Lawrence Livermore Laboratory

[b] From sampling for 2 hr at 3 cfm with a 47-mm filter with atmospheric concentrations taken from W. John et al., Atmos. Environ. 7, 107 (1973).

[c] Not available.

as regards the relative impurity levels of the Gelman and Millipore filters. Washing with deionized distilled water proved highly effective in reducing impurity levels, except for copper, with the Gelman filter, but was less so with the Millipore.

By comparing the impurity levels in the washed Gelman with the approximate minimum expected values shown for ambient air (assuming about 1 m^3 of air sampled per square centimeter of filter) we conclude that, except for copper, the blank corrections would represent less than 30% of the values for the particulate samples. No attempt was made to study batch variations for the Gelman filter.

VII. AN ELECTRON SPECTROSCOPY FOR CHEMICAL ANALYSIS (ESCA) STUDY OF PARTICULATE MATTER ON MF MILLIPORE AND GELMAN GA-1 FILTERS

Since information on the oxidation states of nitrogen was to be sought by ESCA, it was vital that the filter medium either be nitrogen-free or be so covered by particles as to be undetected by this surface technique. To test the latter possibility as a means of avoiding restrictions on filter composition, an MF Millipore and a Gelman GA-1 filter moderately loaded with atmospheric particles were analyzed via ESCA by T. Novakov. Comparison of the results with those from a blank filter indicated that substrate interference was excessive in both cases. We conclude that neither filter can be used for such nitrogen analysis. However, good ESCA spectra for lead, sulfur, and phosphorus oxides in particulate matter were obtained. This implies adequate retention of particles on the filter surface.

VIII. CONCLUSIONS

A 47-mm, 5 μ cellulose ester filter can be employed at 3 cfm as a highly efficient after-filter for the Lundgren impactor. A Gelman GA-1 5 μ cellulose ester filter will permit particulate analysis by XRFA, NAA, and, except for nitrogen compounds, ESCA. To overcome the latter deficiency, a silver membrane filter of 1.2 μ pore size may be useful but would be inappropriate in NAA and XRFA.*

---

*Subsequent studies by T. Novakov confirmed the utility of silver membrane filters for ESCA determinations of N, C, and S.

ACKNOWLEDGMENT

We wish to express our appreciation to Stephen Wall for technical assistance. Partial support was received from Air Resources Board Contract No. 358.

REFERENCES

1. Lundgren, D. A. 1972. J. Coll. Interface Sci. 39, 205.

2. Goetz, A. and Tsuneishi, N. 1951. J. Amer. Water Works A43, 943.

3. Lindeken, C. L., et al. 1964. Health Phys. 10, 495.

4. Haberl, J. B. and Fusco, S. J. GE Tech. Inf. Ser. 70 POD 12.

5. Stern, S. C., Zeller, H. W. and Schekman, A. I. 1960. J. Colloid Sci. 15, 546.

6. Spurny, K. R., et al. 1969. Environ. Sci. Technol. 3, 453.

7. Liu, B. Y. H. and Lee, K. W. 1976. Environ. Sci. Technol. 10, 345.

The Validation of the Lundgren Impactor

JEROME J. WESOLOWSKI, ARTHUR E. ALCOCER, AND BRUCE R. APPEL
Air and Industrial Hygiene Laboratory
Laboratory Services Branch
California State Department of Health
Berkeley, California

I. INTRODUCTION

The Lundgren impactor was used in the Aerosol Characterization Experiment (ACHEX) to provide particle size distributions of elements, mass, and oxidation states of chemical species, with a time resolution of 2 hr, the latter requirement being the principal reason for the choice of this particular impactor. Before the size distributions obtained from any impactor can be considered reliable, specific experiments must be carried out to judge the validity of the data. These include not only calibration of the device but also quantitative determination of such phenomena as wall loss and the bounce-off of particles from the impaction surfaces. This paper describes the results of the laboratory and field work carried out to determine the accuracy of the data collected using the Lundgren during ACHEX.

II. THE IMPACTOR

A schematic diagram of the impactor is shown in Figure 1. There are five stages, each consisting of an aluminum cylinder that can be covered with a removable collection surface. A slit before each cylinder determines the velocity of the aerosol impinging on the cylinder and hence the impaction efficiency for that stage for a particle of given aerodynamic size. Rotation of the cylinders gives time resolution. An after-filter assembly follows the last stage to collect the particles not collected by the impaction surfaces.

A particle striking an impaction surface can stick to the surface, bounce off, break into fragments, or dislodge another

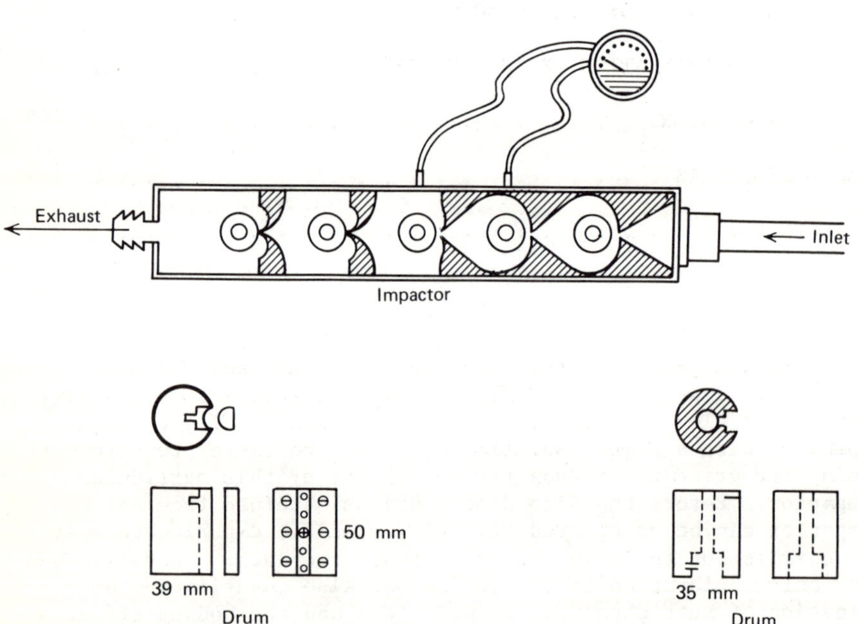

FIGURE 1. Lundgren cascade impactor.

particle previously collected by the surface.* It is difficult to experimentally separate the last three phenomena, and the word "bounce-off" is often used to refer to the general problem. A particle reentering the aerosol stream will either be lost to the walls of the device or be collected by a subsequent stage or the after-filter. The word "sticky" is usually used to refer to surfaces that have good collection efficiency (i.e., cause the particles to stick), but it should be emphasized that their ability to absorb the kinetic energy of the particle may be the governing principle for good collection efficiency, and not necessarily their ability to wet the particles.

The type of impaction surface to be used will depend upon the subsequent chemical analysis required for the aerosol collected. A collection surface covered with a grease (e.g., silicone stopcock grease) would provide good collection efficiency, but would make mass determinations difficult and increase blank values for chemical elements of interest. Furthermore, with analytical techniques such as electron spectroscopy for chemical analysis (ESCA), which analyzes only the upper 30 to 50 Å of surfaces, a medium providing any shielding of the particles must be avoided. Two nonsticky surfaces were employed, 0.0025-cm-thick Teflon and 0.0006-cm Mylar. The former was the preferred substrate for ESCA analyses since particulate carbon data could be obtained. The latter was preferred for alpha-induced X-ray emission analyses. It was also preferred for neutron activation analysis (NAA), since it can withstand more radiation damage than Teflon.

In addition to these substrates a plastic film, approximately 0.006 cm thick and covered with a sticky resin, was provided to the laboratory by Professor Kenneth Noll, University of Tennessee†. Infrared spectroscopy suggested that the film was polyethylene and that the material on the surface was a hydrocarbon resin; it is here referred to as sticky polyethylene (SPE).

---

*The word "reentrainment" is used to refer to this last process, as well as to the loss of previously collected particles by the motion of the air alone. The moving impaction surface of the Lundgren has the advantage of minimizing reentrainment. For example, a cylinder rotated once in 24 hr in an environment with an average 100 $\mu g/m^3$ of particulate matter will accumulate, on the average, less than one monolayer of particles.

†Unfortunately, the material is apparently no longer available commercially, and the original manufacturer cannot be identified.

Although this surface was thicker than desired for alpha-induced X-ray emission analysis, and in general had larger blank values than Mylar and Teflon for many elements, total mass as well as many specific elements could be obtained using this film. Its collection properties were therefore compared to those of uncoated Mylar and Teflon.

The impactor used in this study was improved in a number of ways from the Lundgren impactor employed in the 1969 Pasadena study (1). The improvements included a more convenient system for mounting the impaction surfaces on the cylinders, an improved inlet system, a more accurate and reliable method of rotating the cylinders, redesign of the entrance slits for stages 3 and 4 to reduce wall losses, and a modification to reduce the effects of the bounce-off of large particles. The last improvement was made by introducing an extra stage (called 1B) between stages 1 and 2 with a slit width identical to that of stage 1 (now called 1A) and placing SPE on this new stage. Presumably some of the large particles that bounced off the smooth impaction surface of stage 1A would be captured by the sticky material on 1B. The modified Lundgren impactor should give a slightly different particle size distribution than would an unmodified impactor since 1B would collect some of the particles which otherwise would impact on stage 2. Thus we expect that the summation of 1A and 1B would yield a larger value than that obtained for stage 1 on a four-stage Lundgren, and that stage 2 of the modified Lundgren would receive less material than would be collected on stage 2 of a four-stage unit. However, it is believed that the distortion is minor compated to the effect of bounce-off on the distribution if the four-stage impactor were employed with nonsticky impaction surfaces.

Lundgren has previously compared ambient air mass distributions from sticky (greased) and 1-mil-thick uncoated Teflon surfaces and determined no significant differences (2). However, this was done under conditions of high relative humidity in Riverside, California, where hygroscopic particles can dominate the aerosol composition. Under these conditions bounce-off is less likely. Furthermore, the fact that the mass data showed no appreciable differences does not necessarily imply the equivalence of size distributions for specific elements on sticky and dry surfaces. In contrast to Lundgren's results, Table 1 shows the differences between sticky and "dry" surfaces with dry particles, namely, for coal dust and Arizona road dust (3). Without a sticky collection surface, approximately 70% of the Arizona road dust reached the after-filter, compared to only 15% when the stages were greased. The results with coal dust are similar. The differences exhibited in the laboratory by these extreme cases demonstrate that Lundgren's findings may not apply to

TABLE 1. Comparison of Mass on Sticky and Dry Impaction Surfaces for Lundgren Impactor

Stage[a]	Coal dust (1-5μm)		Arizona road dust (0-15μm)	
	Sticky	Dry	Sticky	Dry
1A	1.0	1.5	3.5	3.2
2	10.9	9.5	13.8	9
3	42.1	10.5	35.7	9.6
4	29.4	6.3	19.5	12.3
After-filter	15.2	72	27.6	66
Sum	100%	100%	100%	100%

[a]Stage 1B was greased with silicon without a supporting film and hence could not be weighed.

ambient aerosols under all conditions. Thus there is a need to perform field comparisons for each experiment to ensure that the apparent size distributions obtained do not result from experimental artifacts.

III. CALIBRATION OF THE IMPACTOR

Design criteria lead to the following 50% cut points at 3 cfm for unit density spheres:

Stage	50% Cut Point (μ)
1A	8.0
1B	8.0
2	4.0
3	1.5
4	0.5
After-filter	<0.5

In field operation the upper cut-off for particles reaching the impactor through the sampling probe depends on the wind conditions, but typically will be on the order of 20 μm. Laboratory calibrations using uranine dye obtained 50% cut points for all units used in the ACHEX that agreed, with one exception, to within 10% of the design criteria when impaction surfaces coated with silicon grease were used. Stage 1B results were lower by

15 to 20%, an expected deviation, as discussed above. A more detailed discussion is given by Appel and Wesolowski (3). For the comparative plots of particle size distributions presented in this report the 50% cut points listed in the above table were used. The concentration on a given stage was plotted at the midpoint between the 50% cutoff for that stage and 50% cutoff for the next larger particle stage. The upper cutoff for the sampling probe was estimated to be 20 μm. The concentration for the largest particle stage was taken simply as the sum of the concentrations on stages 1A and 1B. For the after-filter the concentration is plotted at 0.2 μm. The concentrations are shown per logarithm of the diameter interval.

IV. FIELD EVALUATION OF THE IMPACTOR

The ambient air validation measurements were carried out in the following manner. At each station and for each episode two Lundgren impactors were run side by side, one with the sticky surface described above and the other "dry," that is, with Mylar or Teflon films lacking adhesive coatings. Quite often the latter had a double foil (i.e., both 0.0006-cm-thick Mylar and 0.0025-cm-thick dry Teflon). In these cases the Teflon was placed on the drum first and covered the entire drum, and a Mylar strip of half the width of the drum was placed over the Teflon. Stage 1B always was covered with the SPE. Each impactor had its own after-filter, a 47-mm Gelman GA-1 cellulose acetate membrane filter (4). A second filter of this type sampled the same air but collected all particles entering the sampling probe (i.e., ≤ about 20 μm) and is referred to here as the total filter.

Figure 2 is a schematic showing the sampling system. The efficiency of particle collection by the sampling probe is dependent upon wind speed and direction, especially for large particles. It is difficult, therefore, to design a particle collection system that maintains a constant collection efficiency over a wide range of wind conditions, that is, samples isokinetically. No attempt to do this was made. However, the collection system was so designed that all the separate pieces of aerosol equipment sampled the same air and, as discussed below, sampled isokinetically within the probe. The probe itself extended about 10 ft above the roof of the trailer. A cylindrical wind shield or shroud reduced the horizontal velocity of particles before they entered the probe, resulting in better sampling efficiency for large particles. A 40-mesh stainless steel screen protected the probe entrances from birds, bugs, leaves, and the like. A manifold containing a multitude

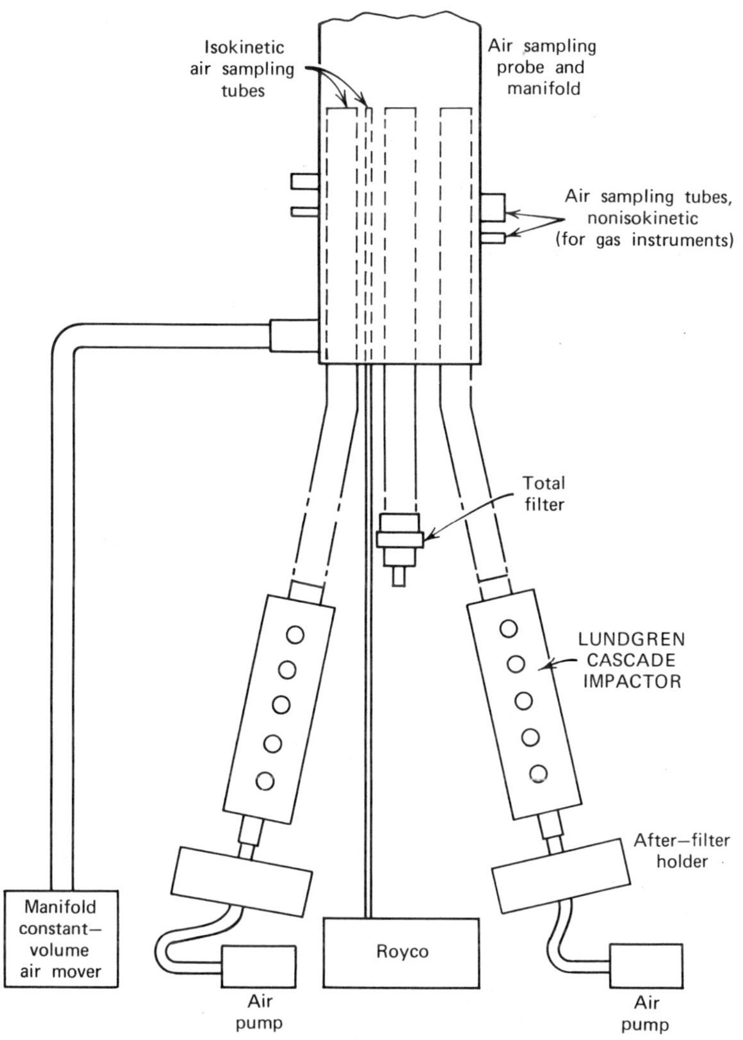

FIGURE 2. Schematic of the aerosol sampling system.

of ports was attached to the probe. Air was moved through the entire system by a constant-volume air mover at 24 cfm. Aerosol equipment was attached to pipes that were concentric with the main manifold. Each piece of aerosol equipment had its own air mover. The crucial design criterion was to make the diameter of each pipe leading to a piece of aerosol equipment of a dimension such that, when combined with the flow rate for that piece of equipment, the linear velocity of air at the pipe entrance equaled the linear velocity of the air moving through the sampling probe. Then all the aerosol equipment sampled isokinetically at this point in the manifold, and thus all pieces sampled the same aerosol.

To experimentally verify the symmetry of the system (i.e., the equivalence of the two impactors), a series of ambient air runs was made using the same impaction surfaces on both Lundgrens, sometimes sticky and sometimes dry. The results are shown in Figure 3. The correlation coefficient is .94, and the results clearly indicate that the system is symmetric.

Next, side-by-side ambient air comparisons were made, using different impaction surfaces, in order to answer four specific questions:

(1) What are the differences in the observed particle size distributions for specific elements for dry Mylar versus dry Teflon?

(2) Does the modification incorporating a sticky stage 1B help eliminate bounce-off?

(3) What are the differences between sticky and dry impaction surfaces?

(4) What are the wall losses, that is, what percentage of the weight of a given chemical species (or of mass) that should have been collected by the impactor stages and after-filter is lost to the walls of the impactor?

Analytical measurements of specific elements were done by R. Ragaini and R. Kaifer at the Lawrence Livermore Laboratory using neutron activation analysis (NAA) (5). Four elements will be discussed in this paper: Al, an indicator of soil particles; Na, an indicator of sea and/or soil particles depending on geographic location; Br, an indicator of automobile-generated aerosol; and V, an indicator of both soil and anthropogenic pollution (6).

In regard to the first question, relevant data were obtained with both synthetic and ambient aerosol. With uranine dye

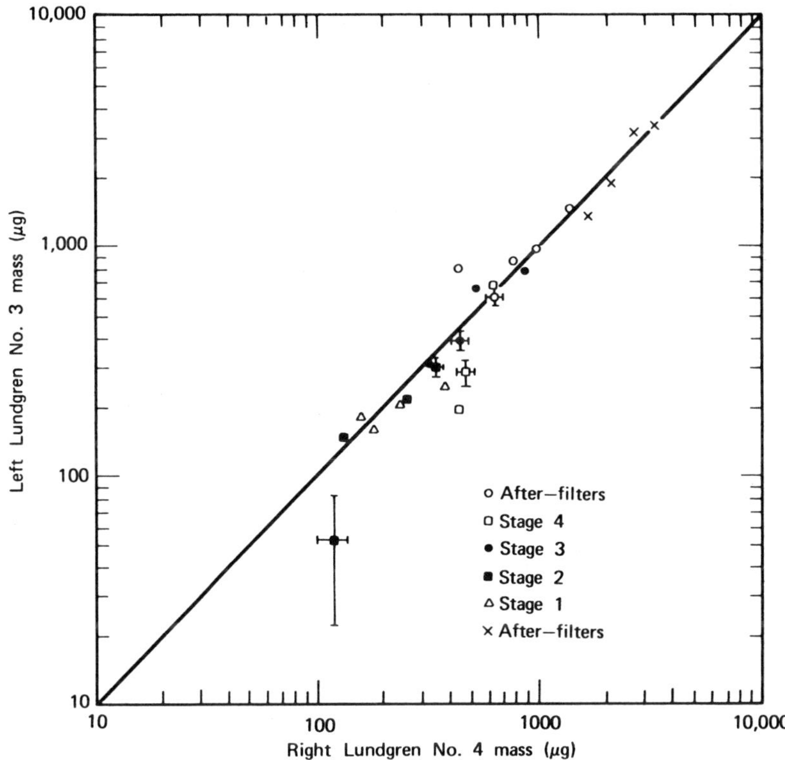

FIGURE 3. Comparison of masses on stages and the after-filters of Lundgren impactors run side by side.

particles laboratory work indicated no significant difference between Teflon and Mylar as collection surfaces (3). Table 2 lists results for ambient air sampling conducted in San Jose, California, on August 25, 1972. With the exception of vanadium, the average Mylar/Teflon ratio is close to unity. It should be noted that the vanadium detected was on the order of tenths of a nanogram per cubic meter and that the errors in some cases are rather large. Clearly the dry Mylar and dry Teflon have similar

TABLE 2. Comparisons of Ambient Air Elemental Concentrations on Dry Mylar and Dry Teflon Impaction Surfaces Collected with the Lundgren Impactor: San Jose, August 25, 1972

Stage	Surface	Element (ng/m^3) NA	Al	V	Br
1A	Teflon	31.0 ± 1.6	15.0 ± 0.8	0.042 ± 0.008	0.76 ± 0.09
	Mylar	22.9 ± 1.9	7.2 ± 0.6	ND	0.64 ± 0.07
	Ratio	0.74	0.48	--	0.84
1B	Sticky	34.3 ± 8.6	205 ± 10	0.34 ± 0.05	3.8 ± 0.1
2	Teflon	75.1 ± 3.8	17.5 ± 0.9	0.078 ± 0.009	1.7 ± 0.1
	Mylar	84.6 ± 4.7	19.9 ± 1.0	ND	2.2 ± 0.1
	Ratio	1.12	1.13	--	1.29
3	Teflon	342 ± 17	80.7 ± 4.0	0.40 ± 0.03	15.3 ± 0.8
	Mylar	365 ± 18	85.8 ± 4.3	0.22 ± 0.09	14.5 ± 0.7
	Ratio	1.06	1.06	0.55	0.95
4	Teflon	179 ± 9	74.9 ± 3.7	0.98 ± 0.05	45.9 ± 2.3
	Mylar	173 ± 9	57.1 ± 2.9	0.64 ± 0.09	41.5 ± 2.1
	Ratio	0.97	0.76	0.65	0.90
	$\overline{R}$ (stages 1A, 2, 3, 4)	0.97	0.86	0.60	1.0

collection characteristics. Also important is the indication that the two halves of the drum collect equivalent amounts of aerosol, as one would expect. The equivalence of Mylar and Teflon is shown graphically in Figure 4.

Table 2 also leads to a discussion of question 2, namely, the effect of incorporating a sticky stage 1B into the Lundgren in order to collect large particles that bounce off stage 1A. We note that for all of the elements studied, stage 1B, which always had SPE on it, collected much more material than stage 1A. For aluminum, a soil-related element expected to most dramatically demonstrate bounce-off, stage 1B collected approximately 10 times more than 1A. When using the impactor with sticky surfaces on <u>all</u> drums, this effect is not observed; in fact, as one would expect, stage 1A collects more material than 1B. This is shown in Table 3. Although stage 1B for the all-sticky system does collect some material, this does not necessarily imply that there is bounce-off from 1A since the finite resolution for particle sizing would demand that 1B collect some particles. From the data in Tables 2 and 3 it is clear that the incorporation of a sticky stage 1B into the Lundgren system, which used dry impactions surfaces on other stages, helped considerably in preventing large particles which bounced off stage 1A from being collected on stages 2, 3, and 4 and the after-filter.

The sodium data merit special comment. We expect Na to come mainly from the sea at locations near the coast and from the soil at locations inland. The hygroscopicity of sea salt should lead to reduced bounce-off relative to that for soil Na. Thus there should be more bounce-off of Na in Fresno than in the coastal stations. Indeed this is so; in Fresno the 1B/1A ratios are essentially the same for Al and Na.

We next address question 3 concerning differences in particle size distributions from sticky and dry impaction surfaces. Three of the four side-by-side runs shown in Figures 4 to 7, those for Berkeley, San Jose, and Pasadena, represent urban, near-coastal locations. Assuming approximately similar production mechanisms, we expect similar particle size distributions and similar amounts of bounce-off for all of the specific elements. Indeed, both appear to be fulfilled. For example, the Br particle size distributions are heavily weighted toward small particles for all four locations. This is consistent with the automobile as the principal source of Br. The V distributions are similar for Berkeley, Pasadena, and San Jose, and have shapes similar to the one for Br, also indicating a combustion source. The V distribution for Fresno, the inland station, however, have large-particle components as well, presumably from a soil source (6). In fact, the V/Al ratio for the first stage is 0.0016, which is close to the value of 0.0018 for average crustal material found

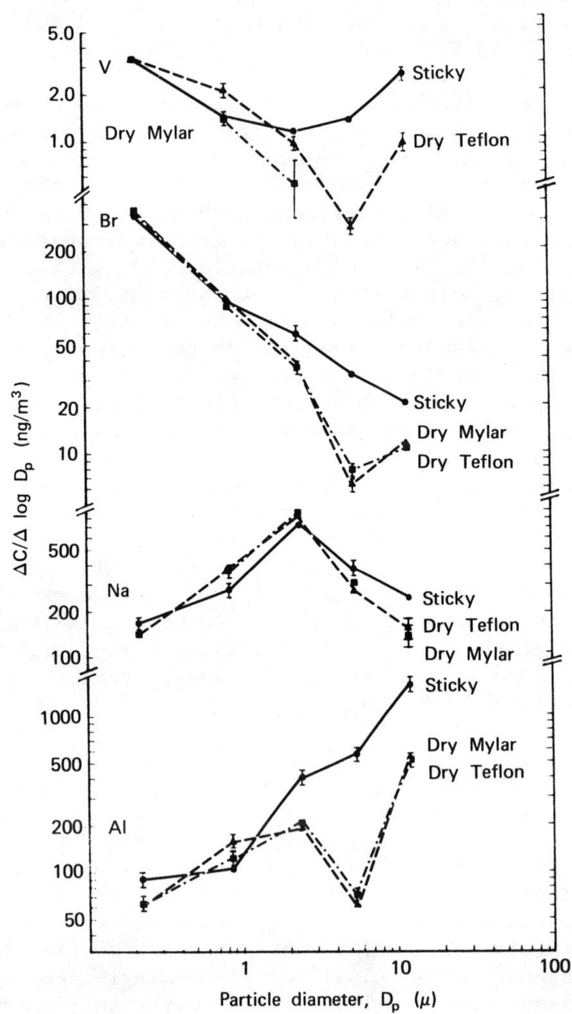

FIGURE 4. Comparison of chemical determinations from aerosols collected on various Lundgren impactor surfaces: San Jose, August 25, 1972.

TABLE 3. Comparison of Concentrations on Stages 1A and 1B from Lundgrens Run Side by Side, One with all Sticky Impaction Surfaces and One with Only 1B Sticky[a]

Location and date	Lundgren	Ratio of 1B/1A			
		Na	Al	V	Br
San Jose, 8/25/72	1B sticky	1.3	18.5	8.1	5.5
	All sticky	0.19	0.21	0.23	0.59
Fresno, 8/31/72	1B sticky	27.1	24.7	15.6	16.1
	All sticky	0.31	0.39	0.35	0.58
Berkeley, 7/28/72	1B sticky	0.50	8.2	1.5	5.9
	All sticky	0.10	0.25	0.17	0.57
Pasadena, 8/24/72	1B sticky	b	18.5	--	9.0
	All sticky	--	0.83	0.019	0.77

[a] The Lundgren with only stage 1B sticky had the drums of the other stages covered half with dry Teflon and half with dry Mylar. An exception was the Berkeley run in which the nonsticky stages were completely covered with dry Teflon.

[b] Dash = no data available.

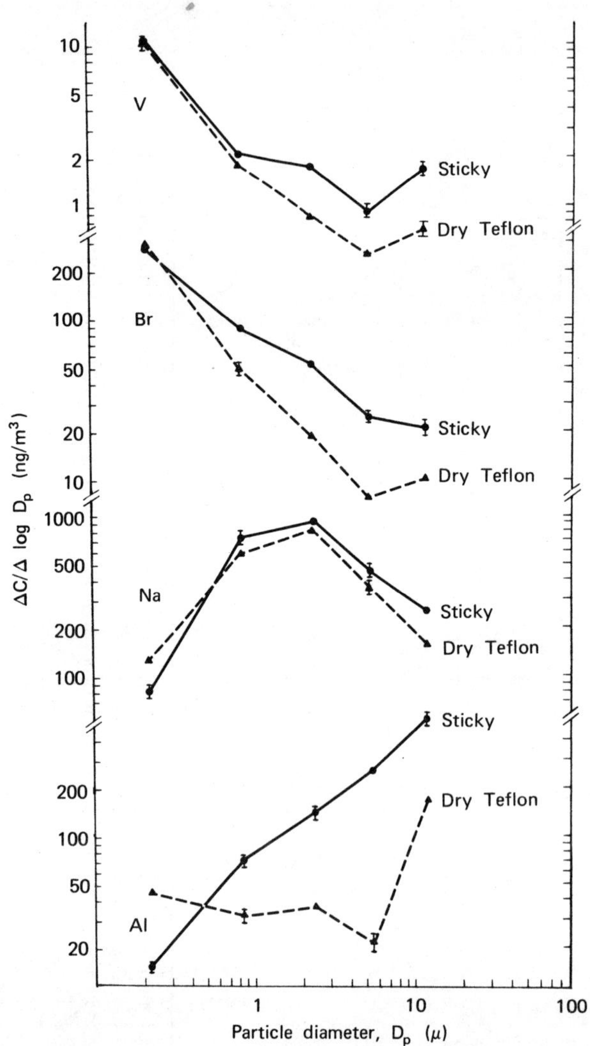

FIGURE 5. Comparison of chemical determinations from aerosols collected on various Lundgren impactor surfaces: Berkeley, July 28, 1972.

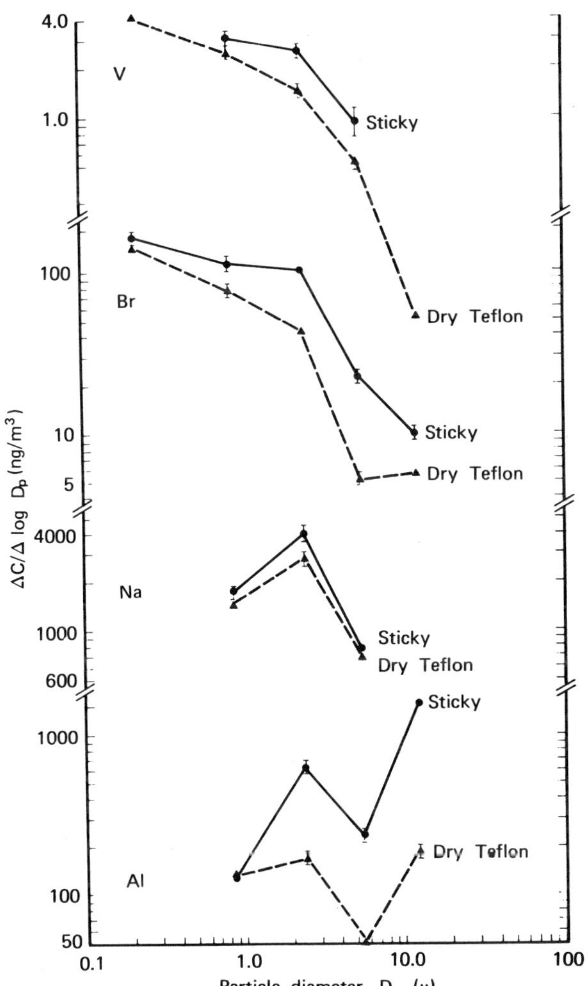

FIGURE 6. Comparison of chemical determinations from aerosols collected on various Lundgren impactor surfaces: Pasadena, August 24, 1972.

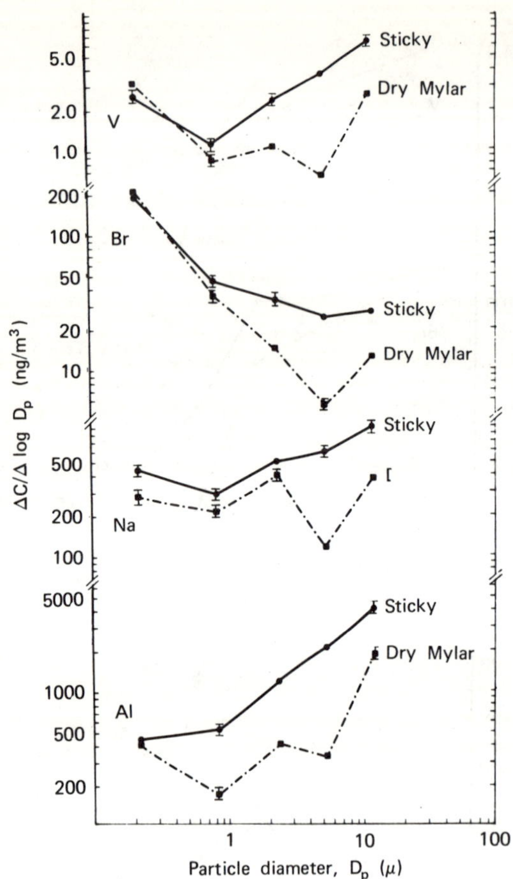

FIGURE 7. Comparison of chemical determinations from aerosols collected on various Lundgren impactor surfaces: Fresno-Visalia, August 31, 1972.

by Goldschmidt (7). The dry impaction surface distributions for Br and V indicate a large amount of bounce-off, particularly for the large-particle stages. Qualitatively, the bounce-off effect is similar at all four stations for Br, and at Pasadena, San Jose, and Berkeley for V. The dry impaction surface V distribution for Fresno shows a larger loss as compared to Br, consistent with soil as an important source of V at this location. The Al distributions from sticky surfaces at all four locations are heavily weighted toward large particles, indicative of a soil source.

Likewise, as one would expect, the bounce-off for this element is the most pronounced. In fact, the sticky and dry distributions are sufficiently different that the use of the two distributions to determine sources could lead to differing conclusions.

The sodium distributions are quite similar for Pasadena, San Jose, and Berkeley, and show a maximum between 1 and 3 μ. According to Junge, 90% of the mass of sea salt particles is carried by "jet" particles, that is, particles greater than 1 μm, which originate upon the bursting of small air bubbles produced by breaking waves, and only 10% by film particles, that is, particles less than 1 μm, which result from the bursting of the bubble film (8).

Thus the Na distributions for Pasadena, Berkeley, and San Jose are entirely consistent with a sea source. On the other hand, the Fresno Na distribution is completely different and has a large-particle component indicative of a soil source. Even more interesting is the minimal difference between dry and sticky impaction surfaces for the coastal stations, whereas the Fresno data display a large difference indicative of bounce-off.

A similar comparison of mass data is shown for Berkeley and Fresno in Figure 8. For Berkeley the mass distributions on sticky and dry surfaces are equivalent within experimental errors,* consistent with Lundgren's work (2). In the case of Fresno, where a large soil component would be expected, there is a difference between sticky and dry surfaces for the large-particle stages. As with the elements, there is little difference between dry Mylar and dry Teflon.

We conclude that the amount of bounce-off is related to the source of the aerosol containing the element in question, rather than to the element itself. Thus bounce-off will vary with location.

We now discuss the last question, that concerning loss of particles to the walls of the impactor. Laboratory studies with uranine dye particles indicated up to 38% loss with the maximum loss for particles in the 4 to 7-μm size range (3). Wall losses with atmospheric aerosols are listed in Table 4. These were obtained by comparing the sum of the concentrations on the stages and after-filter with the total filter run in parallel with the

---

*Caution must be used in interpreting mass data from impactor stages. Small pieces of the thin impaction surfaces can be lost by abrasion during the process of attaching or removing the surfaces from the drums. Although care was taken to minimize such errors, the mass determinations for the stages are accurate only to ±10% for the small-particle stages and ±50% for the lightly loaded large-particle stages.

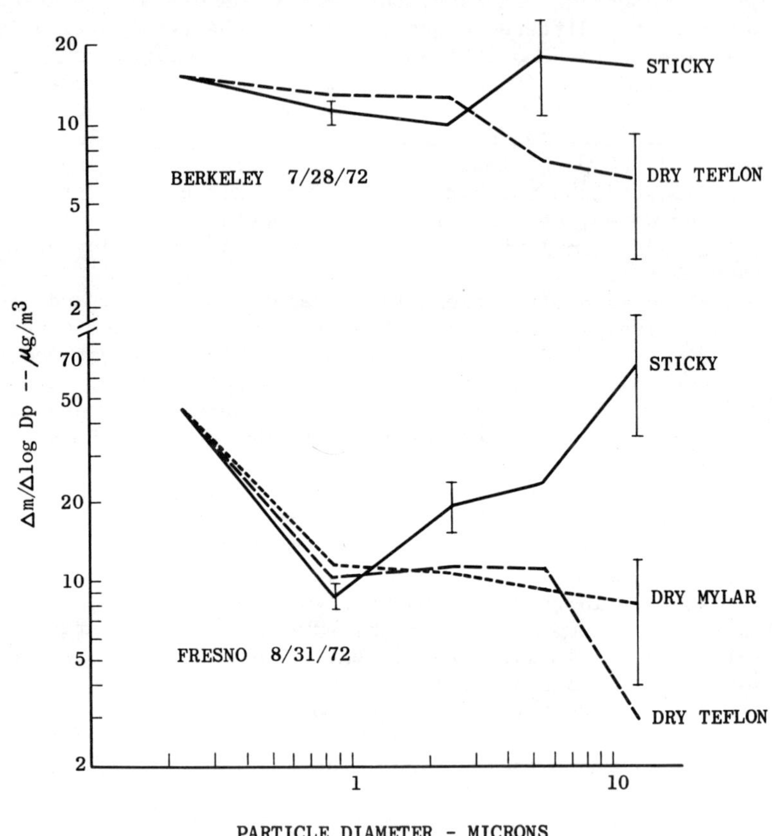

FIGURE 8. Mass particle size distributions.

impactors in the common manifold. We note that there are wall losses for the SPE system but they are smaller than with either the dry Mylar or the dry Teflon system. For V, Br, and Na the wall losses from the sticky system are less than 15%, but Al and mass each have 24% loss. The loss of mass could be high because of the film abrasion problem discussed earlier. The Mylar system shows wall losses greater than the sticky system losses, for both mass and elements, with the greatest loss for Al. The Teflon wall losses are similar to those of Mylar.

TABLE 4. Comparison of Wall Losses for Various Collection Systems[a]

Location and date	Collection system	Mass (%)	Element (%)			
			Na	Al	V	Br
San Jose 8/25/72	Total filter	100	100	100	100	100
	Sticky	78	77	78	84	90
	Mylar	61	79	30	61	86
	Teflon	53	77	32	72	87
Fresno, 8/31/72	Total filter	100	--[b]	--	--	--
	Sticky	60				
	Mylar	36				
	Teflon	35				
Berkeley, 7/28/72	Total filter	100	100	100	100	100
	Sticky	88	99	74	94	82
	Teflon	75	85	26	84	74
Berkeley, 7/30/72	Total filter	100	--	--	--	--
	Sticky	76				
Berkeley, 7/31/72	Total filter	100	--	--	--	--
	Teflon	67				
Berkeley, 8/1/72	Total filter	100	--	--	--	--
	Mylar	91				
	Teflon	69				
Averages	Sticky	76	99	76	89	86
	Mylar	62	79	30	61	86
	Teflon	60	81	29	78	81

[a] Sum of concentrations on stages and after-filters expressed as a percentage of the concentration on the total filter.

[b] Dash = data not available.

It is important to note that the sticky surfaces, though a great improvement over the dry, do not completely eliminate wall loss and bounce-off. This is consistent with the laboratory work of Rao, which showed that the sticky material used in ACHEX had, in the limiting case of hard polystyrene latex spheres, a collection efficiency of only 50% (9).

Recent work by Dzubay et al., using ambient air aerosol, illustrates another consequence of using substrates of poor collection efficiency (10). With two impactors in parallel, one using greased surfaces and one using bare aluminum, they found that the 50% cutpoint shifted from 8 μm for the greased surface to 3.5 μm for the bare aluminum.

For additional information on collection surfaces the reader is referred to a recent article that evaluates collection surfaces using both laboratory-generated aerosol and atmospheric particles (11).

V. CONCLUSIONS

A number of conclusions concerning the Lundgren impactor can be reached from the data discussed above.

(1) There are differences in the particle size distributions for specific elements, as well as mass, collected with Lundgren impactors run with sticky and dry impaction surfaces. The differences are related to the source of the particles containing the element in question, rather than to the element itself. Thus the differences vary with location. In some cases the distributions from the sticky and dry impaction surfaces are sufficiently different that the use of the two distributions to determine sources could lead to differing conclusions.

(2) There are large wall losses for Lundgren impactors using dry impaction surfaces. Though significantly less for sticky impaction surfaces, these losses are not negligible.

(3) Particle size distributions and wall losses for 1/4-mil uncoated Mylar are similar to those for 1-mil uncoated Teflon.

(4) The incorporation of a stage with a sticky impaction surface between the first and second stages, with a slit width identical to that of the first, helps prevent particles that bounce off the dry surface of the first stage from reaching other stages or the after-filter.

(5) Because of the severe wall losses for dry surfaces, summation of the stages and after-filter is not a useful measure of the total weight of the aerosol entering the impactor.

(6) Because of the variation in the particle size distributions for specific elements from location to location, it is doubtful whether a mathematical function can be derived to correct the particle size distributions obtained from dry impaction surfaces.

In view of these conclusions it is recommended that Lundgren impactors with dry impaction surfaces not be used for field measurements unless some method of judging the value of the data similar to that employed above is included as an integral part of the measurements.

This work clearly demonstrates the importance of laboratory and field measurements in assessing the collection errors of size-segregating sampling instruments. It also points to the need for having standardized laboratory methods for generating aerosols that reflect the properties of ambient air aerosols.

ACKNOWLEDGMENTS

We wish to acknowledge our indebtedness to Professor Kenneth Whitby of the University of Minnesota and Carl Peterson of Environmental Research Corporation for the generous use of their aerosol laboratory facilities during the early phases of this experiment before the completion of the aerosol laboratory at AIHL. We thank Richard Ragaini and Robert Kaifer for their cooperation in carrying out the neutron activation analysis. We are especially grateful to SuzAnne Twiss for data reduction, statistical analysis, and her considerable effort in reviewing the manuscript.

REFERENCES

1. Novakov, T., Mueller, P. K., Alcocer, A. E., and Otvos, J. W. 1972. J. Colloid Int. Sci. 39, 1, 225.

2. Lundgren, D. 1972. J. Colloid Int. Sci. 39, 1, 205.

3. Appel, B. R. and Wesolowski, J. J. 1973. "Calibration and Sampling Media Selection Studies with the Lundgren Impactor," AIHL Rep. No. 137, Air and Industrial Hygiene Laboratory, California State Department of Health, February.

4. Appel, B. R. and Wesolowski, J. J. Selection of Filter Media for Particulate Sampling with a Lundgren Impactor, this volume.

5. Ragaini, R., Kaifer, R., and Wesolowski, J. J. 1972. "A Report on Neutron Activation Analysis of Samples," AIHL Rep. No. 103, Air and Industrial Hygiene Laboratory, California State Department of Health, September.

6. Marten, C., Wesolowski, J. J., Kaifer, R., John, W., and Harris, R. 1973. Sources of Vanadium in Puerto Rican and San Francisco Bay Area Aerosol, Environ. Sci. Technol. 7, 817; John, W., Kaifer, R., Rahn, K., and Wesolowski, J. J. 1973. Trace Element Concentrations in Aerosols from the San Francisco Bay Area, Atmos. Environ. 7, 107.

7. Goldschmidt. 1954. Geochemistry. Oxford University Press, London.

8. Junge, C. W. 1972. Our Knowledge of the Physico-Chemistry of Aerosols in the Undisturbed Marine Environment, J. Geophys. Res. 77, 27, 5183.

9. Rao, A. K. 1975. "An Experimental Study of Initial Impactors," Ph.D. dissertation, Particle Technology Laboratory, University of Minnesota.

10. Dzubay, T. G., Hines, L. E., and Stevens, R. K. 1976. Particle Bounce Errors in Cascade Impactors, Atmos. Environ. 10, 229.

11. Wesolowski, J. J., John, W., Devor, W., Cahill, T. A., Feeney, P. J., Wolfe, G., and Flocchini, R. 1977. Collection Surfaces of Cascade Impactors, in X-Ray Fluorescence Analysis of Environmental Samples (T. G. Dzubay, Ed.), Ann Arbor Science, Ann Arbor, Michigan, p. 121.

X-Ray Fluorescence Analysis of ACHEX Aerosols

ROBERT D. GIAUQUE, LILLY Y. GODA, AND ROBERTA B. GARRETT
Lawrence Berkeley Laboratory
University of California
Berkeley

Abstract

The elemental compositions of aerosols collected on filter media and impactor films during the Aerosol Characterization Experiments (ACHEX) were determined by X-ray-induced X-ray fluorescence analysis. Low-power X-ray tubes (<40 W) and semiconductor detector X-ray spectrometers were utilized for the analyses.

Typically, the concentrations of 13 elements (Al, Si, S, K, Ca, Ti, Mn, Fe, Ni, Zn, Br, Sr, and Pb) were routinely ascertained for aerosols collected over 2-hr periods. The concentrations of 22 additional elements determined were usually near or below our detection limits.

I. INTRODUCTION

Elemental analysis of aerosols, coupled with meteorological, particle size distribution, gaseous, and other chemical data, provide information that can be used to estimate the contribution of primary and secondary aerosols to the total airborne particulate matter (1-3). From such data, evaluations of the significance of both natural and anthropogenic sources in the evolution of aerosols may be made. The analytical technique X-ray-induced X-ray fluorescence analysis (XRFA) lends itself to simultaneous determinations of a broad range of elements in airborne particulate matter collected on filter media or impactor films. The technique is quantitative and relatively rapid (minutes), and high sensitivities are obtainable (ng/m^3 of air level). Unfortunately, the method does not permit the determination of elements having very low atomic numbers, such as carbon and nitrogen, which are major constituents of most aerosol specimens. However, the technique is nondestructive and allows subsequent chemical species measurements to be made by other analytical procedures.

II. DISCUSSION OF METHOD

The XRFA method employed for these studies involves the interaction of photons, provided by an X-ray tube either directly or indirectly, with specimen atoms. A fraction of the photons, if of sufficient energy, create inner atomic shell vacancies; subsequent atomic transitions fill the vacancies, and a fraction of these give rise to the emission of characteristic X-rays. These X-rays are measured by a semiconductor detector and are sorted by their energies using pulse-height analysis. The elemental concentrations are determined from the X-ray intensities.

The relative ability to create inner atomic shell vacancies, using various energies of excitation radiations, is determined from the photoelectric cross sections, τ, of the elements. Figure 1 illustrates X-ray photoelectric cross-section curves for Ag Lα (3.0 keV), Mo Kα (17.4 keV0), and Tb Kα (44.2 keV) X-rays (4). As shown, Ag Lα X-rays are over 2000 times more efficient than Tb Kα X-rays for producing photoelectric interactions with the elements Al $\rightarrow$ Cl. Likewise, Mo Kα X-rays are approximately 15 times more efficient than Tb Kα X-rays for producing photoelectric interactions with the elements Al $\rightarrow$ Sr. Tb Kα X-rays

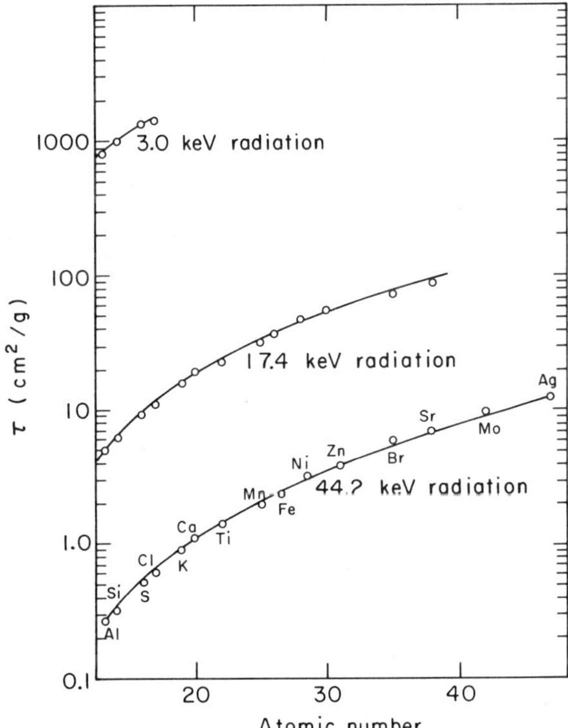

FIGURE 1. Photoelectric cross-section curves for 3.0, 17.4, and 44.2-keV photons.

are efficient for producing photoelectric interactions with the elements Y → La. Consequently, excitation radiations of more than one energy are utilized separately for the determination of a range of elements.

Only a fraction of the vacancies created in a particular energy level are filled by transitions that result in the direct emission of X-rays. Some vacancies are filled by transitions involving the emission of Auger electrons. The fraction of vacancies filled by transitions that directly yield X-rays is known as the fluorescence yield value, ω. Figure 2 shows curves of theoretical values of ω (5, 6) for a wide range of elements for the K and L_{III} energy levels. K X-rays are used for the analysis of elements up to atomic number 57 (La), and L X-rays for elements with higher atomic numbers. The values of ω are

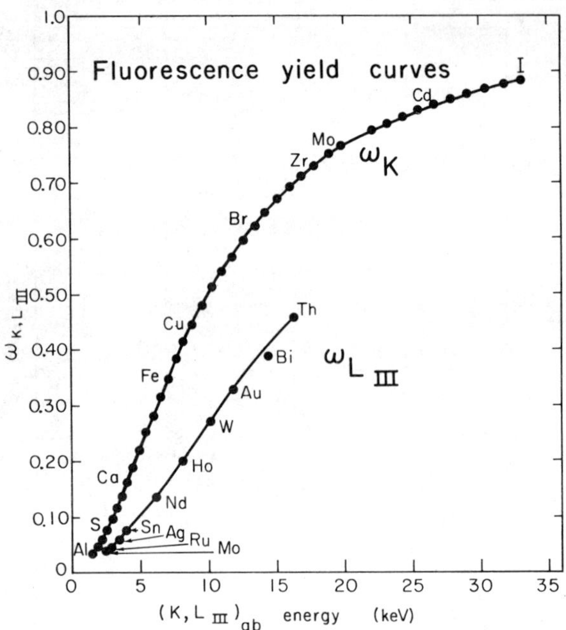

FIGURE 2. Theoretical fluorescence yield curves for the K and L_{III} energy levels.

not influenced by the experimental conditions. Low fluorescence yield values and major absorption effect problems either limit the sensitivities attainable or prohibit determination of elements having very low atomic numbers.

A. Instrumentation and Characteristics

Three different X-ray spectrometers were utilized during these studies to determine the elemental compositions of aerosols collected on filter media and impactor films. The characteristics of these spectrometers, as well as the elements determined, are listed in Table 1. The spectrometers are illustrated in Figures 3 to 6. Both 512 and 1024-channel pulse-height analyzers were used to acquire the X-ray spectra. Corrections for system dead times were made by using gated clocks that measured the total system live times. More complete descriptions of these spectrometers have been reported (7-9).

TABLE 1. Characteristics of Spectrometers Used

	Spectrometer A	Spectrometer B	Spectrometer C
Detector	Grounded guard-ring	Guard-ring reject	Top-hat
Electronics	Pulsed-light feedback	Pulsed-light feedback	Pulsed-light feedback
Resolution (eV) at 5.9 keV	∿225	∿190	∿175
X-ray tube, power (W)	Mo transmission, 20	W, 50	Ag, 50
Excitation method	Direct	Secondary targets	Direct and secondary targets
Total area (cm^2) of specimen analyzed	1-3	5	1
Elements determined	S → Sr, Hg, Pb	Sr → Ba	Mg → Cl

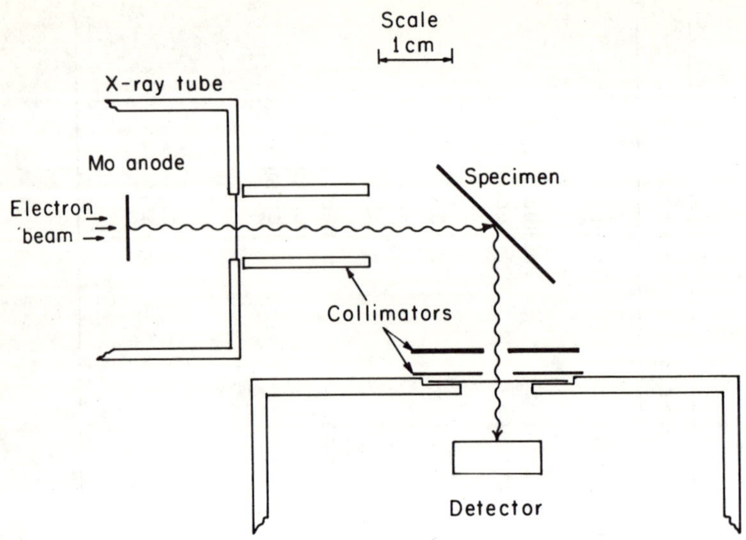

FIGURE 3. Spectrometer A. X-ray fluorescence analysis technique by direct excitation using a Mo transmission X-ray tube.

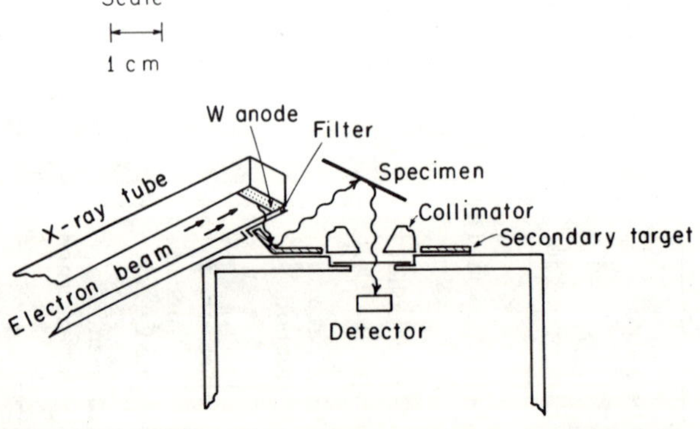

FIGURE 4. Spectrometer B. X-ray fluorescence analysis technique using a W anode X-ray tube and secondary targets to provide the excitation radiation. (Goulding-Jaklevic, EPA-type X-ray spectrometer).

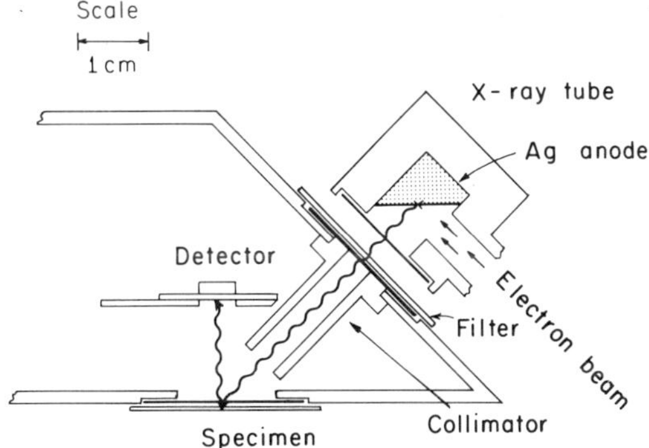

FIGURE 5. Spectrometer C. X-ray fluorescence analysis technique by direct excitation using an Ag anode X-ray tube.

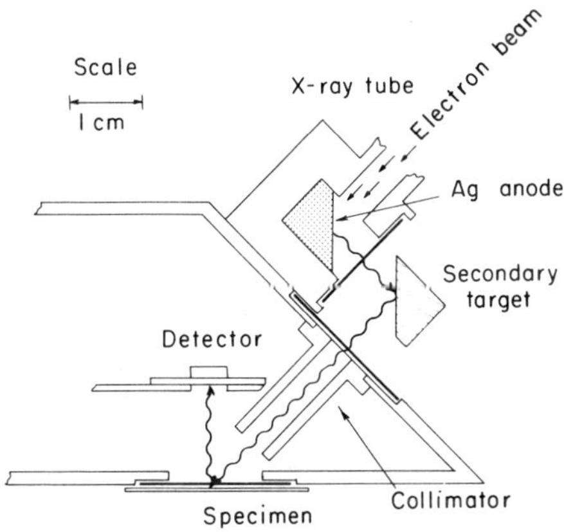

FIGURE 6. Spectrometer C. X-ray fluorescence analysis technique with an Ag anode X-ray tube and secondary targets to provide the excitation radiation.

B. Calibration Methods

Aerosols collected on filter media or impactor films can, in general, be considered as thin specimens. For most analyses the concentration of an element, m_j ($\mu g/cm^2$), is directly proportional to the intensity, I_j, of one of its characteristic K or L X-ray lines and may be expressed as

$$I_j = K_j m_j \tag{1}$$

where K_j is a sensitivity factor for the element j.

For any given geometry and for constant excitation radiation intensity, the relative ability to excite and detect various X-ray lines from "infinitely" thin specimens may be calculated as previously reported (7). Standardization of a spectrometer is achieved using a single-element thin-film standard. The standard is prepared thin enough so that absorption effects are negligible ($\sim$100 $\mu g/cm^2$). Calculated values of K_j for X-ray lines from eight elements have been reported which agree to within several percent of measured values from evaporated thin-film standards (7). In effect, standardization for excitation radiation intensity and total system geometry is done with a single-element thin-film standard, and factors converting counts per second to micrograms per square centimeter for the other elements are calculated.

A thick, pure silicon disk is used to standardize spectrometer C for the analyses of the lower Z elements, since a truly thin uniform standard with negligible absorption effects would be difficult to prepare. For standardization purposes, the mass thickness of a disk is the mass above which an increase in thickness does not measurably increase the fluorescence X-ray line intensity. Mathematically, the mass of a thick element disk may be expressed as

$$m_{thick} = \frac{3.92}{\mu_e \csc \phi_1 + \mu_f \csc \phi_2} \tag{2}$$

where μ_e and μ_f are the total mass absorption coefficients of the element for the excitation and fluorescence radiations. The angles formed by the excitation and fluorescence radiations with the disk surface are ϕ_1 and ϕ_2, respectively. It can be shown that m_{thick} represents the mass for which only 25% of the radiation (excitation x fluorescence) is not attenuated. Hence the mass of the disk for calibration purposes equals $m_{thick}/4.0$. We used thick disks to standardize for aluminum and sulfur and obtained agreements to within 5% of the calculated relative excitation-detection efficiency values normalized to silicon.

C. Data Processing

X-ray spectra of the collected aerosol specimens are recorded on magnetic tape, and the analytical computations are made by a Control Data 7600 computer. Although our programs have been established for the analyses of many types of specimens, much less than 50 K of core space would actually be required for the analyses of aerosols only. The analysis programs are performed in three steps. First, the X-ray spectrum background due to the scattered excitation radiation is removed; second, overlapping X-ray lines are unfolded; and, third, the concentrations of the elements in the original air sampled are calculated from the intensities of the appropriate X-ray lines selected for analysis.

The shape of the scattered excitation radiation background is simply determined by obtaining a spectrum of a blank filter similar in mass to the filters used for the aerosol collections. The scattered X-ray backgrounds within various energy ranges selected for analysis are ratioed to the intensities of the incoherent scattered excitation radiations. This procedure is possible because the typical aerosol loadings contribute only a couple of percent to the total mass of the filter plus collected aerosol. Also, the loadings are similar in average effective atomic number to the collection media.

Fixed sets of channels corresponding to preselected energy ranges are employed for analyses and are verified daily. Minor amplifier adjustments are made, if necessary, to compensate for any slight instrumental deviation such as a baseline or gain shift. To obtain high statistical accuracies, between 70 and 80% of the total, peak areas are utilized for the analyses. The peak fractions used are established from element thin-film deposits. Corrections for overlapping X-ray lines are also determined from the element thin-film deposits by establishing a relationship between individual X-ray peak shapes and intensities. In effect, a series of simultaneous equations are established that compensate for the overlapping X-ray lines occurring within the preselected energy ranges.

The concentrations of the individual elements present in the aerosol are then calculated from the individual X-ray line intensities by using Eq. 3:

$$\frac{\mu g\ (j)}{m^3} = \frac{C_j}{C_s} \times \frac{m_s}{V} \times \frac{1}{K_j} \times A_j \qquad (3)$$

where C_j and C_s are the characteristic X-ray count rates from element j and the standard element of known mass, m_s is the standard mass ($\mu g/cm^2$), V is the volume of air sampled (m^3) per square centimeter of the collection media, K_j is defined by Eq. 1, and A_j is an absorption factor to correct for particle size effects.

For most determinations the value of the term A_j is 1.0. Particle size effects must be considered when analyzing for low atomic number elements such as silicon and aluminum. These elements have low-energy X-rays, <4 keV, and are present principally in the large particles, >2μ. A fraction of the excitation and fluorescence radiations are attenuated in the individual particles containing these elements. The value of the term A_j is determined experimentally using multiple energies of X-ray excitation which act as variable depth probes. Details of the method are reported elsewhere (9).

III. RESULTS

A. Blank Substrates

The different types of collection media utilized during Phase II of the ACHEX program were (a) washed Gelman GA-1 cellulose ester membrane filters for total and after-filters of 2-hr aerosol specimens, (b) sticky polyethylene films for Lundgren impactor surfaces, (c) Whatman 41 cellulose filters for high-volume samples, and (d) glass fiber filters for total and refined aerosol collections for 2-hr intervals. Table 2 lists the concentrations of impurities found in these substrates. Washed GA-1 filters were ascertained to be suitable collection media for analyses of aerosol specimens because of their relatively low contents of impurities. Additionally, they are membrane-surface-type collection substrates. Zinc determinations were not made for aerosols collected on sticky polyethylene because of the high content of this element in the blank. A fraction of the particles collected on Whatman 41 filters are embedded within the substrate. Although Bonner et al. (10) have estimated absorption corrections for particle penetration, determinations for elements with K X-rays of energies 3 keV and less were not made. As shown in Table 2, the glass fiber filters contained substantial amounts of impurities. Consequently, the concentrations of very few elements (Pb, Br, and in some cases Fe) could be determined for aerosols collected on this medium. These filters were used principally to collect aerosols for carbon and extractable organic determinations.

TABLE 2. Blank Filter Impurities ($\mu g/cm^2$)[a]

Element	GA-1	Sticky polyethylene	Whatman 41	Gelman A glass filter
K	ND	ND	ND	42
Ca	0.16	ND	0.05	41
Fe	0.031	0.02	0.02	2.6
Cu	0.058	ND	0.01	ND
Zn	0.004	0.6	0.01	167
Rb	ND	ND	ND	0.41
Sr	ND	ND	ND	0.67
Ba	ND	ND	ND	3.8
Pb	ND	ND	ND	0.12

[a]Values uncorrected for filter absorption effects.

B. Sensitivities

Tables 3 to 5 list the sensitivities (3σ values) for the elements determined with each spectrometer. In each case the values listed are for 10-min counting periods. The sensitivities attainable using spectrometer C have subsequently been improved (9). Figures 7 to 9 show spectra acquired by each of the spectrometers. The values listed are in nanograms per square centimeter.

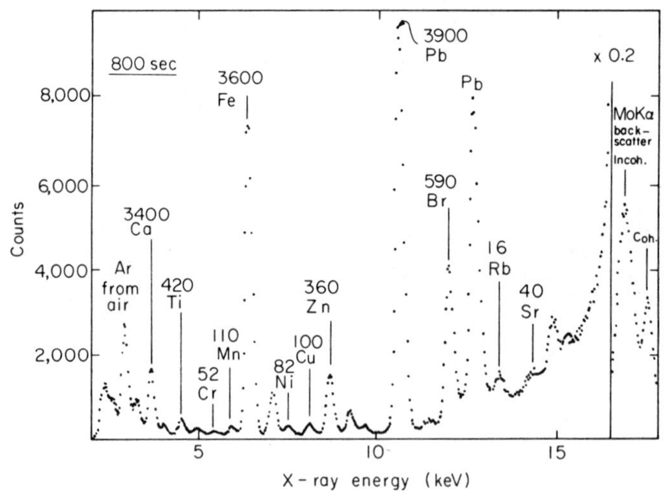

FIGURE 7. Spectrum of a total filter aerosol specimen collected for a 2-hr interval. Spectrum obtained using spectrometer A.

C. Intercomparisons of Analytical Data

An intercomparison of XRFA and NAA results for filter samples indicates that there are losses of aerosol constituents during shipment, handling, and/or analysis. These comparisons are included in another paper entitled "Quality Assurance for the Chemistry of the Aerosol Characterization Experiment," by

TABLE 3. Theoretical Limits of Detection, Spectrometer A

Medium:	Gelman GA-1	Sticky polyethylene
Mass (mg/cm^2):	5.0	5.0
Air volume sampled (m^3/cm^2):	0.75	2.40
Area analyzed (cm^2):	3	1
Element and spectral line	Detection limit (ng/m^3)	
S Kα	900	480
Cl Kα	420	220
K Kα	140	100
Ca Kα	65	55
Ti Kα	26	19
V Kα	21	15
Cr Kα	17	10
Mn Kα	12	8
Fe Kα	13	10
Ni Kα	5	5
Cu Kα	8	5
Zn Kα	5	12
Ga Kα	4	5
As Kα	5	4
Se Kα	5	4
Br Kα	7	5
Rb Kα	10	8
Sr Kα	12	9
Hg Lα	9	7
Pb Lα	9	7
Pb Lβ	23	19

TABLE 4. Theoretical Limits of Detection, Spectrometer B, Tb Secondary Target

Medium:	Whatman 41
Mass (mg/cm^2):	9.0
Air volume sampled (m^3/cm^2):	3.0
Area analyzed (cm^2):	4
Element and spectral line	Detection limit (ng/m^3)
Sr Kα	17
Y Kα	16
Zr Kα	14
Nb Kα	12
Mo Kα	11
Pd Kα	10
Ag Kα	9
Cd Kα	10
In Kα	12
Sn Kα	13
Sb Kα	13
Te Kα	14
I Kα	17
Cs Kα	28
Ba Kα	38
La Kα	56

TABLE 5. Theoretical Limits of Detection, Spectrometer C

Medium:			Gelman GA-1	
Mass (mg/cm^2):			5.0	
Air volume sampled (m^3/cm^2):			0.75	
Area analyzed (cm^2):			1	
Mode:	Zr secondary target	Ag secondary target	Ni secondary target	Direct, Ag filter
Element and spectral line		Detection limit (ng/m^3)		
Mg Kα	--	--	--	320
Al Kα	900	1400	7200	130
Si Kα	560	700	3500	70
S Kα	--	400	1400	40
Cl Kα	--	370	900	40
Ca Kα	--	--	200	--
Ti Kα	--	--	130	--
Fe Kα	--	--	120	--

Appel et al. in this volume. Comparisons of the XRFA data for the Lundgren impactor stages with total filter samples are also included in that paper.

Table 6 lists the elements present in California aerosols, which we determined by XRFA during the ACHEX II program. Additionally, the concentrations and the particle size ranges in which these elements were typically found are listed. The elements enclosed in parentheses were seldom present at measurable levels. Elemental diurnal patterns were usually established for each episode for the elements S, K, Ca, Ti, Mn, Fe, Ni, Zn, Br, and Pb.

D. Elemental Particle Size Distribution Data

Figures 10 and 11 illustrate some of the elemental particle size distribution data obtained during ACHEX II. The Na, Al, and V results were ascertained by instrumental neutron activation

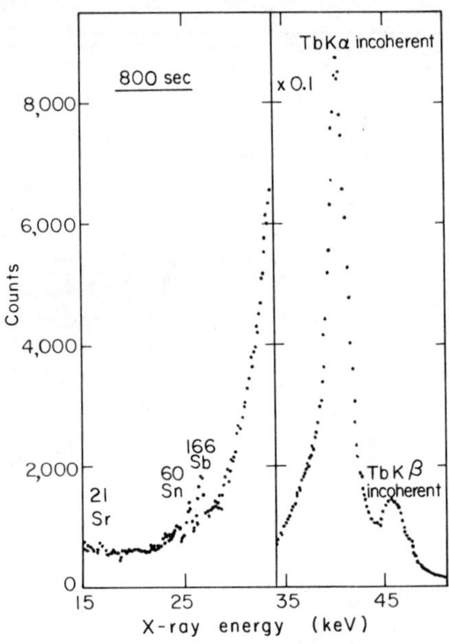

FIGURE 8. Spectrum of a high-volume aerosol specimen collected for a 24-hr interval. Spectrum obtained using spectrometer B.

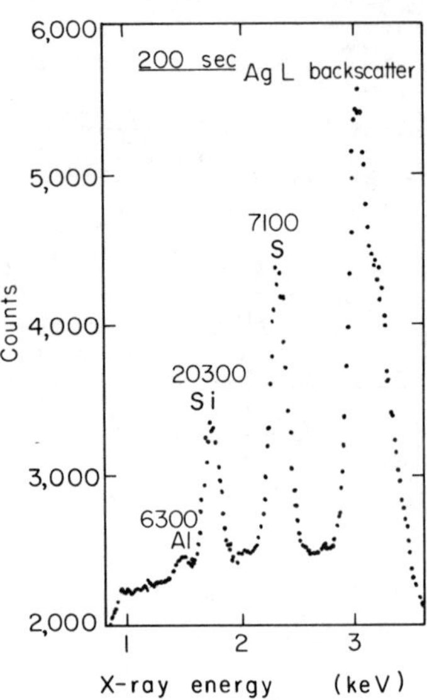

FIGURE 9. Spectrum of a total-filter aerosol specimen collected for a 2-hr interval. Spectrum obtained using spectrometer C in the direct excitation mode.

analysis (INAA). The elements Al, K, Ca, and Fe are predominantly present in the larger particles and are usually thought to originate principally from mechanical processes such as windblown soil dust. The particles containing S, Ni, Br, and Pb are in the submicron size range. These particles originate directly as a result of combustion or indirectly from secondary chemical processes. In California, V in the small particles is generally thought to originate from combustion of fuel oil or from refinery processes. The ratio between V (INAA) and Ni (XRFA) results for 11 after-filter specimens collected at the four sites during ACHEX II was 0.77 ± 0.04. Consequently, nickel may well serve as a tracer element equivalent to vanadium in the Los Angeles South Coast Basin.

TABLE 6. Elements in California Urban Aerosols

Concentration range ($\mu g/m^3$)	Particle diameter	
	>1μ	<1μ
>1	Al, Si, Ca, Fe	S, Pb
0.1-1	K, Ti, Zn, (S)	Ca, Fe, Br, (K)
0.01-0.1	Cr, Mn, Br, Sr, Pb, (Ba)	Ti, Ni, Cu, Zn, (V), (As), (Sn), (Sb), (I)
0.001-0.01	Ni, Cu, (Ga), (Rb)	Mn, (Se), (Cd)

IV. SUMMARY

Analyses of aerosols collected over 2-hr intervals during intensive episode studies, together with other chemical, physical, and meteorological data, were helpful in studying the evolution of aerosols in California. The data were also used to estimate the contributions of natural, primary, and secondary sources to the total airborne particulate matter. By using X-ray fluorescence analysis (XRFA) techniques, 1100 aerosol samples were analyzed. Diurnal and particle size distribution data were obtained for many elements in aerosols collected over 2-hr intervals. These data were of value in accomplishing the objectives of the ACHEX program.

ACKNOWLEDGMENTS

We especially thank G. Hidy, formerly with the Rockwell International Science Center, for supporting our efforts in the ACHEX programs sponsored by the California Air Resources Board. Appreciation is expressed also to F. Goulding, J. Jaklevic, and other members of the Nuclear Instrumentation Group for developing and supplying much of the equipment used for these programs, and to D. Malone for designing the low-energy X-ray spectrometer. N. Brown and D. Gok assisted in developing the computer programs employed. We thank J. Hollander for his encouragement in this work. Finally, we are grateful to J. Wesolowski and B. Appel of the California State Department of Public Health for their support and cooperation.

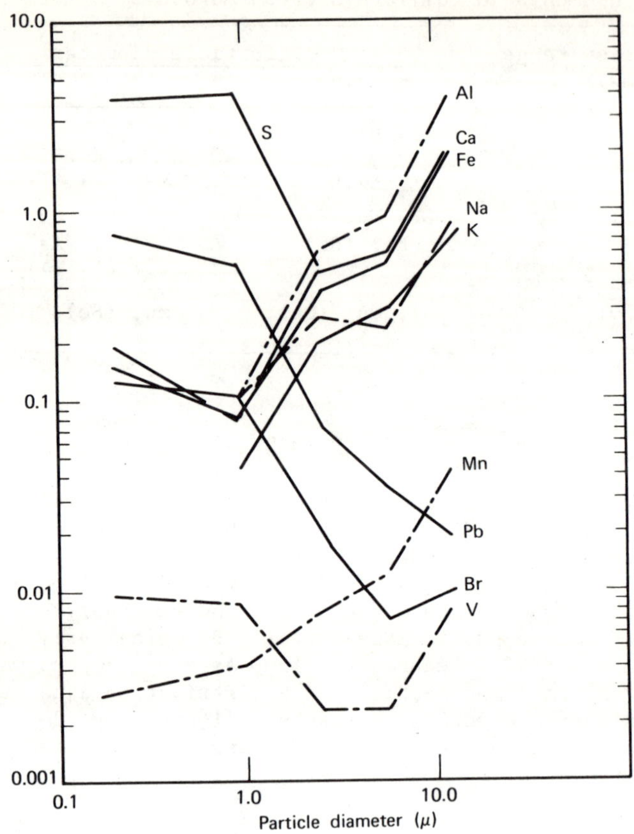

FIGURE 10. Elemental particle size distribution data for aerosol collected at Rubidoux on Spetember 19, 1973, between 1200 and 1400.

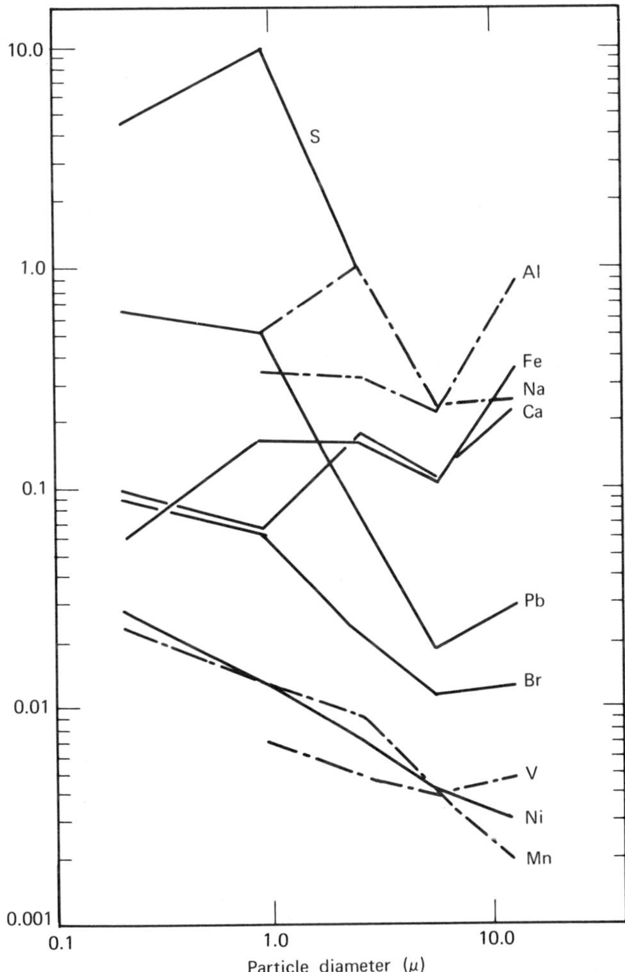

FIGURE 11. Elemental particle size distribution data for aerosol collected at Dominguez Hills on October 5, 1973, between 1000 and 1200.

REFERENCES

1. Lee, R. E., Jr. and von Lehmden, D. J. 1973. Trace Metal Pollution in the Environment, J. Air Pollut. Control Assoc. 23(10), 853-857.

2. Gordon, G. E., Zoller, W. H., Gladney, E. S., and Jones, A. G. 1971. Trace Elements in the Urban Atmosphere, in Nuclear Methods in Environmental Research, Proceedings of the American Nuclear Society Topical Meeting, August, University of Missouri, pp. 30-37.

3. Friedlander, S. K. 1973. Chemical Element Balances and Identification of Air Pollution Sources, Environ. Sci. Technol. 7, 235-240.

4. McMaster, W. H., Del Grande, N. K., Mellett, J. H., and Hubbell, J. H. 1969. "Compilation of X-Ray Cross Sections," University of California, Lawrence Livermore Laboratory Rep. UCRL-50174, Section II, Revision I.

5. Bambynek, W., Craseman, B., Fink, R. W., Freund, H. U., Mark, H., Swift, C. D., Price, R. E., and Venugopala Rao, P. 1972. X-Ray Fluorescence Yields, Auger, and Coster-Kronig Transition Probabilities, Rev. Mod. Phys. 44(4), 716-813.

6. McGuire, E. J. 1971. Atomic L-Shell Coster-Kronig, Auger, and Radiative Rates and Fluorescence Yields for Na-Th, Phys. Rev. A3, 587-594.

7. Giauque, R. D., Goulding, F. S., Jaklevic, J. M., and Pehl, R. H. 1973. Trace Element Determination with Semiconductor Detector X-Ray Spectrometers, Anal. Chem. 45, 671-681.

8. Goulding, F. S., Jaklevic, J. M., Jarrett, B. V., Landis, D. A., Loo, B. Y., and Menge, J. D. 1973. "An Automatic X-Ray Fluorescence Analyzer for Air Particulate Filters," University of California, Lawrence Berkeley Laboratory Rep. LBL-1743.

9. Giauque, R. D., Garrett, R. B., Goda, L. Y., Jaklevic, J. M., and Malone, D. F. 1976. Application of a Low Energy X-Ray Spectrometer to Analyses of Suspended Air Particulate Matter, in Advances in X-Ray Analysis, Vol. 19,

(R. W. Gould, C. S. Barrett, J. B. Newkirk, and C. O. Ruud, Eds.), Kendall/Hunt Publishing Co., p. 305.

10. Bonner, N. A., Bazan, F., and Camp. D. C. 1973. "Elemental Analysis of Air Filter Samples Using X-Ray Fluorescence," University of California, Lawrence Livermore Laboratory Rep. UCRL-51388.

Instrumental Neutron Activation Analysis Techniques for Measuring Trace Elements in California Aerosols

RICHARD C. RAGAINI, H. ROBERT RALSTON, DARREL GARVIS, AND ROBERT KAIFER

Lawrence Livermore Laboratory
Environmental Sciences Division
University of California
Livermore, California

Abstract

Instrumental neutron activation analysis (INAA) done at Lawrence Livermore Laboratory (LLL) played a key role in the 1972-1974 California Aerosol Characterization Experiment (ACHEX). The main purpose of INAA was to measure the particle size distributions and diurnal patterns of various chemical constituents in aerosols collected in California. These data were used to satisfy some of the key objectives of ACHEX.

Secondary uses of INAA were the validation of the Lundgren rotating drum cascade impactors used in the ACHEX, and the validation of other analytical techniques used in the chemical analyses.

These studies reconfirmed previous efforts which determined that techniques using INAA are useful operational methods for chemical analysis of aerosols collected over 2-hr periods in urban air.

I. INTRODUCTION

A key element of the ACHEX was the extensive chemical analysis undertaken on selected aerosol samples during the course of the experiment. Samples for study were chosen on the basis of priority for the intensity of air pollution observed or in terms of other information required to meet the objectives of the program.

Many of the chemical methods used in the ACHEX required levels of sensitivity that exceeded routine procedures used in the past. Therefore considerable development of the sensitivity of methods was undertaken, including neutron activation analysis (NAA), X-ray fluorescence analysis (XRFA), electron spectroscopy for chemical analysis (ESCA), high-resolution spectroscopy, and wet chemical methods for noncarbonate carbon, sulfate, nitrate, and liquid water content. The group at LLL was responsible for carrying out the INAA measurements.

In the ACHEX more than 3000 samples of aerosols were collected during 1971-1973, and INAA was utilized to nondestructively analyze the particulates on 700 samples including Whatman 41 high-volume filters, Gelman 5-μ GA-1 cellulose acetate, low-volume total filters, and filters after Lundgren* rotating drum cascade

*Reference to a company or product name does not imply approval or recommendation of the product by the University of California or the U.S. Energy Research and Development Administration to the exclusion of others that may be suitable.

impactors. Various impactor films were used and analyzed in the experiment: Mylar, Teflon, sticky resin-coated polyethylene, and paraffinated Mylar. The only collection media not studied were glass fiber and silver membrane filters.

The normal procedure was to cut out sections of the substrate and perform INAA as well as XRFA for elemental analyses. Although INAA and XRFA are complementary techniques for trace element analyses, a number of elements are measured in both techniques, such as Fe, Br, V, Mn, and Ca. Interlaboratory comparisons were made to validate the elemental analyses, and the agreements were quite satisfactory. The comparison did point out the importance of validation in an environmental program, especially when chemical analyses are performed by different methods. The data were transmitted to the Air and Industrial Hygiene Laboratory (AIHL) on IBM punch cards for computer data bank storage and later interpretation. The INAA technique played a significant role because of its ability to determine a large number of the chemical constituents of interest. It could detect elements indicative of windblown dust, such as Al and rare earths, elements indicative of marine aerosols, such as Na and Cl, and elements introduced by man, such as Br, V, Zn, and Se.

As discussed in more detail later, two irradiations and four different counts were involved in the analyses. The LLL 3-MW pool-type reactor (LPTR) was used for the irradiations. Gamma-ray spectra were taken with a 40-cm^3 lithium-drifted germanium detector, Ge(Li). For a complete irradiation sequence, 20 to 30 elements were reported. For a short irradiation, about 10 elements were routinely detected. The data transmission time was as short as 1 day and normally 1 week for short-irradiation elements, and was about 1 month for long-irradiation elements.

II. DESCRIPTION OF INSTRUMENTAL NEUTRON ACTIVATION ANALYSIS (INAA)

Since its inception INAA has become increasingly more important as a tool for trace element analysis. This is due to progressive improvements in its sensitivity, made possible by the use of high-resolution, solid-state radiation detectors, improved data reduction techniques and schemes, and increased experience in sample preparation and collection methods.

To be most useful, the INAA technique requires a high thermal neutron flux of the type usually found only in a nuclear reactor. Neutron fluxes of 10^{12} to 10^{14} $n/cm^2 \cdot sec$ can be easily obtained at most reactor facilities.

A. Basic Principles

The procedure used for INAA is to take a sample, expose it to a neutron flux, measure the induced activity as a function of energy, and evaluate the data to quantify the elements present in the original samples (Figure 1). In practice, the samples contain many elements and occasionally several isotopes of the same element. The activation/counting scheme employed must be chosen in such a manner as to optimize the activity of as many elements as is reasonably possible. This involves making two or more separate irradiations for each sample and applying various counting schedules, depending on the type of sample and the elements being sought in the analysis. The irradiation schedule and counting scheme used in this study are discussed in detail below.

The terms Q1 and Q2 refer to the first and second count, respectively, after the short irradiation, and L1, L2, L3 likewise refer to the long irradiation.

For this study the neutron source was the Livermore pool-type reactor (LPTR). The LPTR is a 3-MW thermal research reactor operated for the U.S. Energy Research and Development Administration (ERDA) by the University of California Lawrence Livermore Laboratory.

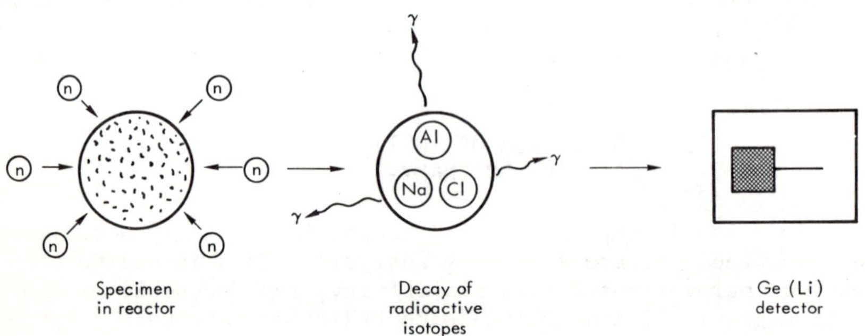

FIGURE 1. Schematic diagram of instrumental neutron activation analysis procedure.

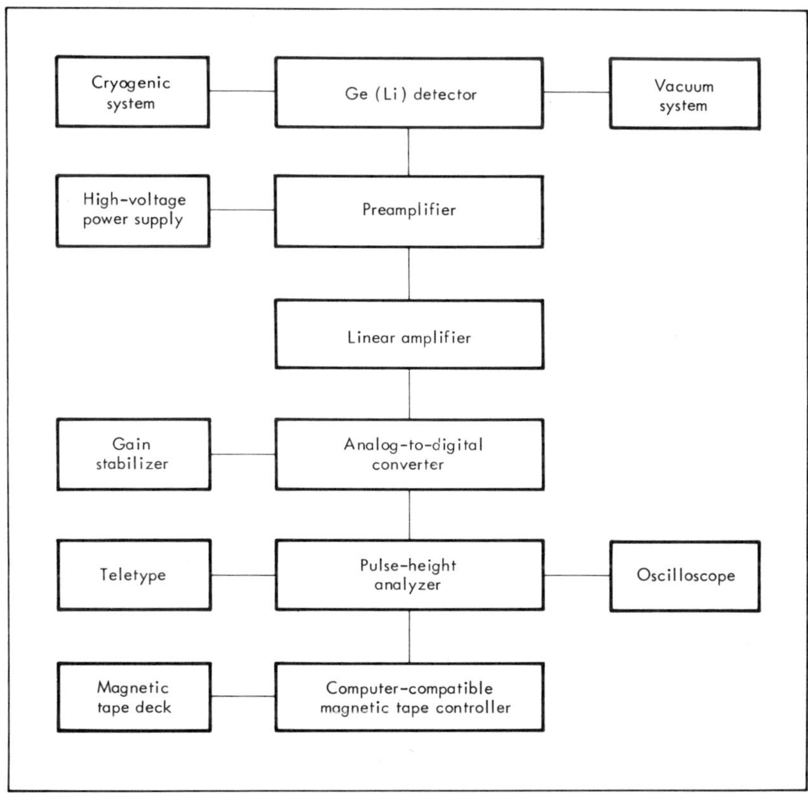

FIGURE 2. Block diagram of INAA detector system.

Figure 2 is a block diagram of the detector system used. It is built around a 40-cm^3 Ge(Li) detector, coupled to a 4096-channel pulse-height analyzer and the appropriate linear electronics.

B. Irradiation and Counting Procedure

The irradiation and counting scheme shown in Table 1 was designed to give the maximum sensitivity for the wide range of elements. The scheme is similar to that used by others for the same type of analyses (1-3).

TABLE 1. Neutron Irradiation and Sample Counting Schedule

Irradiation time, flux	Cooling time	Counting time	Elements detected
2 min, 2×10^{13} n/cm^2·sec	4 min	500 sec (Q1)	Al, V, Cu, Ti, Ca
	20 min	1000 sec (Q2)	Na, Mg, Cl, Mn, Br, I, Ba, In
12 hr, 5.1×10^{12} n/cm^2·sec	20–30 hr	40 min (L1)	Eu, Br, As, W, Ga, Zn, K, Cu, Na, Cd
	6–10 days	80 min (L2)	Sm, Au, Hg, La, Sb
	20–30 days	600–800 min (L3)	Fe, Cr, Co, Zn, Hg, Se, Ag, Sb, Ce, Eu, Sc, Th, Ni, Ta, Hf, Ba, Rb, Zr, Cs, Yb, Tb, Lu

The derived data were recorded on computer-compatible magnetic tape for permanent reference and for subsequent reduction and analysis. The procedure is effective for the identification of approximately 35 elements in a single sample.

Table 2 gives a listing of the parent nuclide, daughter product, half-life, and energy of the gamma rays used in the analysis of the elements of interest. This table shows that ^{65}Zn has a gamma ray with an energy of 1115.4 keV. Thus, if there are large amounts of both Zn and Sc in a sample, it may not be possible to separate the gamma-ray peaks since they will overlap when detectors with currently available resolutions are used. Scandium has another gamma ray with an energy of 889.4 keV, and the only appreciable interference with this peak is the ^{110m}Ag gamma ray at 884.7 keV. By the use of curve-fitting techniques and spectrum stripping, it is possible to resolve large amounts of ^{110m}Ag, ^{46}Sc, and ^{65}Zn in the same sample. A close examination of Table 2 reveals several other interferences of the type discussed above which occur in a typical sample.

The detection limits for 75 elements are given in Table 3 based on a maximum irradiation time of 1 hr, a thermal neutron flux of 10^{13} n/cm^2·sec, and no appreciable interferences (4). These detection limits can, therefore, be thought of as theoretical limits within the state of the art.

Detection limits are shown in Table 4 for the 2-hr samples analyzed in the ACHEX study.

C. Standardization

To generate accurate quantitative information, neutron activation analysis (NAA) must make accurate comparisons of the sample elemental counting information with that for the elemental standard. The elemental standards are individually prepared by dissolving a known amount of the element of interest as a high-purity metal, oxide, or assayed compound, as the chemistry dictates, into triple-quartz-distilled water or dilute quartz-distilled acid. Aliquots of these standards are then spotted onto blank filters of Whatman 41 or Gelman Ga-1, and the filters are processed through the routine NAA procedure described above. These experiments basically provide values of the conversion factor of nanograms per count for each element which are then applied to the sample count data by a computer code to calculate the elemental weights on the filters and impactor surfaces.

The standardization methods differ for short and long irradiations. The main similarity in the two methods is that both utilize premixed solutions containing only those elements of

TABLE 2. Parent, Daughter, Half-Life, and Energy of Observed Gamma Rays for Elements Analyzed by INAA Technique in This Study

Parent element exposed	Daughter isotope observed	Half-life[a]	Gamma-ray energy observed[b]
Ti	$^{51}_{22}Ti$	5.79 m	320.0 (100)
Mg	$^{27}_{12}Mg$	9.46 m	843.8 (100), 1014.5 (37.5)
Cu	$^{66}_{29}Cu$	5.10 m	1039.2 (100)
V	$^{52}_{23}V$	3.75	1434.3 (100)
Al	$^{28}_{13}Al$	2.32 m	1778.7 (100)
Ca	$^{49}_{20}Ca$	8.80 m	3084.4 (100)
Ba	$^{139}_{56}Ba$	82.9 m	165.8 (100)
I	$^{128}_{53}I$	25.0 m	443.3 (100), 526.4 (9.9)
Br	$^{80}_{35}Br$	16.8 m	616.2 (100), 665.6 (17.2)
In	$^{116}_{49}In$	53.7 m	417.0 (40.2), 1097.1 (59.8), 1293.4 (100)
Na	$^{24}_{11}Na$	15.0 h	1368.4 (100), 2754.1 (100)
Cl	$^{38}_{17}Cl$	37.3 m	1642.7 (74.5), 2167.6 (100)
Mn	$^{56}_{25}Mn$	155.0 m	846.7 (100), 1811.2 (29.4)
As	$^{76}_{33}As$	26.4 h	559.1 (100), 657.1 (100)
W	$^{187}_{74}W$	23.9 h	134.2 (31.6), 479.5 (83.1), 618.2 (23.1), 685.7 (100)
Ga	$^{72}_{31}Ga$	14.1 h	629.9 (25.5), 834.0 (100), 2201.6 (27.3)
K	$^{42}_{19}K$	12.4 h	1524.7 (100)

(continued)

TABLE 2 (cont.)

Parent element exposed	Daughter isotope observed	Half-life[a]	Gamma-ray energy observed[b]
Cd	$^{115}_{48}Cd$	53.5 h	492.3 (29.5), 527.9 (100)
Sm	$^{133}_{62}Sm$	48.8 h	103.2 (100)
Au	$^{198}_{79}Au$	64.7 h	411.8 (100)
La	$^{140}_{57}La$	40.2 h	328.8 (21.4), 487.0 (49.4), 1596.6 (100)
Fe	$^{59}_{26}Fe$	45.6 d	1099.3 (100), 1291.5 (77.0)
Cr	$^{51}_{24}Cr$	27.8 d	320.1 (100)
Co	$^{60}_{27}Co$	5.26 y	1173.2 (100), 1332.5 (100)
Zn	$^{65}_{30}Zn$	24.3 d	1115.5 (100)
Hg	$^{203}_{80}Hg$	46.9 d	279.2 (100)
Sc	$^{75}_{34}Sc$	120 d	136.0 (96.0), 264.7 (100)
Ag	$^{108m}_{47}Ag$	5.00 d	434.0 (100), 614.4 (100), 723.0 (100)
Ag	$^{110m}_{47}Ag$	255 d	657.7 (100), 884.7 (79.6), 937.5 (36.5), 1384.2 (27.7)
Sb	$^{124}_{51}Sb$	60.3 d	602.7 (100), 1691.1 (53.2)
Ce	$^{141}_{58}Ce$	32.5 d	145.5 (100)
Eu	$^{152}_{63}Eu$	12.7 y	121.8 (100), 344.2 (85.6), 1408.1 (65.0)
Sc	$^{46}_{21}Sc$	83.9 d	889.3 (100), 1120.5 (100)
Th	$^{233}_{91}Pa$	27.0 d	311.9 (100)

(continued)

TABLE 2 (cont.)

Parent element exposed	Daughter isotope observed	Half-life[a]	Gamma-ray energy observed[b]
Ni	$^{58}_{27}Co$	71.3 d	811.1 (100)
Hf	$^{175}_{72}Hf$	70.0 d	343.3 (100)
Hf	$^{181}_{72}Hf$	42.3 d	133.1 (49.4), 482.2 (100)

[a]Time units are as follows: m = minutes, h = hours, d = days, y = years.

[b]Numbers given in parentheses following the gamma-ray energy are the relative intensities (%) of the gamma rays.

TABLE 3. Detection Limits (Theoretical) of INAA Technique for 75 Elements, Assuming a 1-Hr Irradiation with a Thermal Neutron Flux of 10^{13} n/cm^2·sec and No Appreciable Interferences

Limit of detection (μg)	Elements
1-3 x 10^{-7}	Dy
4-9 x 10^{-7}	Eu
1-3 x 10^{-6}	--
4-9 x 10^{-6}	Mn, In, Lu
1-3 x 10^{-5}	Co, Rh, Ir
4-9 x 10^{-5}	Br, Sm, Ho, Re, Au
1-3 x 10^{-4}	Ar, V, Cu, Ga, As, Pd, Ag, I, Pr, W
4-9 x 10^{-4}	Na, Ge, Sr, Nb, Sb, Cs, La, Er, Yb, U
1-3 x 10^{-3}	Al, Cl, K, Sc, Se, Kr, Y, Ru, Gd, Tm, Hg

(continued)

TABLE 3 (cont.)

Limit of detection (μg)	Elements
$4\text{-}9 \times 10^{-3}$	Si, Ni, Rb, Cd, Te, Ba, Tb, Hf, Ta, Os, Pt, Th
$1\text{-}3 \times 10^{-2}$	P, Ti, Zn, Mo, Sn, Xe, Ce, Nd
$4\text{-}9 \times 10^{-2}$	Mg, Ca, Tl, Bi
$1\text{-}3 \times 10^{-1}$	F, Cr, Zr
$4\text{-}9 \times 10^{-1}$	Ne
1-3	S, Pb
4-9	Fe

TABLE 4. Limits of Detection[a] for Trace Elements in Aerosols with Instrumental Neutron Activation (ng/m^3)

Element	Gelman GA-1	Impactor film[b]
Na	300	200
Mg	100	10
Al	0.3	100
Cl	1	1
Ca	1	10
Ti	10	1
V		0.1
Mn	0.6	
Cu	0.1	1
Br		0.0003
Ln	0.002	0.0004
I		0.02
Ba	2	0.5

[a]Above blank.

[b]Average results for 0.0006-cm Mylar, 0.0025-cm Teflon, and 0.0025-cm polyethylene covered with 0.004-cm sticky resin.

interest. Both have the shortcoming of requiring the preparation of standards of approximately the same concentrations as the samples of interest. This necessitates an estimation of the composition of the sample. This estimation is critical if one is looking at samples which contain large amounts of elements that have interferring gamma rays, such as Ag, Sc, and Zn, as described above, and if any of the short-lived isotopes (e.g., Al, V, Ti) vastly predominate during the Q1 count.

For short irradiations all samples are normalized to a "standard flux monitor." Several different types of flux monitors have been used by different investigators, depending on their specific applications, but each was designed to do the same thing, namely, to normalize a given irradiation back to some known condition or parameter. For this study a titanium flux monitor was used and counted for 60 sec between the Q1 and Q2 counts. Several titanium flux monitors were in use, and all were run for a predetermined time, counted, and found to be statistically identical. The flux monitors were placed in the same position with respect to the sample in the sample carrier during each irradiation. The counts from the flux standard were then ratioed to the master standard count taken with the same flux monitor, and all elemental counts were ratioed up or down by the same factor. The ratioing was performed by the data reduction program, and the program "flags" any standard ratio that deviates from the master standard by more than 10%.

For the long irradiations a standard was prepared for each irradiation. The standard was then irradiated with and counted in the same position and sequence as the sample. Any flux variation that existed from one irradiation to another was thereby taken into account. This procedure effectively calibrated the detector for each sample run.

D. Data Reduction

The data reduction for the INAA portion of this study was performed using the gamma-ray analysis program MIKEGAM. This program is an adapted version of a technique developed by Ralston and Wilcox (5). The accuracy of the code has been verified by interlaboratory and intralaboratory comparison studies which are discussed below.

Basically, the code operates by generating a baseline from the Ge(Li) spectrum, and from this baseline defining and quantitating significant peaks. The program has two major branches: one for processing short-irradiation samples, and the other for processing long-irradiation samples. The errors contributed by counting statistics, sampling, blank contribution, and standards determination are carried along and combined in the final calculations.

E. Specimen-Handling Procedure

Our procedures were planned for the following types of samples: Ga-1 filters, 0.25- by 1-in strips of Mylar, 1-in diameter Whatman 41 high-volumes, sticky film, and parafinnated Mylar. All samples were handled with blunt-ended stainless steel tweezers and inserted into 2-dram polyethylene vials that had been previously cleaned. A vial was then loaded into a reactor polyethylene pneumatic tube "rabbit," along with a titanium foil flux monitor (encased in polyethylene) and a polystyrene disk that acted as a shock absorber. After irradiation the sample was removed from the inner vial after the outer "rabbit" had been discarded, and the sample was inserted into a clean cellophane bag, using clean stainless steel tweezers. The bag-encased sample was then inserted into a Lucite sample holder and placed in front of a Ge(Li) detector for gamma-ray counting (i.e., counts Q1 and Q2). For samples undergoing the long-irradiation procedure, a decay period of at least 3 days was allowed before the long irradiation to allow any radioactivities generated in the short irradiation to die away. For the long irradiation, approximately 12 samples were individually bagged in clean polyethylene and were loaded along with bagged elemental standards, into a polyethylene capsule which rotated during the 12-hr irradiation to ensure flux homogeneity. After the irradiation the samples were run through the L1, L2, and L3 counts. Particular care was taken not to introduce any contaminants into the samples during the short irradiation and subsequent analysis so that the long-irradiation analysis would not be affected.

F. Determination of Procedural Errors

To determine the overall error contributed by various procedures such as weighing and pipetting, and also the geometry in the reactor irradiation facility and the gamma spectrometer, six sodium standards were prepared in the usual manner, irradiated together, and then counted sequentially. The overall variation in obtained activity from each sodium standard showed a fractional standard deviation of less than 1%.

III. RESULTS

A. Blanks

The assignment of average trace quantities in the filter substrate (blank) ranked high among the various factors that had an effect on the precision of trace element analysis by INAA of air particulate matter collected on filter media. This was especially true when a trace element in the blank was determined with a high degree of uncertainty and the concentration of the deposited trace element approached that contained in the blank.

For Phase I various particle collection media were studied. Initially, the effort was directed toward the evaluation of candidate substrates, and after final selection to a more detailed characterization of those chosen. The major candidates were Gelman (GA-1), Millipore, a sticky-surface polyethylene film, Mylar, Teflon film, and paraffinated Mylar. An additional variable was the pretreatment of the media, which consisted of washing either by the manufacturer or by AIHL.

Phase II sample collection was restricted to the use of GA-1 filters for the total and after-filters, sticky polyethylene film for the Lundgren impactor surfaces, and W-41 for the high-volume samples. Phase I blank values for W-41 and sticky polyethylene were used, but since a new washing procedure was used on Phase II GA-1 filters, a separate study was done for them.

Table 5 shows the trace element values in nanograms per square centimeter for each of the substrates used in Phases I and II. The values shown are the grand averages of a varying number of samples of each substrate. They, along with the associated probable errors shown in parentheses, are the best determinations of single values that represent the populations of the filter (or impactor) media used in the sampling under field conditions.

The experience gained during Phase I showed that a better statistical sampling of the pool of washed GA-1 filters was in order for Phase II. Accordingly, the following studies were carried out.

1. Phase II, GA-1, Series 1. A representative group of 13 washed filters which had first been analyzed at LBL was analyzed with the results shown in Table 6. Of particular interest is the extreme variability in the Na values for the 13 filters, the range being an order of magnitude. The 95% confidence interval or two standard deviations are in excess of 200%. Also, Cu, V, and Br show high variability, although not as severe as that of Na.

When these blank values were used in correcting the concentrations for particulate loadings on the after-filters and total

TABLE 5. Grand Averages of Sample Substrate Blank Determinations (ng/cm^2)

Element	Phase II GA-1	Phase I GA-1	Mylar	Teflon	Sticky film	Parafin Mylar	W-41
Na	53 (0.50)	88.8 (0.2)	2.8 (0.2)	1.5 (0.2)	206 (0.08)	25.01 90.08)	94.2 (0.17)
Mg	92 (0.24)	ND	ND	ND	ND	ND	ND
Al	41 (0.45)	29.6 (0.2)	4.0 (0.2)	4.45 (0.2)	17.5 (0.43)	67.4 (0.04)	21.5 (0.20)
Cl	158 0.21	178 0.2	6.8 0.2	2.6 0.2	86.4 0.14	34.5 0.12	407 0.20
Ca	176 (0.41)	158 (0.2)	ND 0	ND 0	285 (0.30)	57.7 (0.83)	70 (0.20)
Sc	0 0	ND ND	ND ND	ND ND	ND ND	0 0	0.01 (0.12)
Ti	ND ND	ND ND	ND ND	3.30 (0.20)	3.82 (0.52)	5.89 (0.52)	ND ND
V	ND ND	ND ND	ND ND	0.016 (0.20)	0.0272 (0.29)	0.112 ().26)	ND ND
Cr	0.99 (0.11)	8.0 (0.28)	0 0	1.85 (0.03)	0.55 (0.12)	0 0	1.46 (0.04)
Mn	0.82 0.74	0.53 0.2	0.03 0.2	0.44 0.2	0.443 0.53	0.722 0.28	0.42 0.20

(continued)

TABLE 5 (cont.)

Element	Phase II GA-1	Phase I GA-1	Mylar	Teflon	Sticky film	Parafin Mylar	W-41
Fe	29.4 (0.27)	56 (0.3)	ND 0	11 (0.44)	80 (0.08)	ND ND	25.8 (0.09)
Co	ND ND	ND ND	ND ND	2.26 (0.03)	ND ND	ND ND	0.21 (0.03)
Cu	16 (0.41)	15.8 (0.20)	ND ND	1.90 (0.20)	3.85 (0.55)	ND ND	13.4 (0.16)
Zn	4.5 (0.15)	ND ND	31.8 (0.07)	1.5 (0.2)	300 (0.01)	ND ND	4.51 (0.09)
Ga	ND ND	ND ND	0.005 (0.005)	0.36 (0.36)	0.10 (0.10)	ND ND	ND ND
As	ND ND	ND ND	0.005 (0.005)	0.36 (0.36)	0.10 (0.10)	ND ND	ND ND
Br	0.38 (0.36)	ND ND	0.44 (0.20)	ND ND	0.85 (0.21)	2.26 (0.03)	13.8 (0.95)
Cd	ND ND	ND ND	0.005 (0.005)	0.36 (0.36)	0.10 (0.10)	ND ND	0.075 (0.32)
Sb	ND ND	ND ND	114 (0.005)	0.02 (0.36)	0.15 (0.10)	ND ND	0.034 (0.12)
I	0.36 (0.10)	ND ND	ND ND	ND ND	ND ND	0.11 (0.70)	ND ND

(continued)

TABLE 5 (cont.)

Element	Phase II GA-1	Phase I GA-1	Mylar	Teflon	Sticky film	Parafin Mylar	W-41
La	ND	ND	0.005	0.36	0.10	ND	ND
	ND	ND	(0.005)	(0.36)	(0.10)	ND	ND
Ce	ND	D	ND	ND	ND	ND	0.11
	ND	ND	ND	ND	ND	ND	(0.18)
Hf	ND	ND	ND	ND	ND	ND	0.82
	ND	ND	ND	ND	ND	ND	(0.10)
Au	ND	ND	ND	ND	ND	ND	2.44
	ND	ND	ND	ND	ND	ND	(0.07)
Hg	0.39	12.6	2.6	1.2	3.0	ND	0.16
	(0.5)	(0.06)	(0.50)	(0.5)	(0.5)	ND	(0.30)
Th	ND	ND	ND	0.007	0	ND	0.025
	ND	ND	ND	(0.35)	0	ND	(0.27)

TABLE 6. AIHL Phase II GA-1 Laboratory Blanks - First Set (ng/cm^2), 13 Samples

Element	$\overline{X}$	σ	FSD
Ti	13.1	1.5	0.11
Cu	54.0	26.2	0.49
Al	170.5	38.7	0.23
Ca	669.8	119	0.18
V	0.156	0.081	0.52
I	0.617	0.195	0.32
Br	2.11	1.16	0.55
Na	210	217	1.03
Cl	472	72.4	0.15
Mn	2.25	0.64	0.28
Mg	197	64	32

filters, it became apparent that they were too high (since very high detection limits were being calculated). Comparison of these values with those obtained for Phase I blanks gave ratios as high as 4.5. No reasonable explanation was found for these discrepancies. It was decided to examine another representative group of blanks.

2. Phase II, GA-1, Series 2. Eight additional GA-1 blank filters were subjected to analysis. The results are shown in Table 7. Again, extremely high variability was seen among the Na values, as well as those for Cu, Al, Ca, and Mn. When these average values were applied to the correction of the concentrations of particulate matter collected on the after-filters and total filters, more reasonable detection limits were calculated.

3. Field Blanks. Data for 28 field blanks were obtained. Again, Na showed extremely high variability. Since these blanks were exposed to field conditions, the variability across any single element cannot necessarily be attributed solely to the innate variability of the unexposed blanks.

TABLE 7. AIHL Phase II GA-1 Laboratory Blanks - Second Set (ng/cm^2), 8 Samples

Element	$\overline{X}$	σ	FSD
Cu	15.6	6.5	0.42
Al	40.8	18.4	0.45
Ca	175.9	71.8	0.41
Na	55.0	70	1.1
Cl	160	34	0.21
Mn	0.54	0.11	0.21

Table 8 is a summary of the field blank data, showing means, standard deviations, and fractional standard deviations. The means for the second series of laboratory blanks are included for comparison.

When the field blank mean is ratioed to the laboratory blank mean, all cases show values greater than unity, with Br exceptionally high (16:1). Since we are dealing with statistical sampling, inspection of these ratios cannot give a satisfactory answer to the question of whether the means are significantly different. If student's t-test is used at the level of $\alpha = 0.05$ (95% confidence), all but Mn, Na, Mg, and I can be said to have different means ($\mu_1 \neq \mu_2$).

Thus we can say that to a confidence level of 0.95 something occurred to the field blanks to cause Cu, Al, Ca, Cl, and Br to be elevated relative to the laboratory blanks.

It should be clear from the above that, even if great effort is put into determining blank values, there is sufficient variability from batch to batch and from filter to filter that concentration measurements near the blank value have large errors associated with them.

B. Quality Assurance

The INAA technique played a prominent part in the quality assurance program of ACHEX. Studies of bromine and nitrate losses from filters as a result of their volatility are discussed earlier, as well as detailed intermethod and interlaboratory comparisons. To avoid duplication this section will discuss only the quality assurance procedures not covered earlier.

TABLE 8. Field Blank Summary - Phase II

Element	Average (ng)	σ	FSD	Average (ng/cm^2)	GA-1 Phase II lab blank average (ng/cm^2)	Field blank/ lab blank	By t-test H_0: $\mu_1 = \mu_2$ at $\alpha = 0.05$
Mg	603	222	0.37	119	92[a]	1.3	1 = 2
Cu	108	54	0.50	21.3	16	1.3	1 ≠ 2
Al	331	124	0.37	65	41	1.6	1 ≠ 2
Ca	1260	365	0.29	249	176	1.4	1 ≠ 2
Ti	ND	ND	ND	ND	ND	ND	
V	0.51	0.22	0.43	0.10	ND	ND	
I	3.1	1.4	0.45	0.61	0.32[a]	1.91	1 = 2
Br	31	16	0.52	6.1	0.38[a]	16.1	1 ≠ 2
Na[b]	402	445	1.11	79.3	55	1.4	1 = 2
Na[c]	304	204	0.67	60	55	1.1	1 = 2
Cl[b]	1106	265	0.24	218	158	1.4	1 ≠ 2
Cl[c]	1051	153	15	207	158	1.3	1 ≠ 2
Mn	5.3	1.7	0.32	1.04	0.82	1.3	1 = 2
Ba	ND	ND	ND	ND	ND	ND	
In	ND	ND	ND	ND	ND	ND	

[a]These values were derived by averaging those out of eight filters which had detectable Br, I, or Mg.

[b]Includes all values. [c]Excludes two highest values.

For the determination of the overall accuracy of the use of INAA for the aerosol characterization study, we had access to three different types of verification. The first was to analyze some standard material and compare our measured values to those certified by the agency issuing the standard. In our case we analyzed orchard leaves as certified by the National Bureau of Standards.

The second approach was to participate in an interlaboratory comparison of trace element determinations on various standardized samples. This was done for coal dust and fly ash in one series, and for gravimetric standard solutions in another.

Finally, direct comparisons of elements determined in the same ACHEX sample by different analytical techniques and laboratories were made during the ACHEX.

1. NBS Orchard Leaves (SRM-1571). Samples of NBS standard reference material - orchard leaves - were analyzed. The values shown in Table 9 are averages of three to five replicates. The NBS certified values are also given. The values in parentheses are for information only and are not certified. The experimental errors are equal to two standard deviations or one-half the range of the data, whichever is greater.

2. NBS-EPA Intercomparison Study of Coal Dust (SRM-1632) and Fly Ash (SRM-1633). A "round robin" intercomparison study was sponsored jointly by the National Bureau of Standards and the Environmental Protection Agency. The standards were coal dust, fly ash, gasoline, and fuel oil. Lawrence Livermore Laboratory participated in the measurements of coal dust and fly ash (6, 7).

Table 10 shows the results reported by other laboratories and LLL.

3. Interlaboratory Comparison of Gravimetric Standard Samples. A comparison of trace element concentrations determined by five different nondispersive X-ray fluorescence techniques used at 11 different laboratories was carried out (8). Standard samples were prepared by Columbia Scientific Industries. The INAA program at LLL and the radioactive, source-excited, nondispersive X-ray fluorescence technique used at LLL were included. Our results, compared with those reported by other investigators, are shown in line 4 of Table 11.

Each set of data reported in Table 11 represented our dried solution deposits and two blanks. The filter substrates were Whatman 41 and Millipore SMWP filters. The eight elements deposited on each sample were Al, K, V, Mn, Fe, Cu, As, and Pb in amounts ranging from 1.7 to 54 $\mu g/cm^2$. All investigators

TABLE 9. National Bureau of Standards SRM-1571 Orchard Leaves ($\mu g/g$)

Element	LLL	NBS
Al	341 ± 39	(350)
Ca	20,400 ± 1000	20,900 ± 300
Ba	42.7 ± 4.7	
Mg	6600 ± 500	6,200 ± 200
Br	13.0 ± 2.6	(10)
V	0.675 ± 0.029	
In	0.17 ± 0.05	
Na	89 ± 9	82 ± 6
Fe	330 ± 40	300 ± 20
Cl	635 ± 47	(700)
Mn	95 ± 3	91 ± 4
As	8.8 ± 3.4	14 ± 2
K	12,200 ± 3600	14,700 ± 300

received separate sets of standards. Lawrence Livermore Laboratory used the same set for X-ray analysis before INAA.

The gravimetric standard solutions deposited on Whatman 41 are referred to as "A" and "C." The values shown for each element in Table 11 are ratios of Whatman 41 and Millipore filter averages to the amount of gravimetric standard solution deposited. Each participant knew only which elements were deposited, not how much. The numbers in parentheses are the standard deviations reported by the various participants.

4. Intercomparison within the ACHEX Program. A number of impactors and filters were subjected to both INAA at LLL and XRFA at Lawrence Berkeley Laboratory (LBL) (9). Earlier discussion lists a series of high-volume specimens showing comparisons for the four elements that were determined by both techniques: Ca, Ti, Mn, and Br. Two determinations were done using X-ray fluorescence. These took place before and after neutron activation. In most cases the second measurement shows a reduction in the amount of each element. Comparison of the first XRFA and

TABLE 10. Comparison of LLL INAA Results for NBS Standard Reference Material Coal and Fly Ash with Results Reported by NBS and by Ondov et al. (7) (μg/g ± 2σ unless % indicated)

Element	NBS Coal SRM No. 1632			NBS Fly Ash SRM No. 1633		
	LLL (INAA)	Ondov et al.	NBS	LLL (INAA)	Ondov et al.	NBS
Al (%)	1.59 ± 0.20	1.85 ± 0.13		12.1 ± 0.5	12.7 ± 0.5	
Sb	3.4 ± 0.1	3.9 ± 1.3		6.9 ± 0.6	6.9 ± 6	
As	5.8 ± 0.3	6.5 ± 1.4	5.9 ± 0.6	64 ± 4	64 ± 4	61 ± 6
Ba	302 ± 8	352 ± 30		2510 ± 160	2700 ± 200	
Br	17.2 ± 2.1	19.3 ± 1.9		6.7 ± 0.6	12 ± 4	
Ca	0.42 ± 0.06	0.43 ± 0.05		4.2 ± 0.2	4.7 ± 0.6	
Ce	18.8 ± 1.0	19.5 ± 0.1		148 ± 6	146 ± 15	
Cl	828 ± 22	890 ± 125		--	42 ± 10	
Cs	1.40 ± 0.08	1.4 ± 0.1		8.6 ± 0.8	8.6 ± 1.1	
Cr	17.6 ± 1.0	19.7 ± 0.9	20.2 ± 0.5	117 ± 7	127 ± 6	131 ± 2
Co	5.51 ± 16	5.7 ± 0.4		39.4 ± 1.2	41.5 ± 1.2	
Eu	0.27 ± 0.02	0.33 ± 0.04		2.39 ± 0.11	2.5 ± 0.4	
Hf	1.00 ± 0.07	0.96 ± 0.05		7.2 ± 0.6	7.9 ± 0.4	
In	0.03 ± 0.02	0.20 ± 0.12		0.27 ± 0.14	0.32 ± 0.10	
Fe	0.84 ± 0.02	0.84 ± 0.04	0.87 ± 0.03	6.8 ± 0.03	6.2 ± 0.3	
La	9.5 ± 0.2	10.7 ± 1.2		81 ± 2	82 ± 2	
Mg	0.25 ± 0.08	0.20 ± 0.05		2.0 ± 0.4	1.8 ± 0.4	
Mn	39.5 ± 0.7	43 ± 4	40 ± 3	505 ± 14	496 ± 49	493 ± 7
K	0.27 ± 0.02	0.28 ± 0.03		1.75 ± 0.18	1.61 ± 0.15	
Rb	19.0 ± 1.5	21 ± 2		105 ± 10	125 ± 10	
Sm	1.72 ± 0.08	1.7 ± 0.2		13.0 ± 0.7	12.4 ± 0.9	
Sc	3.68 ± 0.08	3.7 ± 0.3		27.0 ± 0.6	27 ± 1	
Si	2.9 ± 0.4	3.4 ± 0.2	2.9 ± 0.3	9.0 ± 1.4	10.2 ± 1.4	9.4 ± 0.5
Na	380 ± 12	414 ± 20		3240 + 100	3200 ± 400	
Sr	112 ± 26	161 ± 16		1250 ± 230	1700 ± 300	
Ta	0.23 ± 0.02	0.24 ± 0.04		2.0 ± 0.2	1.8 ± 0.3	
Tb	0.20 ± 0.02	0.23 ± 0.05		1.87 ± 0.15	1.9 ± 0.3	
Th	2.9 ± 0.1	3.2 ± 0.2		23.6 ± 0.8	248 ± 2.2	
Ti	1000 ± 260	1040 ± 110		7150 ± 1200	7400 ± 300	
W	0.65 ± 0.15	0.75 ± 0.17		5.5 ± 1.5	4.6 ± 1.6	
U	1.24 ± 0.05	1.41 ± 0.07	1.4 ± 0.1	10.5 ± 1.0	12.0 ± 0.5	11.6 ± 0.2
V	41 ± 10	36 ± 3	35 ± 3	290 ± 80	235 ± 13	214 ± 0.8
Yb	0.76 ± 0.03	0.7 ± 0.1		5.9 ± 0.3	7 ± 3	
Zn	43 ± 2	30 ± 10	37 ± 4	230 ± 40	216 ± 25	210 ± 20
Zr	28 ± 24			182 ± 76[a]	301 ± 20	

[a]Zr value 182 ± 76 μg/g was corrected for $^{235}U(n,f)^{95}$ contribution. Uncorrected value was 295 μg/g.

TABLE 11. First XRFA Intercomparison Study - Ratio of A and C Averages to Gravimetric Standards

Laboratory code	Sample type	Al	K	V	Mn	Fe	Cu	As	Pb
1	A		1.09(0.21)	0.76(0.06)	0.64(0.06)	0.57(0.20)	0.53(0.08)	0.48(0.07)	0.57(0.07)
	C		0.73 (0.24)	0.68(0.10)	0.68(0.08)	0.65(0.10)	0.60(0.10)	0.69(0.08)	0.74(0.09)
2	A	1.73	0.67	0.72	0.76	0.67	0.79	0.61	0.69
	C		0.91	0.77	0.83	0.74	0.76	0.63	0.69
3	A		0.92(0.07)	0.91(0.07)	0.95(0.07)	0.91(0.03)	0.92(0.04)	0.90(0.04)	0.94(0.03)
	C		0.91(0.06)	0.90(0.06)	0.92(0.04)	0.89(0.04)	0.01(0.03)	0.89(0.03)	0.88(0.03)
LLL	A	0.79(0.01)		0.91(0.01)	0.94(0.02)	0.84(0.01	0.89(0.19)		
INAA	C	0.79(0.08)		0.89(0.01)	0.96(0.02)	0.83(0.01)	1.20(0.19)		
5			0.82(0.02)	1.01(0.02)	0.99(0.03)	0.92(0.03)	1.00(0.03)	0.94(0.03)	1.05(0.03)
			0.81(0.03)	0.95(0.03)	0.95(0.03)	0.91(0.03)	0.96(0.03)	0.94(0.03)	0.99(0.03)
6	A	1.55(0.20)	0.94(0.01)	0.92(0.01)	1.01(0.02)	1.01(0.01)	0.91(0.03)	0.76(0.07)	0.98(0.05)
	C	2.51(0.37)	1.01(0.01)	1.04(0.01)	0.95(0.02)	0.96(0.01)	0.88(0.02)	1.20(0.08)	0.92(0.03)
7	A	1.36(0.12)	0.95(0.06)	0.90(0.06)	1.10(0.07)	1.00(0.05)	0.92(0.04)	0.86(0.06)	1.00(0.05)
	C	1.78(0.11)	1.15(0.07)	1.12(0.08)	1.17(0.08)	1.06(0.04)	1.07(0.06)	0.88(0.05)	1.01(0.05)
8	A		1.06(0.16)	1.01(0.15)	1.12(0.17)	1.11(0.17)	0.97(0.16)	1.03(0.15)	0.96(0.15)
			1.03(0.15)	1.07(0.16)	1.02(0.15)	1.09(0.16)	1.11(0.17)	1.15(0.17)	1.06(0.16)
9	A			0.86	0.90	0.98	1.01	1.00	1.03
	C			1.04	1.10	1.16	1.17	1.11	
10	A	1.66(0.25)	1.28(0.18)	1.15(0.17)	1.21(0.19)	0.93(0.13)	0.74(0.23)	1.10(0.25)	1.13(0.27)
	C	1.76(0.26)	1.25(0.20)	0.91(0.14)	1.11(0.18)	0.86(0.12)	0.86(0.10)	1.05(0.24)	1.09(0.24)
11	A	0.71	0.92(0.06)	1.10(0.06)	1.03(0.08)	1.01(0.008)	0.88(0.09)	0.93(0.12)	0.84(0.08)
	C	0.58	0.92(0.06)	1.17(0.09)	1.09(0.08)	1.09(0.08)	0.92(0.07)	1.06(0.17)	0.81(0.09)
12	A	6.5(2.1)	1.13(0.18)	0.99(0.07)	1.01(0.09)	1.01(0.11)	1.01(0.12)	0.08(0.15)	1.04(0.10)
	C	5.0(1.3)	1.18(0.15)	1.02(0.10)	1.06(0.12)	1.11(0.12)	1.04(0.10)	1.18(0.16)	1.07(0.11)

NAA results shows that both Br and Mn displayed about 15% (on the average) higher values by XRFA. A difference in Br values might be ascribed to volatile loss, while that for Mn might reasonably relate to other causes (e.g., mechanical loss). The reanalysis by XRFA yielded values for Mn and Br that generally agreed better with the NAA results. When results are compared the values for the soil-related elements Ca and Mn are seen to be typically 20 to 30% lower for the second analysis.

This discrepancy can be attributed to handling problems which caused loss of larger particles. This seems to be a reasonable assumption since the specimens were subjected to many shocks both during shipment to Livermore and back to Berkeley and during the irradiation process, which involved the use of a pneumatic sample transport system into and out of the reactor.

Table 12 shows the average ratios of the same four elements (Ca, Ti, Mn, Br) on impactors, foils, and total filters and after-filters.

C. Episodic Data

Figures 3 and 4 illustrate the kind of data produced by INAA and the way they are used in the experimental interpretation. The analyses can be divided into two types: diurnal patterns and particle size distributions. Figures 3 and 4 show the Harbor Freeway episode of September 20, 1972. Lundgren impactor substrates were cut into 2-hr segments, and the Gelman GA-1 after-filters were changed every 2 hr. Samples were collected one block from the freeway. Figure 3 shows the particle size distributions for Br, Na, Cl, and Al for early morning and later afternoon commuter periods. Figure 4 shows the diurnal patterns of Pb, Br, Na, Cl, Al, V, and mass for the after-filters (<5.0-μ particles) for the 24-hr period. The Pb data are taken from Giauque at LBL (9).

In Figure 4 the Cl, Br, and Pb diurnal patterns track each other throughout the day, indicating a common submicron particle source, namely, lead halide emissions from auto exhausts. On the other hand, Al, V, and Na each portray different diurnal patterns and different sources.

The Br distribution in Figure 3 is peaked on submicron particles, typical of an automotive engine combustion source, with 92% (morning) and 82% (evening) on the after-filters. The Pb/Br ratios for the after-filters are 2.1 (morning) and 2.0 (evening) which are typical of fresh exhaust, since the Pb/Br ratio in ethyl gasoline is 2.5.

Although the Cl diurnal pattern is the same as for Br, the Cl is less concentrated on submicron particles than is Br, the

TABLE 12. Comparison of Data Determined by INAA (LLL) and XRF (LBL)[a] (ng/m^3)

Specimen		Ca	Ti	Mn	Br
TB0516TF	XRF-1	2090 ± 80	392 ± 98	93 ± 7	735 ± 29
	INAA	1700 ± 800	455 ± 244	53 ± 3	670 ± 19
UF0670TF	XRF-1	2110 ± 80	395 ± 94	84 ± 7	507 ± 20
	INAA	1890 ± 1636	392 ± 98	58 ± 3	456 ± 13
VH0792TF	XRF-1	3280 ± 130	340 ± 39	69 ± 6	294 ± 12
	INAA	4381 ± 526	375 ± 82	55 ± 3	266 ± 8
VI0806TF	XRF-1	8830 ± 440	815 ± 129	141 ± 8	294 ± 5
	INAA	7506 ± 2764	918 ± 202	119 ± 4	258 ± 7
TD0643TF	XRF-1	1310 ± 50	233 ± 53	42 ± 6	393 ± 16
	INAA	1022 ± 591	154 ± 41	25 ± 2	350 ± 10
WL1096TF	XRF-1	424 ± 37	≤89	26 ± 6	220 ± 9
	INAA	569 ± 273	81 ± 30	15 ± 2	178 ± 5
WK1084TF	XRF-1	693 ± 37	123 ± 30	33 ± 6	214 ± 9
	INAA	1184 ± 494	126 ± 49	27 ± 2	154 ± 5

[a] R. Giauque, January 25, 1974.

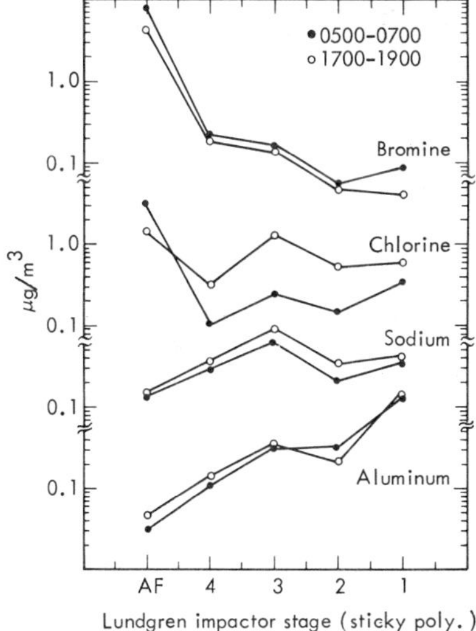

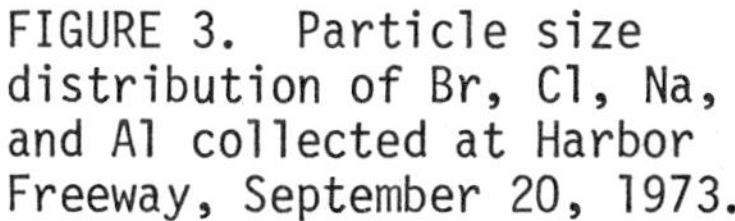

FIGURE 3. Particle size distribution of Br, Cl, Na, and Al collected at Harbor Freeway, September 20, 1973.

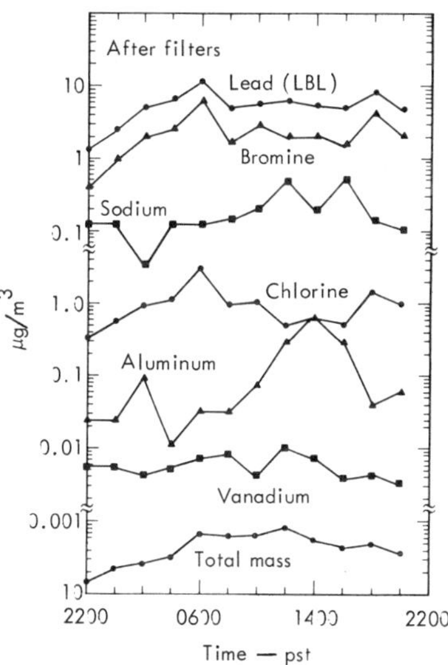

FIGURE 4. Diurnal patterns of selected elements at Harbor Freeway, September 20, 1973.

percentages being 78% (morning) and 35% (evening). The Cl distribution resembles the Na distribution in the micron range. It appears that most of the submicron Cl is from automotive exhaust, while most of the large-particle Cl is from the marine aerosol. Stages 1 to 4 show a deficiency of Cl relative to a sea water Na/Cl ratio of 0.55 at 0500 to 0700: 0.98 (1), 1.4 (2), 2.5 (3), and 2.7 (4). At 1700 to 1900 the Na/Cl ratio is closer to 0.55: 0.72 (1), 0.64 (2), 0.65 (3), and 1.24 (4). On the after-filters the Na/Cl ratios are 0.04 (morning) and 0.1 (evening). These data are consistent with a Cl loss during the morning hours, probably due to oxidizing agents which release Cl to Cl_2 gas.

Aluminum shows a predominance of large particles in Figure 3, typical of soil-derived aerosols, with only 1.6% (morning) and 2.2% (evening) on the after-filter. Sodium is distributed more uniformly in the 1 to 10 range. The peaking of the Na distribution on stage 3 (1.5 to 4.0) is considered to be representative of the marine aerosol. Although we do not include distributions for V, they show this element peaked on submicron particles, indicating a combustion source.

IV. SUMMARY

Instrumental neutron activation analysis (INAA) was one of the techniques used to measure elemental diurnal variations and to determine elemental particle size distributions. It was found to be sufficiently sensitive for a large number of elements, even for 2-hr samples from Lundgren impactor stages.

A secondary role of INAA was the validation of other analytical techniques utilized in the chemical analysis of aerosols. Supplemental experiments were also performed using INAA to validate the collection efficiencies of rotating-stage cascade impactors.

These studies confirmed previous efforts which had determined that rechniques using INAA were useful operational methods for chemical analysis of aerosols samples taken over 2-hr periods in urban air.

ACKNOWLEDGMENT

This work was performed under the auspices of the U. S. Energy Research and Development Administration under Contract No. W-7405-Eng-48. This report was prepared as an account of work sponsored by the United States Government. Neither the United States nor the United States Energy Research and Development Administration, nor any of their employees, nor any of their contractors, subcontractors, or their employees, makes any warranty, express or implied, or assumes any legal liability or responsibility for the accuracy, completeness or usefulness of any information, apparatus, product or process disclosed, or represents that its use would not infringe privately-owned rights.

REFERENCES

1. Rahn, K., Wesolowski, J. J., John, W., and Ralston, H. R. 1971. Diurnal Variation of Aerosol Trace Element Concentrations in Livermore, California, J. Air Pollut. Control Assoc. 21, 405.

2. Dams, R., Robbins, J. A., Rahn, K. A., and Winchester, J. W. 1970. Nondestructive Neutron Activation Analysis of Air Pollution Particulates, Anal. Chem. 42, 861.

3. Zoller, W. H. and Gordon, G. E. 1970. Instrumental Neutron Activation Analysis of Atmospheric Pollutants Utilizing Ge(Li) Gamma-Ray Detectors, Anal. Chem. 42, 257.

4. Guinn, V. P. 1966. Proceedings of the First International Conference on Forensic Activation Analysis, San Diego, California, September, pp. 7-40.

5. Ralston, H. R. and Wilcox, G. E. 1968. "A Computer Method of Peak Area Determinations from Ge-Li Gamma Spectra," University of California Lawrence Livermore Laboratory Rep. UCRL-71210 (preprint).

6. Ondov, J. M., Rancitelli, L. A., Filby, R. H., and Ragaini, R. C. 1974. "Four Laboratory Comparative Instrumental Nuclear Analysis of the NBS Coal and Fly Ash Standards," 7th Materials Research Symposium on Accuracy in Trace Analysis, Gaithersburg, Maryland, October.

7. Ondov, J. M., Zoller, W. H., Olmez, I., Aras, N. K., Gordon, G. E., Rancitelli, L. A., Abel, K. H., Filby, R. H., Shah, K. R., and Ragaini, R. C. 1975. Elemental Concentrations in the NBS Environmental Coal and Fly Ash Standard Reference Materials, Anal. Chem. 47, 1102.

8. Camp, D. C., Cooper, J. A., and Rhodes, J. R. 1974. X-Ray Fluorescence Analysis - Results of a First-Round Intercomparison Study, X-Ray Spectrom. 3, 1.

9. Giauque, R. 1974. Lawrence Berkeley Laboratory, private communications, January 25 and May 2.

An Intermethod Comparison of X-Ray Photoelectron Spectroscopic (ESCA) Analysis of Atmospheric Particulate Matter*

BRUCE R. APPEL, JEROME J. WESOLOWSKI, EMANUEL M. HOFFER, SUZANNE TWISS, AND STEPHEN WALL
Air and Industrial Hygiene Laboratory†
Laboratory Services Branch
California State Department of Health
Berkeley, California

and

SHIH-GER CHANG AND TIHOMIR NOVAKOV
Lawrence Berkeley Laboratory††
University of California
Berkeley, California

*Reprinted from International Journal of Environmental Analytical Chemistry, Vol. 4, page 169, 1976. Reprinted by permission.

†Work supported in part by the California Air Resources Board.

††Portions of this work performed at LBL supported by National Science Foundation-RANN and by U. S. Atomic Energy Commission.

Abstract

A comparison of ESCA with wet chemical and other analytical techniques has demonstrated agreement within about a factor of 2 for sulfate and carbon but gave much larger deviations for nitrate and ammonium in analysis of atmospheric particulate samples collected in California's South Coast Basin. Analyses for nitrate and ammonium were low by a factor of 5 to 10 relative to the wet chemical methods. Loss of volatile ammonium- and nitrate-containing materials under the high vacuum and localized X-ray beam-induced heating conditions of ESCA was the most reasonable explanation of the discrepancy. Temperature programmed ESCA studies support this conclusion and suggest the significance of highly volatile nitrate (e.g., nitric acid) and ammonium compounds in the atmospheric particles.

I. INTRODUCTION

Efforts to characterize atmospheric aerosols often involve determination of aerosol composition as a function of particle size and time of day. This requires analytical methods of high sensitivity since loadings for specific elements in the range of a few nanograms/cm^2 are frequently encountered. As a result, highly sensitive instrumental techniques such as neutron activation analysis (1), X-ray fluorescence analysis (XRFA) (2), and X-ray photoelectron spectroscopy for chemical analysis (ESCA) (3) are being used. The present report focuses on efforts to validate sulfate, nitrate, ammonium, and carbon data as obtained by ESCA in the Aerosol Characterization Experiment (ACHEX).

In addition to high sensitivity, ESCA has other inherent advantages, including the frequent absence of interferences from the substrate (because of its restriction to the sample surface), and its ability to provide, from a single analysis, data on a number of different oxidation states of a given element. The difficulties with ESCA for aerosol analysis include the following:

(1) Peaks in ESCA spectra corresponding to species in the same oxidation state for a given element are often only poorly resolved. For atmospheric samples peak assignments are empirical, being based upon comparison with chemical shifts for a limited set of pure, reference materials.

(2) To obtain quantitative results by ESCA, an element in the sample must be analyzed by another technique and other species determined by ratio to this internal standard. In our study, Pb, determined by XRFA, served as the internal standard. Since XRFA analyzes more deeply into a given particle or surface because of less absorption, Pb values so obtained may not be relevant to the ESCA spectra.

(3) ESCA analyzes only the outer 50 to 150 Å of the exposed particles. So, for example, with a 1-μ spherical particle, ESCA analyzes only the outer 0.5 to 1.5% and thus may not yield analyses representative of the average composition of the particle. On the surface of a filter or impaction medium, particles lying shielded by other particles may not be analyzed due to absorption and scattering. Similarly, particles which penetrate into the pores or beneath the fibers of a filter will probably not be analyzed. Since the degree of penetration, as well as composition, may be a function of particle size, such penetration may lead to biased data.

(4) ESCA requires maintaining the sample in vacuum (10^{-7} to 10^{-9} torr). Furthermore, while the samples are typically maintained at around room temperature as in this intermethod comparison, X-ray bombardment may produce localized heating of the sample. With or without such heating, there is the possibility of loss of volatile species such as nitric or sulfuric acid and ammonium nitrate.

This paper is intended to assess the combined effect of these and other unrecognized sources of error. To determine the accuracy of ESCA in the absence of standard reference materials, data obtained by ESCA were compared with those obtained by other instrumental or wet chemical techniques. Since, at the time of this study, the sensitivity requirements were beyond the range of conventional wet chemical methods, new techniques were developed and validated. These will be outlined briefly.

II. EXPERIMENTAL

A. Sampling and Analysis Strategy

Samples were collected on a 47-mm Gelman GA-1 cellulose acetate filters for sulfate analysis and on 47-mm, 1.2-μ-pore-size silver membrane filters for carbon and nitrogen species analysis. The sampling rate was 3.0 cfm with the membrane filters being changed at 2-hr intervals throughout 24-hr periods. In addition samples were collected on Hi-volume cellulose filters (Whatman 41), sampling continuously during the same 24-hr period.

ESCA analyses were conducted only on the membrane filters, while, depending on sensitivity, wet chemical methods employed either the 2-hr or 24-hr Hi-volume filters.

Intermethod comparisons included, therefore, both direct and indirect comparisons. Direct comparisons involved analyses of sections from the same filters, while indirect comparisons required comparison of calculated 24-hr average values from ESCA analysis of 12 2-hr filters with the wet chemical analysis of the Hi-volume filter.

B. X-Ray Photoelectron Spectroscopy (ESCA)

Extensive reviews of X-ray photoeletron spectroscopy are available (11). Therefore only a brief description of the method will be given here. ESCA measures the kinetic energies of photoelectrons expelled from a sample irradiated with monoenergetic X-rays. The kinetic energy of a photoelectron E_{kin}, espelled from a subshell i, is given by $E_{kin} = h\nu - E_i$, where $h\nu$ is the X-ray photon energy and E_i is the binding energy of an electron in that subshell. If the photon energy is known the determination of the kinetic energy of the photoelectron peak provides a direct measurement of the electron binding energy.

The electron binding energies are characteristic for each element which enables the method to be used for elemental analysis. The binding energies for a given element are modified by the valence electron distribution. Thus the binding energy of an electron subshell in a given atom depends on the atom's chemical environment (e.g., different oxidation states). These differences in the electron binding energies are known as the "chemical shift." The binding energies for oxidized species are greater than for the neutral configuration (a positive shift), while the binding energies show a negative shift for reduced species. The ESCA technique can, therefore, be used for both elemental analysis and for determination of chemical states.

Because of the limited resolution at the time of this study, ESCA spectra were analyzed only for total oxidized sulfur on a routine basis (cf. Figure 1). This includes adsorbed SO_2, sulfite, as well as sulfate (13). Based upon more detailed studies of a limited set of samples, sulfate typically represents at least 70% of this total.

In ESCA analysis for nitrate, nitrate proved to be the only discernible oxidized nitrogen species making total oxidized nitrogen a good approximation for determining this species.

ESCA analysis for ammonium was hampered by the presence of a second reduced nitrogen component. Indeed in the room temperature spectra, NH_4^+ nitrogen sometimes represented as little as 10 to 20% of this second species. This proportion was highly variable, however, making quantitative comparison difficult. Nevertheless, wet chemical ammonium ion determinations have been compared to ESCA values taking the total reduced nitrogen as an upper limit to the true ammonium value.

Intermethod comparison of carbon analysis employed total carbon, making resolution of discrete species or oxidation state unnecessary.

C. Wet Chemical Sulfate Analytical Methods

(1) SRI microchemical (4). In a procedure developed at the Stanford Research Institute samples are extracted with hot water and all soluble sulfur compounds, presumed to be principally sulfate, reduced to H_2S by hydrogenation followed by coulometric titration. The sensitivity was sufficient to permit analysis of aliquots of the same 2-hr filters analyzed by ESCA.

(2) Barium chloride turbidimetric analysis (15). Because of limited sensitivity this method could not be used on 2-hr samples. The method was used to analyze 24-hr Hi-volume samples which were then compared to the average of 12 2-hr low-volume filters collected sequentially throughout this 24-hr period and analyzed by ESCA.

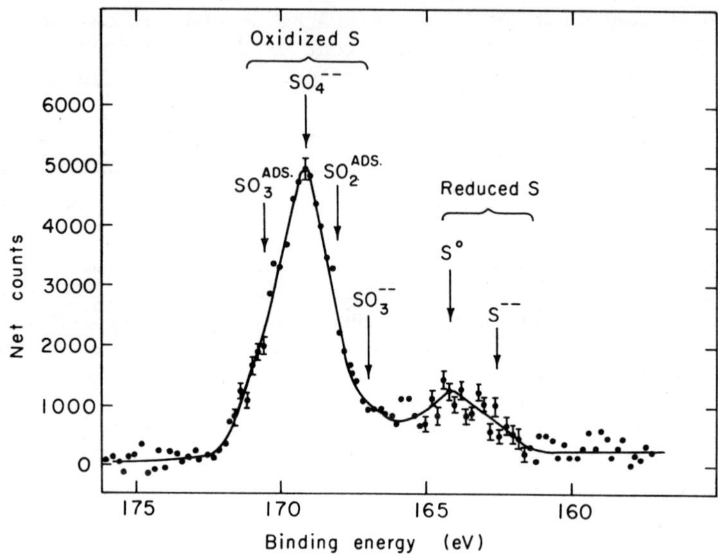

FIGURE 1. Sulfur (2p) photoelectron spectrum of an ambient particulate sample. Expected positions of sulfate, sulfite, neutral sulfur, and sulfide are indicated. In addition, the expected line positions for adsorbed SO_3 and SO_2 are also shown.

(3) AIHL microchemical method (10). Following extraction in 5 ml of water by a micropercolation procedure (7), 2-hr filter samples were analyzed for sulfate colorimetrically by reaction with a barium-dinitrosulfanazo III complex. The precision, accuracy and effects of interferents for sulfate analyses in the range 1 to 13 μg/ml are reported elsewhere (10). This method was applied to sections of the same filters analyzed by ESCA.

D. Wet Chemical Nitrate and Ammonium Analytical Methods

The microchemical procedure for nitrate also permitted analysis of the same 2-hr samples analyzed by ESCA. The conventional 2,4-xylenol method (5) was increased in sensitivity by a 5-fold increase in path length for determining absorbance values and other minor modifications (6). The extracts obtained by the micropercolation procedure (7) were used. A comparison of the micro- and conventional 2,4-xylenol method in the 1 to 20 and 50 to 80 μg/aliquot ranges, respectively, with atmospheric samples yielded average agreement in concentration within 10%. Furthermore, the precision of the micromethod was ±10% or better even at 1 μg/aliquot.

Lacking a suitable micro-method for ammonium ion, ESCA data averaged over 12 2-hr filters were compared to those from the corresponding 24-hr Hi-volume filter collected simultaneously at the same location and analyzed by the idophenol blue procedure (8).

E. Combustion Analysis for Carbon

Samples collected on silver membrane filters were analyzed by combustion to CO_2 with quantitation by gas chromatography as detailed by Mueller et al. (9).

F. Control Studies on Volatile Losses Under ESCA Instrumental Conditions

ESCA measurements were performed as a function of the sample temperature in the range from $-150^{o}C$ to $350^{o}C$. The sample was first cooled in dry nitrogen (at atmospheric pressure) to low temperature. The cold sample was then introduced into the spectrometer vacuum and spectra taken at $-150^{o}C$ and consecutively higher temperatures. The experiments were performed with ammonium sulfate, ammonium nitrate, nitric acid, and a number of ambient air particulate matter samples.

III. RESULTS

A. Sulfate

Table 1 lists the results of the comparisons of ESCA sulfate determinations with those by the SRI and AIHL microchemical methods and by the $BaCl_2$ turbidimetric method. Except as noted, ESCA data were obtained on the Gelman GA-1 filter samples. For the comparison of ESCA and wet chemical methods analyzing sections from the same filter samples, the ratio of means indicated average agreement within a factor of two. The indirect comparison of ESCA and the barium chloride turbidimetric values for sulfate indicates good agreement, on the average. The effect of filter medium was explored by comparing ESCA results for samples

TABLE 1. Sulfate Analyses by ESCA Compared to Microchemical and Turbidimetric Methods

Wet method	No. of samples	ESCA range for sulfate ($\mu g/m^3$)	Wet method range for sulfate ($\mu g/m^3$)	Ratio of wet method to ESCA
SRI microchemical	12[a]	3.3–19	0.6–16	0.5 ± 0.1
AIHL microchemical	8[a]	3.6–30	2.5–9.7	0.7 ± 0.2
$BaCl_2$ turbidimetric	4[b,c]	1.1–6.4	1.0–5.9	1.0 ± 0.4
$BaCl_2$ turbidimetric	4[b,d]	1.8–17.1	4.4–19.0	0.9 ± 0.2

[a] 47-mm Gelman GA-1 cellulose ester filters sampling for 2-hr at 3.0 cfm.

[b] $BaCl_2$ turbidimetric data are from analysis of four 24-hr Hi-volume filter samples.

[c] ESCA values are mean results from 12 2-hr Gelman GA-1 filters in parallel sampling.

[d] ESCA values are mean results from 12 2-hr Selas 1.2-μm Ag membrane filters in parallel sampling.

collected on silver membrane filters with the $BaCl_2$ turbidimetric method. The results for sulfate were not significantly different from those with the Gelman membrane filters. We conclude that ESCA provides sulfate analyses which are correct within a factor of two.

B. Nitrate and Ammonium

Table 2 compares ESCA to wet chemical analyses for nitrate and ammonium employing determinations conducted on the same filters and the indirect procedure outlined above. With nitrate, using both comparative techniques, the ESCA results were lower by a factor of nearly 5.

Using total reduced nitrogen as an approximation to ammonium, the ESCA values appear to agree well with that by the indophenol blue procedure. However, this agreement is fortuitous since the ammonium ion represented as little as 10 to 20% of the total reduced nitrogen making ESCA values low by as much as a factor of 5 to 10. We discuss below a probable cause of the apparent agreement between the wet chemical NH_4^+ and total reduced nitrogen by ESCA.

It is unlikely that the wet chemical methods are subject to large positive interference with these samples. A more likely source of error is the loss of volatile nitrate and ammonium in the ESCA spectrometer. Relevant control studies are detailed below.

C. Carbon

Twenty-nine samples collected on Ag membrane filters were analyzed for total carbon by ESCA and by a combustion technique (9). ESCA values ranged from 8.09 to 146 $\mu g/m^3$ with a mean ratio of combustion to ESCA results of 0.9 ± 0.1. These results suggest average carbon analyses by ESCA to be reasonably accurate. However, the degree of agreement for carbon analysis should also be influenced by the evaporation of the more volatile organic materials under ESCA conditions. The present agreement implies that most of the carbon was nonvolatile.

D. Volatile Losses Under ESCA Instrumental Conditions

Synthetic aerosols of ammonium sulfate were found to be stable in vacuum at $25^{\circ}C$ for time intervals normally used to complete the analysis of air samples; no detectable decrease of either ammonium or sulfate photoelectron peak intensity was

TABLE 2. Ammonium and Nitrate Analysis by ESCA Compared to Wet Chemical Methods

Species	Wet method	No. samples	ESCA range ($\mu g/m^3$)	Wet method range ($\mu g/m^3$)	Ratio of wet method to ESCA
NO_3^-	Micro-2,4-xylenol	9[a]	2.2-26.5	2.4-133	4.6 ± 0.7
NO_3^-	2,4-Xylenol	3[b]	0.9-7.1	5.6-36.4	4.9 ± 1.3
NH_4^+ [c]	Indophenol blue	4[b]	2.8-16.0	2.3-16.4	1.0 ± 0.1

[a]47-mm, 1.2-μ Ag membrane filters run for 2-hr at about 3 cfm.

[b]ESCA values are mean results from 12 2-hr, 1.2-μ Selas Ag membrane filters while the wet chemical analysis for nitrate was conducted on 24-hr Whatman 41 Hi-volume filters collected simultaneously.

[c]With ESCA the NH_4^+ analysis is taken to be 100% of the total reduced nitrogen observed while NH_4^+ could be as low as 10 to 20% of this value. See text.

observed between -150°C and 25°C. However, upon further heating at higher temperatures $(NH_4)_2SO_4$ appeared to decompose to NH_4HSO_4, the ammonium peak decreasing noticeably at ≥80°C, while the sulfate peak remained approximately constant until 150°C. At 150°C, NH_4HSO_4 appeared to be the species present. At higher temperatures both NH_4^+ and sulfate sulfur decreased at the same rate.

Ammonium nitrate proved to be more volatile in vacuum than ammonium sulfate. It decomposed completely at temperatures above about 100°C with equivalent loss of anion and cation. Even at normal sample temperatures of about 25°C a decrease in both nitrate and ammonium peak intensities was observed at a rate of about 15% per 12-hr period. A decrease of much less than 5% was observed for ammonium nitrate between sample temperatures of -150°C and 25°C. Since 12 hr (at room temperature) is the maximum time that an ambient sample was kept in vacuum, an underestimation of ≤15% is to be expected for ammonium nitrate concentration determination.

Nitric acid samples were prepared (14) by exposing polycrystalline graphite to NO_2 and moisture in air.* Intense nitrate photoelectron peaks were observed at -150°C. However, most of the nitric acid was lost when the sample was heated to 25°C.

Figure 2 details typical results of variable temperature ESCA studies conducted with ambient aerosol samples. The ESCA spectrum of the nitrogen 1s region taken with the sample at -150°C is shown in Figure 2a. Nitrate and ammonium photoelectron peaks are seen in addition to a peak, labelled N_x, believed to be due to surface amine and/or amide compounds (3,13). The amount of ammonium, as reflected in the photoelectron peak intensity, exceeds the nitrate peak intensity by a factor of about 6. Therefore at most about 15% of the total ammonium present in the sample at -150°C could be associated with nitrate (i.e., as NH_4NO_3). The remaining 85% of the ammonium could be in the form of ammonium sulfate or other compound.

Figure 2b shows the nitrogen 1s spectrum of the same sample taken after it was heated to 25°C. This spectrum reveals only a trace of the original nitrate and a decrease of about 60% in ammonium peak intensity. The amount of ammonium nitrogen apparently lost by volatilization is about three times the original nitrate nitrogen observed at -150°C. Therefore, the volatile ammonium cannot be principally ammonium nitrate.

Based on these results, the inconsistency between ESCA (at 25°C) and wet chemical determinations for ammonium probably

*Nitrous acid was not observed but would be detectable if present.

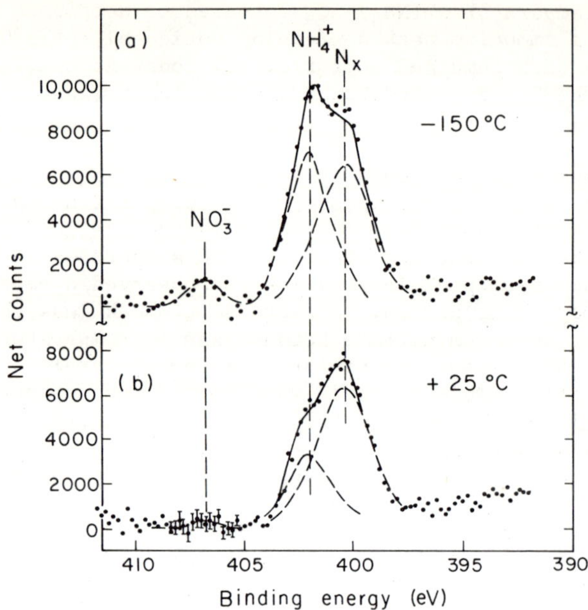

FIGURE 2. (a) Nitrogen (1s) photoelectron spectrum of an ambient particulate sample measured at -150°C. The position of nitrate, ammonium, and N_x species are indicated. (b) Nitrogen (1s) spectrum of the same sample measured at 25°C.

cannot be explained by volatile losses of ammonium sulfate, ammonium bisulfate, or ammonium nitrate. The inconsistency for nitrate can similarly not be explained by the volatilization of ammonium nitrate. The results with nitric acid, however, indicate that only a small fraction of any nitric acid initially present would have been observed by ESCA at 25°C. Thus if a large fraction of nitrate exists as HNO_3 on the filter, this would explain, qualitatively, the low nitrate values by ESCA.

The limited volatility of the ammonium salts and the behavior of the ambient samples suggest that the major fraction of ammonium on aerosol particles is present in a hitherto unrecognized volatile form perhaps produced by chemisorption of ammonia. Indeed, ammonia was found to adsorb on, for example, combustion-produced carbon particles, to form a volatile carboxyl-ammonium salt (12).

It is notable that the volatile ammonium species is found in concentrations similar to the concentration of the other reduced nitrogen species (i.e., N_x), at least for the ambient air samples which have been examined at low temperatures by ESCA. Because the ESCA measurements for intermethod comparisons were performed at 25°C, the total reduced particulate nitrogen is largely in the form of N_x. This might explain the apparent agreement between the results of wet chemically determined concentrations of ammonium and ESCA concentrations of total reduced nitrogen.

E. Precision of ESCA Determinations

The preceding discussion has emphasized the accuracy of ESCA determinations. A qualitative indication of the precision of ESCA is obtained by comparing diurnal patterns for species determined both by ESCA and by alternate procedures. Figure 3 depicts diurnal patterns for sulfate analyses as measured by both ESCA and the SRI microchemical procedure applied to 2-hr low-volume filters. The data is that for a 24-hr moderate smog episode at Pomona, California. ESCA and the SRI procedure yield strikingly similar diurnal patterns suggesting sufficient precision for the ESCA method to permit useful interpretations.

IV. SUMMARY AND CONCLUSIONS

A comparison of ESCA with wet chemical and other analytical techniques has demonstrated agreement within about a factor of 2 for sulfate and carbon but yielded much larger deviations for nitrate and ammonium analysis in the analysis of atmospheric samples from California's South Coast Basin. For these species ESCA analysis performed at 25°C was low by a factor of at least 4.

Detailed ESCA studies with temperature programming from -150°C to 350°C demonstrated that loss of ammonium sulfate, ammonium bisulfate, and ammonium nitrate are probably minor contributors to this discrepancy. The data are consistent with loss of a more volatile nitrate (e.g., nitric acid) and ammonium species the latter perhaps arising from chemisorption of ammonia on particulate matter.

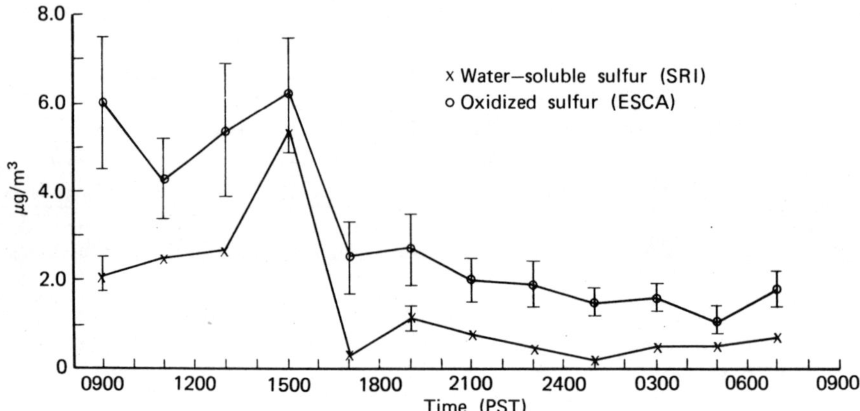

FIGURE 3. Diurnal pattern for sulfate (as sulfur) by ESCA and the SRI microchemical methods; Pomona, California, October 24-25, 1972.

ACKNOWLEDGMENT

We would like to acknowledge the assistance of Messrs. Arthur Alcocer and Paul Bussard of AIHL, Drs. Dale Coulson and James Smith of SRI, and Drs. N. L. Craig and A. B. Harker of LBL for their assistance in this work.

REFERENCES

1. Iddings, F. A. 1969. Environ. Sci. Technol. 3, 132; Zoller, W. H. and Gordon, G. E. 1970. Anal. Chem. 42, 257; Dams, R. et al. 1970. Anal. Chem. 42, 861; Rahn, K., Wesolowski, J. J., John, W., and Ralston, H. R. 1971. J. Air Pollut. Control Assoc. 21, 7.

2. Dittrich, T. R. and Cothern, C. R. 1971. J. Air Pollut. Control Assoc. 21, 716; Giauque, R. D. et al. 1973. Anal. Chem. 45, 671.

3. Novakov, T., Mueller, P. K., Alcocer, A. E., and Otvos, J. W. 1973. J. Coll. Interface Sci. 39, 225.

4. Ludwig, F. L. and Robinson, E. J. 1965. Coll. Sci. 20, 571; Ludwig, F. L. and Robinson, E. J. 1968. Atmos. Environ. 2, 13.

5. Intersociety Committee Methods of Air Sampling and Analysis, American Public Health Association, 1972.

6. Air and Industrial Hygiene Laboratory Method 50, the 2,4 xylenol method for the analysis of sub-milligram amounts of nitrate, 1974.

7. Hermance, H. W., Russell, C. A., Bauer, E. J., Egan, T. F., and Wadlow, H. V. 1971. Environ. Sci. Technol. 5, 781.

8. Weatherburn, M. W. 1967. Anal. Chem. 39, 971.

9. Mueller, P. K., Mosley, R. W., and Pierce, L. B. 1972. J. Coll. Interface Sci. 39, 235.

10. Hoffer, E. M., Kothny, E. L. and Appel, B. R. 1979. Atmos. Environ. 13, 303.

11. Hercules, S. H. and Hercules, D. M. 1972. Environ. Anal. Chem. 1., 1969; Novakov, T. 1973. In Proceedings of the Second Joint Conference on Sensing of Environmental Pollutants, p. 197, published by Instr. Soc. of America, Pittsburgh.

12. Chang, S. G. and Novakov, T. 1975. Atmos. Environ. 9, 495.

13. Craig, N. L., Harker, A. B., and Novakov, T. 1974. Atmos. Environ. 8, 15.

14. Forsythe, W. R. and Giauque, W. F. 1942. J. Am. Chem. Soc. 64, 48.

15. "Selected Methods for the Measurement of Air Pollutants," Public Health Service Publication No. 999-AP-11, 1965.

Microwave Measurements of the Liquid Water Content of Atmospheric Aerosols*

WILLIAM W. HO, GEORGE M. HIDY, AND RONALD M. GOVAN
Science Center
Rockwell International
Thousand Oaks, California

Abstract

A new method is described for the measurement of the dielectric constant of atmospheric aerosols. From these observations the free liquid water content of airborne particles can be deduced. An instrument to measure semi-continuously the dielectric constant of aerosols is discussed, and illustrative observations are given for urban air in southern California and for marine and desert air at nonurban sites. The results taken during the fall of 1972 indicate that a significant fraction of

*From Journal of Applied Meteorology, Vol. 4, page 871, December 1974. Reprinted by permission of the American Meteorological Society.

the marine and urban aerosol observed is free water in the relative humidity range between 40 and 75%. The liquid water content in the aerosol correlates satisfactorily with the extinction coefficient for visible light, as measured with an integrating nephelometer.

I. INTRODUCTION

The degradation of visibility from haze in the atmosphere has long been related to conditions of high humidity. Beyond a relative humidity of ∿70%, the opacity of haze increases markedly. This decrease in visibility is believed to be related to the ability of airborne particles to absorb significant amounts of water vapor. Experiments of Hänel (4) and Covert et al. (1), for example, have demonstrated that atmospheric aerosols are hygroscopic and indeed undergo a substantial increase in mass as the particles equilibrate to increased relative humidity above ∿70%.

Recent work on the chemistry of aerosols, particularly in urban air, has suggested that water plays an important role in particle behavior even at a relative humidity well below 70%. For example, Goetz and coworkers [e.g., Goetz and Pueschel (3)] have reported that liquid particles in the Los Angeles area have been collected that appear to have coatings of organic material. These particles grow on the collection substrate with exposure to water vapor. Indirectly, it appears to be necessary to postulate a liquid water content of at least 10% to account for the total mass composition of filter-collected aerosol equilibrated with air of a relative humidity <50% [e.g., Miller et al. (9)]. Attempts to measure water content of filter-collected aerosols in California suggest a water fraction of ∿10% or greater by mass [e.g., Meyer et al. (8)] for samples collected in air of a humidity range between 40 and 60%.

With the potential of the importance of liquid water to aerosol chemistry, it is desirable to develop methods for its measurement, at least semi-continuously. In this paper, a new method for determining aerosol liquid water content is proposed that makes use of measurements of the dielectric constant of continuously collected particles. The fact that the dielectric constant of liquid water is much larger than other substances expected to be in atmospheric aerosols permits a satisfactory estimation procedure for the water fraction.

The paper first discusses the principle of measurement of the dielectric constant using a microwave system. Then the method is illustrated by presentation of results obtained in urban and nonurban sites in southern California in late summer and fall of 1972. The importance of apparent diurnal changes in liquid water content to aerosol behavior is considered to complete this discussion.

II. EXPERIMENTAL METHOD

A. Principle of Measurement

It is well known that when a small sample with a dielectric constant larger than unity is introduced into a microwave cavity, a shift in the resonant frequency of the cavity occurs which is proportional to the dielectric constant and volume of the sample; this technique has been used extensively in the past to measure dielectric properties of matter. Since chemically unbound liquid water has the property that its dielectric constant in the low-frequency microwave region is an order of magnitude greater than that for other materials, this technique then enables the measurement of liquid water content in small samples with very high accuracy. Water vapor, on the other hand, has a dielectric constant close to unity in the microwave region below 10 GHz so that the technique, in principle, is sensitive only to water in the liquid phase and enables a unique separation of the percentage of liquid water and water vapor in samples of air.

The measurement apparatus consists of a right cylindrical TM_{010} resonant cavity operating in the 3-GHz frequency region with provisions made for insertion of a quartz sample tube along the axis of the cavity (Figure 1). The electric field in this mode is axially symmetric and is a maximum at the center along the axis of the cavity. The behavior of this resonant mode has been investigated extensively in our previous work on precision measurements of the dielectric properties of aqueous solutions [Ho and Hall (6)]. It has the overwhelming advantage over other modes in that its resonant frequency is dependent only on the diameter of the cavity and is not degenerate in frequency with any other resonances. Furthermore, for the S-band frequency region, the dimensions for the cavity can be chosen such that interference from nearby resonances can be kept to a minimum. For the final design in the present case, for instance, the next nearest mode occurs at a frequency whose separation from the dominant mode is several orders of magnitude larger than the width of the resonance. Therefore, no mode mixing exists, and the actual field configuration within the cavity is expected to be very close to that obtained from theory for the boundary conditions given. We expect, then that the observed properties

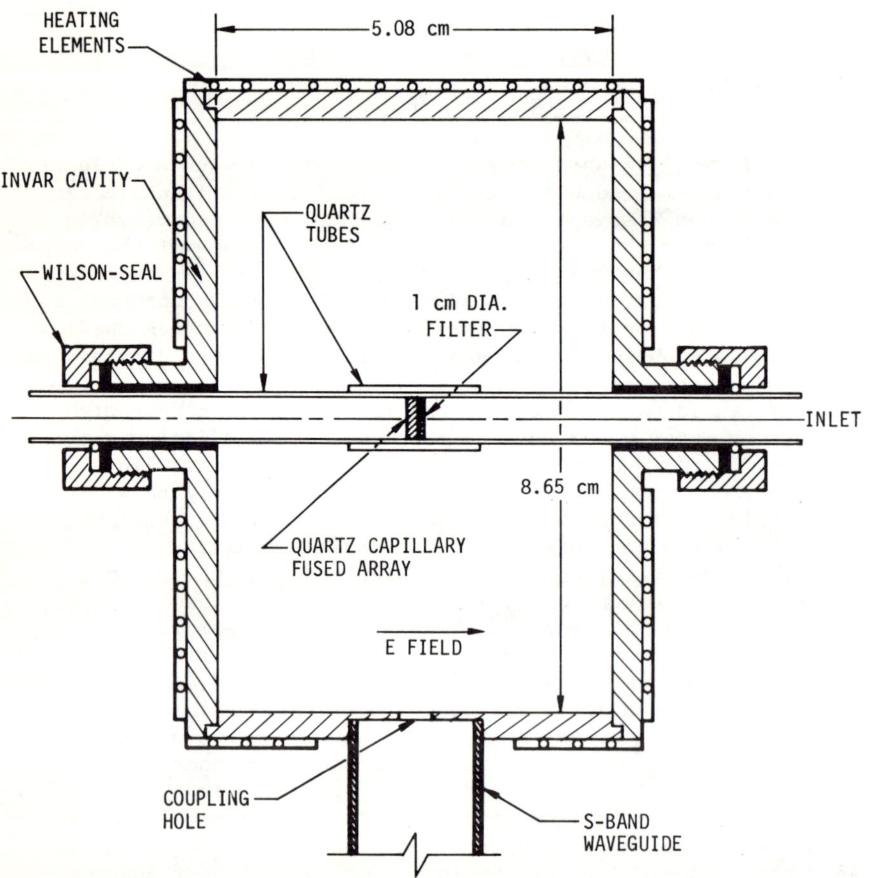

FIGURE 1. Assembly drawing of 2.59-GHz, TM_{010} microwave resonant cavity.

can be predicted quite well from calculations, a conclusion subsequently verified by test measurements.

The electronic and microwave circuits associated with the prototype laboratory instrument are shown schematically in Figure 2. Microwave power is generated by a stable frequency-

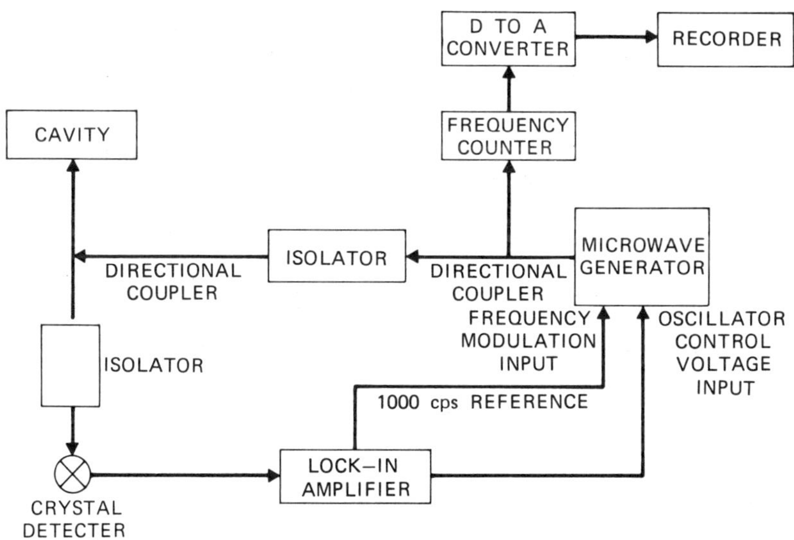

FIGURE 2. Block diagram of electronic circuit.

modulated source whose output frequency is controllable by an external voltage. Microwave power is introduced into the reflection-type cavity resonator by a directional coupler. The cavity is made of Invar and temperature-stabilized to better than 0.05C to minimize frequency drift due to thermal expansion changes in the dimension of the cavity. The reflected power from the cavity is then detected with a crystal and the modulation component of the output is measured with a phase-sensitive detector. When the modulation is small compared to the width of the cavity resonance, the signal is proportional to the derivative of the resonance curve and the output at resonant frequency is zero. The output of the phase-sensitive detector is then used to control the frequency of the microwave oscillator in a feedback loop so that its output is always at the resonant frequency of the cavity. This output is directly measured with a frequency counter. For display purposes and ease of data processing, the measured frequency changes are converted to a signal voltage with a digital-to-analog converter and recorded on a strip chart as a function of time.

The instrument was tested and found to be capable of measuring changes in frequency to better than 100 Hz or 3 parts in

10^7. The long-term stability over a period of about 1 hr is of the order of 200 Hz.

To measure liquid water content in urban aerosols, the instrument must be designed to detect in the order of 10 μg/m^3 of water if the total particulate mass concentration is 100 μg/m^3 or less. To achieve this sensitivity, a preconcentration of particles is required at this stage of development. To collect aerosol particles, a substrate was chosen of Gelman Type A glass-fiber filter. This medium was chosen for its low absorptivity of water vapor and for minimization of the possibility of filter surface wetting by organic liquid and water, as well as for high filter collection efficiency.

The glass-fiber filter was placed normal to the flow in the air sampling tube to collect material over a period of time. Air is passed through the system at a constant measured rate. Since the frequency shift of the cavity is proportional to the total volume of liquid water collected on the filter, a steady decrease in frequency with time should be observed. It is, of course, expected that particulate matter also will be collected on the filter and the total mass of filtered material will in turn produce a shift in addition to that due to the liquid water. This effect, however, can be corrected for by knowing the total weight of the sample collected, which may be measured by direct weighing or using other suitable methods, such as a mass concentration monitor using β-radiation attenuation [e.g., Dresia et al. (12)].

B. Frequency Shift and the Dielectric Constant

The quantitative shift in cavity resonant frequency due to the presence of liquid water is dependent on the geometric configuration of the water as well as its dielectric property and volume. From the results of test cases we have performed, it appears as if the liquid water collected on the glass-fiber filter is mainly in the form of droplets with approximately spherical shape. This is consistent with the findings by direct observations with a scanning electron microscope of filters containing collected material. Therefore, for initial data reduction we have taken as a model the idealization that the collected aerosols are individual dielectric spheres separated from each other by a distance large compared with their average diameter. The applicability of this model is compatible with the glass-fiber filter used in the experiment provided that the filter is not too heavily loaded with airborne particles.

For this model the cavity frequency perturbation is given by the expression

$$-\frac{\Delta f}{f} = \frac{3(\varepsilon_1 - \varepsilon_0)}{(\varepsilon_1 + 2\varepsilon_0)} \frac{1}{2J_1^{\,2}(x_0)} \frac{\Delta v}{V} , \tag{1}$$

where ε_1, the complex dielectric constant of the sample, is given by $\varepsilon_1 = \varepsilon' + i\varepsilon''$, and ε_0 is that for air; $J_1(x_0)$ is the value of the Bessel function $J_1(x)$ at the first zero of $J_0(x)$; and V is the volume of the cavity and Δv the total volume of the sample (no significant departure occurs until the separation between individual spheres and distortions within each sphere become comparable to their diameters). The real part of the expression corresponds to a shift in resonant frequency of the cavity, and the imaginary part to a change in the microwave loss within the cavity due to microwave absorption of the sample material.

As it is much more accurate and convenient to measure the frequency shift of the cavity, we have used it in this experiment as the primary determination of Δv, and the loss measurements were made only on occasions for self-consistency checks.

The expression for the frequency shift due to liquid water droplets in our TM_{010} cavity ($\varepsilon'_1 = 77$, $\varepsilon''_1 = 10$, $\varepsilon'_0 = 1$, $\varepsilon''_0 = 0$, $V = 284.4$ cm^3, $f = 2.6 \times 10^9$ Hz, $2J_1^{\,2}(x_0) = 0.5383$) is given by

$$M_w = 20.3\ (\Delta f)_w , \tag{2}$$

where the mass of liquid water (μg) is obtained from the volume by taking the density as unity, and the frequency shift $(\Delta f)_w$ is in kilohertz.

Tests were performed with the instrument to determine the dielectric constant of dry atmospheric aerosol using the dielectric sphere model, and the results indicated that if the density of typical dry aerosols is assumed to be in the range 1.7 to 2.0 as reported by Whitby et al. (10), then the measured values for ε' ranged from 2.5 to 3.5 with negligible dielectric loss. Consequently, the corresponding expression for dry atmospheric aerosols with a dielectric constant ε' of 3 and an assumed average density of 1.8 is given by

$$M_a = 88.4\ (\Delta f)_a \tag{3}$$

which is approximately four times less sensitive in frequency shift than that for an equivalent mass of liquid water. We note that a change of the value of ε' from 3 to 3.5 only changes the coefficient by 13%.

The total frequency shift for an aerosol sample containing liquid water is then related to the total mass of the sample by the expressions

$$(\Delta f)_T = 4.9 \times 10^{-2} M_w + 1.1 \times 10^{-2} M_a$$
$$M_T = M_w + M_a . \tag{4}$$

The liquid water content is therefore given in terms of the measured parameter $(\Delta f)_T$ and M_T by

$$M_w = 26(\Delta f)_T - 0.29 M_T \tag{5}$$

where the masses are in micrograms and the frequency shift is in kilohertz. This final result is clearly dependent on the dielectric constant and density of the dry atmospheric aerosols. Although the subsequent field measurements are quite consistent with the assumed values of ε' = 3 and 1.8 for density, work is now in progress to further quantify these parameters as well as to determine their possible variation with chemical composition. In addition, work is under way to check the dielectric sphere model for aerosols collected on a filter paper.

We emphasize here that the measurements are sensitive only to free (unbound) water since hydrates have much lower dielectric constants than nonassociated water. The total water concentration in atmospheric aerosols may consequently be somewhat higher than the measurements would indicate.

The above expressions are, strictly speaking, applicable only to aerosols of pure liquid water. However, the dielectric depression for most ionic solutions and acids, i.e., the decrease in ε'_1 due to the presence of ions, is usually small. For instance for alkali halides, the decrease in ε'_1 is only of the order of 10% for concentrations near 1 N. In addition, the dielectric loss term ε''_1 is of the order of three or four times that for distilled water at S-band frequencies. Since the coefficient in Equation 1 is proportional to $[(\varepsilon'_1 - \varepsilon_0) \times (\varepsilon'_1 + 2\varepsilon_0) + \varepsilon''^2_1]/[(\varepsilon'_1 + 2\varepsilon_0)^2 + \varepsilon''^2_1]$, it can be seen that so long as $\varepsilon'_1 \gg \varepsilon_0$ the expression is nearly constant and close to unity, especially when ε'' becomes large. Therefore, the treatment is quite general and the measured results, at least for the spherical model, make no distinction between liquid water and acids or aqueous salt solutions. The coefficient in Eq. 2, for instance, only changes by about 0.5% if the liquid aerosols were composed of 0.5 N NaCl solution and by about 5% for saturated NaCl solutions.

Certain low-molecular-weight organic liquids have dielectric constants comparable to water and can conceivably cause interferences with the measurements. However, they also tend to have high vapor pressures. In the case of urban aerosols, at least for the Los Angeles area, the liquid organic materials present are expected to be high-molecular-weight paraffinic, aromatic, or oxygenated compounds, with carbon numbers in the range of 8 or greater. For these liquids the dielectric constants, in almost all cases, are in the range of 2 to 4. Therefore, one does not expect a significant increase in the dielectric constant of the atmospheric aerosol due to their presence, even though their concentrations in the Los Angeles area are expected to be high, perhaps in the range 20 to 40% by mass. Indeed, this conclusion was verified by the fact that for ambient samples measured under low relative humidity conditions ($\leq$40%) the total mass obtained by direct weighing is in agreement with that calculated from the measured Δf via Eq. 3.

C. Instrument Design

A schematic diagram of the air sampling system is shown in Figure 3. Outside air is introduced into the system at a flow rate of $\sim$1 m^3/hr through a covered air intake. The flow rate is continuously monitored with a Hastings mass flowmeter and maintained at a constant flow rate throughout a sampling period by the adjustment of valve 3. In order to minimize the effect of systematic instrument drift and the effect of changing atmospheric conditions such as variation in barometric pressure, relative humidity, and temperature, the system is calibrated at regular intervals. During the sampling period, which is of the order of 20 min for most cases, valve 2 is open and valve 1 is closed. After each sample run, clean prefiltered ambient air is flowed through the system for the same time interval by closing valve 2 and opening valve 1. The frequency shift of the cavity as a function of time is recorded on a strip chart recorder. A direct comparison of the slope of the line obtained during sampling to that for the calibration case then enables the accurate determination of the frequency shift $(\Delta f)_T$ corresponding to the materials collected on the filter paper. In this method of sampling, the usual difficulty encountered due to absorption

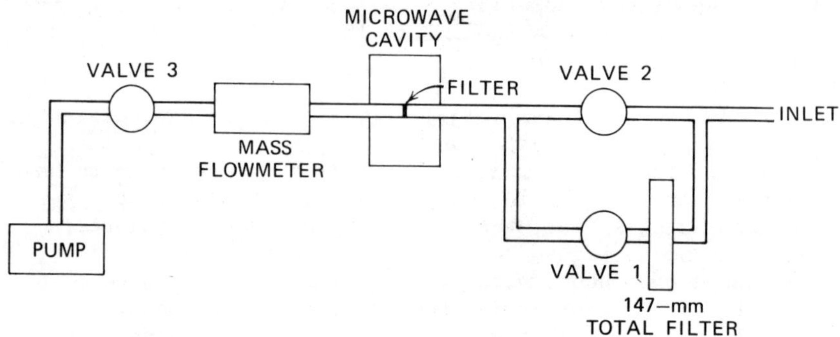

FIGURE 3. Schematic diagram of air-sampling system.

of water vapor on the glass-fiber filter is eliminated so long as the variation in relative humidity of the ambient air is slow compared to the sampling time interval, a condition which is easily satisfied for most cases. Furthermore, slow systematic drifts in the cavity resonant frequency due to temperature variation are also compensated for. As the materials collected on the filter paper are in equilibrium with the ambient air stream, it is not expected that any significant loss in volatiles will occur, provided the change in the partial pressure of the volatiles is slow compared to the sampling time. There is, nevertheless, the possibility of loss of volatiles, especially at very high relative humidity due to the breakup of liquid aerosols upon impaction with the resulting smaller drops having higher probability of evaporation at the same water vapor pressure. However, since the total mass determination, as described in the following section, was also done by filter collection on the same substrate at approximately the same flow rate, the corresponding error would also occur, so that, except in extreme cases, the ratio of liquid water to total aerosol mass would be, to first order, insensitive to this loss.

D. Correction for Total Mass Concentration

Since the analysis resulting in Eq. 5 shows that the observations must be corrected for total mass concentration to achieve quantitative estimates of liquid water content, the total mass concentration must be independently determined continuously.

For purposes of this study, the principal method of measuring total mass concentration of aerosols was via an instrument which determines the mass collected on a glass-fiber filter by β-radiation attenuation. The device used was the Froeseke-Hoepfner Model FH62A Dust Monitor. This instrument was calibrated by collecting atmospheric aerosols on preweighed filters at Pasadena in late summer 1972. The calibration was checked again later in winter 1972-73 for atmospheric aerosols and artificially generated particles, including polystyrene spheres, auto exhaust particles, sodium chloride, and other salt aerosols. The more recent calibration tests indicated that this particular instrument suffers from uncertainties in calibration when heavy metal salt aerosols are used and when the relative humidity exceeds ∿70% [Landis (7)]. However, the instrument can be calibrated and is reproducible for measurement of atmospheric aerosols to an accuracy of ±30% which was adequate for our preliminary tests.

E. Field Observations

To test the feasibility of measuring liquid water content in aerosols, the prototype instrument (called a "waterometer") was used in the field in late summer and fall of 1972. The instrument was operated at Pasadena in the Keck Laboratories of the California Institute of Technology and on a mobile laboratory built and operated under the California Aerosol Characterization Experiment (ACHEX). With the prototype instrument, the mobile laboratory was located (a) next to the Harbor Freeway in downtown Los Angeles, (b) at the Los Angeles County Fair grounds in Pomona, (c) at the Goldstone Tracking Station, and (d) at Point Arguello. The first sites were urban locations ranging from a "source-enriched" site for auto exhaust to "receptor" sites chosen to provide samples of air pollution away from the neighborhood of major sources. The Goldstone site was chosen to be representative of a remote desert location more than 60 mi from any community. The Point Arguello location was next to the Pacific Ocean approximately 20 m above sea level and represents marine air for sampling.

In both the Pasadena station and the mobile laboratory, meteorological instrumentation, trace gas monitors, and integrating nephelometers were in operation at the same time the present instrument was used. Because of the geometrical design and nature of the aerosol sampling system at Pasadena and the mobile laboratory, aerosols $\gtrsim$ 10-μm diameter could not be detected [see, e.g., Hidy (5)].

The air sampling was conducted under conditions ranging from no photochemical smog to light to moderate smog (maximum ozone concentration* > 20 pphm) at the urban locations. On one occasion at Pasadena, drizzle and fog were observed. The weather at Goldstone was warm and dry, with observations taken just after passage of a weak storm front. Conditions at Point Arguello were characterized by moderate to brisk winds with scattered clouds.

Experiments were conducted over periods of 24 hr to investigate the diurnal behavior of water content and total mass concentration. The observations were compared with nephelometer readings and with the total mass concentration as measured with the β-attenuation monitor. Other supporting data included the relative humidity or dew point, the winds, as well as a variety of aerosol data.

III. RESULTS AND DISCUSSION

The aerosol sampling conducted during the 1972 season indicated that significant amounts of water were present in atmospheric aerosols even at relative humidities well below 100%. The hygroscopic nature of airborne particles was well illustrated in the urban air of Los Angeles and near the ocean at Point Arguello. In contrast, the desert air at Goldstone was found to contain essentially no detectable water.

Some typical results from this initial study are shown in Figures 4 to 8, illustrating diurnal changes in urban air. Included with these data are results from the total-mass monitor. In Figure 4 are the diurnal changes measured in Pasadena for conditions of moderate photochemical smog, where the oxidant level exceeded 25 pphm by midday. In this case the total mass concentration of airborne particles was maximum near midday, and was accompanied by a maximum in the apparent water content of the particles. The change in fractional water content of the aerosols follows the change in relative humidity, whose values are noted in the figure.

Figure 5 shows the diurnal pattern in liquid water content for the data obtained at Pasadena on another occasion. Two peaks were observed in the total mass concentration with corresponding increases in liquid water content. The decrease in mass concentration at 1000 local time corresponded to a change in wind direction, and the second peak corresponded to a decrease of wind speed to almost zero.

---

*pphm = parts per hundred million.

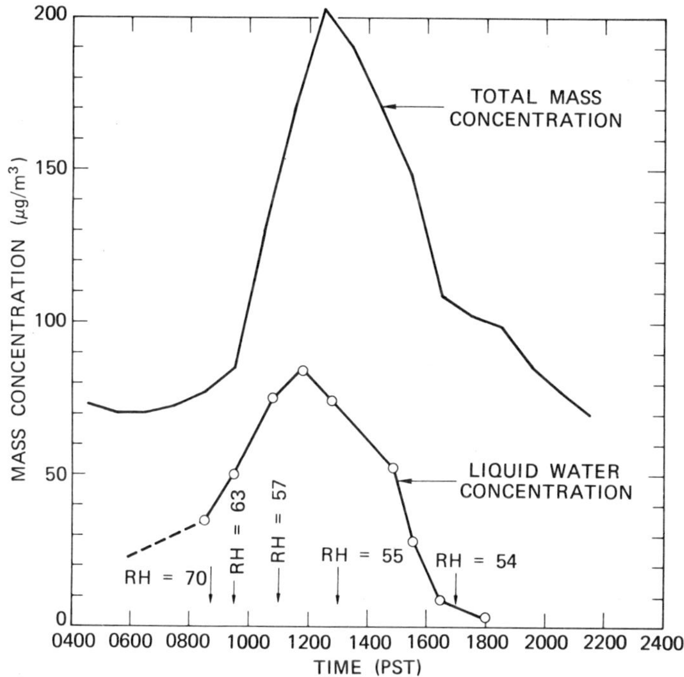

FIGURE 4. Liquid water concentration in atmospheric aerosols at Pasadena, September 15, 1972.

The data in Figure 6 are for another sampling day in Pasadena. Here there was essentially no photochemical smog. The weather was overcast with drizzle and fog through the morning, but clearing in the afternoon. The data from the β-gauge are only semiquantitative because of large errors in mass measurement induced in this instrument at relative humidities ≥70% [Landis (7)]. However, qualitatively, the changes in liquid water content follow well the changes detected in total aerosol mass, as were found for the data in Figure 4. During the morning when fog and drizzle were observed, the liquid water content of material collected was a very large fraction of the total aerosol mass. In some cases the mass fraction of water exceeded 50%.

Another case of measurements of liquid water content in urban air, taken at the Los Angeles County Fairgrounds in Pomona, is shown in Figure 7. This was a condition of light smog where

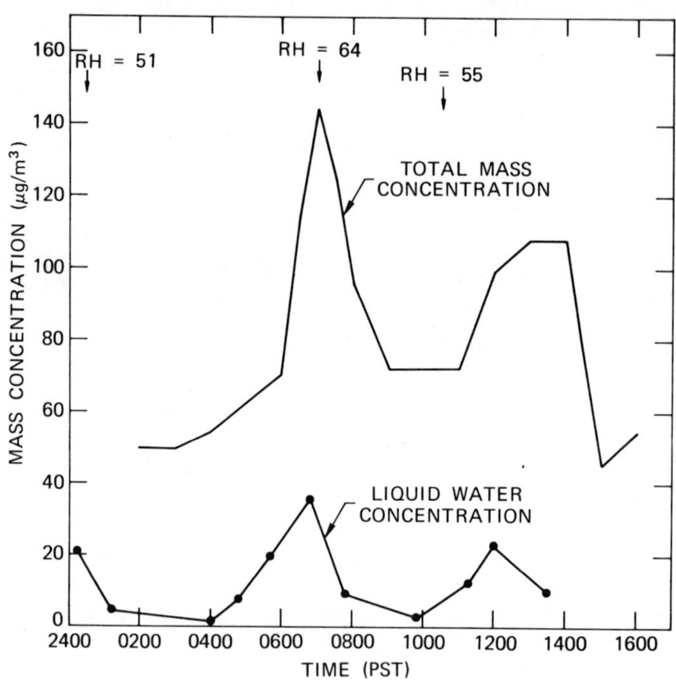

FIGURE 5. Liquid water concentration in atmospheric aerosols at Pasadena, September 20, 1972.

maximum oxidant levels remained below 10 pphm. Again the pattern of change in liquid water content following total mass concentration is observed. However, the maxima in these parameters are observed in the morning hours, in contrast to the Pasadena case in Figure 4. Again, there appears to be a correspondence between changes in relative humidity and water content during the day.

The observation of liquid water content of urban aerosols in the Los Angeles area suggested a fraction of total mass ranging from 10 to 40% for relative humidities from 50 to 70%. With humidity over 70%, the fraction of liquid water can exceed this range significantly.

Measurements of the water in aerosols at the remote desert site are shown in Figure 8. At this location the mass concentration of particles was extremely low and the visibility was

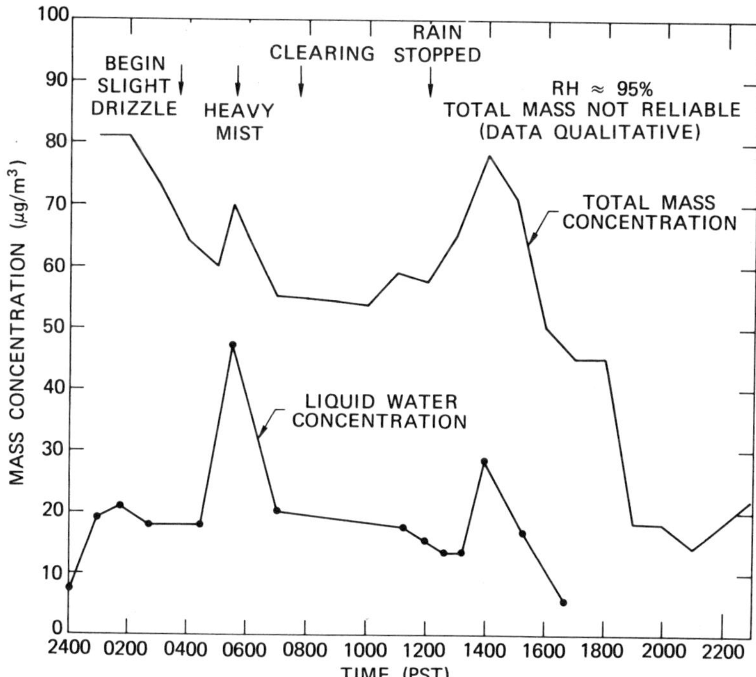

FIGURE 6. Liquid water concentration in atmospheric aerosols at Pasadena, September 9, 1972.

virtually unlimited during most of our visit. Within the limits of detection of the waterometer, essentially no water was present. This conclusion was confirmed by measurement of the water in filter-collected aerosols by a gas chromatographic technique [Meyer et al. (8)].

The observations of liquid water in airborne particle samples at Point Arguello were found to be highly variable. There were marked changes in water content, depending on the wind direction and speed, which were related to nearby spray formation in the surf zone. For most of the observations, the fresh sea salt aerosol being produced at the shoreline appeared to have a water content above 50% water by mass.

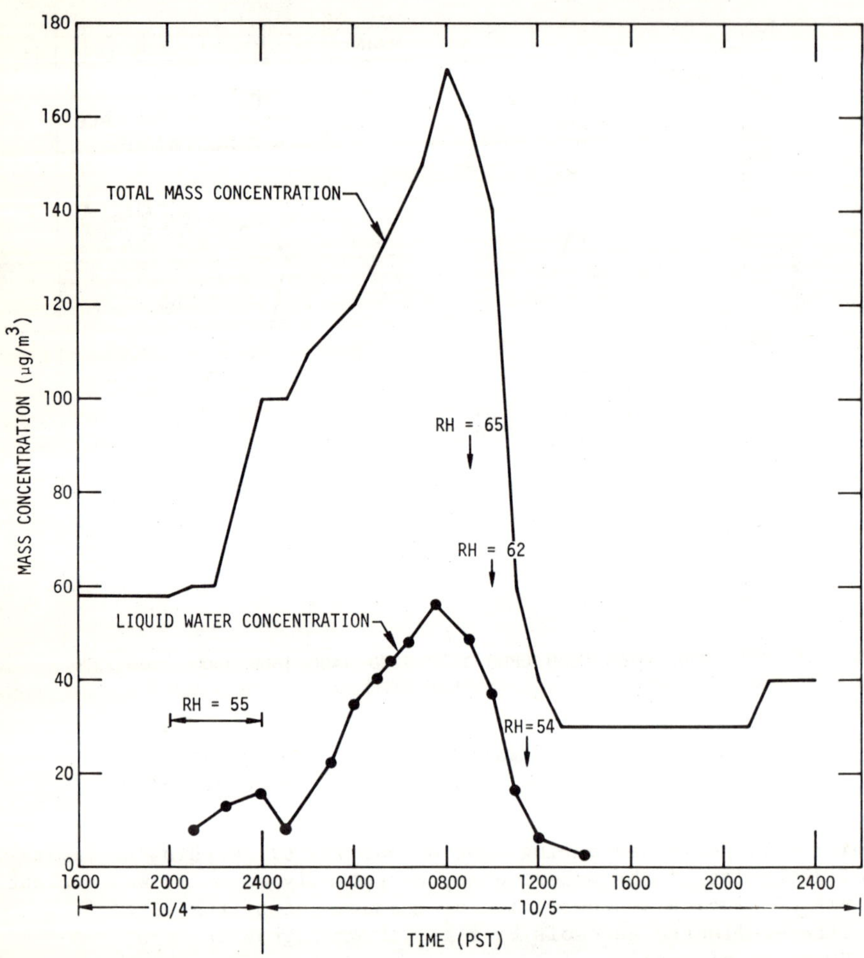

FIGURE 7. Liquid water concentration in atmospheric aerosols at Pomona, October 4-5, 1972.

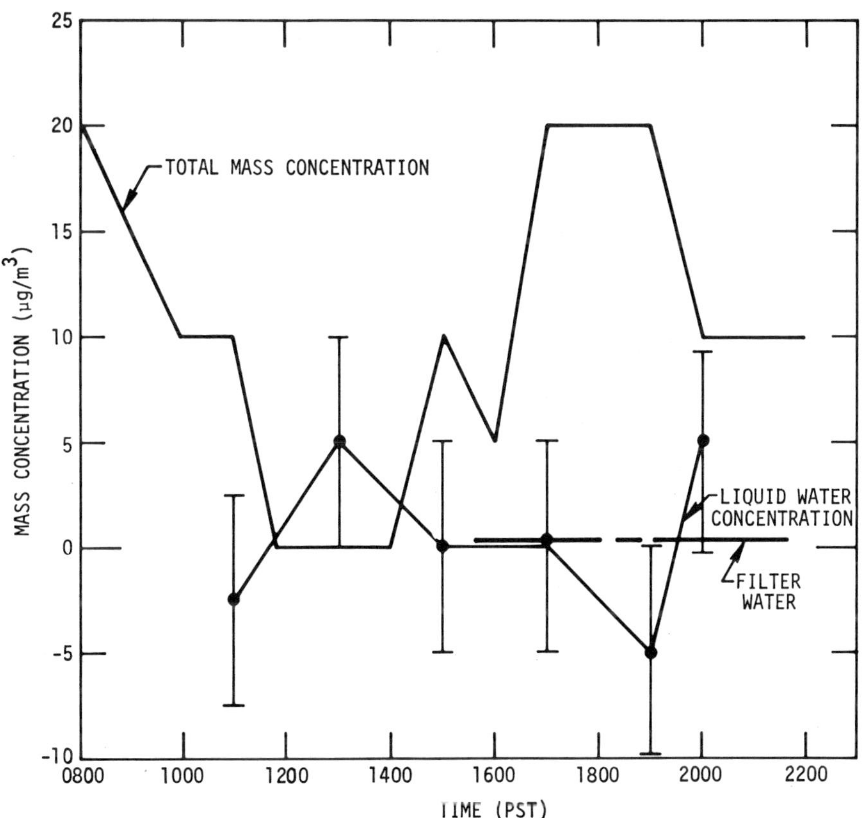

FIGURE 8. Liquid water concentration in atmospheric aerosols at Goldstone Tracking Station, November 2, 1972.

In Figure 9 our measured values of the fractional free liquid water content are plotted as a function of relative humidity and compared with the results of Charlson's Humidified Nephelometer [Whitby et al. (10)]. The data agree fairly well with the humidified nephelometer data except for those associated with very high relative humidity. Our data taken at high relative humidity are not considered reliable due to the previously mentioned difficulty in determining the total aerosol mass as well as the

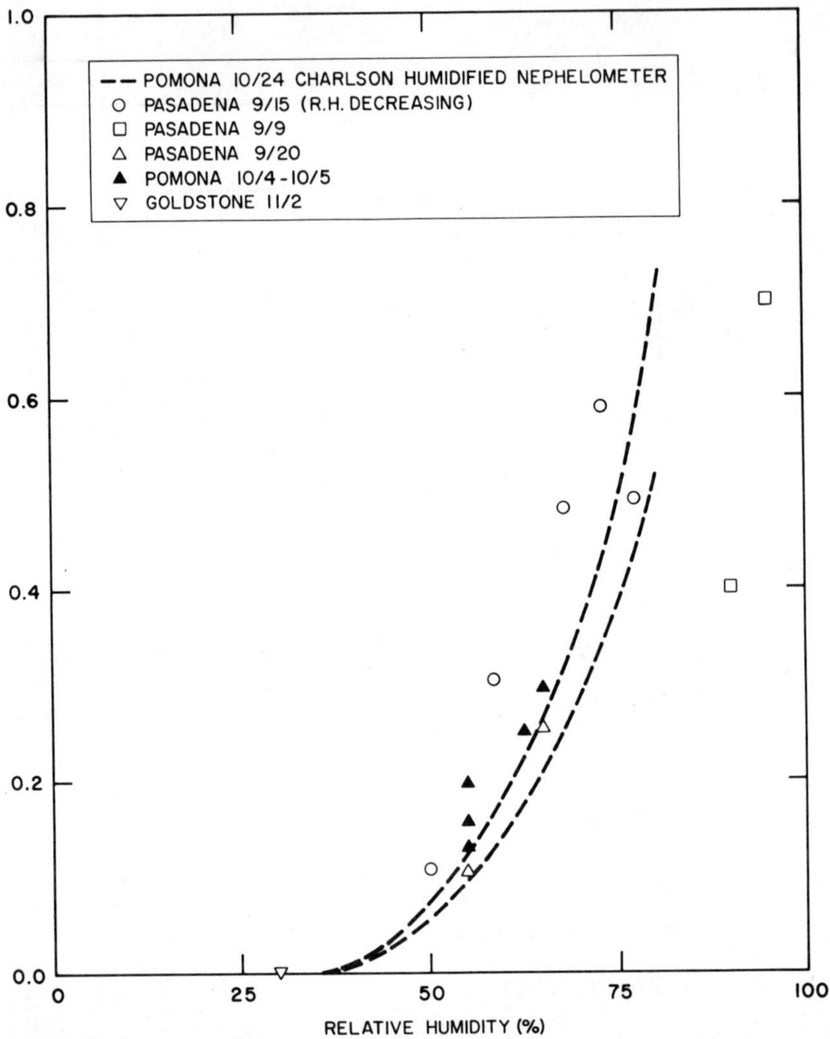

FIGURE 9. Weight fraction of liquid water as a function of relative humidity for the Los Angeles basin area. The data points are results from the present investigation, and the dotted curves are the results obtained by the University of Washington Humidified Nephelometer.

probability of increased water loss in the sampling procedure due to the small pressure drop across the filter. The Pasadena data taken on September 15 under moderate photochemical smog conditions show somewhat higher water content when compared to the rest of the results. This could conceivably be due to differences in the nature of the aerosols present and their hygroscopicity. More likely, it is attributable to the fact that the data were obtained under conditions where the relative humidity decreased fairly rapidly from 77 to 50% during the sampling period, and the aerosols require a fairly long time period before establishing equilibrium with ambient conditions. Similar "hysteresis" effects of this nature have been deduced by Whitby et al. (10) in their study of Los Angeles aerosols with their analysis of the relationship between particle density and changing relative humidity.

Observations during the 1972 study in the Los Angeles area indicated that the $SO_4^=$ and NO_3^- found in the aerosol from 24-hr filter samples could be accounted for stoichiometrically as ammonium salt [Hidy (5)]. Using the measurements of liquid water content, it was found that the salt solutions in the aerosol would be highly concentrated, exceeding 1 molar in many cases. The high concentration of equivalent salt solution would influence the dielectric constant especially in the dielectric loss ε''. However, as discussed in an earlier section, the uncertainty introduced in deducing the liquid water content is comparatively small. The uncertainty in the measurements reported here due to all causes is estimated to be $\pm 30\%$.

The systematic and marked changes in liquid water content with total mass concentration in the aerosols observed in 1972 indicate the importance of water in behavior of the suspended particles. Of particular interest are the diurnal changes found in urban air in the Los Angeles area. At this point it is not certain whether or not the changes in water content are resulting from the advection of more heavily polluted air containing a roughly fixed ratio of water to total mass, or the chemical changes taking place in the aerosol. Other measurements of the chemical composition of aerosols observed at the same time show strong systematic changes in sulfate, nitrate, and organic fraction in the urban aerosol that accompany changes in total mass [Hidy (5)]. The observations of changes in "free" water may be related to the equilibration of the aerosol to moist air as hygroscopic salts such as the nitrates or sulfates are formed by atmospheric chemical transformations. If pure ammonium sulfate is being produced, the substantial absorption of water at relative humidity below 80% cannot be explained. However, if mixtures of ammonium sulfate and nitrate are present, then the aerosols may have much higher affinity for water and therefore are able to retain the water that was absorbed during periods of high relative humidity (usually present during the very early morning hours) for long periods of time even though the relative humidity may decrease substantially during the day (Charlson, private communications, 1973).

A third possibility is that liquid water may actually be produced in chemical reactions on the aerosol particles. Although there are no known inorganic reactions involving atmospheric constituents that can be invoked to rationalize this, there are organic reactions of the decomposition of unstable intermediates from ozone-olefin reactions that could generate liquid water in the airborne particles. This may be of particular interest with the very high organic content found in the Los Angeles aerosol.

There has been indirect evidence available for many years that liquid water in aerosols is a major factor in visibility reduction, particularly at relative humidities >70%. A measure of visibility is generally accepted as being related to the extinction coefficient for visible light, measured by an integrating nephelometer (b_{scat}). To demonstrate the relation between b_{scat} and liquid water content, our observations have been plotted in Figure 10.

Within the experimental error of the measurements ($\pm$30% fractional error) there is a linear relation between b_{scat} and liquid water content. The correlation holds to the Rayleigh light scattering limit, exemplified by the observations at the desert site in Goldstone. This strong relationship over the range of relative humidity of ∿40 to 70% indicates directly, for the first time, the significance of liquid water content to the optical properties of atmospheric aerosols.

It is interesting that the correlation shown in Figure 10 is analogous to the correlation of b_{scat} and total mass concentration. Using the slope of the line in this figure and comparing it with the slope of the b_{scat} versus total mass concentration correlation, one finds qualitatively that the water content in the Los Angeles aerosols is roughly balanced at 30% by mass over the 40 to 70% relative humidity range. This fraction of free water appears to be consistent with estimates from humidified nephelometry observations using the method of Covert et al. (1).

ACKNOWLEDGMENTS

We are indebted to the California Air Resources Board for partial sponsorship of this work under Contract No. ARB-358.

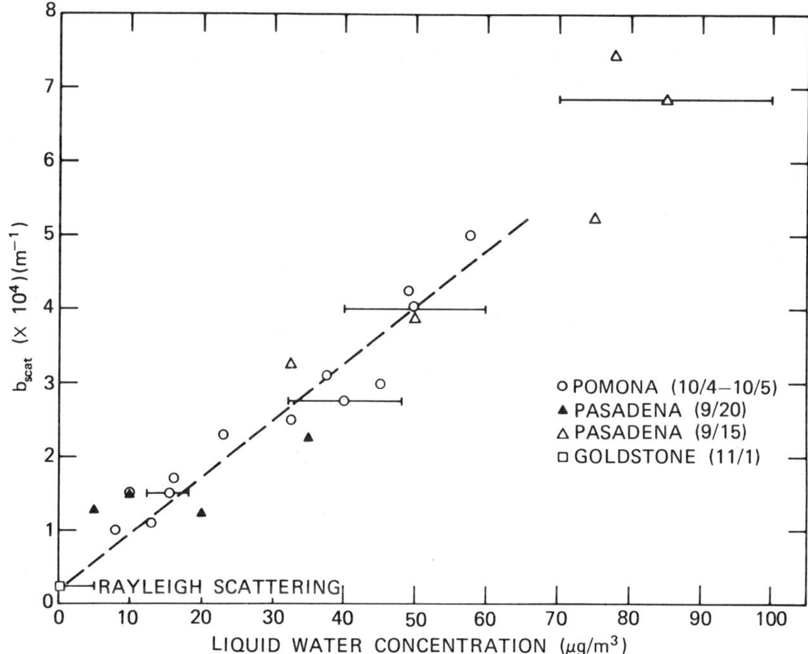

FIGURE 10. Comparison of light-scattering coefficient with the liquid water concentration in atmospheric aerosols.

REFERENCES

1. Covert, D. S., Charlson, R. J., and Ahlquist, N. C. 1972. A Study of the Relationship of Chemical Composition and Humidity to Light Scattering by Aerosols, J. Appl. Meteor. 11, 968-976.

2. Dresia, Von H., Fischötter, Peter, and Feldon, Gerd. 1964. Kontinuierliches Messen des Staubgehaltes in Luft und Abgasen mit Beta Strahlen, V. Deut. Ing. Z. 106, 1191-1195.

3. Goetz, A. and Pueschel, R. 1967. Basic Mechanisms of Photochemical Aerosol Formation, Atmos. Environ. 1, 287-306.

4. Hänel, G. 1970. The Size of Atmospheric Aerosol Particles as a Function of Relative Humidity, Beitr. Phys. Atmos. 43, 119-132.

5. Hidy, G. M. 1973. "Characterization of Aerosols in California," Interim Report for Phase I covering the period October 25, 1971 to April 1, 1973, Rept. SC524.191R, ARB Contract 358, Rockwell International Science Center.

6. Ho, W. and Hall, W. F. 1973. Measurement of the Dielectric Properties of Sea Water and NaCl Solutions at 2.65 GHz, J. Geophys. Res. 78, 6301-6315.

7. Landis, D. A. 1975. "Comments on Calibration of a FH62A Dust Monitor," Atmos. Environ. 9, 1079-1081.

8. Meyer, R. A., Hidy, G. M., and Davis, J. H. 1973. Determination of Water and Volatile Organics in Filter Collected Aerosols, Environ. Lett. 4, 9-20.

9. Miller, M. S., Friedlander, S. K., and Hidy, G. M. 1972. A Chemical Element Balance for the Pasadena Aerosol, J. Colloid Interface Sci. 39, 165-176.

10. Whitby, K. T., Clark, W. E., Marple, V. A., Sverdrup, G. M., Willeke, K., Liu, B. Y., and Pui, D. 1973. "Evolution of the Freeway Aerosols," Rept. No. 209, Particle Technology Lab., University of Minnesota.

Analysis of Particulate Organic Air Pollutants by High-Resolution Mass Spectrometry

RICHARD L. KNIGHTS
Chemistry Department
University of Washington
Seattle, Washington 98195

I. INTRODUCTION

High resolution mass spectrometric (HRMS) analysis of air pollutants was begun in this laboratory in 1970 (1), and results on a limited number of samples were described over the next few years (2, 3). The Aerosol Characterization Experiment (ACHEX) project provided the first attempts at analyzing samples collected for 2-hr periods during the day, in order to follow diurnal variations throughout the day. Tables 1 to 3 present concentrations of organics found in the following samples:

TABLE 1. Identification of Particulate Air Pollutants by High-Resolution Mass Spectrometry

Identifying fragment		Compound name[a]	Relationship to molecular mass[b]	Relative sensitivity factor[c]	Maximum concentration (μg/m^3)[d]
Mass	Formula				
	Hydrocarbons				
71.0860	C_5H_{11}	Total alkanes	M-R	2.5(11)	20.
69.0704	C_5H_9	Total alkenes	M-R	2.5(11)	25.
91.0548	C_7H_7	Monosubst. benzenes	M-R?	17.5	
106.0782	C_8H_{10}	Xylenes, ethylbenzene	M	8.0	
117.0704	C_9H_9	Alkylstyrenes	M-R	7.0	
(103.0546	C_8H_7)	(Substituted styrene)	M-CH_2R	(1.9)	(0.83)
119.0860	C_9H_{11}	Alkyl(C4+)benzenes[e]	M-R	10(2)	0.10
120.0938	C_9H_{12}	Alkyl(C3)benzenes	M	7.8	
134.1094	$C_{10}H_{14}$	Alkyl(C4)benzenes	M	4.2	
148.1252	$C_{11}H_{16}$	Pentamethylbenzene	M	16	0.08
	Polycyclic Aromatic Hydrocarbons				
104.0626	C_8H_8	Tetrahydronaphthalene	M-C_2H_4	6.8	
118.0782	C_9H_{10}	Indan	M	8.4	
128.0626	$C_{10}H_8$	Naphthalene	M	20	0.03
138.1408	$C_{10}H_{18}$	Decahydronaphthalene	M	3.2	
142.0782	$C_{11}H_{10}$	Methylnaphthalenes	M	11	0.01
145.1018	$C_{11}H_{13}$	Dimethyltetrahydro-naphthalene	M-CH_3	8.8	
178.0782	$C_{14}H_{10}$	Anthracene + phenanthrene	M	29	0.015
202.0782	$C_{16}H_{10}$	Pyrene + isomers	M	7.1	
	Oxygenated Aromatics				
93.0340	C_6H_5O	Phenol + NO_2, CHO, COOH? (Mono R-subst. phenols)	M-X	3(0.6)	5.8

(continued)

TABLE 1 (cont.)

Identifying fragment		Compound name[a]	Relationship to molecular mass[b]	Relative sensitivity factor[c]	Maximum concentration ($\mu g/m^3$)[d]
Mass	Formula				
94.0418	C_6H_6O	Phenol	M	10	0.70
110.0368	$C_6H_6O_2$	Dihydroxybenzenes	M	8.4(3)	
108.0576	C_7H_8O	Hydroxytoluene + benzyl alcohol	M	16	0.79
149.0238	$C_8H_5O_3$	Phthalates	M-ROR + H	1	
136.0888	$C_9H_{12}O$	Trimethylphenol + isomers	M	3.5(16)	0.16
	Carboxylic Acids				
60.0210	$C_2H_4O_2$	Total carboxylic acid	M-R+H?	5	3.0
73.0290	$C_3H_5O_2$	Monocarboxylic acids	M-R?	2.0	
86.0368	$C_4H_6O_2$	Pentanedioic acid	$M-CH_2O_2$	0.45(2)	0.64
100.0524	$C_5H_8O_2$	Hexanedioic acid	$M-CH_2O_2$	0.38	1.6
114.0680	$C_6H_{10}O_2$	Heptanedioic acid	$M-CH_2O_2$	1.3(0.45)	0.42
		(Methylhexanedioic acid)	$M-CH_2O_2$	0.45	
105.0340	C_7H_5O	Benzoyl ion		4	
122.0366	$C_7H_6O_2$	Benzoic acid	M	8(3.3)	1.2
136.0524	$C_8H_8O_2$	Methylbenzoic isomers	M	3.5(2)	0.55
150.0680	$C_9H_{10}O_2$	Ethylbenzoic isomers	M	3.5	0.20
164.0836	$C_{10}H_{12}O_2$	Trimethylbenzoics	M	3.5(1.9)	0.37
	Nitrogen Compounds				
43.0058	CONH	Total amides + others		10(1)	0.6
44.0136	$CONH_2$	Amides		22	
40.0186	CH_2CN	Alkylnitriles		1(7)	1.8

(continued)

TABLE 1 (cont.)

Identifying fragment		Compound name[a]	Relationship to molecular Mass[b]	Relative sensitivity factor[c]	Maximum concentration ($\mu g/m^3$)[d]
Mass	Formula				
153.0426	$C_7H_7NO_3$	Hydroxynitrotoluene	M	10(15)	0.01
169.0374	$C_7H_7NO_4$	Toluene oxid. prod.	M	10(15)	0.01
129.0578	C_9H_7N	Quinoline?	M	13	0.06
167.0734	$C_{12}H_9N$	Carbazole	M	12(9)	0.005
179.0734	$C_{13}H_9N$	Acridine + isomers	M	9	0.003
	Secondary Oxidized Compounds				
116.0472	$C_5H_8O_3$	1-Heptene oxid. prod.?	M?	1(2)	1.1
118.0630	$C_5H_{10}O_3$	5-Hydroxypentanoic acid?	M?	1	0.4
112.0524	$C_6H_8O_2$	1-Heptene oxid. prod.?	M?	1(2)	3.
126.0680	$C_7H_{10}O_2$	1-Heptene oxid. prod.?	M?	1(2)	1.
128.0838	$C_7H_{12}O_2$	1-Heptene oxid. prod.?	M?	1(2)	0.9
144.0726	$C_7H_{12}O_3$	?	M?	1	0.4
147.0532	$C_5H_9NO_4$	$HCO(CH_2)_4ONO_2$ + isomers	M	1	
163.0480	$C_5H_9NO_5$	$HOOC(CH_2)_4ONO_2$	M	1	0.2
161.0688	$C_6H_{11}NO_4$	$HCO(CH_2)_5ONO_2$ + isomers	M	1	
177.0636	$C_6H_{11}NO_5$	$HOOC(CH_2)_5ONO_2$	M	1	0.4
189.0636	$C_7H_{11}NO_5$	$HCO(CH_2)_5COONO_2$ + isomers	M	1	
175.0844	$C_7H_{13}NO_4$	$HCO(CH_2)_6ONO_2$ + isomers	M	1	
191.0792	$C_7H_{13}NO_5$	$HOOC(CH_2)_6ONO_2$	M	1	
	Tentative Terpene Products				
154.0994	$C_9H_{14}O_2$	Terpene oxid. prod.?	M	1(0.16)	1.
182.0942	$C_{10}H_{14}O_3$	Terpene oxid. prod.?	M	1(0.16)	0.5
168.1150	$C_{10}H_{16}O_2$	Terpene oxid. prod.?	M	1(0.16)	0.5
184.1098	$C_{10}H_{16}O_3$	Pinonic acid?	M	1(0.16)	0.4

(continued)

TABLE 1 (cont.)

Identifying fragment					
Mass	Formula	Compound name[a]	Relationship to molecular mass[b]	Relative sensitivity factor[c]	Maximum concentration ($\mu g/m^3$)[d]
	Inorganics				
47.9670	SO	Volatile sulfates		1.2	70.
35.9766	HCl	Ammonium chloride	$M-NH_3$	2(6)	1.6
45.9928	NO_2	Ammonium nitrate	$M-NH_4O$	0.57(3)	4.6

[a] Updated 1978 names and values listed, with older 1975 ones used for original ACHEX analyses in parentheses. To calculate a more accurate concentration from older ACHEX alkane data, for example, multiply by 4.5 (= 11/2.5).

[b] M means molecular ion; M-R is loss of alkyl group; and so on.

[c] Relative sensitivity factor is the response of a compound versus quantitative standard orthobromobenzoic acid.

[d] Maxumum concentration found in any 2-hr sample from West Covina on July 24, 1973, the heaviest smog episode analyzed.

[e] Alkylbenzene value formerly reported was based upon C_9H_{11}.

TABLE 2. Concentrations ($\mu g/m^3$) of Air Pollutants by HRMS, Samples at West Covina, California, on July 23-24, 1973

July 24 Time (PDT)	Hydrocarbons							
	C_5H_{11} alkanes	C_5H_9 alkenes	C_7H_7 substituted benzenes	C_8H_{10} xylenes, ethylbenzenes	C_9H_{11} alkyl(C4+)-benzenes	$C_{10}H_{14}$ alkyl(C4)-benzenes	$C_{10}H_8$ naphthalene	$C_{11}H_{10}$ methyl-naphthalene
2220–0220	4.0	4.9	0.53	0.16	0.34	0.18	0.07	0.022
0220–0720	8.9	8.5	0.73	0.30	0.53	0.35	0.14	0.13
0720–0920	13.8	13.8	1.0	0.38	0.64	0.36	0.20	0.14
0920–1120	9.3	12.5	1.0	0.22	0.65	0.39	0.18	0.078
1120–1320	9.8	8.9	1.1	0.56	0.65	0.25	0.14	0.063
1320–1520	10.2	13.4	1.7	0.42	0.67	0.68	0.32	0.23
1520–1720	20.0	24.5	2.0	0.82	1.08	0.58	0.32	0.18
1720–1920	13.8	16.9	1.6	0.67	0.96	0.35	0.36	0.089
1920–2220	7.1	9.8	0.92	0.16	0.42	0.08	0.19	0.035

(continued)

TABLE 2 (cont.)

July 24 Time (PDT)	Oxygenated Aromatics						
	$C_8H_5O_3$ phthalates	C_6H_5O phenol + NO_2, CHO, COOH?	C_6H_6O Phenol	$C_6H_6O_2$ dihydroxy-benzenes	C_7H_6O benzaldehyde + interference	C_7H_8O hydroxytoluene + benzyl alcohol	$C_9H_{12}O$ trimethylphenol
2220–0220	3.0	0.28	0.06	0.10	0.039	0.055	0.21
0220–0720	10.0	0.35	0.11	0.11	0.036	0.076	0.13
0720–0920	5.1	0.79	0.14	0.17	0.011	0.11	0.19
0920–1120	4.5	0.45	0.23	0.27	0.065	0.21	0.22
1120–1320	7.6	0.92	0.23	0.46	0.12	0.23	0.42
1320–1520	7.8	0.98	0.33	0.65	0.047	0.12	0.74
1520–1720	8.8	1.2	0.35	0.58	0.27	0.35	0.19
1720–1920	7.4	0.60	0.20	0.53	0.059	0.23	0.15
1920–2220	5.3	0.30	0.15	0.15	0.078	0.087	0.15

(continued)

TABLE 2 (cont.)

July 24 Time (PDT)	Carboxylic Acids							
	$C_4H_6O_2$ pentanedioic acid	$C_5H_8O_2$ hexanedioic acid	$C_6H_{10}O_2$ heptanedioic acid	C_7H_5O benzoyl ion	$C_7H_6O_2$ benzoic acid	$C_8H_8O_2$ methylbenzoic + phenylacetic	$C_9H_{10}O_2$ ethylbenzoic + phenylpropanoic	$C_{10}H_{12}O_2$ trimethylbenzoic + phenylbutanoic
2220-0220	1.2	2.0	0.39	0.37	0.017	0.042	0.045	0.049
0220-0720	0.61	1.5	0.12	0.58	0.051	0.038	0.087	--
0720-0920	1.0	2.3	0.60	0.72	0.092	0.15	0.022	0.066
0920-1120	3.2	5.7	1.19	1.3	0.18	0.22	0.15	0.024
1120-1320	2.8	5.6	0.98	1.0	0.26	0.24	0.056	0.18
1320-1520	2.9	8.3	1.9	1.4	0.06	0.36	0.17	0.14
1520-1720	4.7	8.5	1.4	1.4	0.32	0.21	0.054	0.045
1720-1920	4.5	7.6	1.16	1.3	0.19	0.26	0.092	0.061
1920-2220	3.4	3.1	0.74	0.98	0.10	0.040	0.026	0.024

(continued)

TABLE 2 (cont.)

July 24 Time (PDT)	Nitrogen Compounds				Possible 1-Heptene Products			
	CONH amides + others	CH_2CN alkyl-nitriles	$C_7H_7NO_4$ toluene oxidn. prod.	C_9H_7N quinoline?	$C_5H_8O_3$ 1-heptene oxidn. prod.?	$C_6H_8O_2$ 1-heptene oxidn. prod.?	$C_7H_{10}O_2$ 1-heptene oxidn. prod.?	$C_7H_{12}O_2$ 1-heptene oxidn. prod.?
2220-0220	0.42	0.78	0.010	0.004	0.087	0.75	0.037	0.043
0220-0720	0.54	1.4	0.005	--	0.29	0.56	0.20	0.060
0720-0920	0.65	1.2	--	0.006	0.58	0.62	0.21	0.32
0920-1120	0.99	1.5	--	0.010	0.47	2.3	0.23	0.21
1120-1320	1.2	1.7	0.008	0.017	--	2.7	1.9	0.15
1320-1520	2.3	4.2	0.037	0.033	2.0	3.1	0.96	1.6
1520-1720	1.3	3.1	0.038	0.084	1.2	5.2	2.0	1.6
1720-1920	0.84	2.6	0.015	0.022	0.46	2.7	0.46	0.15
1920-2220	0.74	2.1	--	0.054	0.62	1.2	0.83	0.97

TABLE 3. Concentrations ($\mu g/m^3$) of Air Pollutants by HRMS, Sampled at Pomona, California, on August 16-17, 1973, and Dominguez Hills, California, on October 4-5, 1973

Pomona August 17 Time (PDT)	C_5H_{11} alkanes	C_5H_9 alkenes	C_7H_7 substituted benzenes	$C_{10}H_8$ naphthalene	$C_8H_5O_3$ phthalates	$C_7H_6O_2$ benzoic acid	$C_4H_6O_2$ pentanedioic acid	$C_5H_8O_2$ hexanedioic acid	CONH amides + others
2110-0120	3.4	2.0	0.20	0.020	1.4	0.75	0.13	1.0	0.67
0120-0620	7.6	1.6	0.17	0.018	2.2	0.66	0.33	0.88	0.21
0620-0910	20.9	28.9	0.43	--	3.7	1.1	1.1	--	0.34
0910-1110	14.7	30.7	0.39	0.023	3.5	1.1	0.64	0.57	0.083
1110-1310	16.9	32.9	0.51	0.085	6.7	1.2	--	--	0.35
1310-1510	12.5	20.0	0.30	0.021	6.4	0.94	--	2.4	0.25
1510-1710	9.3	26.8	0.38	0.025	2.9	0.92	0.32	1.4	0.46
1710-1910	9.8	23.6	0.37	0.011	2.7	0.77	1.5	1.2	0.15
1910-2200	14.2	18.7	0.29	0.048	2.1	0.89	0.74	1.1	0.41
Dominguez Hills October 5									
2300-2300	1.8	2.7	0.13	0.014	2.0	0.17	2.2	3.7	0.11
1210-1410	2.8	3.1	0.20	0.039	1.5	0.74	0.30	0.36	0.16
1410-1610	7.1	6.7	0.39	0.010	3.4	1.2	0.66	0.67	0.24.

Table	Location	Dates	Number of samples
1, 2	West Covina	1973: July 23-24	9
3	Pomona	1973: August 16-17	9
3	Dominguez Hills	1973: October 4-5	3

Conclusions from these HRMS data have been summarized (4) and included in a later chapter of this volume. The ACHEX data have been analyzed in more detail (5, 7) and compared with those for other locations within the United States (5). Results using HRMS have also been reported for more recent samples from the Los Angeles area (6).

II. GENERAL METHODOLOGY

Samples of air pollution aerosol particles were collected on 405 cm^2 of 8 by 10 in. high-volume (at ~80 m^3/hr) glass fiber filters. This complex mixture of organic and inorganic compounds was gradually heated inside the mass spectrometer vacuum system from room temperature to 380°C, at 30°/min. Each compound was thus vaporized at its characteristic temperature range for ionization and fragmentation, and mass analysis. Repetitive scanning of the mass spectrum was necessary for quantitative analysis by following the identifying fragments of the compounds during the time required to vaporize completely, about 30 min.

III. MASS SPECTROMETER

The mass spectrometer used in this work was an AEI (Kratos) MS9, a high resolution, double focusing instrument of Nier-Johnson geometry, using separate electrostatic and magnetic analyzers. A high-resolution instrument was needed to determine the elemental composition of each mass fragment by separating the multiple peaks commonly encountered at each unit mass.

Sample molecules entered the MS9 ion source chamber as gases volatilized from the direct insertion probe. Perfluorokerosene (high boiling) internal mass standard leaked in constantly from a 1-liter ballast volume. The sample molecules were fragmented and ionized by bombardment with a 300-μA beam of electrons of 70-eV energy. The ions of positive charge were accelerated by a potential of 8000 V out of the ion chamber and toward a series of focusing plates and a variable "source slit."

Ions passing through the source slit were separated according to energy in the electrostatic analyzer section and then according to momentum in the magnetic analyzer section. The magnet field strength was variable over the range necessary to focus at the collector slit ions of desired mass (more precisely, of mass/charge ratio m/e, but predominantly of unit charge). Mass spectra were recorded over the mass range 500 to 27. The presence of a beam of ions of a certain mass was detected by passing a focused beam through the collector slit to an electron multiplier, which detected an "ion current" signal proportional to the abundance of that mass fragment. The signal was then amplified with variable multiplier gain before being sent to the computer.

The magnet field strength was decreased logarithmically with time to generate an ion current signal output that was a spectrum of the mass fragments produced by the sample molecules. The magnet was cycled to generate repeated spectra. The experimental results were the changing intensities of masses with time and with temperature (mass thermograms).

A special internally heated sample insertion probe (1, 8) was designed to press one or more 7-mm-diameter punches from the glass fiber filter sample between gold plates. After insertion into the vacuum system, the probe heater volatilized the sample components through a glass capillary into the MS9 ion source chamber.

IV. COMPUTER SYSTEM

A Digital Equipment Corporation PDP-12 laboratory computer controlled the mass spectrometer. Most interaction between computer and operator used the display scope, with operator input by keyboard. Hard copy was output on a Versatec Matrix printer/plotter. Two magnetic tape transports were required for data storage and for program saving and loading. A high-speed disk unit was used for data reduction. A crystal-controlled real-time clock provided a programmable time base.

The PDP-12 used here had a core memory of 8000 12-bit words. Programs required the entire 8000 core for operating instructions and data storage.

V. SAMPLE ANALYSIS

Filter punches of 7-mm diameter were stored under dry ice until just before analysis, when the individual sample container envelope was warmed to room temperature. An aliquot of 1 to 8 μg of the internal quantitative standard, orthobromobenzoic acid (BBA) in methanol, was syringed onto the back of a filter punch. Up to nine punches were then loaded between gold disks and pressed against the heated surface by screwing on the copper end cap. The stainless steel shaft was passed through the fluorocarbon sliding vacuum seal until the glass tip just entered the ion source chamber. The filter was heated at 30°/min from room temperature to 380°C over 15 min. and then held at that temperature for 15 min.

After analysis the probe was cooled to 100°C before removal from the MS9, and then just below room temperature before loading the next sample.

VI. COMPOUND IDENTIFICATION AND QUANTITATION

The compounds most commonly found in urban aerosol samples were identified by the parameters summarized in Table 1. Each reported compound was identified by a unique elemental formula with exact mass representing the molecule or a characteristic fragment. Although HRMS provides an unambiguous indication of the elemental composition of a mass fragment, the possibility always exists of other isomeric compounds contributing to the concentrations reported. For example, the fragment C_5H_9 comes from any alkenes of five or more carbon atoms, and C_9H_{11} can result from any alkylbenzenes having four or more carbons on the alkyl group or groups.

The 30 mass spectra were searched for masses within 0.003 mass units of an identifying fragment, to generate a mass thermogram for each compound. The area under the thermogram (the integrated peak intensities) was proportional to the amount of compound on the filter. The concentration in the air sampled was calculated as follows:

$$\text{Concentration in the air } (\mu g/m^3) = \frac{Ax}{As} \cdot \frac{Ws}{Rx\ F\ V}$$

where Ax is the area under mass thermogram for identifying fragment of compound x, A_s is the area under mass thermogram from BBA, W_s is weight of added internal quantitative standard BBA, R_x is response (sensitivity) factor of compounds relative to BBA = Ax/Wx·Ws/As, F is fractional aliquot of filter used for analysis, and V is volume (m^3) of air sampled through filter.

The largest source of quantitative error was the response factor used to calculate the concentration of each compound. Some factors were the averages of standard runs, but others were literature values or estimates from similar compounds of known responses. A large systematic error from an incorrect response factor would change the reported concentrations, but would not affect relative comparisons among samples. A footnote to Table 1 describes how to calculate a probably more accurate concentration in a case where a response factor has been changed in the light of recent experience.

Repeatability was checked by comparing results of analyses run on different sections of the filter on different days. The coefficient of variation was about 50%, essentially the same for either the half of the concentrations greater than 0.1 $\mu g/m^3$ or the half less than that.

ACKNOWLEDGMENTS

Dennis Schuetzle developed the general MS/computer methodology in this laboratory, obtained Environmental Protection Agency (EPA) support for its development, and is much appreciated for encouraging my entry into the field. Dagmar R. Cronn provided valuable enthusiasm and cooperation in data interpretation. A. L. Crittenden was the principal investigator who tirelessly updated computer software and carried out the monumental task of interpreting the ACHEX and other data into a final EPA report (5). Michael Clancy Higgins was a close friend at the controls of the temperamental MS9 and the frustrating computer. Robert J. Charlson provided useful guidance in air chemistry and its importance as a field of research. Bruce R. Appel provided the samples and financial assistance in processing the data; his patience and interpretation of the results were essential. The U.S. Environmental Protection Agency funded three years of developing HRMS as a viable analytical method. We gratefully acknowledge the assistance of W. E. Wilson and Project Officer R. K. Patterson.

REFERENCES

1. Schuetzle, D. 1972. "Computer Controlled High Resolution Mass Spectrometric Analysis of Air Pollutants," Thesis, University of Washington.

2. Schuetzle, D., Crittenden, A. L., Charlson, R. J. 1973. Application of Computer Controlled High Resolution Mass Spectrometry to the Analysis of Air Pollutants, J. Air Pollut. Control Assoc. 23, 704.

3. Schuetzle, D., Cronn, D. R., Crittenden, A. L., and Charlson, R. J. 1975. Molecular Composition of Secondary Aerosol and Its Possible Origin, Environ. Sci. Technol. 9, 838.

4. Appel, B. R., Wall, S. W., and Knights, R. L. 1975. Characterization of Carbonaceous Materials in Atmospheric Aerosols by High-Resolution Mass Spectrometric Thermal Analysis, this volume. First presented at the Pacific Conference on Chemistry and Spectroscopy, Los Angeles, October.

5. Crittenden, A. L. 1976. "Analysis of Atmospheric Aerosols by Mass Spectrometry," Final Report (Grant No. R 801119), EPA-600/3-76-093, August.

6. Appel, B. R., Hoffer, E. M., Kothny, E. L., Wall, S. M., Haik, M., Knights, R. L. 1979. Analysis of Carbonaceous Material in Southern California Atmospheric Aerosols. 2., Environ. Sci. Technol. 13, 98.

7. Cronn, D. R., Charlson, R. J., Knights, R. L., Crittenden, A. L., and Appel, B. R. 1977. A Survey of the Molecular Nature of Primary and Secondary Components of Particulates in Urban Air by HRMS, Atmos. Environ. 11, 929.

8. Knights, R. L. 1973. Thesis, University of Washington.

PART II

CHEMICAL CHARACTERIZATION

Sources and Elemental Composition of Aerosol in Pasadena, California, by Energy-Dispersive X-Ray Fluorescence*

ROBERT H. HAMMERLE and WILLIAM R. PIERSON
Ford Motor Company
Dearborn, Michigan

---

*Reprinted from Environmental Science & Technology, Vol. 9, page 1058, November 1975.

Abstract

In the fall of 1972 we performed a 29-day study to identify sources of atmospheric particulate matter and assess their contributions to the aerosol in Pasadena, California. Size-fractionated and unfractionated filter samples were collected automatically and were analyzed by an automatic X-ray fluorescence spectrometer. Nine elements proved consistently measurable with reasonable accuracy: Ca, Ti, V, Mn, Fe, Ni, Zn, Br, and Pb. All show orders-of-magnitude fluctuations, with a tendency to follow diurnal patterns related to meteorological factors. Most of the elements can be classified as small-particle elements (Ni, Zn, Br, Pb) or large-particle elements (Ca, Ti, Mn, Fe) with V falling in an intermediate size range. Statistical snalysis of interelement correlations and size distributions indicates that gasoline engine exhaust is the source of the Br and Pb; soil from the basin is the main source of the Ti, Mn and Fe; much of the Ca probably is from cement dust contamination of the soil, and the rest of the Ca is indigenous to the soil. Proportionality and high correlation coefficients ($\sim$0.8) characterize elements from the same sources and correlation coefficients $\sim$0.4 characterize unrelated elements. Using appropriate gravimetric factors, we estimate that the soil contributed 8 $\mu g/m^3$ and the primary exhaust particulate from gasoline engines contributed 5 $\mu g/m^3$, on the average, to the aerosol mass at the Pasadena site during the 29-day period.

I. INTRODUCTION

There is considerable interest in identifying the origins of the atmospheric aerosol. With this as a major objective, a coordinated effort by many investigators was initiated in 1969 in the Los Angeles basin. The results of that study [reported in a series of papers in J. Colloid Interface Sci. 39, 136-304 (1972)] stimulated planning in 1970 for the large-scale California Aerosol Characterization Experiment (ACHEX) (1,2), the observational phase of which was conducted in Pasadena and other sites in August through late October 1972.

During planning of ACHEX, a major difficulty was recognized to be the lack of a means of obtaining chemical information on the aerosol with good time resolution. We believed that X-ray fluorescence on-line with a Si(Li) semiconductor detector could yield elemental analyses for a number of elements concurrently in the aerosol on a reasonably fast time scale. An apparatus embodying this idea was designed and built at the University of

California Lawrence Berkeley Laboratory by Jaklevic et al. (3,4) under the auspices of the U.S. Environmental Protection Agency, with the intent that its first utilization would be in the Pasadena ACHEX, prior to final modifications by LBL and subsequent transfer to EPA.

The present paper reports the results obtained in Pasadena with this system. It represents the first extensive series of measurements, with prompt analysis and readout in the field, of the elemental composition of particulate matter with an automatic X-ray fluorescence spectrometer. It demonstrates not only the value of the superior time resolution but also the advantage offered by the rapid response in permitting decisions to be made during the course of the experiment. Technical difficulties delayed commencement of measurements until the end of October 1972. By this time, the main part of the program involving most of the other ACHEX participants had been truncated owing to unseasonably smog-free weather, and therefore we are able to relate our results to other properties of the basin atmosphere to a less extent than had been anticipated.

Whitby et al. (5) report that the size distribution of aerosols often has two modes separated by a particle-deficient region at approximately 1 to 2 μm. This bimodality may be the result of different formation processes for the two size modes. Any attempt to identify aerosol sources by means of elemental composition should therefore include measurements of composition as a function of particle size. This was done in the present study by measuring the composition of the undifferentiated aerosol and the portion below 1.5 μm concurrently.

Friedlander (6) gives a good description of the basic theory for the use of elemental composition to identify sources. The approach is to solve the material-balance equations for each element measured. For example, the Ca in the aerosol can be written as

$$\mathrm{Ca}(\mu g/m^3) = \sum_i \alpha_i C_i \tag{1}$$

where α_i is Ca content (fraction, by weight) in the particulate matter originating from the ith source, and C_i is concentration ($\mu g/m^3$) of particulate matter sampled that can be attributed to the ith source, subject to the constraint

$$\mathrm{Total}\ (\mu g/m^3) = \sum_i C_i \tag{2}$$

The summations over i are meant to represent summations over all sources, such as soil, sea salt, gasoline engines, oil refining operations, and so forth. The assumption is made that for certain trace species the concentration in the aerosol emitted from a given source is constant in time, both at the source and during transport through the atmosphere. This assumption can be tested from the elemental composition of the many aerosol samples collected sequentially at one location.

In the present study, elemental composition of size-fractionated and unfractionated aerosol was measured at one Pasadena site for a series of 2-hr periods spanning 29 days. Sources are identified and variations in the elemental composition and size distribution of the particulate matter coming from the sources are described statistically. Contributions of the identified sources to the atmospheric particulate loading are deduced.

A preliminary report on the present study has been given elsewhere (7).

II. MODEL

Our results were analyzed in terms of a "single-source model". Let us imagine that, of the many sources of particulate matter, there is only one source which emits elements X and Y. Let us imagine further that the elemental composition of the particulate matter from this latter source remains constant in time, both as it is emitted and as it moves through the atmosphere. Then it follows that the concentrations, X and Y, in the gross particulate matter downwind of the source are always proportional to one another. If a large number of samples are taken at a downwind site and analyzed for X and Y, the best least-squares line $Y = A + BX$ should have the following properties:

(1) The intercept, A, should be zero.

(2) The slope, B, should be equal to the Y/X ratio characteristic of the source.

(3) The variation of Y about the least-squares line should be approximately equal to the error in the measurement of Y, and similarly for X.

(4) The correlation coefficient should approach unity.

The third condition follows from the assumption that the elemental composition of the aerosol is constant, because errors in the measurement of Y and X are then the only sources of variation about the line. The best least-squares line is generally not the regression line, since both Y and X are subject to measurement errors, and also since, as explained by Kendall and

Stuart (8, pp. 375,418), regression is not the correct mode of treatment when a functional relationship between the variables (here, X and Y) is expected.

Let us assume that the concentrations of elements X and Y in the aerosol particles are apportioned among the various particle-size ranges in a manner that remains constant in time and is the same as that found at the source. In that case, the least-squares lines

$$Y(r < r') = A_Y + B_Y Y(\text{total}) \tag{3}$$

and

$$X(r < r') = A_X + B_X X(\text{total}) \tag{4}$$

should each satisfy criteria similar to those listed above for the interelement data.

There may be many atmospheric aerosol systems to which this simple model is inapplicable, most likely owing to multiple sources for the same element, transfrer of material between the gas and particulate phases, coagulation of aerosol particles, or variation in the elemental composition and size distribution at the source.

III. EXPERIMENTAL

A. Sampling Procedure

The sampling equipment was set up in the W. M. Keck Laboratory, California Institute of Technology. Air was drawn from a point 6.7 m above the roof (18.5 m above street level) through a 21-m vertical sampling pipe (7-cm i.d.) described by Whitby et al. (9). The air stream was analyzed continuously by withdrawing parts of it to a number of instruments.

One part of the stream was drawn through a membrane filter (0.8-μm mean pore flow diameter) to collect the aerosol for elemental analysis. This filter is called the total-mass filter.

Another part of the stream was drawn through a similar filter preceded by the first four stages of an Andersen Cascade Impactor. The filter in this case is called the backup filter. The function of the impactor stages was to remove particles larger than ∿1.5-μm aerodynamic diameter ($\rho^{1/2}d$). The backup filter then collected aerosol particles smaller than 1.5-μm aerodynamic diameter. Information about the size distribution of the elements in the gross aerosol can then be obtained by comparing elemental analyses of the total-mass filter and the simultaneously collected backup filter. We verified by means of electron-microscope photographs that the cutoff was 1.5 μm.

The total-mass and backup filters were changed simultaneously every 2 hr by automatic sample changers. These sampling stations were designed and built (4) at the Lawrence Berkeley Laboratory, University of California, for the U.S. Environmental Protection Agency. They can be set to change the sample at any desired interval; the 2 hr generally used in our study represents a compromise between time resolution and sensitivity.

A third part of the stream was drawn through a β thickness gauge (Frieseke & Hoepfner Type FH 62A Dust Monitor) which determined the total particulate mass concentration by measuring the β-ray attenuation in the particulate mass collected on a filter tape. The β gauge has been evaluated and described by Husar (10); stated accuracy is ±25%.

Other parts of the stream were withdrawn for measurements of such items as various gases, light scattering, and condensation-nuclei counts. Carbon monoxide was measured every 10 or 20 min with a Beckman 6800 chromatograph.

Samples were collected continuously (except for short interruptions for maintenance and repairs, and longer interruptions for subsidiary experiments) from October 30 through November 27, 1972. During this 29-day period, 289 total-mass-filter and 223 backup-filter samples were collected. Of the latter, 190 were collected simultaneously with total-mass-filter samples. The β gauge was available only November 16 through November 27; during this period, 114 total-mass-filter samples were collected.

B. Elemental Analysis

The elemental composition of the particulate matter was measured on-site with the automatic X-ray fluorescence spectrometer. The instrument is described in detail elsewhere (3,4). A molybdenum secondary-target X-ray tube was used for the excitation of the sample. A 20-min fluorescence X-ray spectrum of each total-mass filter and each backup filter was accumulated with a Si(Li) semiconductor detector and associated electronics (system resolution 180 eV FWHM at 5.9 keV) and a 1024-channel pulse-height analyzer. The spectrum was automatically analyzed at the end of the accumulation period by a minicomputer system which used a spectrum-stripping program and the stored spectra of 16 elements (K, Ca, Ti, V, Cr, Mn, Fe, Co, Ni, Cu, Zn, As, Se, Br, Sr, and Pb). The results, in ng/cm^2 of filter, were printed by a Teletype minutes after the spectrum was accumulated. Meanwhile, the next filter moved automatically into the spectrometer, and the process was repeated.

Elements lighter than K are relatively difficult to detect because of their low fluorescence yields and especially the strong absorption of the fluorescence X-rays by matter (including the particles themselves). The absorption problem moreover renders the measurement of these light elements highly unreliable. There can also be interference from heavy-element L, M, etc., X-rays; for example, the K lines of S suffer interference from one of the M lines of Pb. Because of these limitations, a number of light elements which would have been of great interest (e.g., Na, Al, Si, S, Cl) are not included in our measurements.

Elements in the $40 < Z < 60$ region are not reported in this study because of the unsuitability of the molybdenum K X-rays for their fluorescence. Capability for excitation of these elements, and also for those slightly below $Z \sim 20$, has been added to the instrument but not in time for the Pasadena study.

IV. RESULTS AND DISCUSSION

There were nine elements for which the average concentration in the 289 total-mass filters was larger than the interfilter error in the measurements: Ca, Ti, V, Mn, Fe, Ni Zn, Br, and Pb. For each of the nine elements, Table 1 shows the average atmospheric concentration; the rms variability in the concentration; the interfilter error, that is, the standard deviation of measurements of a given element on replicate filters; and the intrafilter error, that is, the standard deviation among repeated measurements of a given elment on the same filter (with the filter left in place in the spectrometer for the repeated measurements).

Figure 1 shows the record of atmospheric concentrations of a few elements for the entire sampling period. Similar plots could be constructed for the other elements of the foregoing list.

A. Interelement Correlations and Functional Relationships

The best-fit lines $Y = A + BX$ for possible interelement functional relationships were determined using the method of Bartlett (11), which consists of averaging the points in the first third of the plot of Y versus X, averaging the points in the last third of the plot, and drawing a straight line between the averages. This method ignores the data in the middle third of the plot. We refer to relationships deduced in this way as "best-fit" lines.

The "best least-squares" lines (8,12) were also determined. In this treatment, errors in both X and Y are taken into account (in contrast to the common least-squares procedure that takes

TABLE 1. Mean Concentrations RMS Fluctuation of the Concentrations, and Inter- and Intrafilter Measurement Errors,[a,b] for Elements on Total-Mass Filters

Element	Mean (ng/m^3)	RMS fluctuation (ng/m^3)	Errors: Interfilter (%)[a]	Errors: Intrafilter (ng/m^3)[b]
Pb	2140.0	1851.0	7.0	>6
Br	718.0	649.0	7.2	>3
Fe	268.0	272.0	7.3	>5
Ti	33.2	39.0	28.0[c]	10
Mn	9.3	15.1	28.0[c]	4
Ca	239.0	256.0	52.0	>18
Ni	10.8	11.3	16.0[c]	3
V	8.6	10.9	50.0[c]	5
Zn	75.7	79.1	7.5	>3

[a]Interfilter error ≡ standard deviation of measurements of a given element on replicate filters.
[b]Intrafilter error ≡ standard deviation among repeated measurements of a given element on the same filter, measurements.
[c]The interfilter error for this element is not a constant percentage of the individual determinations; the error itself is essentially constant and is approximately equal to the intrafilter error. The percent of the mean is given.

into account only errors in Y). The errors in X and Y were assumed to be normally distributed and independent of the magnitudes of X and Y. The error ratio between Y and X was assumed to be that given by the interfilter error estimates in Table 1, and the errors in X and Y were assumed to be uncorrelated or only slightly correlated ($r < 0.5$). For most pairs of elements the best least-squares line agrees with the best-fit line. The slopes and intercepts for the best-fit lines with their rms errors are given in Table 2. The variation in Y about the best-fit lines is given in Table 3.

The intrafilter error is not known for the elements Pb, Br, Fe, and Zn. The interfilter percent error in the determination of each of these elements is about 7%, probably due to differences in flow rates and variations in filter positioning in

TABLE 2. The Intercepts, A(ng/m^3) and Slopes, B, of the Interelement Best-Fit Lines, Y = A + BX (Standard deviations, $\pm\sigma$, shown in parentheses).

	X								
V	Pb	Br	Fe	Zn	Ni	Mn	Ti	V	Ca
Ca	A=144(175)	188(158)	-20(14)	36(111)	137(54)	50(22)[a]	17(7)	89(29)[a]	
	B=0.04(0.12)	0.07(0.4)	0.96(0.04)	2.7(3.5)	9(8)	20.3(1.1)	6.7(0.6	17(5)	
V	-1.5(0.5)	0.3(0.5)	0.8(0.7)	-0.9(0.7)[a]	-0.07(0.5)	1.8(0.7)[a]	0.9(0.4)		
	0.0047(0.0002)	0.0117(0.0007)	0.029(0.002)	0.13(0.005)	0.81(0.02)	0.73(0.06)	0.23(0.02)		
Ti	7(5)	10.0(4.5)	-2.2(1.5)	1.7(2.0)	11(3)	6.0(2.5)[a]		8.2(1.3)	3.4(2.0)
	0.012(0.004)	0.03(0.01)	0.132(0.002)	0.41(0.07)	2.0(0.5)	2.91(0.08)		2.9(0.2)	0.124(0.005)
Mn	4(13)	5(15)	-2.4(0.6)	-1.4(0.3)	4.2(3.5)		1.3(66.0)	2.4(1.9)	0.4(0.6)
	0.003(0.004)	0.006(0.013)	0.0435(0.0004)	0.14(0.02)	0.46(0.55)		0.2(0.15)	0.8(0.25)	0.037(0.004)
Ni	2.0(0.7)	4.4(0.6)	4.2(0.4)	2.0(0.7)[a]		5.5(0.8)	4.8(2.5)	3.4(0.45)	5(8)
	0.0041(0.0003)	0.009(0.001)	0.025(0.005)	0.11(0.01)		0.57(0.15)	0.18(0.15)	0.87(0.02)	0.02(0.01)
Zn	33(4)[a]	44(3.5)[a]	18.0(4.7)[a]		28.0(3.5)	31(1.5)	32(18)	35.0(3.5)	33(38)
	0.020(0.003)	0.04(0.02)	0.215(0.008)		4.4(0.5)	4.8(0.7)	1.3(1.0)	4.7(0.45)	0.17(0.09)
Fe	140(60)	167(66)		43(3)	145(20)	74(12)	70(91)	122.0(7.5)	70(38)
	0.060(0.055)	0.14(0.18)		3.0(0.4)	11.3(2.2)	20.8(0.5)	5.9(2.2)	17(2)	0.8(0.3)
Br	-14.6(9.5)		383(51)	258.0(26.0)	330(18)	398(54)	366(40)	332(20)	451(570)
	0.3415(0.00005)		1.25(0.38)	6.1(1.0)	36.0(4.5)	34(10)	10(2)	44.0(3.5)	1.1(0.6)
Pb		117(27)	950(71)	588(115)	865(76)	1061(86)	1009(97)	958(63)	1250(1402)
		2.818(0.0005)	4.4(0.9)	20(3)	118(11)	115(24)	34(6)	136(8)	3.7(1.4)

[a]The best-fit line is different from the best least-squares line.

TABLE 3. Variations About the Best-Fit Line, $S_{y/x}$ (ng/m^3), and the Correlation Coefficients, r (Standard interfilter measurement error is given in parentheses)

	X							
V	Pb	Br	Fe	Zn	Ni	Mn	Ti	V
Ca	$S_{y/x}$=321(125)	325(125)	197(125)	284(125)	312(125)	251(125)	234(125)	271(125)
	r=0.208	0.146	0.795	0.493	0.333	0.560	0.754	0.450
V	6.7(4.4)	7.7(4.4)	8.8(4.4)	9.4(4.4)	6.3(4.4)	8.7(4.4)	7.2(4.4)	
	0.787	0.711	0.595	0.574	0.833	0.552	0.765	
Ti	29.7(9.5)	32.0(9.5)	18.7(9.5)	29.0(9.5)	34.0(9.5)	20.0(9.5)		
	0.623	0.546	0.868	0.637	0.579	0.684		
Mn	14.5(2.7)	14.7(2.7)	10.0(2.7)	13.0(2.7)	13.0(2.7)			
	0.405	0.343	0.866	0.515	0.389			
Ni	8.7(1.8)	10.0(1.8	9.9(1.8)	10.0(1.8)				
	0.579	0.482	0.446	0.554				
Zn	45.0(<5.6)	51.0(<5.6)	37.0(<5.6)					
	0.399	0.313	0.623					
Fe	169.0(<19.8)	220.0(<19.8)						
	0.422	0.347						
Br	84(<52)							
	0.983							

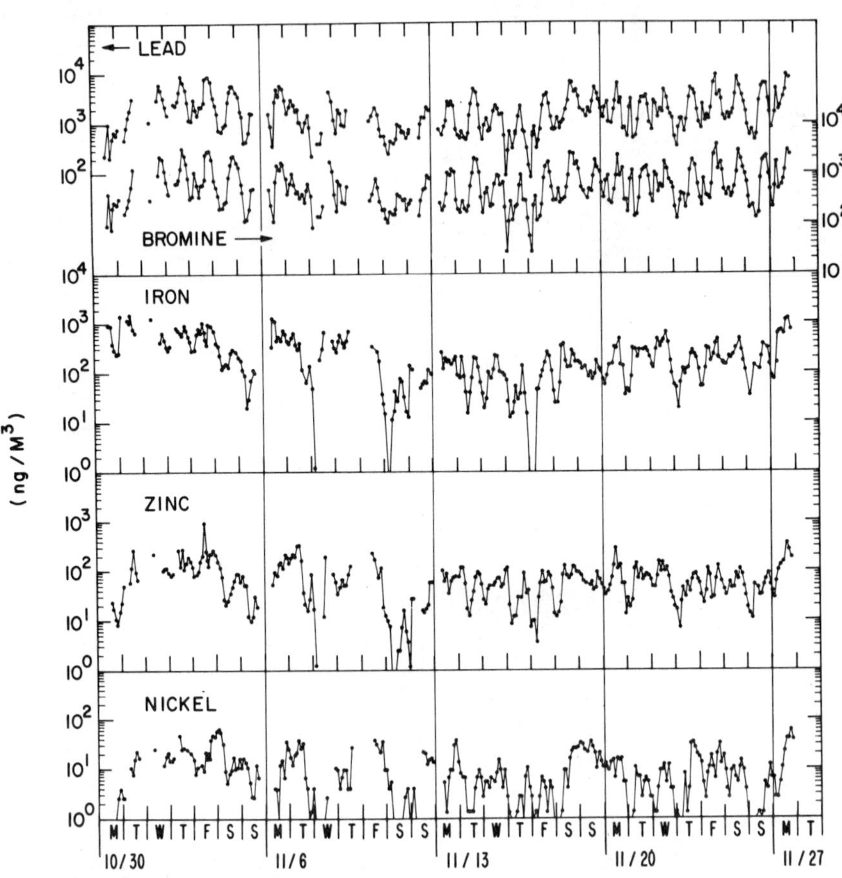

FIGURE 1. Diurnal patterns of Pb, Br, Fe, Ni, and Zn for the entire sampling period. Generally the concentrations of all of the elements maximize at the same time, even though Pb and Br are the only ones shown here with a source in common. Note the close similarity between the Pb and Br patterns.

the spectrometer. These effects cancel in the ratio between two elements on the same filter; hence, the errors in the ratio are less than or equal to the interfilter errors. The interfilter errors (±σ) are also shown in Table 3. The interfilter errors for Ca, V, Mn, Ti, and Ni are precisely what is expected from the statistical error in the determination of the peak area over a background signal. The intrafilter errors for these elements should also be equal to the statistical errors.

The best interelement correlation seen in this study is that between Br and Pb (Table 3; Figure 2). The intercept is zero (within ±2σ). The slope has very small error limits. These facts indicate that the Br and Pb arise from a common source. The correlation of Pb with CO (correlation coefficient: 0.98).

$$\text{Pb } (\mu g/m^3) = 0.888 + 0.365 \times \text{CO (ppm)} \qquad (5)$$

confirms the expectation that all three species, Br, Pb, and CO, originate predominantly from automobiles and other gasoline engines.

The variation of Br about the line in Figure 2 is twice the interfilter standard error. To determine whether this violates criterion #3 of the single-source model, we have analyzed the Br/Pb data in other ways. The Br/Pb ratio for all total-mass-filter samples taken together shows a normal distribution but with a standard deviation (15%) that is broader than expected (10%). An analysis of variance in the Br/Pb ratio shows that the among-day variance (0.0095) is significantly larger than the within-day variance (0.0017). If the Br/Pb ratio varied due to random errors, the two variances would be equal. Hence we conclude that there is a day-to-day variation in the Br/Pb ratio in the atmospheric particulate matter - i.e., the model criterion #3 is not met in the case of Br versus Pb.

Loss of Br from the particulate phase is probably responsible for the variation in the Br/Pb ratio. In support of this, Figure 2 shows that the slope of the regression line is below the Br/Pb ratio (0.386) in tetraethyl fluid. The stoichiometry of the major compounds reported in automobile exhaust particulate matter (13-18) is such that the Br/Pb mass ratio should be approximately equal to the tetraethyl fluid value. Ninomiya et al. (19) find a Br/Pb ratio in automotive exhaust particulate of 0.45.

Loss of Br from particulate matter has been documented by many investigators. Robbins and Snitz (21) reported that the Br/Pb ratio in isolated automobile exhaust particulate decays with a half-life of approximately 15 min; Ter Haar and Bayard (18) reported that most of the Br in isolated automobile exhaust

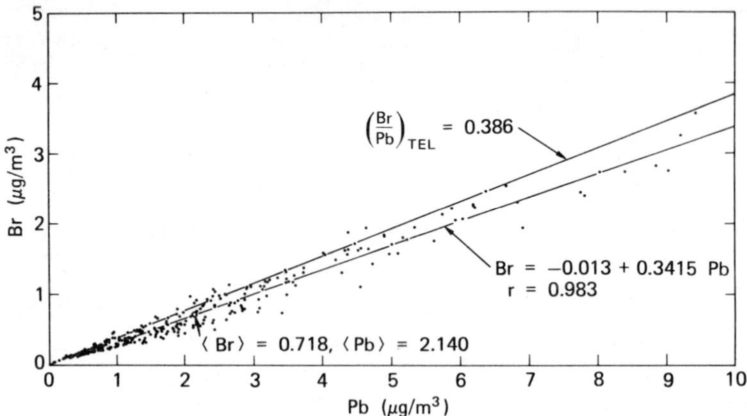

FIGURE 2. Linear regression between Br and Pb, two elements from a common source. The regression line predicted on the basis of the Br/Pb ratio in tetraethyl fluid (TEL) in gasoline is shown. The "best-fit" line (see text) is described by the equation Y=A+BX, where $Y \equiv Br(ng/m^3)$, $A = -0.013 \pm 0.019 (2\sigma) ng/m^3$, $X \equiv Pb(ng/m^3)$, $B = 0.3415 \pm 0.0001$.
Note that the correlation coefficient is high (r=0.983), the intercept is close to zero (A=-0.0131), and the slope (B=0.3415) is close to that expected on the one-source model with gasoline engines as the one source. Comparison between the slope of the regression line and the expected (TEL) value indicates that the vehicle exhaust particles have generally lost about 12% of their bromine.

particulate is lost within 1 hr. Actually, in view of our observed variation in Br/Pb ratio, these decay times seem surprisingly short.

The largest Br/Pb daily averages (0.39) occurred on October 30 and November 11, the two windiest days in the sampling period, presumably because the more rapid advection results in a younger aerosol reaching the site. The lowest Br/Pb daily averages, 0.29, were observed on November 4 and 27, which were days with low wind speeds and which were also the smoggiest days in the sampling period. The day with lowest wind speed, November 21, had a Br/Pb daily average of 0.32, significantly higher than hte minimum, and no smog. The Br/Pb relationship is evidently not straightforward; conditions associated with smog as well as low wind appear to be associated with low Br/Pb ratios.

The Br/Pb ratios in the 2-hr samples are not correlated with Pb concentration. As discussed later, the highest Pb concentration for most days occurs during rush hours. Since fresh automobile exhaust particulate should have the maximum Br/Pb ratio, we might have expected Br/Pb to be positively correlated with Pb. However, the wind speed is also generally lowest during the rush hours, and this will tend to decrease the Br/Pb ratio, perhaps leading to the observed lack of correlation.

The Ti and Fe concentrations are approximately proportional to each other, with a zero intercept (within $\pm 2\sigma$) and a well-defined slope. Apparently, therefore, most of the Ti and Fe originate from a common source. The slope of the best-fit line is within the range (Ti/Fe = 0.12 to 0.24) expected (24) for soil dust. Other sources might be involved, however, since the variation in Ti about the best-fit line is approximately twice the standard measurement error for Ti.

An analysis of variance for the Ti/Fe ratio indicates a significantly larger variance among days (0.885) than within days (0.294). Most of the daily averaged ratios are close to the grand mean, but the daily average for November 11 is more than twice the grand mean. November 11 was a windy, rainy day, and the concentrations of all elements, including Fe and Ti, on the filters were very low as a consequence. Because of the resulting lack of accuracy, it is not certain that the high daily average on November 11 is significant. If November 11 is dropped from the analysis of variance, the among-day variance (0.355) is only 1.4 times the within-day variance (0.252) indicating a single source for Ti and Fe - i.e., soil dust.

Manganese is also expected in soil dust (24). However, the concentrations of Mn and Fe are not proportional to one another in our experiments, though they are well correlated. The nonzero intercept suggests that Fe, to a greater extent than Ti or Mn, is present in sources besides soil.

An analysis of variance for the Mn/Fe ratio shows a few anomalous days when the daily-averaged ratio is substantially different from the grand mean, but generally on those days there was very little particulate matter in the filters and the accuracy of the measurements is poor.

We find that Ti and Mn are not proportional to one another; the intercept is 2.4 times the standard error. The Ti/Mn slope is smaller than the expected 3.6 for soil dust. If one were to force the best fit line to go through the origin, the slope would be 3.6.

In general, then, the high correlation coefficients among the three elements, Ti, Mn, and Fe, and the Ti/Fe and Mn/Fe ratios, support the expectation that all three elements are derived principally from the soil. However, a clear-cut proportionality (zero intercept) exists only between Ti and Fe. The lack of proportionality in the other pairings could be caused by variability in the concentrations of Fe, Ti, and Mn in soil dust, or by other sources of Fe.

Miller et al. (24) showed that the elemental composition of soil varies with location and that the elemental composition of the dust from soil is different from the elemental composition of the parent soil. However, the error limits for some of their analyses are not reported.

Miller et al. (24) and Friedlànder (6) conclude that the Fe in the Pasadena aerosol on July 3, 1969, comes from metallurgical processes. The iron and steel industry in Los Angeles County has been estimated to emit 3.4 tons/day of gross particulate matter, out of a total primary man-made emission of 130 tons/day (25), and hence the possibility of metallurgical operations as a secondary source of Fe cannot be discounted. Our correlations and Ti/Fe and Mn/Fe ratios, however, indicate that soil was the major source of Ti, Mn, and Fe during November 1972. [Strong correlations between Fe, Mn, and other elements prominent in soil, including Al, were used by John et al. (26) to identify the sources of these elements as soil dust in the San Francisco Bay area.]

Calcium is proportional to Fe. The best-fit line satisfies all four criteria of the one-source model. Calcium is almost proportional to Ti; all criteria are satisfied except that the intercept is not zero within 2σ. Calcium is proportional to Mn. The Ca/Fe ratio (the slope of the best-fit line) is 0.96, somewhat larger than that expected (24) for soil dust from a rural site (0.3 to 0.6), but similar to that found (24) for soil dust in the Pasadena urban area [0.6 to 1.4; Miller et al. (24)]. The two highest Ca/Fe values reported for soil in the Pasadena area by Miller et al. (24) were 0.95 and 1.4, which pertain to soil taken near a street and from a construction site, respectively. The Ca enhancement probably stems principally from cement dust. Friedlander (6) believes that cement dust is the principal source of Ca in the Pasadena aerosol. Cement dust is high in Ca, nearly 50% by weight, and nearly devoid of the other elements measured in our study. On the basis of our data and the cited soil compositions, we would estimate that about half of the Ca in our samples is indigenous to the soil and the other half is cement dust (construction and roadway abrasion) mixed into and entrained within the soil.

Calcium is poorly correlated to all the other elements except Ti, Mn, and Fe. This tends to strengthen the foregoing assessment.

Vanadium and Ni appear to be proportional to one another. However, particle-size information (see later) shows that V and Ni actually come from different sources. Nickel is not well correlated with any element other than V.

The measurement error for V is so large that variations about best-fit lines for any of the eight other elements are less than 2.1 times the V measurement error, so that criterion #3 of the model can be satisfied by all elements relative to V. Moreover, V is also well correlated with Ti, Br, and Pb.

Miller et al. (24) attribute both V and Ni to fuel-oil fly ash. However, they did not measure the atmospheric concentration of Ni; they did measure V but did not attempt to compare the concentration against an emission inventory. In the one sample of fuel-oil fly ash studied, they measured the V/Ni ratio to be 3.5, which does not agree with our atmospheric V/Ni ratio, 0.81 ± 0.02. We conclude that Ni is a tag for a source of sources other than vehicles or soil, but we would decline to name fuel-oil fly ash.

The remaining element, Zn, is not proportional to any other element measured, except perhaps V. Figure 3 shows the scatter

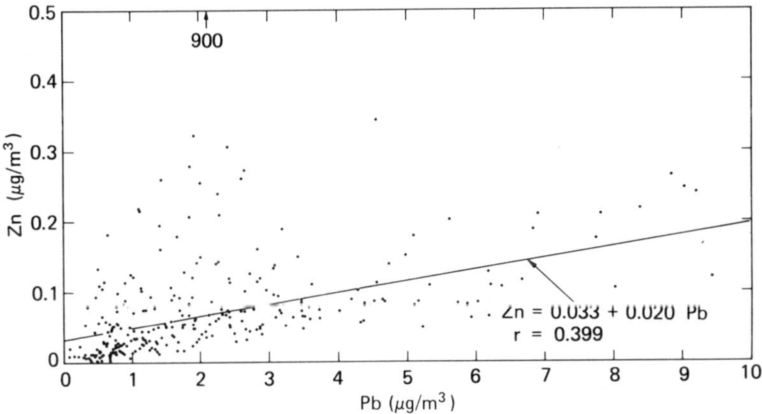

FIGURE 3. Linear regression between Zn and Pb. The "best-fit" line (see text) Y = A+BX, where Y = Zn (ng/m^3), A = 0.033 ± 0.008(2σ)ng/m^3, X = Pb(ng/m^3), B = 0.020±0.006, has a poor correlation (r = 0.399) and large offset from the origin($|A| \neq 0$) characteristic of elements probably originating from independent sources.

diagram for Zn and Pb as an example. The approximate proportionality between V and Zn may be only a consequence of the large uncertainties in the V measurements.

Zinc has rather high correlation coefficients with Fe and Ti. However, there is much more Zn in the aerosol than would be anticipated if the source of Zn were soil dust. This, together with the lack of proportionality between Zn and Fe indicates that Zn does not come from any of the same sources as the other elements measured in this study.

It should be pointed out that almost all of the elements are significantly correlated to one another. A correlation coefficient of 0.25 would be 99% certain of being significant. The average coefficient in this study is 0.56. Undoubtedly this simply reflects that the emission rates from the various sources may be correlated in some cases and that the meteorological conditions, such as inversion heights, modulate all atmospheric concentrations alike. A high correlation coefficient by itself is not a criterion for ascribing two elements to a source in common.

B. Elemental Size Distributions

Whitby et al. (5) report that the size distribution of aerosols, including the aerosol in Pasadena, often has two modes separated by a saddle at approximately 1 to 2 μm. This bimodality implies (and evidence advanced by Whitby et al. seems to corroborate) two classes of formation processes and a lack of mass transfer between the two modes. This being the case, the chemical compositions of the two modes ought to differ and therefore should be looked at separately. This was part of the rationale for measuring compositions demarcated at 1.5 μm. The other reason was to test the size distribution of each element statistically to see if it met the size-distribution criteria of the single-source model as outlined earlier.

In measuring the elemental size distribution, we seek to ascertain the functional relationship between each element on the backup filters and the same element on the simultaneously taken total-mass filters. The intercepts, slopes, and variations about the best-fit line, as determined from our results, are shown in Table 4. In the case of Fe, the slope of the best-fit line (0.235) indicates that 23.5% of the Fe is in particles less than 1.5 μm in diameter.

The error in the ratio of a given element on the backup filter and the same element on the total-mass filter is essentially the interfilter measurement error. These errors are also given in Table 4.

TABLE 4. Best-Fit Lines for Size Distribution Data, X(backup) = A + BX(total).
(The standard errors are given for the slope and intercept. The standard interfilter errors are given in parenthesis under the variation about the line).

Element	Slope	Intercept (ng/m^3)	Variation about line (ng/m^3)
Pb	0.7969 (0.00005)	-3.5 (25.4)	202 (120)
Br	0.7852 (0.00005)	-2.0 (8.4)	67 (41)
Fe	0.235 (0.004)	1.6 (4.2)	39 (5.4)
Zn	0.774 (0.0005)	-1.2 (1.9)	9.9(4.4)
Ni	0.757 (0.0015)	0.4 (0.3)	2.2(1.4)
Mn	0.301 (0.038)	0.4 (0.8)	4.6(2.1)
Ti	0.17 (0.09)	5.3 (2.5)	7.9(4.8)
V	0.56 (0.015)	1.0 (0.6)	3.9(3.4)
Ca	0.06 (0.09)	49 (48)	60 (37)

The concentration of Pb on the backup filter is proportional to that on the simultaneously taken total-mass filter. Similarly, the concentration of Br on the backup filter is proportional to the Br on the total-mass filter. The ratios Pb(backup)/Pb(total) and Br(backup)/Br(total) are equal, to within 1%.

Size distributions of Pb emitted from automobiles have been extensively measured (13-19, 27-37). Except for the Hirschler papers (13,14) (where the collection method may have biased the result), there seems to be general agreement that (once allowance has been made for nonsuspendable particles) the mass median aerodynamic diameter of the airborne exhaust Pb is submicron. Habibi et al. (16,17,33,34) report Pb mass median aerodynamic diameters larger than 1 μm, but Mueller (38) points out that, if the particles larger than 9 m are excluded to approximate the airborne fraction, Habibi's residual distributions also have submicron mass median aerodynamic diameters.

Huntzicker et al. (39) found approximately 80% of the airborne Pb mass to be in particles smaller than 1.5-μm aerodynamic diameter at the Pasadena site in November 1972, in agreement with our values of Pb(backup)/Pb(total) and Br(backup)/Br(total).

Atmospheric Pb size distributions (22,23,39-43) reveal considerable disagreement as to what the mean size is and whether the size distribution varies with time and/or location. Gillette and Winchester (43) report that the Pb size distribution is constant with time and location, at least for sizes greater than 0.2 μm. Our results show that the Pb(backup)/Pb(total) ratio and the Br(backup)/Br(total) ratio were both constant in time during November 1972 at the Pasadena site.

Robbins and Snitz (21) have argued that the loss of Br from automotive exhaust particulate is rate-limited by the diffusion of Br from the interior of the particles to the particle surface. If they are correct, then Br in small particles should be lost much more rapidly than Br in large particles. We find that the Br/Pb ratio - i.e., the slope of the best-fit line, on the backup filters is 0.3426 ± 0.0003, which is almost identical to the Br/Pb ratio on the total filters (0.3415 ± 0.00005). This means that we find no variation of Br/Pb ratio with particle size. Martens et al. (22) have also found the Br/Pb ratio to be independent of particle size. As they point out (22,23), such results argue against the diffusion mechanism for Br loss. The question is still not settled, however (23,44).

The concentration of Fe on the backup filter appears to be proportional to that on the total-mass filter. However, the variance of Fe(backup) about the best-fit line is 7.2 times that expected from measurement errors. An analysis of variance for the ratio Fe(backup)/Fe(total) shows that the among-day variance is significantly larger than the within-day variance. The highest values for this ratio (Fe concentrated in smaller particles than usual) occurred on rainy days (November 14, 16, and 17). The lowest values occurred on dry windy days (October 30 and November 8); this suggests a connection between low ratio and high wind, as would be expected for soil dust. There is one day (November 11) that was both windy and rainy; for that day, the ratio was very close to the grand mean, so that rain might be a factor as well as wind but in the opposite direction. On rainy days the amount of Fe collected on the filter was very low. Systematic errors in the latter situation may confuse the picture, but in general one recognizes in these observations the role of wind in entraining soil dust, and also the role of rain in preventing entrainment and in removing particles by rainout and washout. The shift of the Fe size distribution in favor of smaller particles on rainy days is to be expected if, as is generally accepted (45), rainout and washout preferentially remove the larger airborne particles.

Another explanation for the varying size distribution of the Fe is that the size distribution of the particles as "seen" by the sampling-tube inlet may be altered by the effect of the wind on the trajectories of the largest particles, in a manner analogous to the problem of isokinetic sampling from a moving stream. We found that 37% of the Fe measured by a sampling station on the roof did not reach a second sampling station at the end of the sampling pipe (in contrast to Pb, which shows no significant losses in the sampling pipe). Presumably only the larger Fe-containing particles are lost, so the ratio Fe(backup)/Fe(total) on the roof should be 0.15. The slopes of the Ti and Mn lines are essentially the same as the slope for Fe. The slope of the Ca line is apparently smaller than that of Fe (i.e., the Ca tends, even more than does Fe, to reside in the larger particles).

The concentration of Zn(backup) is proportional to Zn(total). Similarly, the concentration of Ni(backup) is proportional to that of Ni(total).

It is especially interesting that the backup/total ratios for all elements except V cluster about two values, 0.2 and 0.8 (Figure 4), signifying a segregation of particle composition according to particle size. This is consistent with the concept of bimodal particle-size distributions (5) as arising from two categories of particle-generation mechanism - e.g., condensation for the small-particle model and abrasion for the large. In the small-particle class, with 80% of their mass residing in particles smaller than $\sim$1.5 μm, we observe Pb, Br, Zn, Ni, in the large-particle class, with 80% larger than 1.5 μm, we find Ca, Ti, Mn, Fe (Figure 4). All four small-particle elements have the same backup/total ratio within a few percent. Also, the ratio for each of these four is independent of time, which argues against any given one of them having multiple singificant sources. Judging by the particle sizes, Ni and Zn are probably associated with combustion or other high-temperature processes.

Vanadium is anomalous. Though the accuracy in its determination is poor, the slope of the best-fit line for the size distribution is well defined (Table 4). From the size distributions it is evident that V either comes from a source of medium-size particles - i.e., a different source from any of the other elements, or that it comes from two sources, one emitting small particles and the other emitting large ones. The latter alternative is favored by the idea of the bimodal size distribution. However, distribution of the V(backup)/V(total) ratios is a normal distribution with a standard deviation only 50% of the mean (i.e., with two thirds of the radios falling between 0.8 and 0.3); there is thus no indication of two V sources. This evidence contrasts with observations of detailed size distributions and V/Al ratios in the San Francisco Bay area (46) showing the V arising

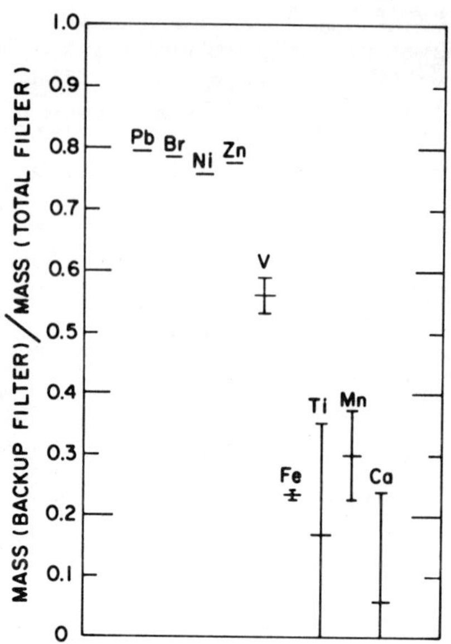

FIGURE 4. Ratio between the mass on the backup filter and the mass on the total-mass filter, for each of the series of nine elements. The errors in the ratio for Pb, Br, Ni, and Zn are too small to be shown on this scale, but the errors in the ratio (2σ) are shown for V, Fe, Ti, Mn, and Ca.

from both small-particle and large-particle sources identified as combustion and soil, respectively.

C. Total Particulate Mass

The average of the total particulate mass loadings for the 12-day period November 16 through 27 at the site in Pasadena, as measured by the β thickness gauge described earlier, was 31.7 $\mu g/m^3$. This average is consistent with values measured with the β gauge at this site in April 1972 (10), but is substantially lower than the 100 $\mu g/m^3$ found at the same site in August and September 1969 (47).

The low particulate loading in our study is probably related to the unusually light smog activity during the period. The alternative possibility of large systematic error in the β-gauge readings can probably be discounted. Beta-gauge measurements of aerosol mass have been shown to agree well with those obtained by gravimetric measurement of filters (10). We have confirmed this to hold for one test sample obtained during November 1972. Random errors with the β gauge are ∿5% for a 10-hr average, or 25% for a 2-hr average. Errors due to atmospheric relative-humidity effects are said to be small (10); moreover, the relative humidity for the latter part of November 1972 was quite constant.

Each element measured on the total-mass filter correlates well with the β-gauge readings. With the exception of Ca and V, all of the correlation coefficients are approximately 0.75.

D. Source Strengths

The strengths of the sources identified in this study are calculated from our experimental tag-element atmospheric mean concentrations, together with the source compositions in Table 5 given by Friedlander (6). The results are shown in Table 6 and compared there to Friedlander's estimates. We find that, at the Pasadena sampling site during November 1972, gasoline-engine exhaust particulate matter was on the average 5.35 $\mu g/m^3$, and the average soil-dust concentration was 8.35 $\mu g/m^3$. During the latter part of November when the gross-mass measurements were available, gasoline-engine exhaust particulate matter was on the average 5.6 $\mu g/m^3$ or 18% of the total particulate mass, and soil dust was 6.4 $\mu g/m^3$ or 20% of the total.

E. Nature of the Sources

There are a number of indications that the principal automotive contribution to the aerosol at our test location during November 1972 came from traffic close to the sampling site, and not from the Los Angeles basin in general.

First, Br has been reported to be lost from isolated automotive exhaust particulate matter on a rapid time scale (see the section "Interelement Correlations and Functional Relationships"). Our average Br/Pb ratio was 0.34 with about 5% accuracy, as compared to 0.39 expected for fresh automobile exhaust, suggesting an effective aerosol age on the order of minutes, not hours - i.e., too young to have been advected from distant parts of the Los Angeles basin.

Second, the weekday diurnal pattern of Pb concentration consistently shows two peaks (Figure 1), viz., a rather weak peak at about 0800 hr and a strong peak at about 1700 to 1900 hr. The Pb making up these peaks appears to come from a local source, since the wind speed is zero, or nearly so, during these morning and evening peaks. If the wind speed is zero at the time of

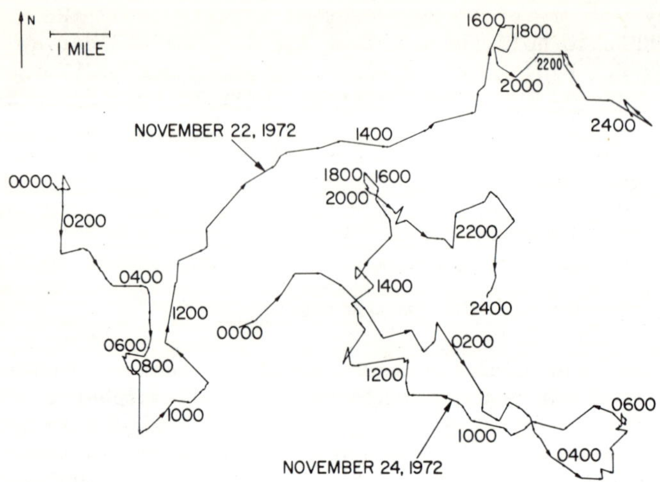

FIGURE 5. Wind traces for November 22 and 24, 1972, as measured on the roof of the Keck Laboratory. Each arrow specifies the wind direction (by orientation) and speed (by arrow length corresponding to distance traveled in 10 min) for 10-min periods throughout the day. The arrows are placed end-to-end, giving the apparent displacement for the day.

the Pb peak, the Pb in the peak must be either from a local source, or from a distant source whose plume just reaches Pasadena when the wind stalls. The implausibility of the latter alternative becomes obvious upon consideration of the wind patterns - for example, on November 22 and 24 (Figure 5). It is improbable that these two different patterns should bring the plume of a distant Pb source or sources to Pasadena just as the wind stalls and give afternoon Pb peaks with similar times and intensities for both days (Figure 1).

Third, the weekday Pb peaks occur at the same time as the weekday rush hours. It is reasonable that the increased traffic is partly responsible for the Pb peaks. If the Pb were advected from a distant part of the basin, the peak should occur a few hours after the traffic rush. (Some of the strong evening peaks do show a trailing edge which may indeed be a vestige of this, except that they seem to last only as long as the calm does).

Fourth, the slope of the regression line of Pb on CO (0.365 $\mu g/m^3$)/ppm is characteristic of low vehicle speeds. Airborne Pb emissions increase with speed (16,17,35), while CO

TABLE 5. Compositions of Various Sources of Particulate Matter (%). (Taken from Friedlander[a]).

	Sea salt	Soil dust	Auto exhaust	Fuel-oil fly ash	Portland cement
C compounds			40.3[b]		
Na	30.6	2.5		5	0.4
Mg	3.69	1.4		0.06	0.48
Al		8.2		0.8	2.38
Si		20		1	10.7
S	2.56				
Cl	55.0	6.8			
K	1.1	1.5		0.2	0.53
Ca	1.16	1.5		1.3	46.0
Ti		0.4		0.06	0.144
V		0.006		7	
Cr				0.1	
Mn		0.11		0.06	
Fe		3.2	0.4	6	1.09
Co		0.002		0.2	
Ni		0.004		2	
Cu		0.008		0.2	
Zn		<0.01	0.14	0.02	
Br	0.19		7.9		
1	1.4x10				
Ba		0.06		0.1	
Pb		0.02	40.0	0.07	

[a]Ref. 6.
[b]As a tarry substance assumed 90% C.

TABLE 6. Average Source Contributions to Aerosol Sampled 18.5 Meters Above Street Level at One Site in Pasadena

Source	Friedlander[a] 9/3/69, μg/m^3(%)	This work 11/16-27/72 μg/m^3 (%)
Soil dust	11.4(11.4)	6.4(20)
Gasoline-engine exhaust particulate[b]	8.2(8.2)	5.6(18)
Fuel-oil fly ash	0.1(0.1)	≤0.09(≤0.3)[c]
Cement dust	1.7(1.7)	0.1(∼0.3)
Total mass	100.0	31.7

[a]Ref. 6.
[b]Primary particulate only-i.e. does not include aerosol formed by smog reactions.
[c]The upper limits require the assumption that the observed V can be attributed to fuel-oil fly ash. The assumption appears unwarranted at present (see text).

emissions decrease (48,49); judging from figures given by Habibi (16,17) for Pb and by Duprey (48) for CO, our slope corresponds to a speed of ∿27 mph. Freeway traffic, comprising some 37% of the vehicle-miles driven in the basin (50), would be characterized by slopes of 1 to 1.4, approximately (51), corresponding to high speeds. It would be difficult to argue that the freeways, and hence the basin at large, are contributing substantially to the Pb and CO observed at our site. (The nearest freeway is the end of the Pasadena Freeway, 1+ miles away.)

Colucci et al. (51) have reported higher slopes, 0.8, for the regression line of Pb on CO in Pasadena. This indicates that more distant traffic may at other times contribute to the observed PB levels in Pasadena. The slope reported by Huntzicker et al. (39) for the same site and approximately the same time as our studies was only slightly higher than the value observed by us.

We have examined meteorological data for a number of weather stations in the basin for the period of our study. Consideration of lapse rates (thus, mixing depths), wind speeds, and other data indicates that the Pb peaks, and concurrent peaks in the other elements, are primarily a consequence of meteorological factors. The peaks are brought about by wind stagnation. (In the case of Pb, the peaks are reinforced by injection at the same time.) The morning peak is brought to an end by the breakup of the night-time inversion, and the evening peak is terminated by the breakup of the early-evening calm. The peaks are most intense when they occur during an inversion.

Soil dust, judging by the observed Ca/Fe ratios, appears to come from the Los Angeles urban area (see the secion "Interelement Correlations and Functional Relationships"). Like the Pb, the soil dust loading (as judged by the Fe concentration) is at a maximum during the morning and evening calms (Figure 1); this is reflected in the negative correlation between wind speed and Fe loading. (The correlation coefficient is -0.16, which is just significant at the 95% confidence level.) The argument for the dominant role of meteorological factors in modulating the concentrations of all the elements appears even more convincing in the case of Fe (where injection during the peak period is less likely) than in the above-cited case of Pb.

If, on the other hand, the peaks in the soil dust loading represent dust stirred up by vehicle traffic, one would expect Pb and Fe to be highly correlated. But Fe is more highly correlated to every other element measured than it is to Pb and Br (Table 3).

Rahn et al. (52) have demonstrated the dominant role of meteorology in the diurnal variations of aerosol elements in the Livermore Valley. While their patterns diverge from ours in important respects, the point is qualitatively clear that the meteorlogy is crucial and can induce very great concentration fluctuations.

It is interesting to consider the contribution of roadway abrasion to the aerosol loading at Pasadena. If we ignore the lack of correlation between Pb and Ca, and adopt the extreme view that all of the airborne Ca comes from roadway abrasion, we find that the upper limit on airborne particulate matter from roadway abrasion is about one-tenth as much as the amount of exhaust particulate, or at most a few tenths of milligrams per vehicle-mile - a very low figure, probably consistent with the expected size of particles from such a source.

V. SUMMARY

The more important conclusions from this study are listed below.

Gasoline-engine exhaust is the chief source of the Br and Pb. It appears to have come primarily from local street traffic during our sampling period. Particulate matter present in gasoline-engine exhaust contributed 5.35 $\mu g/m^3$ on the average to the aerosol mass during our sampling period at the Pasadena site. (This figure does not include contributions resulting from reactions involving exhaust gases in the atmosphere.)

Soil is the main source of the Ca, Ti, Mn, and Fe. The soil dust is mainly from within the Los Angeles basin. About half of the Ca in the soil dust results from contamination of the basin soil by cement dust. The soil contributed 8.35 $\mu g/m^3$ on the average to the aerosol mass during the sampling period.

Vanadium, Ni, and Zn are tags for at least three further sources not identified.

All elements measured show strong concentration fluctuations (orders of magnitude) on a short ($\leq$2-hr) time scale.

Elemental concentrations tend to show a diurnal pattern with morning and evening peaks evidently dominated by meteorological factors.

The average correlation coefficient for elements from a source in common was 0.8. Elements unrelated by source are still correlated, with an average coefficient of 0.4; hence, the existence of a correlation between elements does not suffice to establish a common origin.

The size distributions of Ni, Zn, Br, and Pb are similar - i.e., predominantly in small particles and constant in time.

The size distributions of Ca, Ti, Mn, and Fe are similar - i.e., predominantly in large particles. The size distribution of Fe varies with wind speed and rain.

The size distribution of V differs from all the other elements measured; its MMED is about 1.54 μm, intermediate between the two main size categories.

The average Br/Pb mass ratio was 0.34 with a RMS variation of ±0.05.

The daily average Br/Pb ratio varies significantly from day to day. The windiest days had the highest Br/Pb averages (0.39); the smoggiest days had the lowest Br/Pb averages (0.29).

The Br/Pb ratio is the same for small particles as for large particles at any given time.

We have accounted for approximately 40% of the particulate mass in Pasadena during November 1972.

The results obtained in this study demonstrate the power of the new experimental technique employed. Its chief limitation is the difficulty of measuring the light elements (S, Cl, and lighter). A much greater range of elements would have been attainable if several X-ray excitation energies had been available as is the case now. Sensitivity can be gained at any time during an experiment at the expense of time resolution and vice versa. The method therefore seems advantageous for continuous on-line sampling.

ACKNOWLEDGMENTS

We are pleased to acknowledge F. S. Goulding, J. M. Jaklevic, and B. V. Jarrett of the Lawrence Livermore Laboratory, University of California, and R. K. Stevens of the National Environmental Research Center, U.S. Environmental Protection Agency, Research Triangle Park, N.C., for their development of the X-ray fluorescence sepctrometer used in this study and for their interest and participation in the study. We thank Prof. S. K. Friedlander, J. J. Huntzicker, and the group at the California Institute of Technology, especially R. B. Husar (now at Washington University, St. Louis), for their interest and participation in the study and for their hospitality while we were in Pasadena. We are grateful to T. G. Dzubay, of the National Environmental Research Center (EPA), for assistance in planning and in data collection. We wish to thank A. M. Greenberg, of the Air Pollution Control Division, Wayne County Department of Health, for his patient teaching of the art of reading meteorological records. Finally, we thank Prof. K. Rengan, of Eastern Michigan University, for his help with the extensive neutron activation analysis work.

REFERENCES

1. Hidy, G. M. 1973. "Characterization of Aerosols in California", Interim Report for Phase I (October 25, 1971 to April 1, 1973) to the Air Resources Board, State of California (ARB Contract No. 358), December 15.

2. Hidy, G. M., Appel, B. R., Charlson, R. J., Clark, W. E., Friedlander, S. K., Hutchison, D. H., Smith, T. B., Suder, J., Wesolowski, J. J., and Whitby, K. T. 1974. Paper No. 74-119, 67th Annual Meeting, Air Pollut. Control Assoc., Denver, Colo., June 9-13.

3. Jaklevic, J. M., Goulding, F. S., Jarrett, B. V. and Meng, J. D. 1973. Lawrence Berkeley Laboratory Report LBL-1743.

4. Goulding, F. S. and Jaklevic, J. M. 1973. "X-Ray Fluorescence Spectrometer for Airborne Particulate Monitoring", Office of Research and Monitoring Report EPA R2-73-182 (National Technical Information Service Report PB-225038). April.

5. Whitby, K. T., Husar, R. B. and Liu, B. Y. H. 1972. J. Colloid Interface Sci. 39, 177-204.

6. Friedlander, S. K. 1973. Environ. Sci. Technol. 7, 235-240.

7. Hammerle, R. H. and Pierson, W. R. 1974. 167th ACS National Meeting. Los Angeles, Calif., March 31-April 5.

8. Kendall, M. C. and Stuart, A. 1961. The Advanced Theory of Statistics, Vol. 2. Hafner Publishing Co., New York.

9. Whitby, K. T., Liu, B. Y. H., Husar, R. B. and Barsic, N. J. 1972. J. Colloid Interface Sci. 39, 136-164.

10. Husar, R. B. 1974. Atmos. Environ. 8, 183-188.

11. Bartlett, M. S. 1949. Biometrics 5, 207-212.

12. Acton, F. S. 1966. Analysis of Straight-Line Data, Dover edition. Dover Publications, New York; Wiley, New York. 1959.

13. Hirschler, D. A., Gilbert, L. F., Lamb, F. W. and Niebylski, L. M. 1957. Ind. Eng. Chem. 49, 1131-1142.

14. Hirschler, D. A. and Gilbert, L. F. 1964. Arch. Environ. Health 8, 297-313.

15. Moran, J. B., Manary, O. J., Fay, R. H. and Baldwin, M. J. 1971. "Development of Particulate Emission Control Techniques for Spark-Ignition Engines," Office of Air Programs Report APTD-0949 (National Technical Information Service Report PB-207312). July.

16. Habibi, K. 1970. Environ. Sci. Technol. 4, 239-248.

17. Habibi, K. 1973. ibid. 7, 223-234.

18. Ter Haar, G. L. and Bayard, M. A. 1971. Nature 232, 553-534.

19. Ninomiya, J. S., Bergman, W. and Simpson, B. H. 1970. Paper EN-10G, Second International Clean Air Congress of the International Union of Air Pollution Prevention Association, Washington, D.C., December 6-11.

20. Pierrard, J. M. 1969. Environ. Sci. Technol. 3, 48-51.

21. Robbins, J. A. and Snitz, F. L. 1972. ibid. 6, 164-169.

22. Martens, C. S., Wesolowski, J. J., Kaifer, R., and John W. 1973. Atmos. Environ. 7, 905-914.

23. Martens, C. S., Wesolowski, J. J., Kaifer, R., and John W. 1974. ibid. 8, 1340-1342.

24. Miller, M. S., Friedlander, S. K. and Hidy, G. M. 1972. J. Colloid Interface Sci. 39, 165-176.

25. Lemke, E. E. 1971. "Profile of Air Pollution Control", County of Los Angeles, Air Pollution Control District, Los Angeles, Calif.

26. John, W., Kaifer, R., Rahn, K. and Wesolowski, J. J. 1973. Atmos. Environ. 7, 107-118.

27. Mueller, P. K. Helwig, H. L. Alcocer, A. E., Gong, W. K., and Jones, E. E. 1962. ASTM Special Publication No. 352, Symposium on Air-Pollution Measurement Methods, Fourth Pacific Area National ASTM Meeting, Los Angeles, Calif., pp. 64-77, October 5, (published 1963 by ASTM).

28. Mueller, P. K. 1967. J. Air Pollut. Control. Assoc. 17, 583-584.

29. Moran, J. B. and Manary, O. J. 1970. "Effect of Fuel Additives on the Chemical and Physical Characteristics of Particulate Emissions in Automotive Exhaust", Office of Air Programs Report APTD-0618 (National Technical Information Service Report PB-196783), July.

30. Moran, J. B., Baldwin, M. J., Manary, O. J. and Valenta, J. C. 1972. "Effect of Fuel Additives on the Chemical and Physical Characteristics of Particulate Emissions in Automotive Exhaust", Office of Research and Monitoring Report EPA-R2-72-066, June.

31. Lee, R. E., Jr., Patterson, K., Crider, W. L. and Wagman, J. 1971. Atmos. Environ. 5, 225-237.

32. Bergman, W. 1971. "Characterizing and Measuring Automotive Particulate Emissions with Two Improved Sampling Techniques", presented at Central States Section of the Combustion Institute, Ann Arbor, Michigan, March 23 and 24.

33. Habibi, K. 1971. SAE Paper No. 710638.

34. Habibi, K., Jacobs, E. S., Kunz, W. G., Jr., and Pastell, D. L. 1970. "Characterization and Control of Gaseous and Particulate Exhaust Emissions from Vehicles", presented at the Fifth Technical Meeting of the West Coast Section, Air Pollution Control Association, San Francisco, October 8 and 9.

35. Ter Haar, G. L., Lenane, D. L., Hu, J. N. and Brandt, M. 1971. Paper No. 71-111, 64th Annual Meeting, Air Pollution Control Association, Atlantic City, N.J., June 27-July 2; J. Air Pollut. Control Assoc. 22, 39-46 (1972).

36. Sampson, R. E. and Springer, G. S. 1973. Environ. Sci. Technol. 7, 55-60.

37. Ganley, J. T. and Springer, G. S. 1974. ibid. 8, 340-347.

38. Mueller, P. K. 1970. ibid., 4, 248-253.

39. Huntzicker, J. J., Friedlander, S. K. and Davidson, C. I. 1973-1975. 166th ACS National Meeting, Chicago, Ill., August 1973; 167th ACS National Meeting, Los Angeles, Calif., March 1974; Environ. Sci. Technol. 9, 448-456 (1975).

40. Robinson, E. and Ludwig, F. L. 1967. J. Air Pollut. Control Assoc. 17, 664-669.

41. Daines, R. H., Motto, H. and Chilko, D. M. 1970. Environ. Sci. Technol. 4, 318-322.

42. Lundgren, D. A. 1970. J. Air Pollut. Control Assoc. 20, 603-608.

43. Gillette, D. A. and Winchester, J. W. 1972. Atmos. Environ. 6, 443-450.

44. Moyers, J. L. and Colovos, G. 1974. ibid. 8, 1339-1340.

45. Junge, C. E. 1963. Air Chemistry and Radioactivity, pp. 131-141, 289-98, Academic Press, New York.

46. Martens, C. S., Wesolowski, J. J., Kaifer, R. and John, W. 1973. Environ. Sci. Technol. 7, 817-820.

47. Lundgren, D. A. 1972. J. Colloid Interface Sci. 39, 205-210.

48. Duprey, R. L. 1968. "Compilation of Air Pollution Emission Factors", Public Health Service Publication No. 999-AP-42.

49. Kircher, D. S. and Armstrong, D. P. 1972. "An Interim Report on Motor Vehicle Emission Estimation". U.S. Environmental Protection Agency, Office of Air Quality Planning and Standards, APTD-1471, October (Revised January 12, 1973).

50. Roth, P. M., Roberts, P. J. W., Liu, M.-K., Reynolds, S. D., and Seinfeld, J. H. 1974. Atmos. Environ. 8, 97-130.

51. Colucci, J. M., Begeman, C. R., and Kumler, K. 1969. Air Pollut. Control Assoc. 19, 255-260.

52. Rahn, K., Wesolowski, J. J., John, W. and Ralston, H. R. 1971. ibid 21, 406-409.

Material Balance for Automobile-Emitted Lead in Los Angeles Basin

JAMES J. HUNTZICKER,[†] SHELDON K. FRIEDLANDER,[††] and CLIFF I. DAVIDSON[†††]
W. M. Keck Laboratories of Environmental Engineering
California Institute of Technology
Pasadena, California

[†] Present address: Department of Environmental Science, Oregon Graduate Center, 19600 NW Walker Road, Beaverton, OR 97005.

[††] Present address: Chemical, Nuclear, and Thermal Engineering Department, University of California at Los Angeles, Los Angeles, CA 90024.

[†††] Present address: Department of Civil Engineering, Carnegie Mellon University, Pittsburgh, PA 15123.

ABSTRACT

The flow of automobile-emitted lead through the Los Angeles basin has been estimated from measurements of particle size distributions, atmospheric concentrations, and surface deposition of lead at various sites around the basin. Approximately 24 metric tons/day of lead as gasoline antiknock additives are consumed. Of this, about 18 tons/day are exhausted to the atmosphere, and 6 tons/day are retained in the cars. Of the exhausted lead, two thirds deposits over the land area of the basin, and one third is advected out of the basin. The lead blown out of the Los Angeles area is the major source of atmospheric lead for regions immediately downwind. Automobile-emitted lead also accounts for more than half of the anthropogenic lead input to the Los Angeles coastal waters. Most of the lead is accounted for by independent estimates of the separate transport processes, but uncertainties in certain pathways are discussed.

I. INTRODUCTION

In this paper we report on a material balance carried out on automobile-emitted lead in the Los Angeles basin. Estimates are made of the daily consumption of lead by automobiles and the amounts exhausted to the atmosphere, deposited on the land and roadways, and advected out of the basin. We also show how automobile-emitted lead contributes to the lead influx to the coastal waters. The mass flows are based on new measurements of atmospheric lead concentrations, particle size distributions, the surface deposition of lead, and lead in surface water runoff as well as data from the literature. The area for which the calculation is made includes the major population centers of Los Angeles and Orange Counties (Figure 3).

For a conservative species, such as lead, the terms appearing in the material balance are well defined with respect to a given geographical area. As examples, the total lead emitted to the atmosphere, lead deposition on surfaces, and the quantity advected by the winds past the borders of the region are all exact quantities for any given time period.

Such quantities can be estimated as shown below, but their values are necessarily approximate at this time. The estimates will, however, be useful in a number of possible applications. Relative contributions by Los Angeles and by local sources to atmospheric lead concentrations in regions downwind of the basin can be estimated. Similar calculations can be made on the lead input to coastal waters from various sources. Such results should be of value in making policy decisions related to nondegradation of air quality and to water quality standards. It will also be possible to estimate the flow of certain other trace pollutants from the lead balance.

II. CONSUMPTION OF LEAD

Although detailed information on the consumption of alkyl lead additives is not available, the lead consumption rate for Los Angeles can be estimated as follows: During 1972 the average distribution rate of taxable gasoline for the State of California was 104 million liters/day (1). During the same year Los Angeles and Orange Counties accounted for 41.4% of the automobiles, motorcycles, and trucks in California (2). If it is assumed that gasoline consumption is proportional to the number of motor vehicles, the consumption rate for the Los Angeles region is 42.9 million liters per day. The average concentration of lead in Southern California gasoline for the winter 1971-72 was 0.56 ± 0.06 g/liter (see Appendix A). This gives an average 1972 lead consumption rate of 23.7 ± 2.4 metric tons/day. (For bookkeeping purposes, we will often carry one more significant figure than justified.)

The Los Angeles County Air Pollution Control District (3) estimates that in 1972 350 tons/day or about 5 x 10^5 liters/day of gasoline were lost by evaporation from automotive fuel systems and during gasoline handling operations. If the tetraethyllead fraction does not change during evaporation, then approximately 0.3 ton/day of lead in tetraethyllead vapor are emitted into the atmosphere, and the remainder, 23.4 tons/day, are consumed by automobiles.

III. NATURE OF LEAD EMISSIONS

Hirschler, et al. (4,5) showed that the size and amount of lead-containing particles were sensitive functions of driving mode. At cruising speeds, the exhaust fraction varied between

14 and 54%. During full throttle acceleration, however, large amounts of lead were reentrained from the exhaust system giving exhaust fractions up to 200% of the input. Subsequent studies by Mueller et al. (6), Ter Haar et al. (7), and Habibi (8,9) confirmed Hirschler's general results. Ter Haar et al. (7) found that for automobiles with new exhaust systems, the exhaust fraction was small but increased with age, indicating that a break-in period of several thousand miles was necessary before the exhaust system deposits stabilized. Thus, an accurate picture of typical lead emissions under consumer conditions can only be obtained by monitoring lead emissions for many thousands of miles.

Hirschler et al. (4,5) studied three automobiles over periods which included both uncontrolled surburban driving and programmed chassis dynamometer tests. They found that three cars with 27,000, 19,300 and 9800 accumulated miles retained 21.2, 27.5, and 23.1% respectively, of the input lead. Ter Haar et al. (7) conducted a similar study on one car but also attempted to construct a complete mass balance for the lead by collecting the exhausted lead in a cyclone separator and total filter. After 12,000 miles on a mileage accumulation route, 30% of the input lead remained in the oil, oil filter, and exhaust system, and 54% had been exhausted. Ter Haar et al. (7) speculated that the remaining 16% was lost during handling of the filter and exhaust system.

For an automobile operated under simulated consumer test conditions, Habibi (9) found an emission rate of 89% between 20,000 and 33,000 accumulated miles. However, Habibi's results for these mileages are probably not representative of the total vehicle population, since they do not take into account the break-in periods for new cars and cars with new mufflers.

To estimate the amount of lead exhausted, the retention factors (i.e., percent of lead remaining in the car) of Hirschler et al. (4,5) and Ter Haar et al. (7) are assumed representative of the total vehicle population. A retention factor of 30% is used for the Ter Haar results without trying to account for the missing 16%. This is a consistent application of the Hirschler and Ter Haar data since Hirschler measured only the lead remaining in the car and did not construct a total mass balance. The average retention factor for the four automobiles sampled is 25 ± 4%. Thus of the 23.4 ± 2.4 tons/day of lead which are burned, 5.8 ± 1.1 tons/day are retained in the car, and 17.6 ± 2.6 tons/day (by difference) are exhausted to the atmosphere. (Because of the changing nature of gasoline consumption, this exhaust rate applies strictly only to 1972). The use of the Hirschler and Ter Haar retention factors implies muffler and exhaust system changes at about 27,000-mile intervals although the actual interval may be longer. Such a change will restore the automobile to a low emission state and will entail another break-in period. Firm data are not available to evaluate these effects.

Brief (10) has shown that auto exhaust lead contains both a particulate fraction and an organic vapor phase fraction. Measurements on eight pre-1961 European and English cars showed that the vapor phase component was about 12% of the particulate component. Because of his sampling scheme, however, the particulate lead measured by Brief probably consisted primarily of particles smaller than about 9 μm, which, according to Habibi (9) (see below), make up 43% by weight of the particulate exhaust. Thus if Brief's results can be applied to Los Angeles automobiles, approximately 0.9 ton/day of organic vapor phase lead and 16.7 tons/day of particulate lead are exhausted. Recent measurements of particulate and vapor phase organic lead near roadways by Skogerboe (11) also indicate a significant organic lead contribution. The fate of the vapor phase lead is discussed in the section "Airborne Lead: Removal by Wind". Lead emissions from nonautomotive industrial sources are small (∿0.3 ton/day) (12) and are not included in the mass balance.

Of the lead which remains in the car, one third to one half is in the oil and oil filter (4,6,7). The fate of used motor oil is poorly understood. Some is poured into the sanitary sewer system, some poured onto the ground, and some reclaimed (13).

IV. EXPERIMENTAL

The material balance method developed in this paper is based on measurements of atmospheric particle size distributions, surface deposition fluxes, and atmospheric lead and carbon monoxide concentrations. Our measurements were made to provide data relatively contemporaneous with 1972, our reference year, and based on a consistent set of experimental and analytical techniques. Earlier measurements of particle size distributions (14), airborne lead concentrations (15-18), lead deposition on *Avena sativa* (wild oats) (19), and carbon monoxide concentrations (3,17,18) have been reported for the Los Angeles area.

Lead deposition was measured at Pasadena for 10 different 1- or 2-week sampling periods between November 1972 and February 1974. A 1-week synoptic measurement of deposition was made at Pasadena and five other locations in the Los Angeles basin during August 1973. Deposition measurements were also made at three locations near a freeway in May 1973, at two coastal islands, and at four sites each along the coast, in the Mojave Desert, in the San Gabriel-San Bernardino Mountains, and in the Coachella Valley during the summer of 1973. During each of these measurement periods, lead deposition in Pasadena was also measured. Size distribution measurements were made at Pasadena and a freeway. Size distribution and atmospheric lead concentrations were also measured at five sites in the Los Angeles basin during the 1972 and 1973 phases of the California Aerosol Characterization Experiment (ACHEX) (20,21).

The substrates for the deposition measurements were roughened FEP Teflon disks (0.5 mm thick) with exposed areas of 71 cm^2. The sample disk was placed on top of a larger FEP disk. Both disks were secured to a stainless steel disk (6 mm thick) by a TFE Teflon ring bolted to the steel disk. The assembled collector presented a relatively low profile to the wind. All Teflon parts were cleaned once in cold, concentrated HNO_3, twice in hot, concentrated HNO_3 (G. Frederick Smith redistilled), followed by a rinse in double distilled water. The Teflon pieces were wrapped in Saran Wrap until the deposition collectors were assembled just prior to a sampling period. All glassware used for these measurements was cleaned and stored in a similar manner.

At the end of the collection period, the sample disks were transferred to crystallization dishes. For those samples taken at locations other than Pasadena, the transfer was performed in the field inside a clean Lucite box which protected the samples from contamination. After the samples had been returned to the laboratory in Pasadena, 20 ml of concentrated HNO_3 (G. Frederick Smith, redistilled) were added to the crystallization dishes and the samples digested at near the boiling point for about 1 hr. The Teflon disks were then removed and the solutions evaporated to dryness over a period of several hours. The samples were redissolved in pipeted volumes of 5 or 10 ml of 4 $\underline{N}$ HNO_3 and transferred to cleaned (as above) polyethylene or FEP Teflon sample bottles.

These solutions were analyzed for Pb by flame atomic absorption spectroscopy using the 217 nm line. For some samples, H_2 continuum lamp measurements for nonatomic absorption was also made. These corrections were always very small. Lead blanks were measured by subjecting a clean Teflon disk to the same digestion-concentration procedure as the samples. In all cases the blanks were small - usually at or below the limit of detection. All of the processed urban samples contained greater than 1 μg Pb/ml of solution and gave signal-to-noise ratios greater than 10. The samples from the remote locations contained less lead and gave poorer signal-to-noise ratios.

The collection properties of different surfaces were also investigated. The deposition fluxes on the normal Teflon collector, a Teflon collector coated with paraffin oil, and a 125 mm diameter crystallization dish filled with water to within 1 cm of the top were 35, 40 and 40 ng/cm^2·day, respectively. In a separate experiment, the water collector gave a deposition rate of 38 ng/cm^2·day and the Teflon collector 42 ng/cm^2·day. Thus, the physical nature of relatively smooth surfaces does not affect their particle collection properties.

Size distribution measurements were made with Andersen cascade impactors. The impaction substrates were FEP Teflon disks (0.25 mm thick) which rested on clean quartz or stainless steel plates. The Teflon disks were cleaned as above. The after-filters were 0.45 μm pore-size (Millipore HAWP 04700) filters cleaned by soaking for several hours in 6N HCl followed by rinsing for several hours in double distilled water. The chemical analysis for the Teflon disks was as above. The filters were dissolved in hot concentrated HNO_3 and analyzed in the same manner as the other samples. Filter blanks were determined by dissolving five clean filters and analyzing for lead content. To check the filter analysis procedure, a known amount of lead was added to a filter and the analysis conducted as usual. The ratio of the measured concentration to expected concentration was 1.08, which is within the estimated experimental error. For one of the Pasadena-size distributions, lead concentrations were also measured by C. Patterson and Y. Hirao, using isotope dilution mass spectroscopy. We have used the average of the atomic absorption and mass spectroscopic data. The particle size calibration of the impactor was checked by a gravimetric technique using monodisperse polystyrene latex particles and agreed closely with the manufacturer's specifications.

Organic vapor-phase lead was measured by bubbling filtered ambient air through concentrated HNO_3 at about 1 liter/min (11). Lead analyses were by flameless (carbon rod atomization) atomic absorption.

High-volume filter (Whatman 41), low-volume filter (Gelman GA-1), and Lundgren impactor (polyethylene impaction substrates) samples were taken during the ACHEX study. All lead analyses were performed by R. Giauque of the Lawrence Berkeley Laboratory, using a high-resolution X-ray fluorescence technique (22,23). Carbon monoxide measurements were made with Beckman GC 6800 air quality chromatographs, calibrated against span gases, which in turn had been calibrated by the Air and Industrial Hygiene Laboratory of the California Department of Public Health.

In this paper, experimental errors are expressed as one standard deviation about the mean unless noted otherwise. These are related only to variations about the mean and not to individual uncertainties in each separate measurement. Uncertainties for derived quantities are compounded in the usual manner.

V. REMOVAL OF LEAD FROM THE ATMOSPHERE: DEPOSITION

Of the 16.7 tons/day of particulate lead which are exhausted, most is in the form of lead halide particles, although other chemical species have been identified. Depending on the particle size, a number of environmental pathways are available. For example, very large particles (D_p > 10 μm) settle rapidly. (Unless otherwise specified, all particle diameters are for the aerodynamically equivalent unit density spheres.) Smaller particles may remain airborne for a longer period, but some eventually deposit in the urban area by convective diffusion and other mechanisms. The mechanisms for particle removal have been discussed by Chamberlain (24,25) and Sehmel (26) for relatively well-defined smooth and rough surfaces. A knowledge of the particle size distribution is important in evaluating these effects.

The size distribution of auto exhaust lead aerosol has been studied by many investigators (4-9,27-30). Only Habibi (9) and Ter Haar et al. (7) attempted to simulate actual driving conditions, and of these, the study by Habibi was most detailed. Consequently, much of the following discussing is based on Habibi's results.

Habibi (8) sampled auto exhaust at the end of a 12-meter-long wind tunnel with an isokinetically operated cascade impactor. Coarse particles deposited in the wind tunnel were also measured and assigned to the first impactor stage. For an automobile operated solely on a chassis dynamometer programmed to the 1968 federal mileage accumulation schedule, the mass median diameter of the particulate lead increased from about 1 μm at 5000 accumulated miles to greater than 15 μm at 28,000 miles. At this latter mileage, 57% of the mass of lead was associated with particles larger than 9 μm in diameter. A car which had been driven on the road under typical consumer conditions for 15,000 miles followed by about 18,000 miles on the programmed chassis dynamometer also gave a lead aerosol with 57% by weight in particles larger than 9 μm in diameter and a mass median diameter greater than 15 μm.

Ter Haar et al. (7) tested 26 cars with 17,000 to 92,000 miles of service and concluded that 55% of the emitted lead was in "coarse" particles (i.e., particle diameters greater than about 5 μm). For an automobile operated for 12,000 miles on a mileage accumulation route, Ter Haar et al. also found that 58%

of the emitted lead was as coarse particles. (As noted above, however, 16% of the input lead could not be accounted for.) Ninomiya et al. (29) measured the size distribution of auto exhaust aerosol and found that approximately 20 to 30% by weight of the lead was as "coarse material" (500 μm < D_p < 5000 μm) for a dynamometer cycle consisting of a cold start followed by four federal test-procedure driving cycles.

For our analysis, we assume that the particle size distribution of auto exhaust lead at 33,000 accumulated miles measured by Habibi (9) is typical of Los Angeles cars. [From the known age distribution of California automobiles and the mileage accumulation rate of Los Angeles automobiles (Lees et al., 31), an average accumulated mileage of 57,000 is estimated.] This distribution is plotted in Figure 1 along with lead size distributions taken at a receptor site in Pasadena which was not in the immediate influence of traffic. In the receptor site distributions

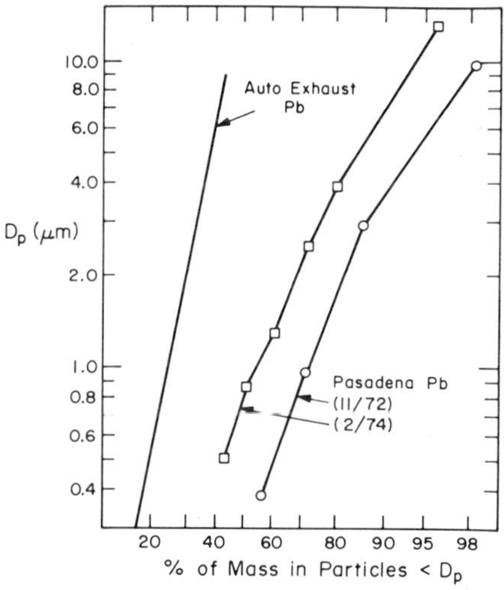

FIGURE 1. Cumulative mass distributions for lead aerosol in auto exhaust (9) and at Pasadena, a receptor site. Abscissa is a log-normal scale, and D_p is the aerodynamic, unit density particle diameter.

only 2 to 7% of the mass is in the greater than 9 μm fraction as compared to 57% for the auto exhaust distribution. Size distributions at receptor sites in the ACHEX study (21) also show very little mass (<5%) above 9 μm. In Figure 2, differential mass distributions for lead aerosol at a site 1 meter from a freeway and at the Pasadena receptor site show that the large-particle mode (D_p < 7 μm) at the freeway is severely attenuated at the receptor site. Size distribution measurements by Daines et al. (32) also show a decrease in large-particle lead with increasing distance from a highway.

The difference between the source and receptor size distributions is due in part to the rapid deposition of very large particles near the roadway. Habibi (8,9) found that approximately one half to two thirds of the greater than 9 μm fraction deposited within 7 meters of the automobile exhaust pipe in the wind tunnel experiments. In our analysis we assume that all of the greater than 9 μm fraction deposits on or near the roadway, and we label this fraction "near" (source) deposition. Although a cutoff at 9 μm is somewhat arbitrary, Heichel and Hankin (33) found that lead-containing particles deposited on trees adjacent to a heavily traveled road ranged in aerodynamic diameter from 7 to 32 μm with a mean of 17 μm. The near deposition fraction is thus 57% of the exhausted lead to which we assign an uncertainty

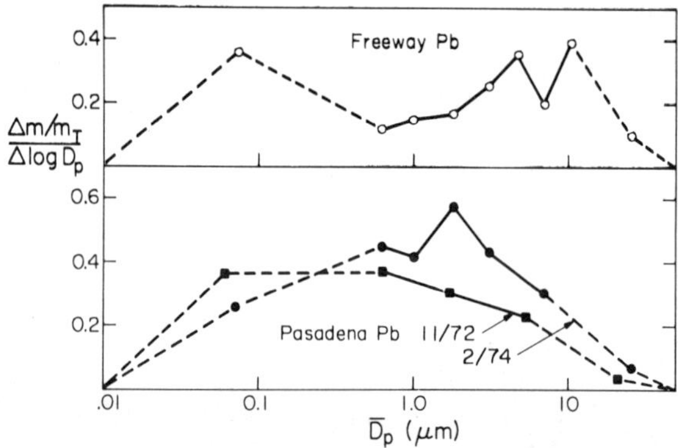

FIGURE 2. Differential mass distributions for lead aerosol at 1 meter from a freeway (May 1973) and for Pasadena (November 1972 and February 1974). Abscissa is the log mean diameter, and the dashed lines assume that the smallest particle size is 0.01 μm and the largest 50 μm; m_T is the total mass loading.

of ±10% to account for variations among automobiles and muffler changes at various mileages. This amounts to 9.5 ± 2.2 tons/day.

Deposition measurements at 1, 30, and 150 meters from a freeway, at which the traffic flow was unidirectional and slightly downhill, gave a factor of 15 decrease between the sites at 1 and 30 meters. The integrated deposition over this 150-meter strip could account for only about 10% of the near deposition expected from average emissions based on the above considerations. However, Habibi (34) has pointed out that most of the large-particle emission probably occurs at locations such as freeway on-ramps or uphill stretches where heavy accelerations occur frequently. We do not have deposition data near such sites. Other studies (32,35) have shown that considerable variation in the transport of large-particle lead away from the freeway can occur, depending on road characteristics and wind conditions.

It is convenient to designate lead which deposits at large distances from the source as "far" deposition. The average Pasadena deposition for 10 different periods between November 1972 and February 1974 was 45 ± 11 ng/cm^2/day. The deposition fluxes measured at different sites during the synoptic urban sampling period have been normalized to the Pasadena flux measured during the same period and are shown in Figure 3. These normalized deposition fluxes will be designated "deposition factors". The average deposition factor for the basin, excluding coastal sites, was 1.0 ± 0.4.

Related data, such as the atmospheric concentrations of lead and the concentrations of lead on the tops of wild oats (*Avena sativa*), can be used to check whether our deposition measurements are characteristic of the basin. [Studies by Motto et al. (36), Dedolph et al. (37), and Rabinowitz (19) have shown that lead in the tops of grass comes primarily from lead aerosol, while the soil contributes only 2 to 3 ppm (dry weight) of lead.] Deposition on horizontal surfaces and on oats depends primarily on the atmospheric concentration of lead and is relatively insensitive to the particle size distributions and wind conditions in the size range of 0.05 to 2 μm (24-26). Thus deposition factors based on normalized concentrations of atmospheric lead and lead on the tops of wild oats should be equivalent to our flat surface deposition measurements.

Rabinowitz (19) measured lead concentrations in wild oats between December 1971 and January 1972 for many sites in the Los Angeles area. These data represent the accumulation of lead over the whole growing season. [Rains and Thornton (38) have shown that lead is not washed off *Avena* by rainfall.] Tepper (15) has

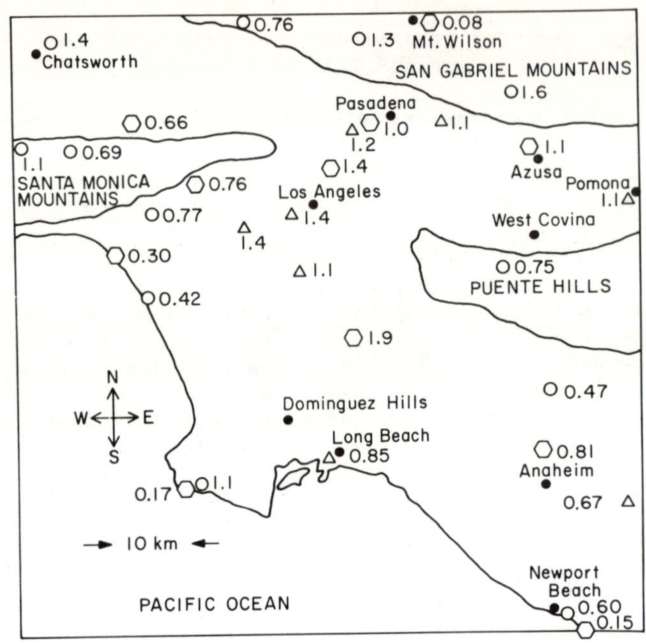

FIGURE 3. Distribution of deposition factors in the Los Angeles Basin. Hexagons correspond to surface deposition measurements from this work, the open circles to the wild oats (Avena sativa) data (19), and the triangles to the airborne data (15, 16, 20). The solid circles indicate specific locations (e.g., Pasadena). The southern boundary of the basin is taken to be the Pacific coast and the northern boundary the first crest of the San Gabriel Mountains (e.g., Mt. Wilson).

reported airborne lead concentrations of eight sites in the Los Angeles area for 1968-69 and the National Air Surveillance Networks (16) for four sites for 1969. Airborne lead concentrations for two sites in the basin during 1972 have been measured by ACHEX (21). The oats and airborne data have been normalized to the respective Pasadena values and are shown in Figure 3. In the case of the oats data, no measurement was made in Pasadena. Thus, these data were first normalized to central Los Angeles and renormalized to Pasadena on the basis of the airborne lead data.

The distribution of deposition factors in Figure 3 indicates that our deposition measurements are consistent with other measurements. The deposition flux can be calculated from the product of the average Pasadena deposition (45 ± 11 ng/cm^2/day), the basin-wide deposition factor (1.0 ± 0.4), and the area of the basin (4430 km^2). The resultant far deposition is 2.0 ± 1.0 tons/day.

This value assumes that the Los Angeles basin is a smooth surface. Vegetation and man-made structures will alter this number, but the magnitude of the effects is difficult to evaluate. The deposition on grassy surfaces can be estimated from the deposition velocities of Sehmel et al. (39) for 0.7-cm grass and an 8-km/hr wind. If the average atmospheric concentration is taken to be 2.4 μg/m^3 (the average for all measurements in this study) and the Pasadena size distribution assumed to be typical of the basin, then the estimated lead deposition for grass is about 40 ng/cm^2 (of land area)·day, which is not significantly different from our experimental value for the Teflon surface. Deposition on vertical surfaces will also contribute to the total far deposition. The vertical surface of buildings is about 25% of the total land area, and only about one fourth of this is exposed to the wind at a given time. Thus the additional deposition on buildings is probably small. Trees will also increase the far deposition term, but this effect is difficult to estimate. In general, more research is necessary to clarify these questions.

VI. AIRBORNE LEAD: REMOVAL BY WIND

Airborne lead and carbon monoxide, both of which result almost completely from automobile exhaust in the Los Angeles area, are known to be positively correlated (17,18). On the time scale of the air flow through the Los Angeles basin, carbon monoxide is unreactive and therefore conserved. Only on very smoggy days with high oxidant concentrations will CO be removed at rates of the order of 1%/hr by reaction with OH (40,41). To estimate the amount of lead blown out of the basin, we consider the basin to be a continuously stirred flow reactor. In this approximation, the input and output flows of carbon monoxide are equal, and the lead output flow can be estimated from the relationship

$$q_{Pb} = q_{CO}\overline{[Pb/CO]} \qquad (1)$$

where q_i is the mass flow rate of species i out of the basin and $\overline{[Pb/CO]}$ is the average value of the ratio of lead to carbon monoxide at receptor sites in the basin. This approximation implies a constant proportionality between lead and carbon monoxide throughout the atmospheric mixing layer. Although vertical profiles of lead concentrations are not available, this assumption is reasonable because the dominant mechanism for the mixing of both species is eddy diffusion which will disperse both the gas and the aerosol (of which about 70% by weight is in the submicron fraction) in a similar manner.

Lead and carbon monoxide concentrations at five different sites in the basin are available for 1972, 1973, and 1974 from this work and from the ACHEX measurements. With the exception of one sampling period, all the lead measurements were made by high-volume sampling with Whatman 41 filters. Simultaneously, lead concentrations were sampled at 2-hr intervals by low-volume sampling with Gelman GA-1 membrane filters. For the 1973 ACHEX measurements, the ratio of "high volume" lead to "low volume" lead was 0.86 ± 0.08 where "low volume" lead is the lead concentration determined by summing and averaging the low-volume filters over the high-volume sampling period. The high-volume filter data have been used in our calculations because of the greater possibility of systematic error (e.g., from contamination) involved in the handling of the low-volume filters. Within the accuracy of the treatment in this paper, no significant difference results from the choice of one or the other set of data however.

The average Pb/CO ratios for the sites are given in Table 1, and the average for all sites is 0.69 ± 0.18 $\mu g/m^3$ (Pb)/ppm (CO). The Pomona site is at the eastern border of our study region, and the Rubidoux site is farther east and somewhat outside the study region. Consequently Pb/CO ratios at these sites should be representative of air leaving the Los Angeles metropolitan area since the major air flow is to the east and north. The close agreement between Pb/CO at these sites and $\overline{\text{Pb/CO}}$ implies that the continuously stirred flow reactor formalism is a reasonable approximation.

When the $\overline{\text{Pb/CO}}$ ratio is adjusted for nonautomotive carbon monoxide and converted to a dimensionless weight ratio, the ratio is $6.2 \pm 1.6 \times 10^{-4}$. On the basis of seven-mode cycle data, the Los Angeles Air Pollution Control District (12) estimates that 7200 tons of carbon monoxide were emitted each day by automobiles during 1972. Orange County adds an additional 20% (42), bringing the total to 8600 tons/day. We have assigned an uncertainty of ±50% to this emission, although the uncertainty may be even larger (12). The resulting rate of removal of a particulate lead by advection is then 5.3 ± 3.0 tons/day.

TABLE 1. Pb/CO at Various Sites in Los Angeles Basin

Site	Pb/CO (μg/m^2·ppm)
Pasadena (10)	0.46
Pomona (7)	0.74
West Covina (5)	0.79
Dominguez Hills (2)	0.81
Rubidoux (3)	0.65
Average	0.69 ± 0.18

The numbers in parentheses indicate the number of sampling days at each site.

In addition to the particulate airborne lead, there is also a vapor phase, organic component. Purdue et al. (43) found the organic component to be about 10% of the total for six American cities. Skogerboe (11) measured organic fractions ranging between 4 and 12% at a receptor site in Fort Collins, Colorado. For a 3-day period in June 1974, we measured the organic component to be 6 ± 1% of the total Pasadena airborne lead. If this is typical of the Los Angeles region, 0.3 ton/day of vapor-phase lead are removed by advection. The difference between the input of organic lead (1.2 tons/day) and the output is due to photolysis of the organic lead vapor to produce a lead-containing aerosol (44). Eventually all of the vapor phase lead will decompose to an aerosol.

A mass balance can now be constructed, and, as shown in Table 2, the agreement between input and output routes is good. (This mass balance applies to dry weather, the situation for more than 90% of the days in Los Angeles). Because each of the output terms in the balance was independently derived, the agreement between input and output indicates that all major environmental pathways have been considered. This conclusion must be tempered, however, by the relatively large uncertainties in each of the terms. Improvements in the flow estimates will require better source characterization and a better theoretical and experimental understanding of particle removal processes.

The three output routes listed in Table 2 are not necessarily sinks. For example, lead which deposits on the streets can be washed off by rain into storm sewers which empty into the ocean, or lead which is blown out of the basin serves as an input to other geographical regions. In the remainder of this paper,

TABLE 2. Mass Balance for Automobile-Emitted Lead

Input (tons/day)			Output (tons/day)		
Evaporation of tetraethyllead		0.3	Near source deposition		9.5 ± 2.2
Auto exhaust		17.6 ± 2.6	Far deposition		2.0 ± 1.0
Aerosol	16.7		Removed by wind		5.6 ± 3.0
Organic vapor	0.9		Aerosol	5.3	
			Organic vapor	0.3	
		17.9 ± 2.6[a]			17.1 ± 3.9

[a] Inclusion of industrial emissions (0.3 ton/day) increases the total lead emissions to 18.2 tons/day.

we consider two subsystems of environmental pathways: the input of lead to the Los Angeles coastal waters and the fate of the lead blown out of the urban area.

VII. LEAD INPUT TO THE COASTAL WATERS: ATMOSPHERIC DEPOSITION

The input routes to the coastal waters are conveniently divided into atmospheric deposition, rainout-washout, rainy weather runoff, dry weather runoff, discharge of treated sewage, and direct discharge of untreated wastes. We shall treat all but the last. With the exception of sewage discharge and direct discharge, all the routes originate in the atmosphere.

We have no data on the particle collection properties of the ocean surface, and consequently our estimate of the atmospheric deposition on the coastal waters is based on dry deposition measurements on the coastal islands. As noted above, the lead deposition rates on our standard Teflon collector and on a crystallization dish filled with water were the same. Of course, the dynamics of the ocean surface are quite different from those of the crystallization dish, and wave action may alter the particle deposition velocity at the ocean surface.

The deposition fluxes of lead on Santa Catalina Island and San Clemente Island during the summer of 1973 are plotted as a function of distance from the coast in Figure 4. Rabinowitz (19)

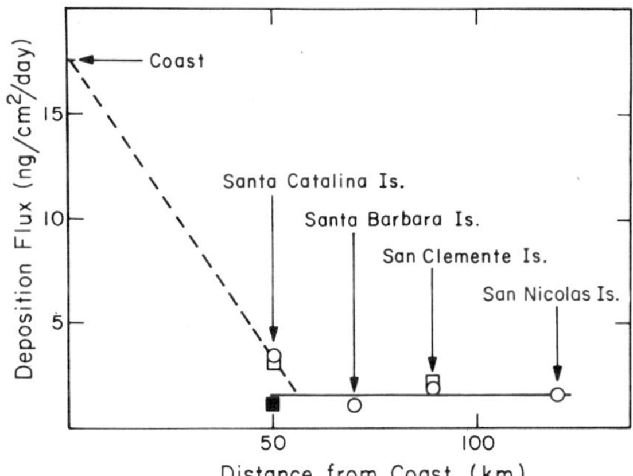

FIGURE 4. Deposition of lead as a function of distance from the coast. Open squares are deposition measurements from this work, the solid square the deposition measurement by Patterson and Settle (45), and the circles the equivalent flat surface deposition fluxes from the *Avena sativa* (19). The significance of the solid and dashed lines is explained in the text.

has measured the concentration of lead in the tops of wild oats on San Nicolas, San Clemente, Santa Barbara, and Santa Catalina Islands and along the coast. From deposition factors calculated from these data, equivalent flat surface deposition fluxes have been derived and are also plotted in Figure 4. A soil contribution of 2 ppm (dry weight) for the oats data was subtracted. The coastal deposition value in Figure 4 is the average of the surface deposition and oats data.

Patterson and Settle (45) have measured the surface deposition of lead on Santa Catalina Island for the 2-week period immediately following our measurement and found a flux of 1.4 ng/ $cm^2 \cdot day$ in comparison with our value of 3.3 $ng/cm^2 \cdot day$. They also measured the $^{206}Pb/^{207}Pb$ ratios for the Catalina lead and lead depositing in Pasadena and found a significant difference (1.171 and 1.193, respectively). Since leads are different if the $^{206}Pb/^{207}Pb$ ratios differ by 0.3%, they concluded that the lead depositing on Catalina during their measurement did not originate in Los Angeles air. This is consistent with the local meteorology. The prevailing wind pattern during the year is onshore flow during the day. At night, weak offshore drainage winds converge with the normal onshore flow. The western edge of the convergence zone is poorly defined, but may extend as far offshore as the eastern edge of Santa Catalina Island (46). Thus Santa Catalina can occasionally receive Los Angeles air, but except under Santa Ana conditions (east winds from the mountains and deserts), the islands which are farther offshore do not receive Los Angeles air.

Hidy et al. (47) have measured an average atmospheric concentration of lead of 120 ng/m^3 on San Nicolas Island (about 120 km from the coast) during the summer and fall of 1970. This corresponds to a surface deposition rate of about 2 ng/cm^2·day, in reasonable agreement with the Patterson and Settle value of 1.4 ng/cm^2·day for Santa Catalina and our value of 2.4 ng/cm^2·day for San Clemente Island. The somewhat higher value of Santa Catalina deposition, which we measured, may represent a slight penetration of Los Angeles air, although no isotopic ratio measurements were made to confirm this. Early morning east winds on Santa Catalina were recorded on 5 days during our measurement, but not at all during the Patterson-Settle measurement. The deposition of cadmium on Santa Catalina was also somewhat higher than on San Clemente Island (48).

We interpret the coastal islands data by assuming that the influence of Los Angeles beyond Santa Catalina Island is negligible. Because deposition data between the coast and Catalina are not available, a simple linear interpolation (the dashed line in Figure 4) is used to estimate the decrease in deposition over this region. The "background" or non-Los Angeles lead is represented by the solid line in Figure 4 and is the average of Patterson and Settle's Catalina flux and the fluxes for the outer islands. The dashed line defines a region of influence of Los Angeles extending to about 55 km from the shore which, for a linear coastal distance of 80 km, corresponds to an area of 4400 km^2. The average (dry) deposition flux over this region is 0.35 ton/day or 120 tons/year assuming 346 "dry" (i.e., rainfall less than 2.5 mm/day) days (49). This estimate is likely to be an upper limit because of the interpolation and, as such, does not depend critically on the somewhat enhanced deposition on Catalina. Further measurements are necessary to define the origin of the "background" lead.

Chow et al. (50) and Bruland et al. (51) have measured lead accumulation rates in the ocean sediment of 4.7 ng/cm^2·day in San Pedro Basin (∿30 km from the coast), 2.5 ng/cm^2·day in Santa Monica Basin (∿50 km from the coast), and 5.7 ng/cm^2·day in Santa Barbara Basin (∿30 km from the coast). (Santa Monica and San Pedro Basins are directly off Los Angeles, while Santa Barbara Basin is off Santa Barbara, about 100 km northwest of Los Angeles). Although these accumulation rates are of the same order as the deposition fluxes shown in Figure 4, their interpretation is complicated by various factors: the influence of sewage outfalls, a combination of advective ocean transport and settling, and dissolution during settling. Thus, a direct comparison between these accumulation rates and our deposition measurements is not possible, but the general agreement lends support to our interpretation of the deposition data.

The lead input to the coastal waters associated with rainout-washout can be calculated from the average annual rainfall over the coastal waters and the lead content of the rainfall. The average rainfall at three coastal sites and Santa Catalina Island is 30 cm/year (49). In 1966-67, Lazrus et al. (52) measured the flux of rainfall lead on Santa Catalina to be 25 $ng/cm^2 \cdot cm$ of rain. Thus, the lead input to the region of influence defined by the deposition measurements is 30 tons/year.

Although positive identification of the origin of the rainfall lead has not been made, meteorological data strongly suggest Los Angeles as the source. During the period of heaviest rain, November through February, the majority of the precipitation winds are from the quadrant centered on E-ENE with an average speed of about 18 km/hr (46). Such winds will carry Los Angeles air well out over the ocean. In fact, the region of influence of precipitation winds from Los Angeles may be larger than the region of influence determined from the deposition measurements. No data are available, however, to anwer this question.

VIII. LEAD INPUT TO THE COASTAL WATERS: RUNOFF AND SEWAGE

In addition to the direct input by rainout-washout, rain storms will wash lead-containing particles into the storm sewers and ultimately into the coastal waters. Lead in runoff results primarily from street dirt, as soluble lead depositing on soil is immobilized by sorption on soil particles (53) and is not subject to significant washoff. During the winter of 1971-72, the Southern California Coastal Waters Research Project (54) measured trace metal concentrations in storm water runoff flowing in the concrete-lined rivers, which act as storm sewers for the region. These rivers carry significant amounts of water only during storms. The samples were obtained by a depth-integrated method at various times throughout the storms. The chemical analysis included both the dissolved and particulate fractions for the metals. There was only one major storm during this period, however, and it is difficult to arrive at a firm number for the runoff input. Because a firm data base is lacking for this environment pathway, we have developed a simple model for estimating the flux of runoff lead. This model is discussed in detail in Appendix B. From the model it is estimated that approximately 140 tons/year are washed off the streets into the coastal waters during the rainy season (November to April). During the dry season, a small amount of water flows in the streets as a result of such domestic activities as lawn watering. Runoff during this season adds another 10 tons/year (Appendix B).

The final contributor to the anthropogenic lead burden of the coastal waters is the discharge of treated sewage, which amounts to 230 tons/year or 0.64 ton/day (55-57). The coastal water inputs are summarized in Table 3. The four "atmospheric" routes (dry deposition, rainout-washout, and dry and wet season runoff) account for more than 50% of the total, indicating that effective control of lead emissions to the coastal waters can be achieved only when both sewage and "atmospheric" routes are controlled. The runoff and sewage routes are essentially point sources of contaminants to the coastal waters, whereas the deposition and rainout-washout routes act over a larger area of influence. It is likely that these two types of routes will have different ecological effects.

TABLE 3. Input of Anthropogenic Los Angeles Lead to the Coastal Waters

Input route	Tons/year
Storm runoff	140
Dry deposition	120
Rainout-washout	30
Dry runoff	10
Municipal sewage	230

IX. TRANSPORT OF WIND-BLOWN LEAD OUTSIDE THE BASIN

To investigate the fate of the 5.6 tons/day of lead blown out of the basin, lead deposition at four sites each in the San Gabriel-San Bernardino Mountains north of Los Angeles and in the Mojave Desert north of the mountains was measured. The average deposition factors in these regions were 0.07 and 0.04, respectively, in general agreement with factors calculated from wild oats data (19) for the same regions. These factors, uncorrected for local sources, represent the maximum amount of Los Angeles lead depositing in these regions. If these deposition factors are applied to a 10,000-km^2 area corresponding to the mountain and desert part of the quadrant centered at northeast and extending approximately 150 km from downtown Los Angeles, a deposition flux of 0.2 ton/day is obtained. Although the source of this lead cannot be unequivocally attributed to Los Angeles, the general wind patterns and the sparse local traffic density strongly suggest Los Angeles as the source.

Directly to the east of Los Angeles is a semiurban region (Riverside-San Bernardino) of about 2500 km^2 in area. Atmospheric lead concentrations measured at two sites in this region during the 1972-73 ACHEX study were typical of lead concentrations in the Los Angeles urban area. From the previous discussion of lead deposition, the maximum deposition factor for Los Angeles lead in this region can be set equal to the average Los Angeles deposition factor. (We specify maximum, because local sources as opposed to lead blown out of Los Angeles will also contribute to the deposition.) This gives a flux of 1.1 tons/day. When the deposition in the mountains, deserts, and coastal waters is included, we can account for only 1.7 out of the 5.6 tons/day which leave Los Angeles. The difference, about 4 tons/day, is transported beyond a radius of 150 km from the city and exceeds the total lead emissions in the surrounding semiurban and nonurban areas [as estimated on the basis of the respective motor vehicle population and carbon monoxide emissions (42)].

X. SUMMARY

The flow of automobile-emitted lead is summarized in Figure 5. In this flow diagram, the flow rates are daily averages - i.e., the flow rates are the yearly fluxes divided by 365. We do not have data for the rainout-washout of lead over the land area of the basin. Thus, the values for far deposition and wind removal are for dry weather only, the prevailing situation for about 90% of the year in Los Angeles. Of the 18 tons/day which are exhausted, approximately two thirds deposits within the basin while one third is blown out. Most of the lead which deposits on the land is immobilized in the soil, but that which settles on streets can be washed into storm sewers and ultimately into the coastal waters during the rainy season. The four "atmospheric" routes of wet and dry season runoff, dry weather deposition, and rainout-washout account for more than 50% of the input of Los Angeles lead to the coastal waters, implying that a control strategy for coastal water lead must consider both automobile-emitted lead as well as wastewater discharges. The lead which remains airborne is primarily in submicron particles and can be

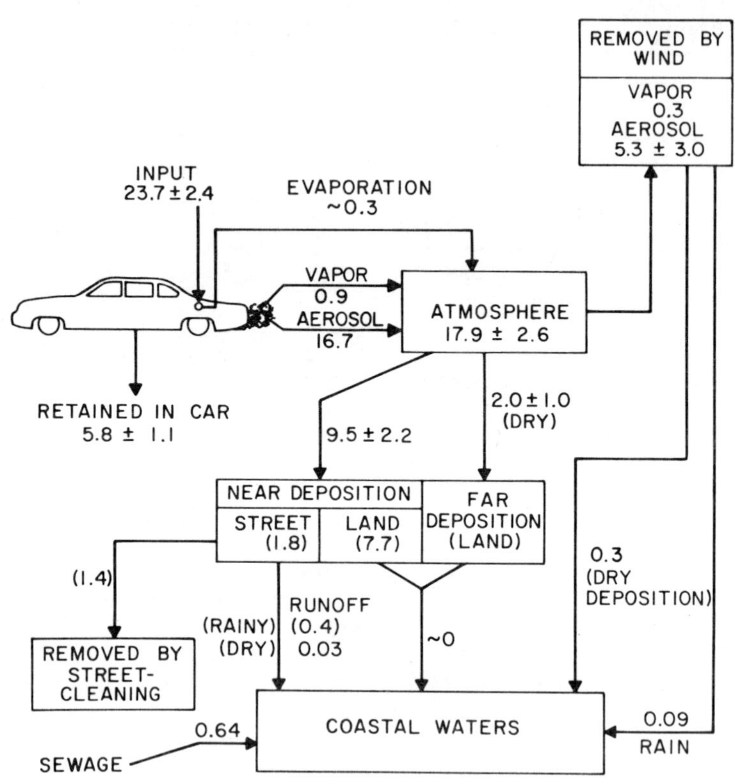

FIGURE 5. Flow of automobile-emitted lead through Los Angeles Basin. With the exception of the far deposition and wind removal fluxes, which are for dry weather only, the values are daily averages calculated by dividing the yearly totals by 365. All fluxes are in metric tons per day. Numbers in parentheses are model-dependent and in need of a more extensive data base.

transported well beyond the boundaries of the urban region. This lead is the major source of atmospheric lead for the regions immediately downwind of Los Angeles, and control of atmospheric lead in these regions will depend upon control of lead emissions in Los Angeles.

The material balance-flow pathway approach is general and can be applied to other environmental pollutants. Although the method does not reveal the details of pollutant disperson in the environment, the requirement of mass balance demands that all important environmental pathways be identified and quantified. This is potentially a powerful tool in assessing the environmental impact of a pollutant. Unfortunately, however, large uncertainties exist for both input and output terms for the lead balance and will exist for most other species for which the data bases are relatively meager in comparison with lead. Refinement of this approach will require more detailed source characterizations and a better theoretical and experimental understanding of particle removal processes in the atmosphere.

APPENDIX A. LEAD CONTENT OF LOS ANGELES GASOLINE

For the winter of 1971-72 the U.S. Bureau of Mines (58) reported that for gasoline sold in Southern California, the average premium gasoline contained 0.73 g of Pb/liter, the average intermediate-grade gasoline 0.56 g/liter, and the average regular grade 0.40 g/liter. In addition, the Bureau of Mines also reports a composite value for so-called "third-grade" gasolines. We have subdivided this category into "low-lead" gasoline with 0.12 g/liter and "unleaded" with 0.01 g/liter. Because of the lack of uniformity with respect to labeling of the various grades, our use of "low-lead" and "unleaded" may not correspond to what one buys at a gasoline station. For example, many actual "low-lead" gasolines are probably classified by the Bureau of Mines as "regular".

According to the Los Angeles Air Pollution Control District (12) 62.9% of the gasoline sold in Los Angeles County in the summer of 1972 was premium. (Because of the increasing fraction of low-compression vehicles on the road and the relative increase in the consumption of lower lead gasoline, this percentage applies strictly only to 1972.) The remainder is divided among intermediate, regular, "low-lead" and "unleaded" grades. The lead concentration for the intermediate grade falls within the range of concentration found in regular gasoline, and little uncertainty is introduced by neglecting this grade. The sale of "unleaded" gasoline in 1972 was small and can be neglected because very few automobiles were built to run on unleaded fuel. Thus, we consider two cases which represent upper and lower limits for lead consumption. In the first we assume 62.9% of the gasoline is premium and 37.1% regular and in the second 62.9% premium and 37.1% "low-lead". The average lead content from these two cases is 0.56 ± 0.06 g/liter, where the uncertainty in this case refers to the two possible extremes.

APPENDIX B. ROADWAY DEPOSITION OF LEAD AND RUNOFF

The accumulation of lead in street dirt, the source of runoff lead, can be approximated by a simple differential equation:

$$\frac{dm}{dt} = Q - Rm \tag{B1}$$

the solution of which is:

$$m = \frac{Q}{R} + e^{-R\ \Delta t}\left(m_0 - \frac{Q}{R}\right) \tag{B2}$$

where m is the mass of lead on the street, Q is the deposition rate minus the amount blown off the roads by wind, R is the fraction removed per unit time by streetsweeping, Δt is the time in weeks since the last rain storm, and m_0 is the amount of lead remaining on the streets at the end of the last storm. Equation B2 is exact only when deposition and sweeping are continuous and constant with repect to time, but is sufficiently accurate for this analysis. The amount of lead washed off the streets is the product of m in Equation B2 and the washoff efficiency ε, which is a function of the rainfall intensity.

During the first major storm (December 21-22) of the 1971-72 rain season, 9.8 tons of lead were washed off the Los Angeles River and Ballona Creek drainage basins (54). These drainage basins account for 39% of the traffic density in the Los Angeles area (59) and an extrapolation on the basis of traffic density to the whole Los Angeles area gives a washoff of 25 tons. Because of the weak nature of the preceeding storms, it is likely that the mass loading of lead on the streets prior to the December 21-22 storm was close to the steady state value (Q/R in Equation B2). Thus $\varepsilon Q/R$ is equal to 25 tons.

Annual washoff can be estimated in the following manner. The rainy season in Los Angeles extends from November to April with 8 days of rainfall greater than 12 mm (49) or an average of about one major storm per month. (Actual rainfall patterns are complex and variable). For a model rain year comprised of six major storms at 4-week intervals, t in Equation B2 is 4 weeks for all storms except the first. Pitt and Amy (60) found that the removal efficiency of lead by streetsweepers was 51% for a single pass. If the streetsweeping frequency is once per week, R is 0.51 per week. Using these values of R, t, and Q/R and neglecting m_0 for all storms, we obtain a washoff in the first storm of 25 tons and in subsequent storms 22 tons/storm for a total of about 140 tons/year. This estimate is independent of Q/R and is only weakly dependent on R and m_0. The critical assumptions were the representative nature of the December 21 storm with respect to washoff efficiency and the extrapolation to area-wide runoff on the basis of traffic density. Considerable variations in the runoff lead flux can occur, however, and more data are needed for a better understanding of this environmental pathway.

The deposition rate of lead on the streets can be estimated by setting the washoff efficiency for the December 21 storm to one. If, as assumed above, the streets are swept once a week with a 51% efficiency for lead removal, the roadway deposition rate is 1.8 tons/day or 660 tons/year. The remainder of the near deposition, 7.7 tons/day, deposits on the land adjoining the roadways. Because of the model dependency and the assumptions involved, these numbers must be regarded as approximate.

There is also a small runoff contribution during the dry season. We have measured the total lead content of dry weather runoff in the Ballona Creek drainage basin. If the result is extrapolated on the basis of traffic density to the whole urban area, a daily flow of 0.03 ton/day or 10 tons/year is obtained. The total input to the coastal waters via runoff is therefore 150 tons/year. The difference between the roadway deposition and the runoff flux is 510 tons/year or 1.4 tons/day. This is the amount removed by streetsweeping. In Los Angeles, this lead is dumped in sanitary landfills and is immobilized in the same manner as soil lead.

ACKNOWLEDGMENT

We thank C. C. Patterson, Y. Hirao, D. Settle, J. R. Spencer, G. Gartrell, P. J. W. Roberts, D. Young, J. Rath, S. Garcia, and L. Hashimoto for technical assistance and advice.

This work was supported in part by grants from the Rockefeller Foundation, the California Air Resources Board, and the National Institute of Environmental Health Sciences, Grant No. 3 T01 ES00080-07S1.

REFERENCES

1. California State Board of Equalization. Motor Vehicle Fuel (Gasoline) Distributions, Department of Business Taxes, 1972.

2. Carey, R. T. (Ed.). 1973. "California County Fact Book 1973", (County Supervisors Association of California, California County Government Education Foundation, 1973). The actual percentage of California vehicles in the metropolitan parts of Los Angeles and Orange Counties is somewhat less than 41.4 because not all of the two counties are included in the study region. The excluded areas are sparsely populated, however, and do not contribute significantly to the automobile population.

3. Lemke, E. E. 1971. "Profile of Air Pollution Control", (Air Pollution Control District, County of Los Angeles.

4. Hirschler, D. A., Gilbert, L. F., Lamb, F. W., and Niebylski, L. 1957. Ind. Eng. Chem. 49, 1131.

5. Hirschler, D. A. and Gilbert L. F. 1964. Arch. Environ. Health 8, 297.

6. Mueller, P. K., Helwig, H. L., Alcocer, A. E., Gong, W. K., and Jones, E. E. 1964. ASTM Special Tech. Publication No. 352, 60.

7. Ter Haar, G. L., Lenane, D. N., Hu, J. N. and Brandt, M. 1972. J. Air Pollut. Contr. Assn. 22, 39.

8. Habibi, K. 1970. Environ. Sci. Technol. 4, 239.

9. Habibi, K., 1973. ibid. 7, 223.

10. Brief, R. S. 1962. Arch Env. Health 5, 527.

11. Skogerboe, R. K. 1974. Private Communication.

12. Los Angeles County Air Pollution Control District. 1972-1974. ibid.

13. Waste Oils Pose Disposal Dilemma", Environ. Sci. Tech. 6, 25 (1972).

14. Robinson, E. and Ludwig, F. L. 1967. J. Air Pollut. Contr. Assn. 17, 664.

15. Tepper, L. B. 1971. "Seven-City Study of Air Pollution Lead Levels: An Interim Report", Environ. News, Environmental Protection Agency, June 4.

16. National Air Surveillance Networks data supplied by R. Ireson. Environmental Protection Agency, San Francisco, Calif. 1973.

17. Public Health Service, Division of Air Pollution, U.S. Department of Health, Education and Welfare. 1965. "Survey of Lead in the Atmosphere of Three Urban Communities", PHS 999-AP-12.

18. Colucci, J. M., Begeman, C. R. and Kumler, K. 1969. J. Air Pollut. Contr. Assn. 19, 255.

19. Rabinowitz, M. 1972. Chemosphere 1, 175.

20. Hidy. G. M. 1973. "Characterization of Aerosols in California" (Interim Report for Phase I. State of California Air Resources Board Contract No. 358).

21. Hidy, G. M. 1974. Unpublished data from ACHEX (Phase II).

22. Giauque, R. D., Goulding, F. S., Jaklevic, J. M., and Pehl, R. H. 1973. Anal. Chem. 45, 671.

23. Giauque, R. D., Goda, L. Y. and Brown, N. E. 1974. Environ. Sci. Technol. 8, 436.

24. Chamberlain, A. C. 1967. Proc. Roy. Soc. A296, 45.

25. Chamberlain, A. C. 1967. Contemp. Phys. 8, 561.

26. Sehmel, G. A. 1972. In Assessment of Airborne Particles. T. T. Mercer, P. E. Morrow, W. Stöber, Ed., p. 48. Charles A. Thomas, Springfield, Ill.

27. Moran, J. B. and Manary, O. J. 1970. "Effect of Fuel Additives on the Chemical and Physical Characteristics of Particulate Emissions in Automotive Exhaust", Interim Technical Report to the National Air Pollution Control Administration, Dow Chemical Co., APTD 0618, PB 196 783.

28. Lee, Jr., R. E., Patterson, R. K., Crider, W. L. and Wagman, J. 1971. Atmos. Environ. 5, 225.

29. Ninomiya, J. S., Bergman, W. and Simpson, B. H. 1971. In Proceedings of the Second International Clean Air Congress, M. Englund and W. T. Beery, Ed., p. 663. Academic Press, New York.

30. Ganley, J. T. and Springer, G. S. 1974. Environ. Sci. Technol. 8, 340.

31. Lees, L., et al. 1972. "Smog - A Report to the People", Environmental Quality Laboratory, California Institute of Technology, Pasadena, Calif.

32. Daines, R., Motto, H. and Chilko, D. 1970. Environ. Sci. Technol. 4, 318.

33. Heichel, G. H. and Hankin, L. 1972. Ibid 6, 1121.

34. Habibi, K. 1973. Private communication.

35. Cahill, T. A. and Feeney, P. J. 1973. "Contribution of Freeway Traffic to Airborne Particulate Matter", University of California, Davis UCD-CNL 169, Final Report to the California Air Resrouces Board.

36. Motto, H., Daines, R., Chilko, D. and Motto, C. 1970. Environ. Sci. Technol. 4, 231.

37. Dedolph, R., Ter Haar, G., Holtzman, R. and Lucas, H., Jr. 1970. Ibid, p. 217.

38. Rains, D. W. and Thornton, S. 1970. "Vegetation as an Indication of Long-Term Lead Pollution, Its Extent and Effect", University of California, Davis, Calif.

39. Sehmel, G. A., Sutter, S. L. and Dana, M. T. 1973. "Dry Deposition Processes", Battelle Northwest Laboratory Rep. 1751 PT1.

40. Weinstock, B. and Niki, H. 1972. Science 176, 290.

41. Wilson, W. E., Jr. 1972. J. Phys. Chem. Ref. Data 1, 535.

42. State of California. 1972. "Implementation Plan for Achieving and Maintaining the National Ambient Air Quality Standards", January.

43. Purdue, L. J., Enrione, R. E., Thompson, R. J. and Bonfield, B. A. 1973. Anal. Chem. 45, 527.

44. Milde, R. L. and Beatty, H. A. 1959. In Metal Organic Compounds. Adv. Chem. Ser. No. 23, 306, ACS, Washington, D.C.

45. Patterson, C. and Settle, D. 1974. "Contribution of Lead via Aerosol Deposition to the Southern California Bight", J. Rech. Atmos., Special Publication, "Internat. Symp. on Chemistry of Sea/Air Particulate Exchange Processes" (Nice, France), July-Dec. 1965.

46. DeMarrais, G. A., Holzworth, G. C. and Hosler, C. R. 1965. "Meteorological Summaries Pertinent to Atmospheric Transport and Dispersion over Southern California", Tech. Paper No. 54, U.S. Weather Bureau, Washington, D.C.

47. Hidy, G. M., Mueller, P. K., Wang, H. H., Karney, J., Twiss, S., Imada, M., and Alcocer, A. 1974. J. Appl. Meteor. 13, 96.

48. Davidson, C. I., Huntzicker, J. J. and Friedlander, S. K. 1974. Division of Environ. Chem., 167th Meeting, ACS, Los Angeles, Calif., March.

49. U.S. Department of Commerce. 1968-1973. Climatological Data California, National Oceanic and Atmospheric Administration, Environmental Data Service.

50. Chow, T. J., Bruland, K. W., Bertine, K. K., Soutar, A., Koide, M. and Goldberg, E. D. 1973. Science 181, 551.

51. Bruland, K. W., Bertine, K., Koide, M. and Goldberg, E. D. 1974. Environ. Sci. Technol. 8, 425.

52. Lazrus, A. L., Lorange, E. and Lodge, J. P. Jr. 1970. Ibid. 4, 55.

53. Zimdahl, R. L. 1972. In "Impact on Man of Environmental Contamination Caused by Lead", H. W. Edwards, Ed., p. 98, Colorado State University, NSF Grant GI-4 Interim Rep.

54. Southern California Coastal Water Research Project (SCCWRP). 1973. "The Ecology of the Southern California Bight: Implications for Water Quality Management", (SCCWRP TR104): The runoff of lead was calculated by L. Hashimoto from raw data supplied by D. Young of SCCWRP.

55. Parkhurst, J. E. 1973. "Technical Report: Waste Discharge to the Ocean", Sanitation Districts of Los Angeles County, January.

56. City of Los Angeles, Department of Public Works. 1973. "Technical Report: Waste Discharge to the Ocean", January.

57. County Sanitation Districts of Orange County. 1973. "Technical Report for Water Quality Control Plan Ocean Waters of California", January.

58. Shelton, E. M. 1972. "Motor Gasolines, Winter 1971-72". Mineral Industry Surveys, Petroleum Products Survey No. 75, U.S. Department of the Interior, Bur. Mines, June.

59. Roberts, P. J. W., Roth, P. M. and Nelson, C. L. 1971. "Contaminant Emissions in the Los Angeles Basin", Appendix A of "Development of a Simulation Model for Estimating Ground Level Concentrations of Photochemical Pollutants", Rep. 71 SaI-6, Systems Applications Inc., Beverly Hills, Calif.

60. Pitt, R. E. and Amy G. 1973. "Toxic Materials Analysis of Street Surface Contaminants", Environmental Protection Agency Rep. EPA-R2-73-283.

Sulfate and Nitrate Data from the California Aerosol Characterization Experiment (ACHEX)*

BRUCE R. APPEL, EVALDO L. KOTHNY, EMANUEL M. HOFFER, AND JEROME J. WESOLOWSKI
Air and Industrial Hygiene Laboratory
Laboratory Services Branch
California State Department of Health
Berkeley, California

*Published, in part, in Environmental Science & Technology 12, 418 (1978).

Abstract

Sulfate and nitrate data obtained in the summers of 1972 and 1973 with five fixed and one mobile sampling laboratory within California's South Coast Air Basin are presented. Included are 24-hr average concentrations, diurnal variations, and size distributions obtained using a Lundgren cascade impactor. In addition, results for ammonia, ammonium, and other chemical species are given. Two-hour average concentrations of nitrate and sulfate up to 149 and 70 μg/m^3, respectively, were observed. The mass median diameters for sulfate were 0.3 to 0.4 μm, and for nitrate, 0.3 to 1.6 μm, the latter varying with locations in the basin. Ratios of ionic constituents and ambient ammonia levels suggest that the sulfate and nitrate are present principally as ammonium salts.

I. INTRODUCTION

The chemical analysis of aerosol samples collected in the Aerosol Characterization Experiment (ACHEX) have provided data on many of their constituents (1). Of special importance have been nitrate and sulfate determinations. Although such data were obtained for aerosols collected throughout California, this paper focuses on data obtained in the South Coast Air Basin (SCAB). Although both sulfate and nitrate levels are elevated in most urban areas, California's SCAB exhibits particularly high values of nitrate as compared to other parts of the United States. Thus caution must be exercised in extrapolating the following discussion to other areas.

Sulfate and nitrate are of concern because of possible adverse health effects as well as visibility reduction. The possible harmful effects of these materials are enhanced because they are present in particles that are predominantly in the respirable size range. Light scattering, with the attendant reduction in visibility, is especially important for particles in the 0.1 to 1-μm size range. Sulfate and nitrate are among the most important contributors to atmospheric aerosols in this size range.

This paper includes the observed 24-hr average concentrations of suspended sulfate and nitrate in the SCAB, diurnal patterns for sulfate and nitrate as compared to other pollutants, and size distributions of sulfate and nitrate on a 24-hr average basis. Sampling in the SCAB was done during the periods September-October 1972 and July-October 1973. In 1972 sulfate and nitrate analyses were obtained only on 24-hr high-volume samples, while in 1973 short-term, low-volume samples were also analyzed for these species. Accordingly, this paper will emphasize the 1973 results.

II. EXPERIMENTAL

A. Aerosol Collection and Analysis Strategy

Particles less than about 20 μm were sampled isokinetically in the aerosol manifold system of the ACHEX mobile laboratory by three units: a five-stage Lundgren cascade impactor and two filters collecting all particles $\leq$ ca. 20 μm ("total filters"). In addition, aerosols were sampled on a 24-hr basis with a conventional high-volume sampler using Whatman 41 (cellulose) filters. The inlet for all samplers was from about 5 to 13 m above ground level. The total filters were 47-mm Gelman GA-1 cellulose acetate and 47-mm Gelman A glass fiber filbers, which permitted sampling at 3.0 cfm while providing >99% removal of particles (measured with room air dust with a condensation nuclei counter) (2). The cellulose ester and cellulose filters were used for determining inorganic aerosol constituents, and the glass fiber filters, for carbonaceous material. Cellulose acetate filters were also employed as the fifth stage ("afterfilter") of the Lundgren impactor. Total filters and afterfilters were changed with the same frequency, usually 8 times per 24-hr period, with 2-hr changes during daylight hours. In addition to the mobile laboratory, total-filter samples with the same time resolution were collected at five satellite stations during the 1973 summer sampling periods. Figure 1 shows the locations sampled by the mobile laboratory and satellite stations.

The Lundgren impactor has been previously described (3). The present units were modified by addition of a second stage 1 to remove more completely dry soil particles >8.0 μm in aerodynamic diameter (4) and were calibrated using laboratory-generated aerosols. The 50% cutoff points for unit density spheres, based on design criteria, agreed well with experimental calibrations and were 8.0, 4.0, 1.5, and 0.5 μm. The impaction surfaces were strips of polyethylene coated with a sticky hydrocarbon resin. Both laboratory and field evaluations of this surface indicate that substantial particle bounce-off should be expected with dry particles (4,5). This effect should be of lesser importance with sulfate and nitrate particles, however, which are expected to be liquid or sticky solid particles. The impaction strips for stages 1 to 4, representing a 24-hr period, were extracted in water to obtain 24-hr average size distributions for sulfate and nitrate. The extraction procedure

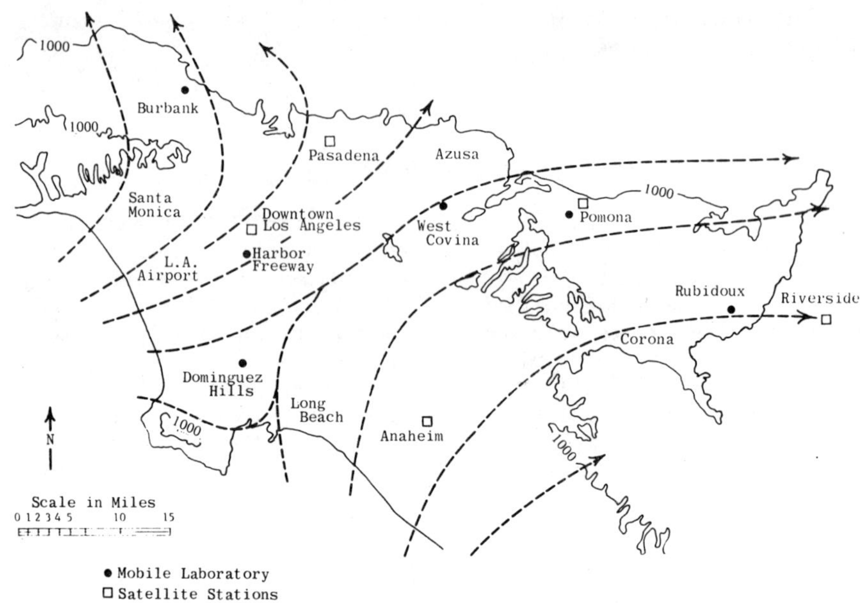

FIGURE 1. Sampling sites and typical summer daytime winds, 1972-1973 ACHEX.

consisted of immersion in water at $80^{\circ}C$ for 2 hr in sealed test tubes and yielded $\geq$92% removal of water-soluble sulfate and nitrate. For determining detailed 24-hr average size distributions, the 24-hr average sulfate and nitrate <0.5 μm was calculated from the serially collected after-filters, and the results used together with those for stages 1 to 4.

B. Aerosol Analytical Methods

Sulfate in low-volume and high-volume samples were analyzed by the Air and Industrial Hygiene Laboratory (AIHL) microchemical procedure (6) and the $BaSO_4$ turbidimetric procedure (7), respectively. Both yielded a precision, expressed as a coefficient of variation, CV, of about 5%. Comparative analysis of high-volume filter sample extracts by the procedures named and, additionally, by the automated methylthymol blue technique (8) and a modified Brosset procedure (9) yielded agreement within 10%. The accuracy of the AIHL procedure, established by extraction and analysis of glass fiber filter strips spiked with known amounts of sulfate, averaged 90%. Nitrate analysis was done by the 2,4-xylenol procedure with minor modification (10,11). The CV for these analyses was 13%. In addition, x-ray-induced, energy-dispersive x-ray fluorescence analysis (XRFA) provided the total sulfur in the aerosol for these same filter samples (12). Ammonium analyses were done by the indophenol blue procedure (13), for which a CV of 5% was found.

C. Total Carbon

Samples collected on Gelman A glass filter filbers, preextracted with ethanol to reduce the carbon blank, were analyzed by combustion to CO_2 with quantitation by gas chromatography using the procedure of Mueller et al. (14). The CV for this method was 2%.

D. Pollutant Gas Analysis

Ozone was monitored continuously at the mobile laboratory with a REM Model 612 chemiluminescent analyzer calibrated against 2% buffered KI. Recent work has demonstrated that this calibration procedure yields values which are high by about 29% relative to values determined by the current reference method, ultraviolet absorption (15). The results are uncorrected for this difference. Nitric oxide and nitrogen dioxide was monitored continuously at the mobile laboratory with a Bendix Model 8101B chemiluminescent analyzer. Sulfur dioxide was determined using a Varian Model 1460 analyzer (chromatographic separation with a flame photometric detector). The continuous pollutant gas analyzers were recalibrated after relocation of the mobile laboratory to a new sampling site.

Ammonia determinations employed Gelman A glass fiber prefilters and bubbler sampling with 0.1N H_2SO_4, followed by analysis with the indophenol blue procedure (13). Samples were stabilized against bacterial decomposition with chloroform and analyzed within 2 weeks of collection. Collection efficiency for ammonia in the range 40 to 80 $\mu g/m^3$ was 65 ± 10% (16). Ammonia results have been corrected for this reduced collection efficiency. At lower concentrations it is likely that collection efficiencies decreased.

E. Quality Assurance

Separate reports, available on request, were prepared covering efficiencies of filter media(2), cascade impactor performance (4,5), and comparison of results obtained on different filter media and by different analytical methods (17). The ratio of the means of the analyses of 400 cellulose acetate filter samples for total sulfur by XRFA and for sulfate by the AIHL microchemical method was 1.01 ± 0.02.

III. SPATIAL DISTRIBUTION OF SULFATE AND NITRATE

Table 1 shows 24-hr average high-volume data from the ACHEX obtained at different times during the 1972-1973 period. Nitrate and sulfate together often comprised more than 15% of the total aerosol mass in the SCAB. Sulfate was less variable than nitrate throughout the basin, both on a microgram per cubic meter and on a percentage basis. The data in Table 1 are listed by location, moving from west to east. The map included as Figure 1 shows these locations. The nitrate level generally increased,

TABLE 1. ACHEX Data for Nitrate and Sulfate in the South Coast Air Basin, 1972-1973 (24-hr Average Values)[a]

		NO_3^-		SO_4^{2-}	
Sampling location	No. of Episodes	($\mu g/m^3$)	(% of Mass)	($\mu g/m^3$)	(% of Mass)
Dominguez Hills	2	4.1	3.1	17.8	13.6
Harbor Freeway	2	6.3	5.6	5.7	5.1
Pasadena	6	6.5	9.0	7.0	9.7
West Covina	5	8.2	4.5	21.8	12.0
Pomona (1972)	5	19.8	14.6	9.6	7.1
Pomona (1973)	2	7.9	4.8	11.5	7.0
Rubidoux	3	31.5	12.5	10.7	4.2
Riverside	9	15.2	13.4	7.7	6.8

[a]Analyses on high-volume samples collected on Whatman 41 filters during July-October periods.

moving eastward. No such trend was seen for sulfate. Table 2 shows data from the National Air Surveillance Network (NASN) for 1968, based on 26 sampling periods at each location (18). The NASN data display trends similar to those shown in Table 1, the nitrate/sulfate concentration ratio increasing from 0.5 at Long Beach to 1.2 at Riverside.

TABLE 2. NASN Network Data for Nitrate and Sulfate in the South Coast Air Basin for 1968 (Annual 24-hr Average Values)[a]

Sampling location	NO_3^- ($\mu g/m^3$)	NO_3^- (% of mass)	SO_4^{2-} ($\mu g/m^3$)	SO_4^{2-} (% of mass)
Long Beach	5.7	5.0	12.7	11.1
Los Angeles	7.7	6.0	10.2	7.9
Ontario	9.1	7.9	9.0	7.8
Riverside	10.2	8.8	8.3	7.2

Source: Reference 18.

[a]Data based on geometric mean of 26 24-hr sampling periods.

IV. CATIONS ASSOCIATED WITH SULFATE AND NITRATE

Ammonia was monitored because of its probable significance in the formation and neutralization of particulate sulfates and nitrates. Figure 2 summarizes the data obtained in the SCAB, based on samples collected with 4-hr time resolution. The diurnal plot for Pomona was obtained by compositing results for similar time intervals for 4 days.

Observed NH_3 concentrations ranged from 2 to 74 $\mu g/m^3$. Sites in the western part of the basin (Pasadena and Harbor Freeway) showed lower concentrations than those further east (Pomona and Riverside). A morning peak was evident at the Harbor Freeway, at Pomona, and for one of the two Riverside data sets. An evening or nighttime peak was also seen at Pomona and Riverside. The marked diurnal variations observed appear consistent with Miner's conclusions (19) that "the major sources of urban-produced ammonia are the internal combustion engine, combustion of fuels for heating, and the incineration of wastes." Observed ammonia levels may be compared to the ammonia concentrations required to neutralize the sulfate and nitrate aerosol, determined assuming their initial formation as the corresponding acids. Results based upon 24-hr average values, given in Table 3, indicate that the NH_3 was usually more than sufficient to neutralize these acids if present. This table also compares SCAB NH_3 concentrations with those observed at two "background" sites, Point Arguello and Goldstone. The former is on the Pacific Coast, about 100 miles north of Los Angeles; the latter, in

TABLE 3. Twenty-four-Hour Average Ammonia Relative to Sulfate and Nitrate Concentrations ($\mu g/m^3$)[a]

Site	Date	NH_3[a]	SO_4^{2-}	NO_3^-	Obs./calc. NH_3[b]
Harbor Freeway	9/27-9/28/72	15.3	7.1	4.6	4.1
Pasadena	10/4-10/5/72	3.5	3.7	6.9	1.1
Pomona	10/24-10/25/72	35.4	19	36.4	3.2
Riverside	9/27-9/28/72	58.3	8.5	16.5	7.7
Point Arguello	11/9-11/10/72	15.0	6.5	1.0	5.9
Goldstone	11/1-11/2/72	6.5	1.2	1.5	8.5

[a] Means for sequentially run 4-hr samples. Sulfate and nitrate results are from 24-hr high-volume samples collected on Whatman-41 filters.

[b] Calculated NH_3 equals the concentration of NH_3 required to neutralize SO_4^{2-} and NO_3^-, assuming their presence as H_2SO_4 and HNO_3.

TABLE 4. Comparison of Expected and Observed Ammonium Values for SCAB and Background Sites[a]

Site	Number of Samples	Expected NH_4^+ based on NO_3^- and SO_4^{2-} present ($\mu g/m^3$)[b]	Observed NH_4^+ as percent of expected value
Harbor Freeway	2	4.0	103
Pasadena	6	4.0	82
West Covina	4	10.3	76
Pomona - 1972	5	8.9	94
Pomona - 1973	2	6.6	75
Riverside	8	7.9	93
Rubidoux	3	15.5	73
Dominguez Hills	2	7.9	62
Point Arguello	1	2.7	131
Goldstone	1	0.9	79

[a] Analyses of 24-hr high-volume samples collected on Whatman 41 filters.
[b] Assuming the composition $(NH_4)_2SO_4$ and NH_4NO_3.

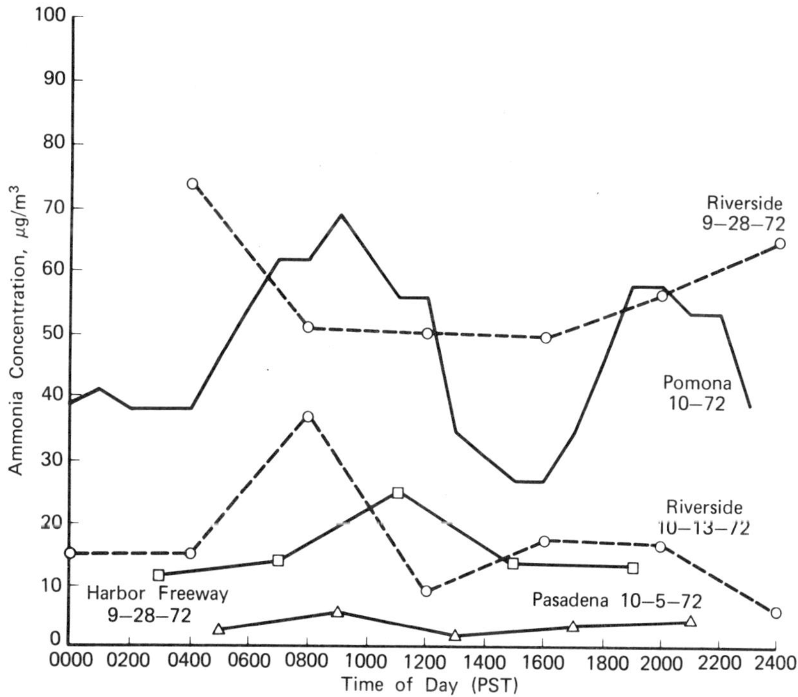

FIGURE 2. Diurnal variation of ammonia at sites within the SCAB.

the Mojave Desert. The levels in the eastern part of the SCAB are much larger than those found at these sites.

The ammonia data suggest that particulate sulfates and nitrate collected in 24-hr samples should exist as the ammonium salts, assuming their initial formation as the corresponding acids. Table 4 compares the observed levels of ammonium in the particulate (in $\mu g/m^3$) with the amounts calculated assuming the species present were $(NH_4)SO_4$ and NH_4NO_3. For sites in the SCAB, values range from 62% at Dominguez Hills, a site enriched in sulfur oxides, to 103% adjacent to the Harbor Freeway. These data imply that, except near sulfur oxide sources, the ammonium salts are the dominant species. This technique does not, however, distinguish between neutralization of acidic constituents before and after collection. For example, H_2SO_4, if present, might be neutralized after collection on a filter by ammonia. Furthermore, it does not necessarily reflect the composition existent in the particulate matter; adsorbed nitric acid coexistent with a roughly equivalent amount of ammonia adsorbed on other aerosol constituents would be indistinguishable from ammonium nitrate by this technique. Recent studies employing electron spectroscopy for chemical analysis of filter samples suggest that such alternatives may be significant in some cases (20).

V. DIURNAL PATTERNS FOR SULFATE AND NITRATE

Figures 3 and 4 compare the diurnal patterns for sulfate with those for other pollutants at West Covina and Pomona. The diurnal plots for sulfate are quite similar to those for SO_2, ozone, and total carbon. The peak concentration occurred at about 1500 hr (PST) for both sites, which are situated about 16 km apart. The maximum 2-hr average sulfate concentration exceeded 50 $\mu g/m^3$ at both locations.

In contrast to sulfate, the diurnal pattern for nitrate often exhibited its maximum during the morning, close to maxima for gas-phase NO_x. This is illustrated in Figure 5 for West Covina, where the peak occurred during the 0600 to 0800 PST period. Figure 6 shows the diurnal variation of nitrate and sulfate at West Covina for two particle-size fractions. The morning peak for nitrate was associated predominantly with particles >0.5 μm, while small-particle nitrate became increasingly important during the afternoon.

Tables 5 and 6 compile the daily maximum concentrations and corresponding times for sulfate and nitrate, respectively, at all sampling sites. The sulfate results are notable for the lack of consistency in times of diurnal maxima; examples of morning, afternoon, and nighttime maxima are present. No clear trend is evident for relating time of sulfate maxima with location. The range in concentration maxima is similar at all locations with 2-hr values up to 69.5 $\mu g/m^3$.

The nighttime maxima on October 3-4 are notable for their coincidence in time and similar concentrations at four locations in the SCAB. The nitrate maxima given in Table 6 indicate 2-hr average concentrations up to 149 $\mu g/m^3$. In contrast to sulfate, over 60% of the nitrate maxima occurred during the period 0600 to 1200 hrs. Except for the October 3-4 episodes, the higher nitrate maxima occurred at the more eastern sites in the basin. As with sulfate, the nitrate maxima on October 3-4 occurred during evening or nighttime hours.

TABLE 5. Diurnal Sulfate Maxima ($\mu g/m^3$) and Times for Maxima (PST) at Mobile Laboratory and Satellite Stations

	Site								
Date[a] (Time)	Dominguez Hills	Downtown Los Angeles	Pasadena	Anaheim	West Covina	Pomona (mobile lab)	Pomona (satellite)	Rubidoux	Riverside
7/23-24/73 (2100-2100)					69.5 (1400-1600)				
7/24-25/73 (2300-1600)					42.6 (1400-1600)				
7/26/73 (0500-1830)					22.9 (1200-1400)				
8/8-9/73 (2100-2130)		11.8 (1230-1430)	15.5 (1500-1700)	9.9 (1500-1700)	16.1 (1600-1800)		16.8 (0800-1000)		15.1 (1200-1400)
8/16-17/73 (2100-2100)		50.8 (1000-1200)		27.4 (1030-1230)		51.8 (1400-1600)	29.1 (1200-1400)		26.6 (0930-1130)
8/22-23/73 (2100-2100)						11.1[b] (1600-1800)			
9/5-6/73 (2300-2300)								36.7 (1600-1800)	
9/18-19/73 (2000-2000)							49.4 (0830-1030)	48.2 (1000-1200)	57.1 (1130-1330)
9/24-25/73 (2300-1800)								14.8 (1000-1200)	
10/3-4/73 (2000-2000)		67.5 (2230-0030)	65.8 (2200-0000)	57.2 (2000-2400)[c]			35.0 (0930-1330)		30.3 (2130-2330)
10/4-5/73 (2100-2100)	50.3 (1000-1200)								
10/10-11/73 (2100-2100)	11.8 (1200-1400)								

[a] Sampling intervals varied slightly between sites.
[b] Calculated for XRFA sulfur value.
[c] Constant for two successive 2-hr periods.

TABLE 6. Diurnal Nitrate Maxima ($\mu g/m^3$) and Times for Maxima (PST) at Mobile Laboratory and Satellite Stations

	Site								
Date[a] (Time)	Dominguez Hills	Downtown Los Angeles							
7/23-24/73 (2100-2100)					32.2 (0600-0800)				
7/24-25/73 (2300-1600)					50.2 (0800-1000)				
7/26/73 (0500-1830)					36.1 (0800-1000)				
8/8-9/73 (2100-2130)		13.4 (0800-1000)	9.8 (1030-1230)	9.4 (1500-1700)	22.7 (1000-1200)		26.9 (0800-1000)		38.5 (1830-2130)
8/16-17/73 (2100-2100)		5.6 (0130-0530)		9.5 (0600-0800)		64.0 (0800-1000)	56.1 (1400-1600)		57.6 (0930-1130)
9/5-6/73 (2300-2300)								81.8 (2000-2300)	
9/18-19/73 (2000-2000)							51.7 (0830-1030)	149.0 (1000-1200)	25.5 (0930-1130)
9/24-25/73 (2300-1800)								55.8 (1000-1200)	
10/3-4/73 (2000-2000)		62.3 (2030-2230)	25.9 (1800-2000)	56.4 (1730-1930)			53.2 (0530-0930)		91.8 (2130-2330)
10/4-5/73 (2100-2100)	16.1 (2100-0100)								
10/10-11/73 (2100-2100)	27.3 (0800-1000)								

[a]Samping intervals varied slightly between sites.

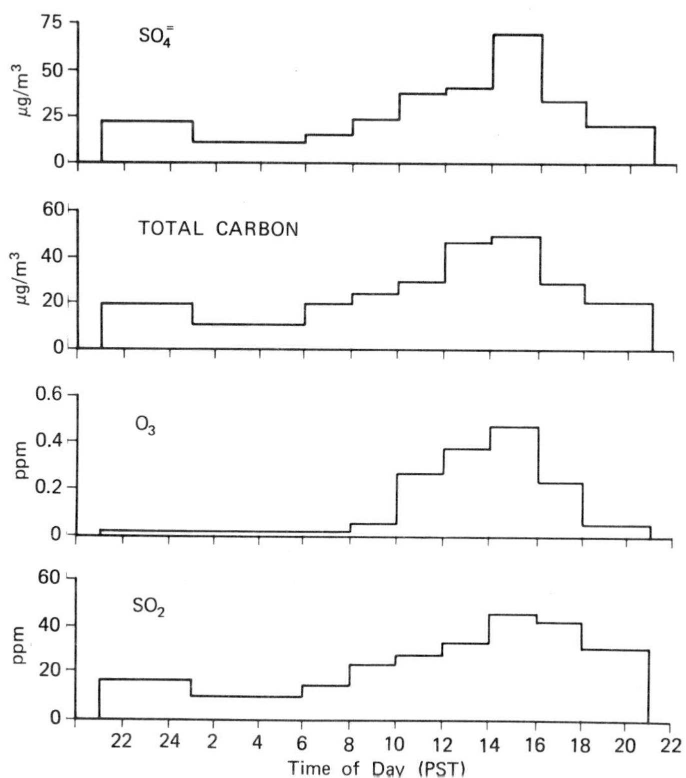

FIGURE 3. Diurnal variations for sulfate and selected pollutants at West Covina, July 23-24, 1973.

VI. SIZE DISTRIBUTIONS WITH THE FIVE-STAGE LUNDGREN IMPACTOR

Figures 7a to c show size distributions for SO_4^{2-} and NO_3^- at a number of sites. Particle size is plotted against Δ mass/Δ log (particle diameter), so that the area beneath the curve is proportional to the mass of the chemical species in that size interval.

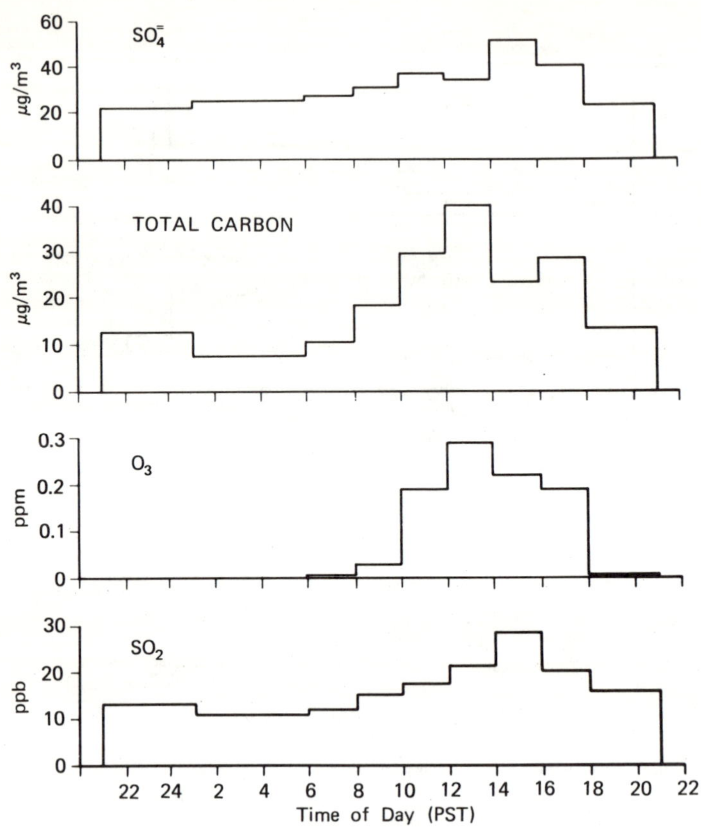

FIGURE 4. Diurnal variations for sulfate and selected pollutants at Pomona, August 16-17, 1973.

In Figure 7a the sulfate distributions at the more easterly sites are notable for their similarity; all show maximum sulfate contents on stage 4 (geometric mean of the aerodynamic diameter was 0.87 μm). Figure 7b shows the distributions for both sulfate and nitrate at a source-enriched site, Dominguez Hills. On October 4-5 the sulfate distribution strongly resembles that for the more easterly sites. On October 10-11 the sulfate level was relatively low, and the distribution shifted somewhat toward small particle size. However, the mass median diameter (mmd) was not appreciably altered.

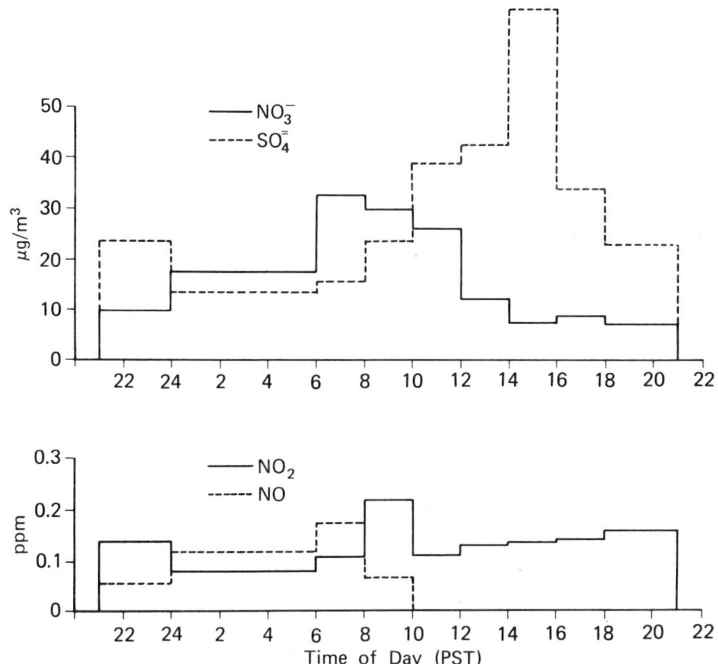

FIGURE 5. Diurnal variations for nitrate, sulfate, and NO_X at West Covina, July 23-24, 1973.

The 24-hr average particle size distributions for nitrate (Figure 7c) display a remarkable change from a nearby flat distribution at West Covina to a sulfate-like size distribution at Rubidoux. The observed size distribution may reflect the relative importance of small- and large-particle nitrate formation mechanisms (e.g., a water-phase and a photochemical, gas-phase process) and, since nitrates are somewhat hygroscopic, the ambient humidity.

Table 7 lists the calculated mmd values for nitrate and sulfate for the episodes analyzed. The sulfate value is relatively constant and, with the exception of the episodes at Rubidoux, exhibits a consistently smaller diameter than does nitrate.

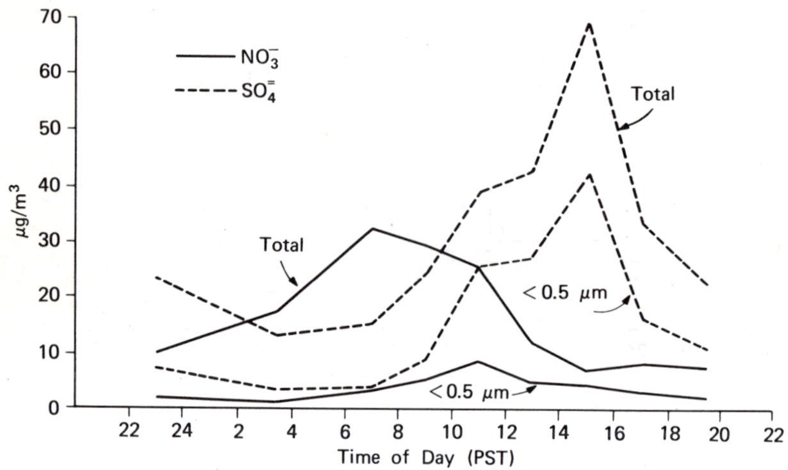

FIGURE 6. Diurnal variations in size distributions for sulfate and nitrate at West Covina, July 23-24, 1973.

The size distributions obtained for sulfate are in agreement with previously reported South Coast Air Basin and San Francisco Bay area values obtained with a Goetz aerosol spectrometer sampling for 8-hr daylight periods (21). The reported mmd's ranged from 0.24 to 0.55 μm.

Lundgren reported an average mmd for sulfate of 0.3 μm, sampling in Riverside, with values ranging from 0.1 to 0.6 μm (22). These California results compare to 0.4 and 0.3 μm obtained for Cincinnati and Chicago, respectively, by Nader and coworkers employing an Andersen sampler (23).

Available data for nitrate size distributions are limited; Lundgren reported for Riverside an average mmd of 0.8 μm, with values ranging from 0.6 to 1.2 μm (22). These are in the range found in the present study for the entire SCAB but are somewhat greater than the ACHEX values for Rubidoux.

VII. SUMMARY

Particulate nitrate and sulfate in the South Coast Air Basin have been shown to differ significantly in diurnal variations, spatial distributions, and, often size distributions. Two-hour

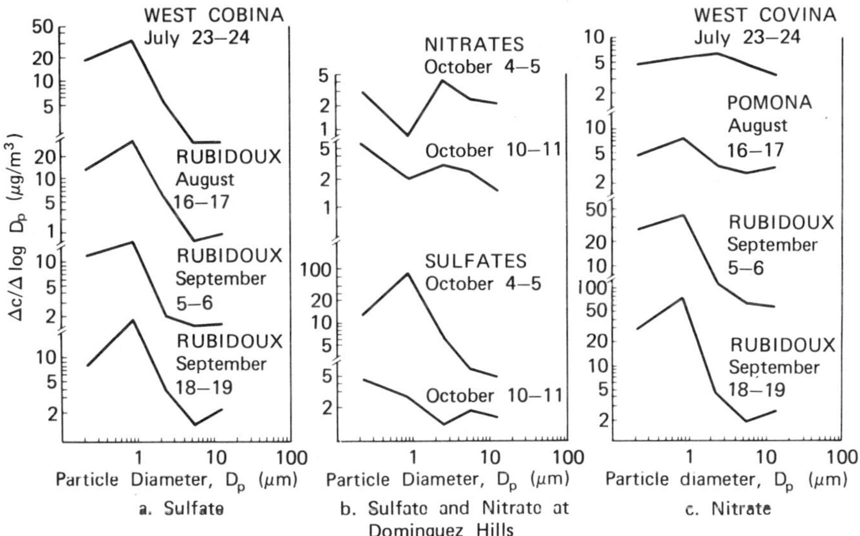

FIGURE 7. Twenty-four-hr average size distributions for sulfate and nitrate (1973 sampling period).

average nitrate and sulfate concentration up to 149 and 70 μg/m^3, respectively, were observed. Nitrate maxima often occurred in the morning hours with generally higher maxima at quarterly sites. Sulfate maxima were more variable in time with no apparent spatial pattern.

The data suggest that both sulfate and nitrate exist primarily as ammonium salts, with some evidence in support of unneutralized sulfates in the vicinity of sulfur oxide emissions. Mass median diameters of 0.3 to 0.4 μm for sulfate and with a broader range, 0.3 to 1.6 μm, for nitrate were measured, with the larger sizes in the western part of the basin.

ACKNOWLEDGMENTS

We wish to acknowledge the assistance of S. Twiss, S. Wall, and Y. Tokiwa in preparing this paper. This work was supported in part by the California Air Resources Board, Division of Technical Services, Research Section.

TABLE 7. Mass Median Particle Diameter: South Coast Air Basin, 1973[a]

Sampling location	Nitrate (μm)	Sulfate (μm)
Dominguez Hills		
October 4-5	1.64	0.43
October 10-11	0.72	0.42
West Covina		
July 23-24	1.13	0.34
July 26[b]	0.62	<0.22
Pomona		
August 16-17	0.68	0.39
Rubidoux		
September 5-6	0.33	0.33
September 18-19	0.34	0.43

[a]Based, except as noted, on 24-hr average size distributions obtained with the five-stage Lundgren impactor.

[b]A 13-hr (0500-1800) average.

REFERENCES

1. Hidy, G. M. et al. 1972. J. Air Pollut. Control Assoc. 25, 1106. Copies of the data bank, on magnetic tape, containing approximately 30,000 determinations are available from the Research Section of the California Air Resources Board.

2. Appel, B. R., and Wesolowski, J. J. Selection of Filter Media for Particulate Sampling with a Lundgren Impactor, this volume; John, W., and Reischl, G. P. 1978. "Measurements of the Filtration Efficiencies of Selected Filter Media", AIHL Rep. No. 184, Air and Industrial Hygiene Laboratory, and in Atmos. Environ. 12, 2015.

3. Lundgren, D. A. 1967. J. Air Pollut. Control Assoc. 17, 225.

4. Wesolowski, J. J., Appel, B. R., and Alcocer, A. 1973. "Preliminary Report on the Use of the Lundgren Impactor for Particle Size Measurements", AIHL Rep. 138, Air and Industrial Hygiene Laboratory, California State Department of Health, and this volume.

5. Wesolowski, J. J., John, W., Devor, W., Cahill, T. A., Feeney, G., Wolfe, G., and Flocchini, R. 1977. Collection Surfaces of Cascade Impactors, in X-ray Fluorescence Analysis of Environmental Samples, Ann Arbor Science Publishers, Inc. Rao, K. 1975. Ph.D. Thesis, The University of Minnesota, Particle Technology Laboratory.

6. Hoffer, E. M., Kothny, E. L. and Appel, B. R. 1979. Atmos. Environ. 13, 303.

7. "Selected Methods for the Measurement of Air Pollutants," U.S. Public Health Service Pub. No. 999-AP-11, 1965.

8. Technicon Industrial Method 118-71W.

9. Brosset, C., and Ferm, M. 1974. "Air Quality-Sulfur Dioxide Concentrations in Air Analysis by the Thorin (Spectrophotometric) Method", International Organization for Normalization, ISO/TC 146/WG 1/TG 4N18.

10. Intersociety Committee Methods of Air Sampling and Analysis, American Public Health Association (1972).

11. Hoffer, E. 1974. "Nitrate Method for Low-Volume Filters," AIHL Method No. 50, Air and Industrial Hygiene Laboratory, California State Department of Health.

12. Analyses by Giauque, R., Garrett, B., and Goda, L., Lawrence Berkeley Laboratory, University of California.

13. Harwood, J. E. and Kuhn A. 1970. Water Res. 4, 805. Rommers, P. J., and Visser, J. 1969. Analyst. 94, 653.

14. Mueller, P. K., Mosley, R. W., and Pierce, L. B. 1972. J. Colloid Interface Sci. 39, 235.

15. Unpublished studies by the Ozone Calibration Committee, W. DeMore, Chairman, jointly sponsored by the Environmental Protection Agency, the California Air Resources Board, and the Los Angeles Air Pollution Control District, 1975.

16. Kothny, E. L., Tokiwa, Y., and de Vera, E. R. 1973. "Atmospheric Ammonia and Ammonium in Particulate Matter," AIHL Method No. 29, Air and Industrial Hygiene Laboratory, California State Department of Health.

17. Appel, B. R., Wesolowski, J. J., Alcocer, A., Wall, S., Twiss, S., Giauque, R., Ragaini, R., and Ralston, R. 1974. "Quality Assurance for the Chemistry of the Aerosol Characterization Experiment", this volume.

18. Air Quality Data for 1968 from the National Air Surveillance Network and Contributing State and Local Networks, Environmental Protection Agency, Research Triangle Park, N.C., August, 1972.

19. Miner, S. 1969. "Air Pollution Aspects of Ammonia", NTIS Rep. PB-188 082.

20. Appel, B. R., Wesolowski, J. J., Hoffer, E., Twiss, S., Wall, S., Chang, S. G., and Novakov, T. 1976. Int. J. Environ. Anal. Chem. 4, 169.

21. Ludwig, F. L., Coulson, D. M., Robinson, E., and Cavanagh, L. A. 1964. Interim SRI Report for Contract No. PH86-64-54.

22. Lundgren, D. A. 1970. J. Air Pollut. Control Assoc. 20, 603.

23. Roesler, J. F., Stevenson, H. J. R., and Nader, J. S. 1965. J. Air Pollut. Control Assoc. 15, 576.

ADDENDUM

Subsequent to this work Gelman GA-1 cellulose acetate membrane filters were found to collect gaseous nitric acid forming artifact particulate nitrate.* Accordingly, the ACHEX nitrate results are probably too high in some cases. The error is probably especially large for samples collected at sites in the eastern portion of the SCAB. However, no quantitative assessment of this error can be made.

*Appel, B. R., Wall, S. M., Tokiwa, Y., and Haik, M. 1979. Interference Effects in Measuring Particulate Nitrate in Ambient Air, Atmos. Environ. 11, 873.

Analysis of Carbonaceous Materials in Southern California Atmospheric Aerosols*

BRUCE R. APPEL, PAUL COLODNY,† AND JEROME J. WESOLOWSKI
Air and Industrial Hygiene Laboratory
California State Department of Health
Berkeley, California

Abstract

A technique is described for estimating the contributions of elemental carbon and primary organic and secondary organic

---

*Reprinted from Environmental Science & Technology, Vol. 10, p. 359, April 1976.

†Present address: MCA Disco-Vision, Inc., 1640 N. 228th Street, Torrange, California 90511.

materials to atmospheric particulate matter collected in California's South Coast Air Basin. The technique employs a combination of solvent extractions and carbon determinations; primary organics are estimated from the carbon solubilized in cyclohexane and secondary organics by the carbon solubilized by successive extraction with benzene and methanol-chloroform minus the primary organics. At the same time an upper limit estimate of the elemental carbon is obtained from the carbon remaining insoluble after the two-step extraction. Support for this technique is provided for samples collected at four locations.

I. INTRODUCTION

The carbonaceous material present in atmospheric aerosols is a combination of elemental carbon, organic (including polymeric), and inorganic compounds (e.g., carbonate salts). When considering the origins of atmospheric aerosols, the organic fraction may be further divided into "primary" and "secondary" materials (1). The former term indicates material introduced into the atmosphere directly in the particle state. Materials which, because of their low vapor pressure, condense shortly after introduction into the atmosphere from an elevated temperature source are also considered "primary." "Secondary" refers to particles formed as the result of homogeneous or heterogeneous reactions in the atmosphere.

While considerable research effort has been devoted to determining the concentrations and origins of organic particulate constituents known to be hazardous (e.g., polynuclear aromatics), less effort has been devoted to characterizing the nature and origins of carbonaceous material more generally. Even the proportion of the elemental carbon present remains a matter of speculation. Previous studies of carbonaceous matter have primarily examined organic solvent-soluble fractions and include the work of Mader et al. (2), Dubois et al. (3), Ciaccio et al. (4), and Cukor et al. (5). The latter group analyzed solvent extracts from suspended particulate matter collected near a major intersection in New York City. They concluded that the organics strongly resembled used automobile lubricating oils (i.e., were "primary" in origin).

Mueller et al. reported a technique for determining carbonate and noncarbonate carbon (6). Their work in analyzing aerosols from Pasadena, California, demonstrated that carbonate carbon was consistently less than 5% of the total carbon present and that noncarbonate carbon (including both elemental and organic carbon) represented from 18 to 44% of the total suspended particulate matter.

Recent work includes studies by Grosjean and Friedlander (7) and Grosjean (8). Grosjean compared the ability of various solvents and solvent mixtures to extract organic carbon. For atmospheric particulate matter collected in Pasadena, California, the extraction efficiency of the polar solvents, ethanol and acetone, increased with the ozone concentration measured during the time of particle collection, while for the solvents isooctane and methylene chloride, no change was observed. Although Grosjean did not use these results to quantitate primary and secondary organics, it is clear from these results that solvent extraction might be so used.

As part of the work sponsored by the California Air Resources Board in the Aerosol Characterization Experiment (ACHEX), we have sought methods to determine ambient air concentrations of elemental carbon as well as primary and secondary organics. As in Reference 6, carbonates were again shown to represent less than 5% of the total carbon present and have, therefore, been ignored. Numerous techniques were explored, including column chromatography, high-resolution mass spectroscopy, thermal analyses, and multiple solvent extraction with total carbon analyses of the various solvent extracts. We present here a discussion of the letter technique since it has proved to be the most useful.

The basis of this technique is to equate primary organics in atmospheric aerosols to those organics soluble in cyclohexane and total organics, to those solubilized by successive extraction with benzene followed by methanol-chloroform. An upper limit estimate to the elemental carbon present is obtained from the carbon remaining insoluble after the two-step extraction. The secondary organics are obtained by difference between total and primary organics. All determinations are based upon carbon analyses rather than total weight to avoid errors resulting from solubilization of inorganic salts (e.g., NH_4NO_3).

The techniques for estimating primary and secondary organics will be supported by comparisons with ozone data obtained during sample collection and used to indicate the extent of photochemical (i.e., secondary) organic particle production expected.

The technique was applied to samples collected on filters at four locations within California'a South Coast Air Basin. Sampling locations for the study are shown in Figure 1. These include a site adjacent to a complex of chemical plants and refineries, Dominguez Hills (DH), and three receptor sites with respect to photochemical smog, West Covina (WC), Pomona (PO), and Rubidoux (RB). As indicators of levels of air pollution on the days sampled, Table 1 lists maximum values observed for ozone and the light-scattering coefficient (b_{scat}) (9), as well as the total suspended particulate matter and particulate carbon

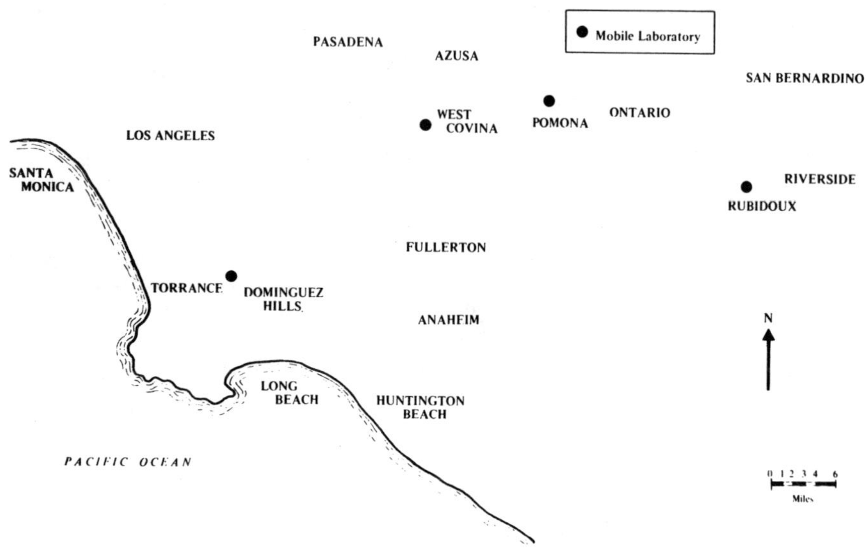

FIGURE 1. Sampling sites for the ACHEX.

averaged over the sampling period shown. The episodes include days of light, intermediate, and heavy photochemical smog.

II. EXPERIMENTAL

A. Materials

Cyclohexane, benzene, and 1:2 v/v methanol-chloroform were Eastman Spectrograde having a nonvolatile residue less than 0.001%. For the 85 ml of each solvent used for extractions, this represents <0.8 mg.

B. Particulate Sample Collection

Gelman Type A 8 x 10-in. glass fiber filters were used after preextraction for 18 hr in refluxing ethanol to reduce the carbon blank. The extractable carbon from a precleaned filter amounted to 2 to 7% of the weight of the total organic carbon extracted from a loaded filter, including the solvent residue. The carbon

TABLE 1. Sampling Episodes in the South Coast Air Basin

Sampling site	(date)	Sampling time (PST)	Max. O_3 (ppm)	Max. b_{scat} x 10^{-4} m^{-1}	Total suspended particulate (μg/m^3)	Carbon[a] (% of TSP)
West Covina (WC)	(7/11-7/12/73)	2100-2100	0.197	7.0	137	10.5
	(7/23-7/24/73)	2100-2110	0.542	16.0	211	11.4
	(7/24-7/25/73	2300-1605	0.449	10.2	241	9.50
	(7/26/73)	0500-1826	0.230	8.38	194	6.60
	(8/8-8/9/73)	2100-2100	0.177	9.51	123	8.70
Pomona (PO)	(8/16-8/17/73)	2100-2100	0.321	7.75	180	9.11
	(8/22-8/23/73)	2100-2100	0.194	3.47	148	7.36
Rubidoux (RB)	(9/5-9/6/73)	2300-2300	0.239	11.0	262	5.76
	(9/18-9/19/73)	2300-2300	0.256	21.8	284	5.39
	(9/24-9/25/73)	2300-1800	0.188	8.49	211	3.8
Dominguez Hills (DH)	(10/4-10/5/73)	2100-2105	0.141	7.35	148	7.16
	(10/10-10/11/73)	2100-2100	0.119	5.38	114	12.2

[a]Carbon obtained by analysis of a refined filter sample, expressed as a percent of the total suspended particulate matter obtained in simultaneous sampling with a conventional high-volume sampler (i.e., without size segregation).

blank by combustion (6) of a 1-in. disk of the filter was 4 μg and represented about 1% of the carbon found with a similar section from a loaded filter. Filter samples were collected for the time periods shown in Table 1, employing a high-volume sampler equipped with a cyclone for removal of particles greater than 3.5 μm (10). Comparing total carbon determinations for the resulting refined samples with those for low-volume glass fiber filters simultaneously sampling 0 to 20-μm particles, the mean ratio refined/total particulate carbon was 1.04 ± 0.09. Thus, on average, all of the carbonaceous material was present in the refined particle samples. Immediately following collection, the filter samples were stored at dry ice temperature with protection from atmospheric moisture, and were brought to room temperature just prior to analysis.

C. Solvent Extraction

The filters were cut in half, and one-half extracted with cyclohexane. The other half was extracted first with benzene and then with 1:2 v/v methanol-chloroform. The extraction period was from 6 to 8 hr with each solvent, using a Soxhlet extractor. Solvent was removed at steam-bath temperature under a stream of nitrogen. An independent experiment employing a Du Pont Thermal Evolution Analyzer demonstrated, for about 400 samples from the same locations, <0.5% volatilization of carbon up to 100°C relative to the total evolved at up to about 500°C. Thus, steam bath treatment of organic extracts is not expected to produce significant loss of organic aerosol constituents. Following solvent removal, the residue was dried to constant weight over silica gel and the weight of the extract determined with a precision of ± 0.1 mg in 25-ml Erlenmeyer flasks. By weighing empty flasks as controls, variations in weight due to changes in buoyancy effects or other factors were used to obtain corrected extract weights.

D. Carbon Analysis

Aliquots (100 to 600 μg) of each solvent extract were weighed into disposable aluminum sample boats with a Cahn Model 4100 electrobalance with a precision of ± 1 μg. The sample was combusted to CO_2 and determined by gas chromatography, using the technique detailed by Mueller et al. (6). The blank correction for the boats was 8.3 ± 1.9 μg C. The resulting carbon determination was compared to that on a section of the original filter prior to extraction (the total carbon). The difference permitted calculation of the carbon remaining insoluble in the extracting solvent.

E. Ozone Determination

Ozone was monitored continuously throughout the sampling periods by a REM chemiluminescent analyzer calibrated against buffered 2% KI (11). Recent work has demonstrated this calibration procedure to yield values high by about 29% relative to ozone values determined by the reference method, uv absorption (12).

F. b_{scat} Determination

The light-scattering coefficient was obtained continuously with an integrating nephelometer (13) calibrated against Freon 12 and particle-free air.

III. RESULTS AND DISCUSSIONS

A. Comparison of Solvent Extraction Efficiencies

Table 2 summarizes the results of the extractions with cyclohexane and of successive extractions with benzene and methanol-chloroform. The results are expressed as $\mu g/m^3$. Also shown for comparison are the weight percent carbon contents of each extract.

On the average, the benzene/cyclohexane solubles weight ratio was 1.8 which compares to the values 1.6 and 1.2 reported by Gordon (14) and by Grosjean (8), respectively. The methanol-chloroform solvent mixture extracted an additional amount nearly equal, on a mass basis, to the prior benzene extraction except for the epixodes at RB where the second extraction yielded from three to five times the mass extracted in benzene. In all cases the cyclohexane and benzene extracts had the texture of a tacky gum while the methanol-chloroform solubles were brittle and, in the cases of the RB episodes, had a definite crystalline appearance. The latter was subsequently identified by infrared spectroscopy and X-ray diffraction to be predominantly ammonium nitrate.

The carbon content of both the cyclohexane and benzene extracts was about 60%. As expected from the solubilization of inorganics and the prior extraction with benzene, the methanol-chloroform extracts exhibited much lower carbon contents, especially for samples collected at Rubidoux; the overall mean percent carbon with the mixed solvent extract was about half the value for cyclohexane and benzene.

TABLE 2. Solvent Extraction of Refined[a] Particulate Matter Samples[b]

Sampling site	(date)	Cyclohexane extractables ($\mu g/m^3$)	Cyclohexane extractables (% carbon)[c]	Benzene extractables ($\mu g/m^3$)	Benzene extractables (% carbon)	$CH_3OH-CHCl_3$ extractables (after benzene extraction) ($\mu g/m^3$)	$CH_3OH-CHCl_3$ extractables (after benzene extraction) (% carbon)
WC	(7/11-7/12/73)	4.9	--[d]	11.0	62.6	12.0	39.3
	(7/23-7/24/73)	--	--	16.6	61.7	16.4	33.0
	(7/24-7/25/73)	8.8	55.0	18.1	65.8	23.8	43.7
	(7/26/73)	6.6	60.4	10.9	67.4	16.3	34.5
	(8/8-8/9/73)	7.4	61.4	7.6	65.2	10.0	32.3
PO	(8/16-8/17/73)	5.4	67.8	11.0	72.2	13.3	38.3
	(8/22-8/23/73)	6.1	61.0	9.0	62.0	11.0	29.6
RB	(9/5-9/6/73)	7.9	59.3	13.7	61.5	60.0	9.3
	(9/18-9/19/73)	5.3	56.9	10.5	60.2	52.2	6.6
	(9/24-9/25/73)	3.8	62.9	5.3	59.8	15.8	17.0
DH	(10/4-10/5/73)	3.8	50.9	7.8	43.6	7.8	36.6
	(10/10-10/11/73)	9.0	58.5	11.5	60.5	9.0	18.2
Mean:		6.3 ± 1.8	59.4 ± 4.6	11.1 ± 3.7	61.9 ± 6.8	20.6 ± 17.2	28.2 ± 12.3

[a] Indicates particles ≤ 3.5 μm.

[b] Coefficients of variations for all determinations range from 4 to 12% except for WC (7/11-7/12/73) for which the range was 9 to 23%.

[c] Weight percent carbon in the solvent extract.

[d] Not determined.

B. Elemental Carbon Estimation

To demonstrate the technique employed to estimate elemental carbon, Table 3 compares the efficiency of cyclohexane, benzene, and benzene plus 1:2 v/v methanol-chloroform for extraction of carbon from the particulate samples expressed as a percent of the total carbon present in the original filter sample. Clearly, with the use of the nonpolar solvents cyclohexane and benzene, on the average more than 50% of the carbon remained unextracted. By extraction with methanol-chloroform, about half of the carbon remaining after benzene extraction is solubilized and is, therefore, not in elemental form. These results are quite similar to those of Grosjean and Friedlander (7), who found up to 40% additional organics solubilized by a polar solvent following nonpolar solvent extraction.

If the remainder after successive benzene and methanol-chloroform extraction is equated to elemental C, then, on the average, only 22% of the total carbon present exists in this form. Since all of this insoluble fraction need not be elemental carbon, this technique provides only an upper limit to the level of elemental carbon present.

C. Estimation of Primary and Secondary Organics

Based in part on the work of Cukor et al. (5), we believed that primary organic particulate matter in urban areas has as its principal source partially oxidized lubricating oils. To further demonstrate this, a comparison was made of the cyclohexane-soluble organics in atmospheric particles collected at 12 locations throughout California with cyclohexane extracts from automobile exhaust particulate matter. The exhaust sample was collected on a glass fiber filter by the staff of the California Air Resources Board's Vehicle Emission Control Division, El Monte, California, using a 1970 Chevrolet equipped with a 350-$in.^3$ displacement V-8 engine and driven over a 300-mile route. Employing column chromatography over silica gel and successive elution with isooctane, benzene, and 1:1 v/v methanol-chloroform to obtain an aliphatics, aromatics, and polar organics ("polars") fraction, we found that cyclohexane extracts from 46 ambient air samples yielded an average composition of 24% aliphatics, 15% aromatics, and 61% polars. This compares to the values 31% aliphatics, 8% aromatics, and 61% polars found for the organics extracted from the exhaust aerosol, suggesting a similar composition. In these cases, therefore, the cyclohexane-soluble organics appeared to be primary in origin.

TABLE 3. Comparison of Extraction Efficiency of Cyclohexane, Benzene, and Benzene Plus CH_3OH-$HCCl_3$ (% of Total Carbon)

Sampling site (date)		Cyclohexane-extracted C	Benzene-extracted C	Benzene + (CH_3OH-$HCCl_3$)-extracted C
WC	(7/11-7/12/73)	--	41	81
	(7/23-7/24/73)	--	43	65
	(7/24-7/25/73)	21	52	97
	(7/26/73)	31	57	100
	(8/8-8/9/73)	43	46	77
PO	(8/16-8/17/73)	22	48	79
	(8/22-8/23/73)	34	51	82
RB	(9/5-9/6/73)	31	56	93
	(9/18-9/19/73)	20	41	63
	(9/24-9/25/73)	30	40	74
DH	(10/4-10/5/73)	18	32	59
	(10/10-10/11/73)	38	50	62
Mean:		28.8 ± 8.3	46.4 ± 7.3	77.7 ± 13.9

The demonstration that cyclohexane-soluble organics from the South Coast Air Basin are not secondary products would provide more direct support for the use of cyclohexane-soluble carbon to estimate primary organic carbon. To this end we assume that photochemical reactions are the principal sources of secondary organics in the South Coast Air Basin, and that such reactions occur predominantly during daylight hours, but the resulting secondary (aerosol) products are collectable throughout longer periods. Accordingly, Figure 2 plots cyclohexane-soluble carbon

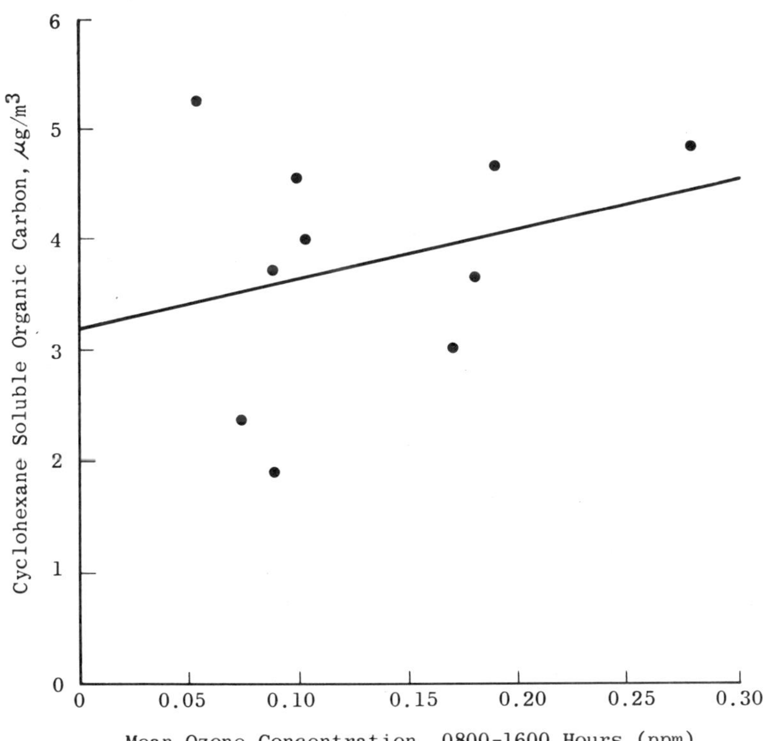

FIGURE 2. Cyclohexane-soluble organic carbon versus mean ozone concentration for sampling sites in the Los Angeles Basin.

for 10 sampling episodes against the mean daytime ozone level. The resulting scatter diagram, with least-squares slope not significantly different from 0 at the 95% confidence level, and the large intercept suggest the lack of dependence of cyclohexane-soluble organics on photochemical reactions leading to secondary organics.

We next examine the organic carbon solubilized by benzene which does not dissolve in cyclohexane. If cyclohexane solubilizes all the primary organics, then the additional soluble organics should be secondary reaction products. Figure 3 plots the difference in carbon solubilized by the two solvents versus mean daytime O_3. In contrast to the cyclohexane solubles, a correlation is obtained with Spearman's $\rho = 0.81$ and intercept near zero, thus supporting our hypothesis that cyclohexane solubilizes principally primary organics.

Finally, the correlation of mean O_3 against total secondary organic carbon, defined as the total soluble C less cyclohexane-soluble C, is shown in Figure 4. The correlation coefficient $\rho = 0.95$ and the near zero intercept for the least-squares fit regression line suggests that it can account for nearly all of the photochemically related secondary organics.

We rationalize the difference in solubility behavior of primary and secondary organics by noting the importance of relatively low-molecular-weight difunctional compounds as likely secondary aerosol constituents (15, 16). The substantial dipolar and hydrogen bonding interactions possible between such molecules and other insoluble aerosol constituents suggest the requirement for solvents of higher polarity to solubilize secondary organics. It should be emphasized, however, that the observed correlations between solubility behavior and aerosol origin may only be valid for the Southern California area.

Using this approach and approximating elemental carbon by insoluble carbon as discussed above, we summarize in Table 4 the composition of the carbonaceous material at the four Southern California locations. By this approach, secondary organics dominated at all of the photochemical smog receptor sites while elemental carbon and primary organics were of greater importance at Dominguez Hills.

Further work is planned to verify the analytical approaches used here and to refine these techniques to permit analysis of samples collected over shorter time periods.* Hopefully, conducting such analyses with better time resolutions will provide more information on sources, formation mechanisms, and transport for suspended carbonaceous material. Such information is a necessary prerequisite to a successful control strategy.

*These studies are reported by Appel et al., Environ. Sci. Technol. 13, 98 (1979).

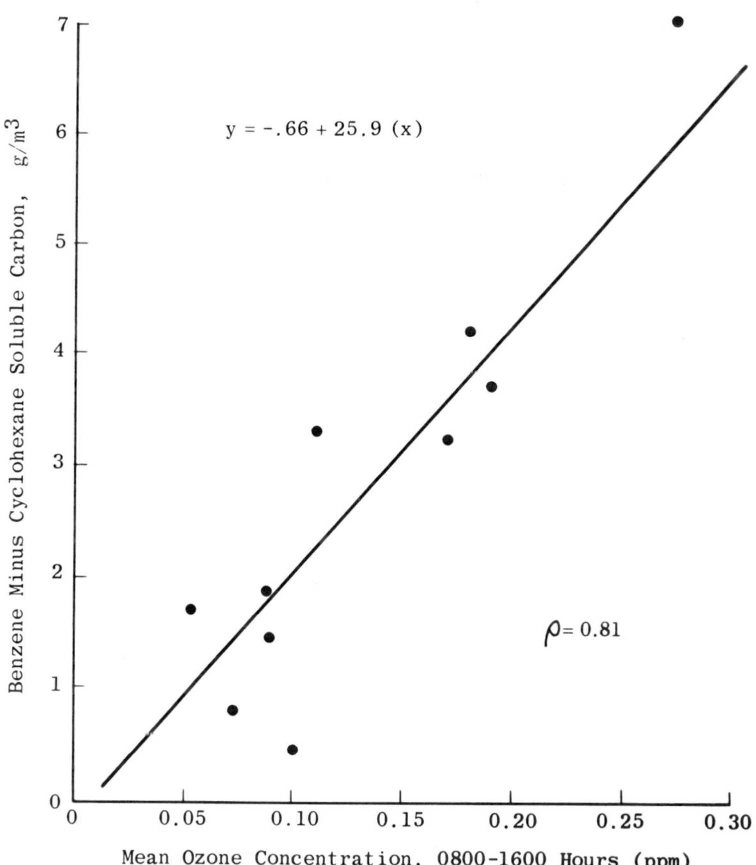

FIGURE 3. Benzene minus cyclohexane-soluble organic carbon versus mean ozone concentration for sampling sites in the Los Angeles Basin.

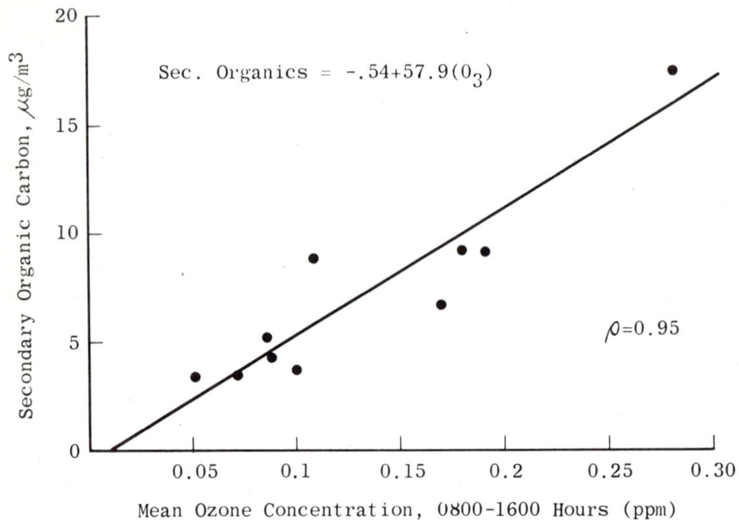

FIGURE 4. Calculated secondary organic carbon versus mean ozone concentration for sampling sites in the Los Angeles Basin.

ACKNOWLEDGMENTS

The authors wish to express their appreciation to P. K. Mueller for guidance during the initial stages of the work, to D. Grosjean of the California Institute of Technology for useful discussions, to A. Bockian of the California Air Resources Board for the automobile particulate matter sample, to E. M. Hoffer for development of some of the laboratory procedures used in this study, and to S. Twiss for statistical evaluation of the data.

REFERENCES

1. Stern, A. C. 1968. Air Pollution, Vol. I, p. 149, Academic Press, New York.

2. Mader, P. P., et al. 1952. Ind. Eng. Chem. 44(6), 1352.

TABLE 4. Composition of Carbonaceous Material

Sampling site (date)		Percent secondary organic C[a]	Percent elemental C[b]	Percent primary organic C[c]
WC	(7/11-7/12/73)	--	19	--
	(7/23-7/24/73)	--	34	--
	(7/24-7/25/73)	76	3	21
	(7/26/73)	70	0	30
	(8/8-8/9/73)	34	23	43
PO	(8/16-8/17/73)	57	21	22
	(8/22-8/23/73)	48	18	34
RB	(9/5-9/6/73)	62	7.3	31
	(9/18-9/19/73)	44	37	19
	(9/24-9/25/73)	44	26	30
DH	(10/4-10/5/73)	41	41	18
	(10/10-10/11/73)	24	38	38

[a] $\frac{(C_6H_6 + \text{MeOH-CHCl}_3)\text{sol C} - C_6H_{12}\text{sol C}}{\text{total C}} \times 100.$

[b] $\frac{\text{Total C} - (C_6H_6 + \text{MeOH-CHCl}_3)\text{sol C}}{\text{total C}}$. This is probably a maximum value.

[c] $\frac{C_6H_{12}\text{sol C}}{\text{total C}} \times 100.$

3. Dubois, L., et al. 1970. Atmos. Environ. 4, 199.

4. Ciaccio, L. L., et al. 1974. Environ. Sci. Technol. 8(10), 935.

5. Cukor, P., et al. 1972. Ibid. 6, 633.

6. Mueller, P. K., Mosler, R. W., and Pierce, L. B. 1972. J. Colloid Interface Sci. 39, 235.

7. Grosjean, D. and Friedlander, S. K. 1974. Paper No. 74-154, 67th Annual Meeting Air Pollution Control Association, Denver, Colorado, June.

8. Grosjean, D. 1975. Anal. Chem. 47, 797.

9. Williamson, S. J. 1973. Fundamentals of Air Pollution, Ch. 2, Addison-Wesley, Menlo Park, California.

10. Supplied by Pacific Environment Service, Santa Monica, California.

11. AIHL Method 2A and Rep., "A Guide for the Evaluation of Atmospheric Analyzers," P. K. Mueller et al., EPA Contract 68-02-0214, Ch. 5 and 6.

12. Unpublished studies by the Ozone Calibration Committee, W. DeMore, Chairman, jointly sponsored by the Environmental Protection Agency, the California Air Resources Board, and the Los Angeles Air Pollution Control District, 1975.

13. Supplied by Meteorological Research Inc., Altadena, California.

14. Gordon, R. J. 1974. Atmos. Environ. 8, 189.

15. O'Brian, R. J., et al. 1975. Environ. Sci. Technol. 9, 577.

16. Schuetzle, D. et al. 1975. Ibid., p. 838.

Paper presented in part at the Pacific Conference on Chemistry and Spectroscopy, San Francisco, October 1974 and based on studies funded by the California Air Resources Board Research Section.

The statements and conclusions in this report are those of the authors and not necessarily those of the California Air Resources Board. The mention of commercial products, their sources, or use in connection with material reported herein is not to be construed as either an actual or implied endorsement of such product.

Characterization of Carbonaceous Materials in Atmospheric Aerosols by High-Resolution Mass Spectrometric Thermal Analysis

BRUCE R. APPEL, STEPHEN M. WALL, AND RICHARD L. KNIGHTS*
Air and Industrial Hygiene Laboratory
Laboratory Services Branch
California State Department of Health
Berkeley, California

I. INTRODUCTION

As detailed in other chapters of this volume, three classes of atmospheric aerosol constituents have emerged of special concern because of potential health hazards and visibility reduction, namely, sulfates, nitrates, and carbonaceous material. This paper describes the results of an investigation of the volatile fraction of the carbonaceous material in aerosols sampled in California's South Coast Air Basin and analyzed by high-resolution mass spectrometric thermal analysis (HRMSTA).

The experimental technique and the rationale for its usefulness have been discussed in detail elsewhere in this volume. Briefly stated, particulate matter samples, supported on

*Chemistry Department, University of Washington, Seattle, Washington.

prewashed glass fiber filters, are introduced into a temperature-programmed inlet system of a high-resolution mass spectrometer. By use of an internal standard introduced in known amount, and measured or estimated response factors, the system has achieved a precision of $\pm 50\%$ and is considered accurate within about a factor of 2.

This paper will address three questions:

(1) What types of compounds are expected in organic aerosols present in the South Coast Air Basin atmosphere?

(2) What are the principal constituents in the aerosol identifiable by HRMSTA?

(3) Are the diurnal variations and concentrations of the probable primary and secondary organics consistent with expectations?

II. SOURCES OF ORGANICS IN AEROSOLS

Particle phase organics may be classified as being primary or secondary. "Primary" refers to the organic compounds introduced into the atmosphere directly in the particle state. The most important known source of primary organic aerosols is the aerosolized lubricating oil present in motor vehicle exhaust (1-3). Such oil consists principally of alkanes and cycloalkanes and partially oxidized materials derived from them.

"Secondary" organics consist principally of materials which condense to the particulate state following chemical reactions in the atmosphere to produce materials of low vapor pressure. Such aerosols include materials that are clearly associated with photochemical smog products. Based on laboratory studies, Table 1 lists an aerosol-forming reactivity scale for the gaseous precursors under simulated photochemical smog conditions without added SO_2. Of the materials shown, cycloalkenes are the most efficient precursors of organic aerosols.

The compositions of the aerosol products from such smog chamber studies have been reported in several cases. For example, cyclohexene forms difunctional acids and various mixed difunctional species in the aerosol products, as shown in Figure 1. The aerosols from cyclohexene were found to be quite similar in composition to those from xylene. Dicarboxylic acids were also identified as products from smog chamber studies with toluene, 1-heptene, and 1,7-octadiene, as well as cyclohexene

TABLE 1. Smog Chamber Aerosol Formation Potential of Hydrocarbons (4-8) (without added SO_2)

Very low	Low	Intermediate	High
Alkanes	Alkenes $\leq C_6$	Benzene	Diolefins $\geq C_6$
		Monoalkylbenzene	Cyclic alkenes
		Dialkylbenzene	
		Trialkylbenzene	
		Alkenes $> C_6$	

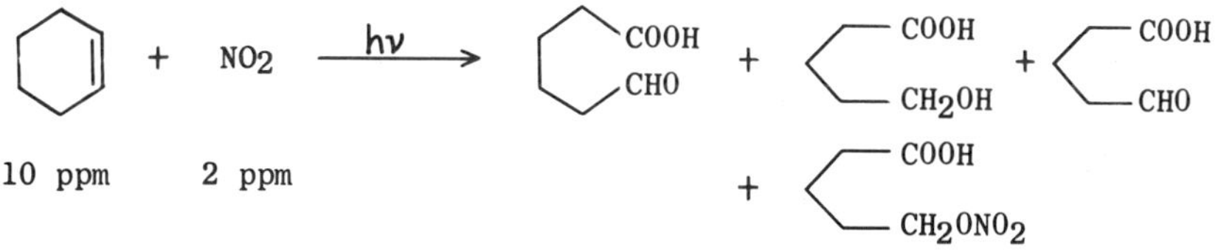

W.E. Schwartz, P.W. Jones, C.J. Riggle and B.F. Miller "Organic Characterization of Cyclohexene/NO_x Aerosol," submitted for publication to Environmental Science and Technology.

Figure 1. Identifiable photochemical aerosol constituents from cyclohexene. Results from Reference 10.

(9). On the basis of such studies, cycloalkenes, heavier aliphatic alkenes, and aromatics are considered the most likely secondary organic aerosol precursors in the atmosphere.

III. AMBIENT AIR RESULTS

Sampling for this study was conducted at three locations, shown in Figure 2 - Dominguez Hills, West Covina, and Pomona - on different days over a 4-month period. Dominguez Hills is in the western part of the basin adjacent to large, stationary pollution sources. Ozone levels are typically low in this area,

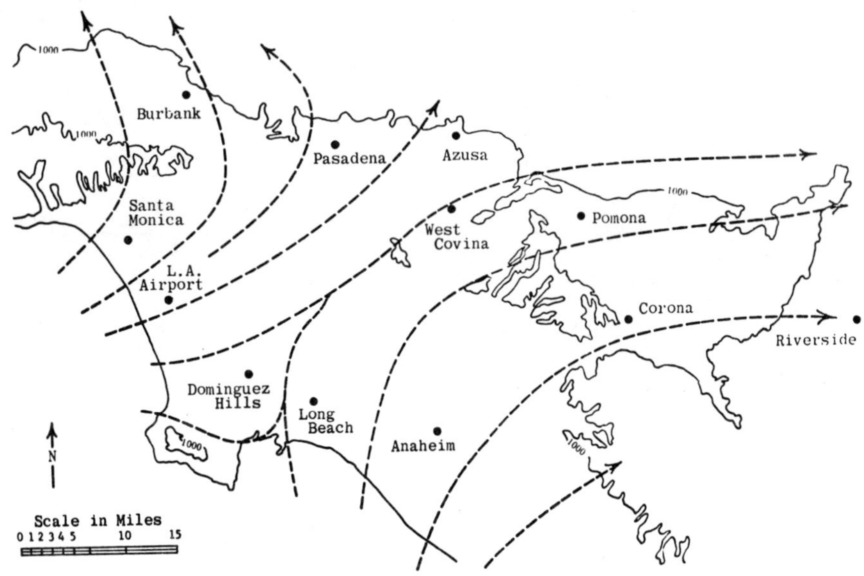

FIGURE 2. Wind Patterns Over the South Coast Air Basin on July 24, 1973, 1600 hr PST.

and, accordingly, relatively low concentrations of secondary organic aerosols are expected there. Figure 2 also shows the 1000 ft elevation contour lines and wind patterns for a particular day during the sampling program. Such patterns are typical of daytime summer winds in the South Coast Air Basin. Thus locations such as West Covina, Pomona, and Riverside can receive pollutants from stationary and mobile sources transported over relatively long distances. Such pollutants have had ample time for reaction to form secondary products. West Covina and Pomona are, therefore, considered photochemical smog receptor sites. Accordingly, the composition of the particulate organic was examined for consistency with measures of smog intensity at these three sites (e.g., the ozone concentration).

Concerning the principal constituents in the aerosol identifiable by HRMSTA, Table 2 lists the seven most abundant organic compounds or classes of compounds found at all three locations, the fragments used for their identification, and their concentration ranges. Of the seven listed, three are aliphatic

TABLE 2. Most Abundant Compounds or Classes in Organic Aerosols in the South Coast Air Basin

Compound	Identifying fragment(s)	Concentration range ($\mu g/m^3$)
Alkanes plus alkenes	C_5H_{11} + C_5H_9	5 - 10
Phthalates	$C_8H_5O_3$	1.4 - 10
Hexanedioic acid (adipic acid)	$C_5H_8O_2$	0.4 - 9
Pentanedioic acid (glutaric acid)	$C_4H_6O_2$	0.1 - 5
Heptanedioic acid (pimelic acid)	$C_6H_{10}O_2$	0.1 - 2
Substituted benzenes	C_7H_7	0.1 - 2
Alkyl(C_{4+})benzenes	C_9H_{11}	0.03- 1.1
Dihydroxybenzenes	$C_6H_6O_2$	0.02- 0.7

dicarboxylic acids such as were formed from cyclohexene, toluene, 1-heptene, 1,7-octadiene, and possibly xylene in smog chambers. Total alkanes and alkylbenzenes are almost surely primary in origin. The concentration of phthalates is higher than that due to experimental artifacts; the apparent ambient air levels may be reflective of the widespread use of phthalates as plasticizers. Acid nitrates are measurable, but their concentration is generally more than a factor of 10 below that of the corresponding diacids.

Next we consider the third question: Are the diurnal variations and concentrations of the probable primary and secondary organics consistent with expectations? Figure 3 compares the diurnal variations of hexanedioic (adipic) acid, total alkanes plus alkenes, gaseous hydrocarbons (HC), and lead at West Covina. Lead is an automotive exhaust indicator and displays a maximum between 0800 and 1000 PST. The diurnal variation of gaseous hydrocarbons is rather similar to that of lead and suggests an automobile origin. By contrast, hexanedioic acid displays a single midafternoon maximum coinciding with the ozone maximum consistent with a secondary origin. Total particulate alkanes plus alkanes display morning and midafternoon maxima. The first is probably automobile related. The second is of unknown origin,

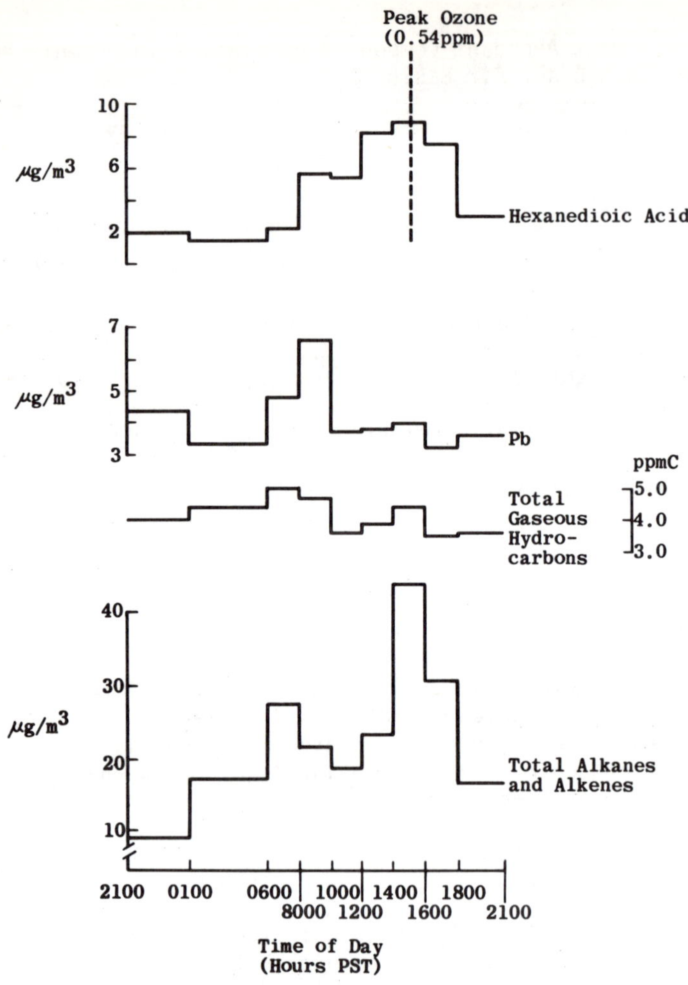

FIGURE 3. Diurnal variations of selected pollutants: West Covina, July 23-24, 1973.

but may reflect transport of primary organics from the industrial complex in the vicinity of Dominguez Hills, upwind on this day.

Figure 4 shows similar results obtained for Pomona. Again hexanedioic acid and ozone peak together. Lead here shows the usual morning maximum. Just as at West Covina, total alkanes

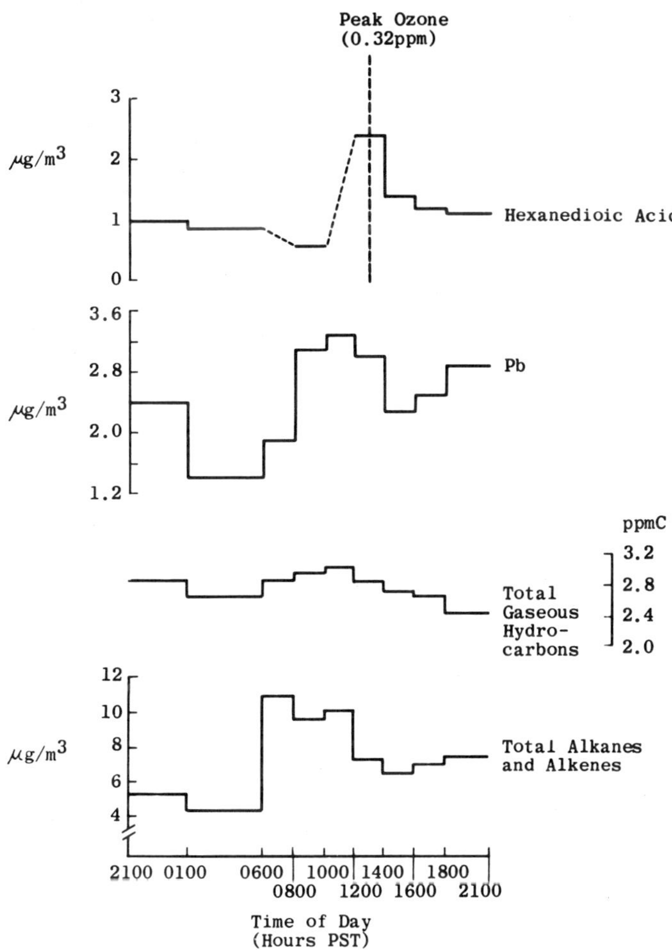

FIGURE 4. Diurnal variations of selected pollutants: Pomona, August 16-17, 1973.

plus alkenes appear to have both an automobile and an additional source.

The concentrations of observed acid nitrates and aldehyde nitrates at West Covina, based upon their molecular ions, are shown in Figure 5 relative to total alkanes plus alkenes. The afternoon increase in concentration of these difunctional materials is greater than that for the primary aerosol indicator,

TABLE 3. Concentrations of Organics During Period of Peak Ozone Concentration Relative to Values at Dominguez Hills

Organics	Dominguez Hills (1000-1200)[a]	West Covina (1400-1600)	Pomona (1200-1400)
Hexanedioic acid	1.0	24.7	6.7
Pentanedioic acid	1.0	15.7	N.D.
Total alkanes and alkenes	1.0	7.6	1.3
Hexanedioic acid total alkanes and alkenes	1.0	3.3	5.4

[a]Values in brackets are times, (PST).

total alkanes plus alkenes. It may be significant that the acid nitrate peak follows that for aldehyde nitrate; the acid nitrate is possibly formed by oxidation from the aldehyde nitrate.

For comparison of the aerosol at the three sampling sites, Table 3 lists the concentrations of the likely secondary and primary materials at the three sites relative to those at Dominguez Hills during the 2-hr time period in which the ozone peak occurred. Clearly the secondary materials are present in much higher concentrations at the receptor sites. For total alkanes plus alkenes, the difference in the three sites is much reduced, while hexanedioic acid, relative to total alkanes plus alkenes, is much higher at the receptor sites.

In an effort to establish a correlation between the intensity of photochemical smog and secondary aerosol formation, Table 4 lists correlation coefficients between ozone, hexanedioic acid, acid nitrates, aldehyde nitrates, and, for comparison, alkanes plus alkenes. Data are shown for West Covina and Pomona only, since results from Dominguez Hills were insufficient. The findings indicate no significant correlation between ozone and the model primary organic, total alkanes plus alkenes. However, correlation coefficients between ozone and the difunctional organics ranged from 0.7 to 0.9.

Figure 6 plots ozone against hexanedioic acid concentrations at all sites. The data for West Covina and Pomona are best fit by separate lines with differing slopes. Although Dominguez Hills results are shown, they are insufficient to define a line.

TABLE 4. Correlation of Ozone Concentration with Selected Primary and Secondary Organic Species.[a]

Station	Species			
	Total alkanes and alkenes	Hexandioic acid	$HOC(CH_2)_nONO_2$ (n = 5,6)	$HOOC(CH_2)_nONO_2$ (n = 4,5)
West Covina	N.S.[b]	.91	.70	.76
Pomona	N.S.	.79	–[c]	–
West Covina and Pomona	N.S.	.67	–	–

[a]Spearman's rank order correlation coefficients.

[b]Correlation coefficient is not significantly different from zero at the 95% confidence level.

[c]Insufficient data.

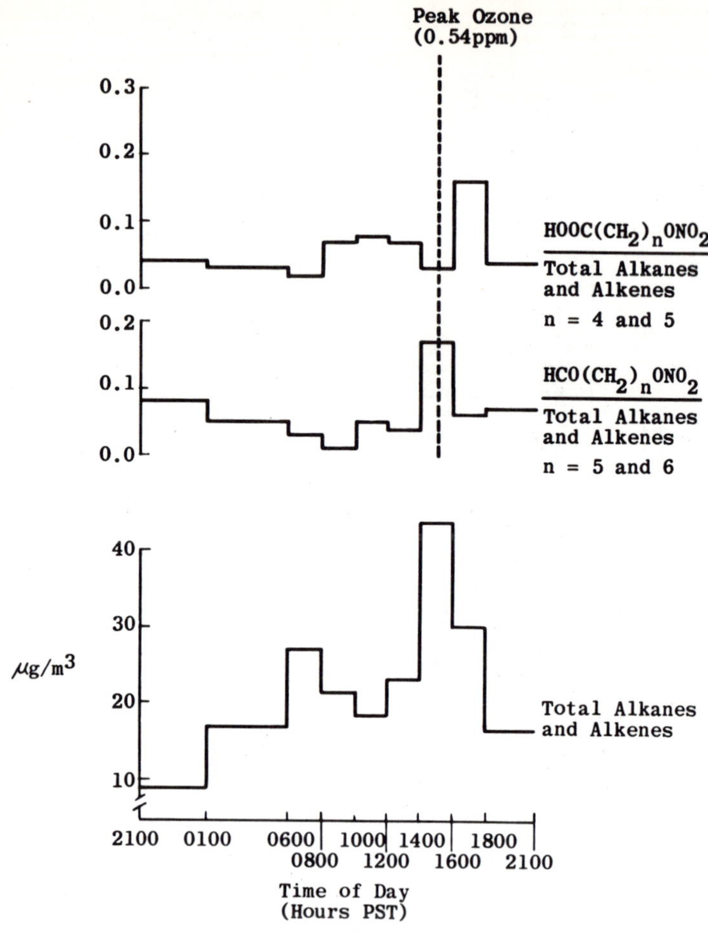

FIGURE 5. Diurnal Variation in Concentration of a Model Secondary Organic Relative to a Model Primary Organic; West Covina, July 23-24, 1973.

IV. SUMMARY

Organic aerosols are expected to consist of primary constituents such as aerosolized lubricating oils and secondary components including organics bearing carboxyl, nitrate, aldehyde, and alcohol functional groups. Difunctional compounds should be especially abundant in aerosols because their low vapor pressure, relative to monofunctional organics, favors condensation.

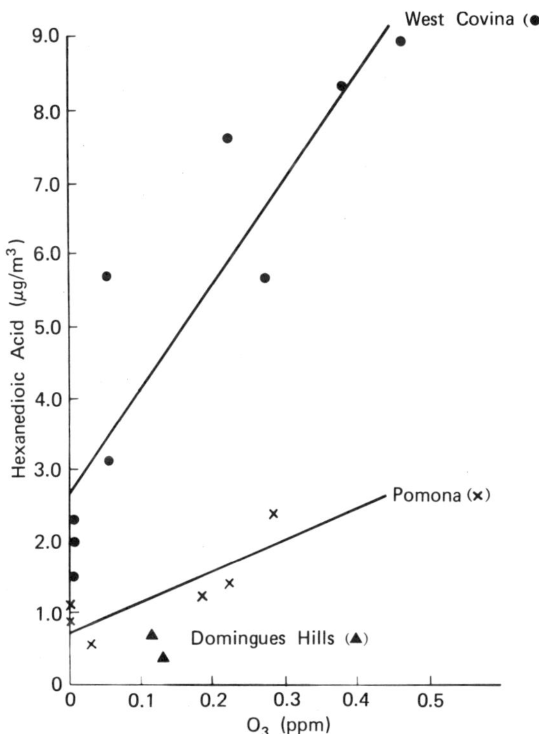

FIGURE 6. Apparent Correlation of Hexanedioic Acid and Ozone Concentrations.

Consistent with these expectations, the major identifiable organic constituents in the aerosols include alkanes plus alkenes, probably primary in origin, and C_5 to C_7 dicarboxylic acids, probably of secondary origin.

Studies conducted at three locations in the South Coast Air Basin indicate an increasing concentration of selected difunctional organics at more easterly sites, whereas total alkanes plus alkenes remain more nearly constant over the Basin. Concentrations of difunctional acids, acid nitrates, and aldehyde nitrates appear to correlate with ozone levels, while those of total alkanes plus alkenes do not. This is consistent with the formation of these difunctional organics in the atmosphere.

ACKNOWLEDGMENTS

We wish to acknowledge the assistance of A. L. Crittenden and Dagmar Cronn in working up the HRMSTA results and J. J. Wesolowski for helpful comments. The Environmental Protection Agency provided partial support for personnel at the University of Washington for HRMSTA.

REFERENCES

1. Ciaccio, L. L., et al. 1974. Environ. Sci. Technol. 8 (10), 935.

2. Cukor, P., et al. 1972. Environ. Sci. Technol. 6, 633.

3. Appel, B. R., et al. 1976. Environ. Sci. Technol. 10, 359.

4. Wilson, W. E., Jr., et al. 1972. "Haze Formation - Its Nature and Origin", Battelle Report to the EPA under Contract No. CPA 70-Neg. 172.

5. Keng, E. Y. H., et al. 1972. "Formation of Nonvolatile Particles from Organic Vapors", Final Technical Report, Project B-358, Georgia Institute of Technology.

6. Groblicki, P. J. and Nebel, G. J. 1969. "The Photochemical Formation of Aerosols in Urban Atmospheres", General Motors Research Publ. GMR-957.

7. O'Brian, R. J. 1975. Environ. Sci. Technol. 9, 577.

8. Bockian, A. H., et al. 1972. "Photochemical Aerosol Studies: A Status Report", California Air Resources Board.

9. Knights, R. L. and Crittenden, A. L. 1975. "Information on Photochemical Air Pollutants Obtained from Smog Chamber Experiments", presented at the First Chemical Congress of the North American Continent, Mexico City, December.

10. Schwartz, W. E., et al. 1974. "Chemical Characterization of Model Aerosols," NTIS Report No. PB 238 557.

ADDENDUM

More recent studies have demonstrated that, during periods of stagnation when pollutants remain trapped in the Basin from one day to the next, the diurnal variations of secondary organics and ozone may show relatively poor correlation.*

---

*Appel, B. R., Hoffer, E. M., Kothny, E. L., Wall, S. M., Haik, M., and Knights, R. L. 1979. Environ. Sci. Technol. 13, 98.

Formation of Photochemical Aerosol from Hydrocarbons: Atmospheric Analysis

ROBERT J. O'BRIEN[*], JAMES H. CRABTREE, JOHN R. HOLMES,
MARGARET C. HOGGAN, AND ALBERT H. BOCKIAN
California Air Resources Board
El Monte, California

Abstract

Atmospheric aerosol samples from various locations in California (particularly the Los Angeles area) were partially analyzed for their organic content. Emphasis was placed on photo-

---

[†]Present address: Department of Chemistry, Environmental Sciences Doctoral Program, Portland State University, Portland, Oregon 92707.

chemically produced material and in particular the identification of known products of organic aerosol precursors from environmental chamber experiments. The presence of organic nitrates in atmospheric aerosols gives a good indication of their photochemical origin. Mono- and dicarboxylic acids made up a significant portion of the organic material in aerosol samples taken in the Los Angeles area under smoggy conditions. These products have been shown previously to result from the photooxidation of olefins and oxides of nitrogen. The highly oxygenated nature of Los Angeles particulate was quite different from primarily emitted auto exhaust aerosols.

I. INTRODUCTION

Some areas of California, particularly the South Coast (Los Angeles) Air Basin, experience severe visibility reduction resulting from light scattering by suspended particulate matter. The State air quality standard for visibility (visibility not to be reduced to less than 10 miles for relative humidity less than 70%) is exceeded on approximately two days out of every three. During periods of intense photochemical activity, prevailing visibilities are sometimes reduced to one-half mile over wide areas of the Basin.

The physical mechanisms leading to the occurrence of heavy loadings of light-scattering aerosol have been elucidated to a large extent (1, 2). The bulk of the visibility reduction seems to be due not principally to directly emitted particulate matter, but rather to condensed species formed via the photochemical reaction of gaseous primary pollutants. The aerosols produced in this fashion accumulate preferentially in the size range 0.1 to 1.0 μm, which is the important range of particulate diameters for the scattering of visible light.

The chemical composition of California atmospheric particulate matter in general and of these secondary aerosols in particular is only partially known. It has been estimated that in the South Coast Air Basin approximately two thirds is anthropogenic in origin and approximately half of this or one third of the total is formed chemically in the atmosphere (3). Most analytical investigations into the composition of California particulate matter have been devoted to the inorganic constituents; very little effort has been made on the organic material.

Inorganic products of photochemical or other secondary reactions are relatively well understood--sulfate and nitrates are known to be produced in the atmosphere by the oxidation of SO_2 and NO_2. Ammonium sulfate, for instance, is a major constituent of particulates in the eastern part of the United States. Relatively well-developed techniques exist for the analysis of these inorganic ions, and for metallic elements as well.

Considerable information is available on the partial composition of atmospheric particulate matter in California (4). Data from the National Air Surveillance Network (5) on aerosol composition in California's South Coast Basin are summarized in Table 1. Measurements for three inorganic ions associated with secondary aerosols--sulfate, nitrate, and ammonium--are reported, as well as values for the total benzene-soluble organic portion. It is seen that, on a basin-wide average, the amounts of nitrate, sulfate, and benzene solubles are approximately equal and account (with ammonion ion) for about 28% of the total mass loading.

However, this figure does not accurately represent the contribution of these aerosol constituents to the overall problem. The benzene-soluble fraction underestimates the total organic content of the aerosol significantly because much of the organic material present in the aerosol is highly polar and insoluble in benzene. Moreover, these aerosol constituents are to a great extent concentrated in the small particle size range and thus make a contribution to visibility reduction, or potential health hazard, which is disproportionately large relative to their mass fraction.

In this paper, we describe a partial analytical scheme for the organic portion of atmospheric aerosol, estimate the relative contribution of primary and secondary sources to the organic material, and assess the overall contribution of organic material to the total atmospheric mass loading. Particular emphasis has been placed on identifying in the atmosphere known organic products of photochemical reactions (6).

II. EXPERIMENTAL

A. Sampling of Atmospheric Aerosol

Aerosol samples were collected with standard high-volume air samplers and Gelman Type A glass fiber filters (without binder). Sampling times were usually 24 hr, but in some cases were shorter or longer.

Preliminary work indicated that commercial glass fiber filters, although labeled as flash fired, contained considerable extractable organic material which interfered with the chemical analysis (see below). Consequently, all filters were prewashed in a large Soxhlet extractor for two days with absolute alcohol and subsequently dried at room temperature. The filters were

TABLE 1. National Air Surveillance Networks Data for 1968, California South Coast Basin (Maximum and Arithmetic Mean of Particulate Constituents, g m^{-3})

	Suspended particulate		NO_3^-		SO_4^{2-}		NH_4^+		Benzene soluble mean
	Max.	Mean	Max.	Mean	Max.	Mean	Max.	Mean	
Burbank	340	122	26	8	39	11	10	1	14
Glendale	216	98	21	7	29	11	4	1	12
Long Beach	290	132	18	7	52	15	31	3	13
Los Angeles	272	135	21	9	41	13	9	2	15
Ontario	389	135	72	12	18	10	6	1	8
Pasadena	195	113	17	7	41	12	8	1	12
Riverside	255	130	38	13	37	10	14	2	10
San Bernardino	213	108	22	9	36	10	10	1	7
Average		122		9		12		2	11
% of total average		--		7		10		2	9

pre- and postconditioned at 21° and 55% relative humidity before the initial and final weighings.

B. Recovery of Organic Aerosol

Soxhlet extraction of the glass fiber filters was employed for recovery of the organic material. Generally, one quarter of the 8 x 10-in filter was extracted for 3 or more hours in a 250-ml Soxhlet. The solution containing the sample was reduced in volume to about 5 ml by boiling, transferred to an aluminum weighing pan, and reduced to dryness in a stream of warm N_2. The sample was weighed to the nearest 0.1 mg, redissolved in the pan in 1 or 2 ml of solvent, and transferred to a vial. The sample was again dried with warm N_2 and then made up as a 10% (wt/vol) solution in the extracting solvent. Finally, the aluminum pan was reweighed to determine the amount of residue.

Most studies of atmospheric aerosol in the past have employed either benzene, methylene chloride, or cyclohexane extractions of organic material from the sample. Due to the presence of very highly polar organic compounds in these aerosol samples, however, it was necessary to use alcohol (95% ethanol, 5% propanol) as a solvent. For example, a single filter sample extracted in a Soxhlet successively with cyclohexane, benzene, methylene chloride, and alcohol yields appreciable organic material in all stages of the extraction. The alcohol extract also contains considerable inorganic material, particularly nitrates, thus complicating the analysis in some cases.

The efficiency of the alcohol extraction in removing organic material was tested on four samples collected in different areas of the South Coast Air Basin. The filters were extracted in the usual fashion in alcohol and then total noncarbonate carbon remaining on the filters was determined by an independent laboratory. The results showed 68, 80, 84, and 87% of the organic material had been removed in the extraction. These percentages were calculated as:

$$\%\ \text{removed} = 100 \times \left[1 - \frac{\text{total carbon residue}}{\text{total extract wt} - NH_4NO_3\ \text{wt} + \text{total carbon residue}}\right]$$

Most samples in this study were extracted from individual quarters of the filter in benzene and in absolute alcohol (95% ethanol, 5% propanol).

As mentioned above, the glass fiber filters were found to contain very appreciable amounts of oxygenated organic material. A blank 8 x 10-in filter (Gelman Type A) extracted in a Soxhlet with benzene yielded 2.4 mg of material. Extraction with ethanol removed 14.8 mg. A prewash with alcohol lowered these background organics to acceptable levels, less than 2.0 mg per filter with alcohol. Solvents were predistilled to remove any nonvolatile material.

C. Infrared Analysis of Extracts

Infrared spectra of aerosol extracts were obtained with a Perkin-Elmer 521 equipped with a beam condenser. Samples were prepared for analysis by placing an aliquot of the aerosol solution (generally 1 to 10 μl) on a small agate mortar. When the solvent had evaporated about 10 to 20 mg of KBr were added and the mixture was ground to intimacy. A portion of the mixture was pressed into a 1.5-mm diameter micropellet. In this way, high-resolution, full-scale spectra could be obtained from less than 100 μg of aerosol.

It was not generally possible to keep traces of water out of the KBr pellets in the process of grinding and pressing. Thus, the characteristic water bands at 3400 to 3500 cm^{-1} and at 1630 cm^{-1} were always present. In addition, a band at 1385 cm^{-1} was sometimes present in the blank. Since these bands were variable in magnitude, interpretation of these spectral regions was made with caution.

D. Chromatographic Separation of Organic Constituents

Previous work (6) has indicated that carboxylic acids are a major aerosol product of the photochemical reaction of olefinic hydrocarbons. Further, preliminary analysis of atmospheric aerosol samples by infrared spectroscopy showed the presence of strong carbonyl bands attributable to carboxylic acids. Consequently, a paper chromatographic technique for separating these acids was developed (7).

Whatman No. 41 paper was cut into 9-in. square sheets and washed in a large Soxhlet with absolute alcohol to remove organic impurities. The paper was dried and then washed in the Soxhlet with azeotropic 2% aqueous acetic acid to prevent tailing of the acid spots. Ten percent solutions of the extracts were spotted in one corner, and the paper was coiled and placed in a glass jar for elution. The solvent front was allowed to move 15 cm past the spot. Two-way chromatography was employed. The solvent in the first direction was alcohol-water-ammonia in the proportions 190-10-2. The second-direction solvent was ethoxyethanol-butylether-acetic acid-water, 84-84-29-4. The first-direction solvent moved the acids as their ammonium salts, and in the second direction they were converted to free acids by the acetic acid. Although optimized for acids, this system also adequately separated less polar materials and inorganic salts as well. The results of separations of several reference compounds, as well as some synthetic photochemical aerosols, have been presented previously (6).

The chromatogram was allowed to dry overnight to remove the acetic acid. Nonvolatile acidic compounds were then detected by spraying with bromcresol purple [40 mg in 100 ml of formaldehyde-ethanol (1:5) adjusted to pH 10 with NaOH]. Exposing the chromatogram to ammonia vapors helps to visualize the spots. This spray reagent detected mono- and dicarboxylic acids, acidic or basic salts, and amines. Long-chain fatty acids were not detected. The detection limit for mono- or diacids in the C_5 to C_{10} range was about 0.1 μmol.

In the solvent system used, many inorganic salts apparently underwent a metathetical reaction with aqueous ammonia. Thus, a salt such as $NaNO_3$ gave two spots: a blue (basic) spot due to NaOH and a yellow (acidic) spot due to NH_4NO_3. Ammonium nitrate gave a single acidic spot. In addition to the chromatographic analysis, inorganic nitrates were measured by specific ion electrode and aliphatic aldehydes by MBTH test (8). Total carbon and molecular weight determinations were performed by the Schwarzkoph Laboratory, Woodside, New York.

III. RESULTS AND DISCUSSION

The ir spectra of the extracts of a typical aerosol collected on a smoggy day in downtown Los Angeles are shown in Figure 1. In this case, the filter sample was extracted successively in benzene followed by absolute alcohol. (The usual procedure was to extract separate portions of the filter in these two solvents.) The necessity of an alcohol extraction to more completely remove the organic material is illustrated in Figure 1. The benzene extraction removed 12% of the total amount of material on the filter and the subsequent alcohol extraction removed an additional 20% of the total. It can be seen from Figure 1a that the benzene extract is characterized by strong carbonyl and organic nitrate bands. The alcohol extraction removed further organic material, as seen by the CH and carbonyl bands, and dissolved a large amount of ammonium nitrate as well. Apparently all of the organic nitrate, but only part of the organic acid, was removed in the benzene extraction. A single extraction in ethanol would remove close to the total 32%, however.

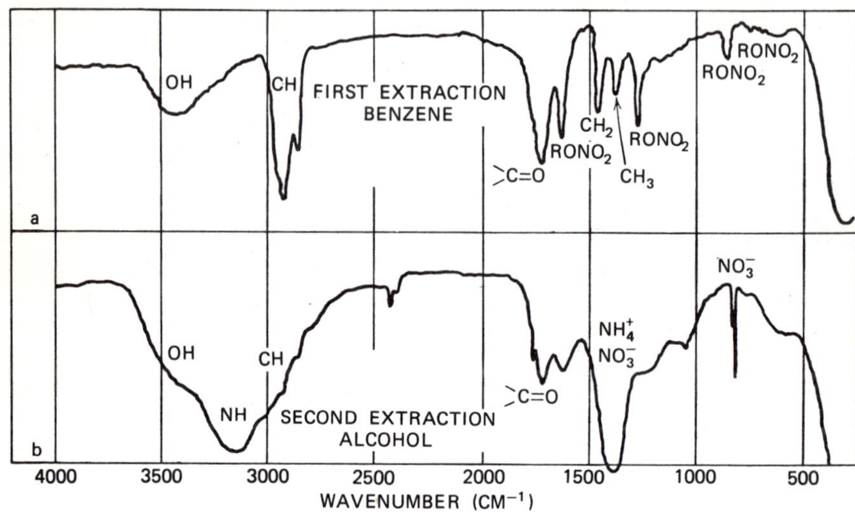

FIGURE 1. Infrared spectra of two successive solvent extractions of the same filter sample. Eight-hour sample taken 8:30-16:30, 9/14/72, at downtown Los Angeles.

The benzene extract, Figure 1a, gave a relatively simple spectrum similar to those obtained from the aerosols generated from mono- or diolefins under simulated atmospheric conditions (6). All five bands, characteristic of organic nitrates are present (700, 755, 860, 1280, 1630 cm^{-1}). Sharp bands indicating the aliphatic organic nature of the aerosol exist in the CH_n regions at 1385, 1460, and about 2900 cm^{-1}. Carboxylic acid is indicated by the strong carbonyl bands at 1720 cm^{-1} and the broad OH band about 3450 cm^{-1}. There is no indication of any aromatic content in this sample, but the ir does not conclusively rule this out.

The alcohol extract, Figure 1b, consists primarily of ammonium nitrate. Note the characteristic doublet at 830 cm^{-1} and the bands at 1385 and 3150 cm^{-1}. Overtones are present at 1760 and 2420 cm^{-1}. However, appreciable organic material is present as indicated by the carbonyl and C-H bands. Paper chromatography indicated the presence of appreciable amounts of dicarboxylic but no detectable monocarboxylic acid in the alcohol extract.

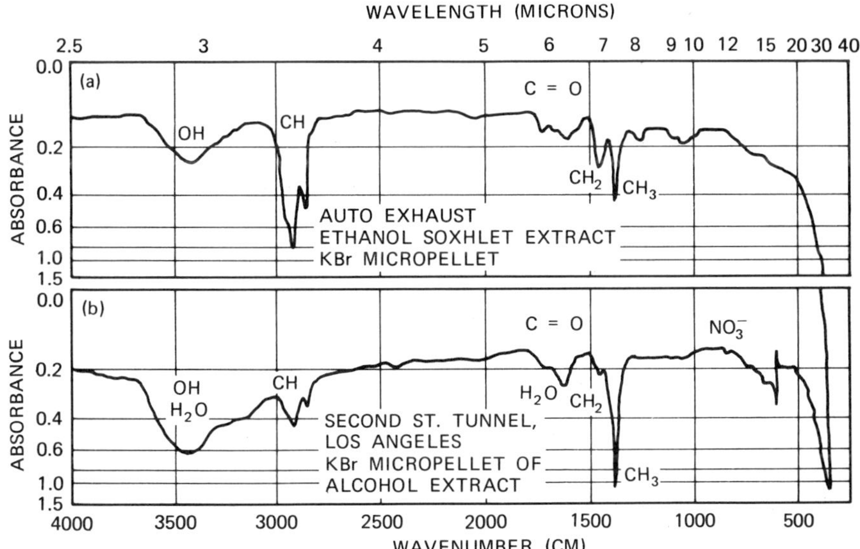

FIGURE 2. Infrared spectra of organic fraction of (a) auto exhaust particulate, and (b) particulate from vehicular tunnel in downtown Los Angeles.

In redissolving the extracts in 1 or 2 ml of alcohol, much of the inorganic salt was lost, as these are of limited solubility. Most organic material did redissolve.

In contrast to this ambient smog aerosol, the ir of directly emitted auto exhaust particulate is shown in Figure 2. Motor vehicles contribute 42% of all directly emitted particulates in the Los Angeles Basin, according to data of the Los Angeles County Air Pollution Control District (LAAPCD, 1971).

Figure 2a shows the infrared spectrum of an alcohol extract of an auto exhaust sample. This sample was taken during a seven-mode (California) cycle on a dynamometer from a 1964 vehicle burning low-lead fuel. The aerosol is primarily aliphatic in nature and exhibits only a small carbonyl band (relative to the C-H intensity). The aliphatic nature is also reflected in the fact either hexane or alcohol would dissolve most of the material present.

Figure 2b shows a 2-hr hi-vol sample taken during morning rush hour in a tunnel in downtown Los Angeles. During this period (6 to 8 a.m.) secondary aerosols should not yet be present to a large extent. The enrichment of this sample by directly emitted auto exhaust is evidenced by the fact that the aerosol mass loading in the tunnel was three times the mass loading obtained concurrently outside the tunnel. Here again the aliphatic organic nature is apparent, with a small carbonyl to C-H band intensity ratio, about 0.3 to 1. In addition to the organic material, ammonium nitrate is present which is probably from the ambient atmosphere.

As part of this preliminary characterization study, atmospheric aerosol samples were obtained from the South Coast Air Basin and elsewhere in California. These samples were extracted in alcohol and in benzene and were analyzed by infrared spectroscopy, paper chromatography, and for inorganic nitrate by specific ion electrode. The results of these analyses are presented below. In addition, aliphatic aldehydes were determined by MBTH test but were found to be negligible (<1% of the organic fraction) in all samples.

One 72-hr sample from downtown Los Angeles was chosen for detailed analysis by paper chromatography. This sample was taken beginning 9:00 a.m., 9-15-72, during moderate smog conditions. Figure 3a gives the ir spectrum of the bulk alcohol extract, which amounted to 32% of the total filterable mass. This sample is seen to have the characteristic bands due to carbonyl and organic nitrate groups and ammonium nitrate. Figure 4 shows a paper chromatogram of this sample sprayed with the acid-base indicator to visualize the spots. Spots #4 and #5 are the characteristic spots attributable, respectively, to dicarboxylic acid and (multifunctional) monocarboxylic acid. See the previous paper (6) for the chromatograms of a number of reference compounds.

Fifteen separate aliquots of this sample were chromatographed and each spot cut out. The collected spots were grouped and extracted from the paper. The total amount of aerosol involved was about 26 mg of which about 4 mg was recovered from spot #5. Spot #1 was not cut out, and the other spots had too little material to weigh accurately. Infrared spectra were obtained for all the spots. Spots #2 and #3 gave spectra similar to spot #4 which is shown in Figure 4b. This spectrum is fairly simple and is in agreement with that of a dicarboxylic acid. There are no bands due to organic nitrate.

Spot #5 which should contain monocarboxylic acid and multifunctional monocarboxylic acid is shown in Figure 4c. The chief bands are carbonyl and organic nitrate, the latter much enhanced

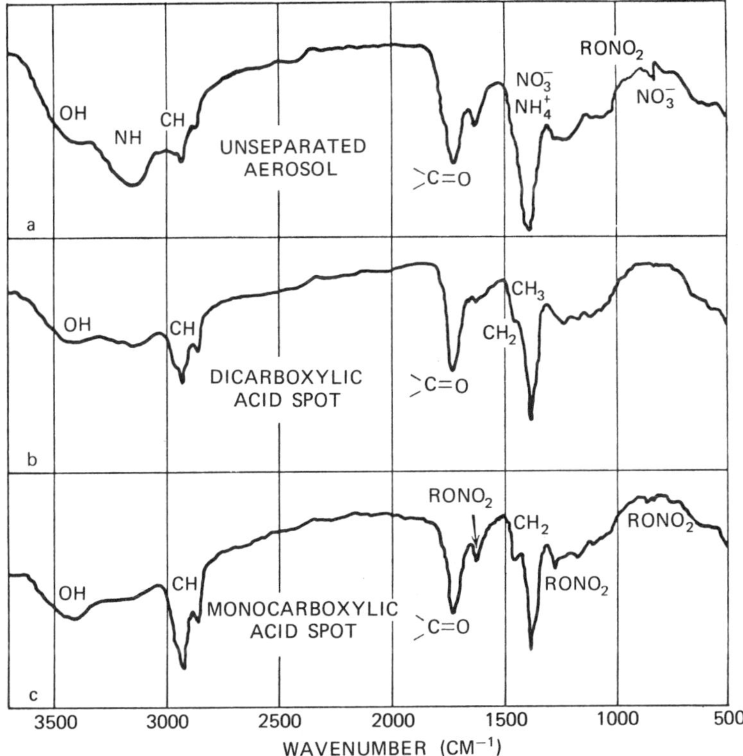

FIGURE 3. Infrared spectra of aerosol sample taken in downtown Los Angeles. Seventy-two-hour sample taken commencing 9/15/73. Chromatographic separation illustrated in (b) and (c), which refer to Figure 4.

over the unseparated aerosol. On the basis of the spot location, we infer that the organic nitrate groups must be joined to monocarboxylic acids. Nonacidic nitrates would move to the upper right corner of the paper. There is a small band at 830 cm^{-1} due to inorganic nitrate. It is believed that this band may be due to reaction of the organic nitrate with KBr. Growth of the inorganic nitrate band has been observed as a KBr pellet ages.

A molecular weight determination by osmometry was performed on the material isolated from spot #5. This would give an

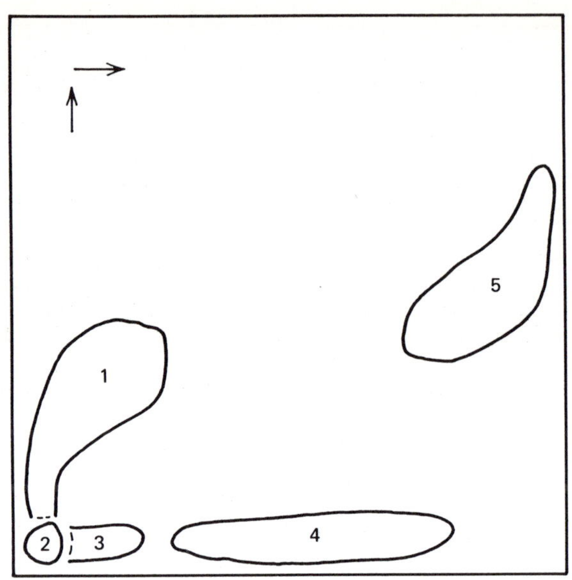

FIGURE 4. Paper chromatographic separation of aerosol sample of Figure 3. Solvent moves vertically and then to the right. (1) Inorganic salts, and (2) and (3) polar dicarboxylic acids, (4) dicarboxylic acids, (5) monocarboxylic acid.

average molecular weight for the compounds present in this spot. The molecular weight was 322. This corresponds to a 20-carbon monocarboxylic acid or to a 15-carbon monocarboxylic acid with a nitrate group.

A rough quantitative estimate of the proportion of nitrate to carboxylate groups, as well as the carbon chain length, can be obtained from the infrared spectra and band strengths of pure compounds prepared for this purpose. Hexanoic acid-6-nitrate was used as a reference compound to determine the relative band strengths of C-H, C=O, and $RONO_2$. For this reference compound the C-H:C=O absorbance ratio was in good agreement with published values. Using these band strengths, we concluded that in the monocarboxylic acid spot there are about nine carobyxlate groups for each nitrate group--that is, one acid molecule in nine contains an organic nitrate group. The carbon chain length (C-H band relative to the C=O band) is about 20, in good agreement with the molecular weight determination.

Although stearic acid (20 carbons long) would have the proper retention to appear in spot #5, it is not acidic enough to give a yellow spot with the indicator used. There must be some shorter chain acids or multifunctional acids present in order to yield the yellow spot color.

Analyses of a number of atmospheric samples obtained at various sites in California are summarized in Table 2. The net weight of the sample is given, as well as the percentage of the sample soluble in benzene and in absolute alcohol (95% ethanol, 5% propanol). Most hi-vol samplers ran for 24 hr at about 50 cfm so the total mass for these may be converted to the approximate mass loading by the formula: total weight (mg)/2 = $\mu g/m^3$.

Benzene removes very little inorganic material. Alcohol, on the other hand, will remove some salts. Nitrates are fairly soluble in alcohol, while sulfates are almost insoluble. Some bromides and chlorides are soluble as well. Extraction of known amounts (15 mg) of NH_4Cl and NaBr from a filter paper for 1 hr resulted in 86% and 96% recovery, respectively, of these salts. Probably the major inorganic portion of the alcohol extract would be nitrates, as nitrates and sulfates comprise the bulk of the salts in urban particulate. NH_4NO_3 gave 100% yield on extraction and $(NH_4)_2SO_4$ gave 5%.

Total inorganic nitrate was measured by specific ion electrode. Assuming a molecular weight of inorganic nitrate of 82 (NH_4NO_3 = 80, $NaNO_3$ = 85) the total percentage of inorganic nitrate is also expressed in Table 2. Since the alcohol extraction will remove most or all of the benzene solubles, adding the benzene-soluble and nitrate percentages and subtracting from the alcohol-soluble percentage will give the approximate percentage of organic aerosol insoluble in benzene. This percentage is often higher than the percentage soluble in benzene.

A rough relative estimate of the concentration of mono- and dicarboxylic acid and of the strength of the carbonyl and organic nitrate infrared bands is also given in Table 2. The acid concentrations are determined from paper chromatograms on the basis of spot size and intensity of color. This is admittedly a crude estimate, since it depends on retention on the paper for different acids and on acid strength for spot color. The infrared band intensities for carbonyl and nitrate are measured relative to the C-H band strength at 2925 cm^{-1}.

Infrared spectroscopy gives a useful semiquantitative means of distinguishing between NH_4NO_3 and other nitrate salts. Thus while the specific ion electrode measurements gave total inorganic nitrate, on the basis of the ir spectrum, this nitrate could be identified further. In general, the concentration of

TABLE 2. Analysis of Atmospheric Aerosol Samples

Date	Location	Sample time (hr)	Net weight (mg)	Benzene extract (%)	Alcohol extract (%)	Inorganic nitrate (%)	Diacids	Monoacids	Organic nitrates	Carbonyl
6-8-72	Alturas	24	131	8	13	3	--	--	O	L
6-3-72	Inyo County	24	121	6	12	4	--	--	O	L
6-2-72	Tahoe City	24	165	8	12	3	--	--	O	L
6-8-72	Petaluma	24	125	8	21	3	--	--	O	L
6-2-72	San Luis Obispo	24	127	5	23	15	--	--	O	O
6-8-72	Sacramento	24	119	1	27	6	O	O	O	L
6-3-72	Madera	24	260	3	16	8	L	O	M	L
6-2-72	Modesto	24	213	1	30	9	L	O	L	L
6-2-72	Redding	24	201	4	22	10	M	O	M	L
6-7-72	Santa Barbara	24	136	2	25	16	O	O	O	L
6-2-72	Merced	24	280	3	16	8	L	O	L	L
6-2-72	Alturas	24	143	2	10	4	--	--	O	L
6-8-72	Santa Maria	24	114	3	20	9	--	--	O	L
7-7-72	Coalinga	12	263	2	12	4	L	M	O	L
7-7-71	Mira Loma	24	588	3	12	3	H	O	H	H
8-2-71	Mira Loma	24	599	2	15	11	--	--	M	H
9-8-72	Mira Loma	48	946	4	18	9	--	--	H	H
10-6-72	Mira Loma	24	882	2	14	8	--	--	M	M
6-25-71	Mira Loma	76	1202	2	15	7	L	O	M	M
6-26-72	Exit of Tepee Burner		610	38	78	1	--	--	O	M
7-11-72	Los Angeles	48	450	11	33	10	H	H	H	H
9-12-72	Los Angeles	8	98	--	37	12	M	O	L	L
9-13-72	Los Angeles	8	127	--	41	11	H	O	M	M
9-14-72	Los Angeles	8	101	12	33	11	M	M	H	H
9-15-72	Los Angeles	72	647	--	32	--	H	H	M	H
9-18-72	Los Angeles	7	73	--	28	--	H	H	H	H
9-21-72	Los Angeles	8	148	--	27	6	M	H	H	H
9-22-72	Los Angeles	8	179	--	40	17	M	H	H	H
9-21-72	Riverside	24	263	--	32	13	O	H	H	H
9-22-72	Riverside	72	472	6	42	17	H	O	H	H

O, L, M, H refer to the relative strength of the particular species as analyzed by paper chromatography (di- and monoacids) or infrared spectroscopy (organic nitrates and carboxyl group). O = absent, L = low, M = medium, H = high.

TABLE 3. Mass Percentages of Components on High-Volume Filters (Parentheses Indicate Number of Samples Analyzed, ± Indicates Range of Values. In b Quantitative Values are only Approximate)

	Central California	LA, Riverside	Mira Loma
a. Mass Percentages of Components on Hi-Vol Filters			
Ethanol extract	19 ± 8 (14)	32 ± 7 (10)	15 ± 2 (5)
Redissolved	12 ± 5 (14)	27 ± 7 (10)	13 ± 3 (5)
Organic fraction (see b, below)	14 ± 6 (7)	19 ± 4 (6)	11 ± 1 (2)
Nitrates (SIE)	8 ± 5 (14)	12 ± 4 (7)	8 ± 3 (5)
Benzene extract	4 ± 2 (14)	9 ± 4 (2)	3 ± 1 (5)
Redissolved	4 ± 2 (14)	6 ± 4 (2)	2 ± 1 (5)
b. Mass Percentages of Components in Organic Fraction			
Aldehydes (MBTH)	0.9 ± 0.8 (7)	0.3 ± 0.3 (6)	0.4 ± 0.2 (2)
Monoacids (as benzoic)	2 ± 5 (&)	22 ± 14 (6)	0 ± 0 (2)
Diacids (as homophthalic)	2 ± 3 (&)	10 ± 6 (6)	7 ± 8 (2)
Total acids (ir)	--	55 ± 10 (6)	--
Organic nitrates (ir)	--	17 ± 4 (6)	--

NH_4^+ ion was highest in the Los Angeles Basin and low elsewhere in the state. For instance, the Santa Barbara sample, taken within a mile of the ocean, was about 16% inorganic nitrate but showed no NH_4^+ ion.

The samples are divided into the regions: "nonurban" California, LA/Riverside, and Mira Loma in Table 3. In Table 3 the solvent extractables and inorganic nitrate are given. The organic fraction is the portion of the total sample soluble in alcohol, less the concentration of nitrates. Table 3 further breaks down the organic fraction into some of its constituents. These quantities are only rather crude estimates because the techniques employed are still only semiquantitative. Nevertheless, they give some idea of the mass balance in the organic fraction.

In general, the LA/Riverside samples had more organic material and more inorganic nitrates. The Mira Loma samples had much wind-blown sand and dust so the relative organic and nitrate portion was less. The Mira Loma samples were approximately a year old when extracted. The "nonurban" California samples were extracted within a week or two of being taken and the LA/ Riverside samples generally within one or two days.

IV. CONCLUSIONS

The number of sample analyses presented in Table 2 is too small to draw any statistically valid conclusions about sources or the magnitude of local problems. However, seven significant conclusions may be drawn on the basis of these preliminary results:

The very highly oxygenated nature of the organic fraction of particulates collected in the South Coast Basin in August is quite different from that of directly emitted auto exhaust particulate or particulate from any other known source. This is an indication of the occurrence of secondary conversion processes in the atmosphere.

Organic nitrates, an uncommon chemical for which there is no known emission source, serve as a good indicator of the presence of secondary conversion processes leading to aerosol formation. Organic nitrates may be readily identified by their characteristic ir spectrum. These compounds are quite evident in samples taken in August in the South Coast Basin of California, and have been found to a lesser extent in some samples taken elsewhere in the state. However, they may not be important on a relative mass basis.

Mono- and dicarboxylic acids made up an appreciable portion of the organic content of particulate matter under photochemical smog conditions in the South Coast Basin. These are very likely to be of photochemical origin.

Benzene or alkanes are not suitable solvents for extraction of organic material from airborne particulate, especially in areas of photochemical activity where highly polar aerosol compounds are formed. Alcohol, a better solvent, will extract about 80% of the organic carbon. The alcohol solubles, less inorganic salts, amount to 20 to 30% of the total mass loading of several samples taken in the South Coast Basin in moderate smog conditions. This percentage is smaller in less polluted areas of the state.

Aliphatic aldehydes make up a negligible portion of the organic material in the particulate samples examined. This is in agreement with vapor pressure considerations.

Inorganic nitrate, of photochemical, or at least secondary origin, makes up about 10% of the mass loading of the samples analyzed. The reaction of NO_2 and NaCl to yield $NaNO_3$ is probably an important source in coastal areas as evidenced by high NO_3^- percentages in samples from Santa Barbara. The prevalent form of nitrate in the South Coast Basin is NH_4NO_3; ammonium ion is less prevalent elsewhere in the state (based on ir evidence).

No indication of aromatic constituents could be detected in the ir of atmospheric aerosols. While not conclusive evidence for their absence, this at least suggests that the aerosols are not primarily aromatic. It should be noted that the ir spectrum of a synthetic aerosol generated from indene (an aromatic olefin) and from mesitylene did not exhibit particularly intense aromatic ir bands (6).

As mentioned above, the presence of organic nitrates is a very good indication of the existence of photochemically generated aerosols. The sample analyses presented in this report thus indicate the widespread presence of photochemically produced aerosols in a variety of locations in California. The actual magnitude of this aerosol problem and its impact have yet to be determined.

Carboxylic acids also are known from chamber experiments to be products of aerosol-forming reactions. These have been found in the atmosphere in this study. In addition, both mono- and dicarboxylic acids have recently been identified in Los Angeles particulate matter by a mass spectral technique (9). Monocarboxylic acids can be formed from either mono- or diolefins but dicarboxylic acids have been found formed only from diolefins or probably cyclic olefins (6). Since both types of acids have been found in these samples, it is likely that both types of olefins are present as aerosol precursors in the atmosphere. In a limited number of chamber studies (6), aromatic compounds were shown to form only minor amounts of acids, thus further indicating the importance of some types of olefins as atmospheric aerosol formers. Potential sources of these diolefins, cyclic olefins, or long-chain monoolefins have not yet been identified. Conceivably, direct emission sources for carboxylic acids may exist in addition to their formation by secondary processes. The presence of organic nitrate groups on acid molecules isolated from the atmosphere does show the formation mechanism to be significant.

When present in the atmosphere as a gaseous aerosol precursor, as well as when present in the aerosol itself, carboxylic acids or their salts may act as emulsifying agents or soaps, drawing together chemically dissimilar species in a single droplet and enhancing aerosol formation and growth.

This work has been of the nature of a pilot study and is not intended to be a statistically valid assessment of the magnitude of the secondary aerosol problem in California. This is because a relatively limited number of samples has been analyzed to date, and because of the limitations inherent in the analytical techniques. However, the experimental techniques developed as well as the preliminary information gained should lead the way to more definitive work. This will require a more comprehensive sampling program followed by more thorough and automated sample analysis. Such a program is currently under way in this laboratory. Once the identity of the actual aerosol components is established, knowledge of aerosol formation mechanisms should allow us to identify their precursor hydrocarbons and, ultimately, the sources of these hydrocarbons.

REFERENCES

1. Whitby, K. T., Husar, R. B., and Liu, B. Y. H. 1972. J. Colloid Interface Sci. 39, 177.

2. Husar, R. B. and Whitby, K. T. 1973. Environ. Sci. Technol. 7, 241.

3. Hidy, G. M., and Friedlander, S. K. 1971. The Nature of the Los Angeles Aerosol, in Proceedings of the Second International Clean Air Congress, p. 391, Academic Press, New York.

4. Los Angeles County Air Pollution Control District. 1971. "Profile of Air Pollution Control."

5. Environmental Protection Agency. 1972. Air Quality Data for 1968 from the National Air Surveillance Network and Contributing State and Local Networks, APTD-0978.

6. O'Brien, R. J., Holmes, J. R., and Bockian, A. H. 1975. Environ. Sci. Technol. 9, 568.

7. Hais, I. M., and Macek, K. 1963. "Paper Chromatography," Czechoslovak Acad. Sci., Prague.

8. Sawicki, E., Hauser, T. R., Stanley, T. W., and Elbert, W. 1961. Anal. Chem. 33, 93.

9. Schuetzle, D., Crittenden, A. L., and Charlson, R. J. 1973. J. Air Pollut. Control Assoc. 23, 704.

This work was presented in part at the Western Regional American Chemical Society Meeting, October 12-16, 1972, San Francisco, California.

Atmospheric Aerosol Formation by Chemical Reactions

GEORGE M. HIDY*
California Institute of Technology
Pasadena, California

and

C. S. BURTON†
Science Center, Rockwell International
Thousand Oaks, California

*Present address: Environmental Research & Technology, Inc.
Westlake Village, California
†Present address: Systems Applications, Inc.
San Rafael, California
‡Reprinted in revised form from International Journal of Chemical Kinetics, Symposium No. 1, p 509-542, 1975.

Abstract

This paper is a review of several aspects of aerosol formation processes that occur in the Earth's atmosphere. Important contributors to atmospheric aerosol chemistry are the sulfates, nitrates, and organic compounds which are considered to be formed mainly by reactions of trace reactive gases in air. The phenomenology of the evolution of aerosols by such chemical processes is illustrated by recent observations taken in Los Angeles smog. The results of this program show the strong relationships between sulfate, nitrate, and organic carbon formation and gas-phase processes of photochemical smog.

Suspected mechanisms of secondary aerosol production are summarized, with consideration of both homogeneous and heterogeneous processes. These mechanisms are examined in the light of laboratory simulations and the knowledge of atmospheric behavior to deduce the potential importance of certain classes of reactions for explaining aerosol evolution. Such considerations illustrate well the complexities of gas-particle interactions in atmospheric chemistry.

I. INTRODUCTION

Over the past century hundreds of studies have been made on visibility reduction in the atmosphere which is associated with aerosols making up haze. It has been known for sometime that haze involves light scattering from tiny submicron particles suspended in air. Yet only recently has there been widespread recognition that such particles play an active role in air chemistry. Through their production and growth, workers have come to link them with the atmospheric removal of certain reactive gases such as the sulfur and nitrogen oxides and hydrocarbon vapors. The chemical mechanisms of aerosol formation in the atmosphere remain uncertain, but it is likely that many processes play a role, including important photochemical reactions.

Since Tyndall's nineteenth century classical experiments on aerosol optical behavior, it has been known that exposure of reactive gases to light can produce copious quantities of airborne particles. More recent observations have led investigators to believe that sunlight-induced chemical reactions are responsible for a substantial fraction of the intense haze observed over cities such as those in southern California. The relation between haze and photochemical reactions in polluted air was deduced many years ago from the work of Haagen-Smit and co-workers, and photochemical aerosol production was explored by many investigators, including Stephens, Hanst, Renzetti, and Doyle [e.g., Leighton (1)]. The possibility of widespread haze formation by sunlight-induced reaction of background nitrogen oxides and natural terpenoid compounds from vegetation was suggested several years ago by Went (2). Recent research on the stratospheric aerosol has indicated that the major constituent by mass of airborne particles at high altitude is sulfur. One explanation for the presence of such material is via oxidation of traces of sulfur dioxide above the troposphere [e.g., Friend et al. (3)].

With increasing concern for the influence of human activities on the atmosphere, there is interest in improving the air quality levels for aerosols generated from atmospheric pollutants. Thus there has been a substantial increase in research activity devoted to atmospheric aerosol chemistry, particularly photochemistry. At the same time that a variety of new laboratory experiments have been updated, major new field studies have been implemented to characterize atmospheric haze formation in more detail. Perhaps best documented of these is the 1969 Pasadena Smog Aerosol Study, whose methods and results were reported, respectively, by Whitby et al. (4) and in the volume Aerosols and Atmospheric Chemistry, edited by Hidy (5). This project was an exploratory one, but recently much larger investigations have been initiated, including the California Aerosol Characterization Experiment (ACHEX) [Hidy et al. (6)] and a portion of the Regional Air Pollution Study (RAPS).

The key to improved characterization of photochemical aerosol behavior in the atmosphere is the quantitative description of the physical and chemical properties of these aerosols as well as the kinetics of their formation. A combination of observations, atmospheric simulation, and laboratory experiments is required to elucidate the complexities of particle evolution. The purpose of this paper is to review the current knowledge in these areas and to synthesize them into suggested key processes of interest in the atmosphere. First, several important new results from the study of photochemical aerosols are outlined. Then conclusions from simulation experiments are related to recent basic investigations to elucidate relevant physicochemical mechanisms applicable to the troposphere and the stratosphere. Finally, a current picture of aerosol formation in the atmosphere is discussed in relation to the combination of evidence from field observations and knowledge of physical chemistry.

II. ATMOSPHERIC OBSERVATIONS

A. The Troposphere

Photochemical aerosols in the atmosphere require an operational or phenomenological definition. One definition might come from a comparison of the nature of suspended material from atmospheric samples and that generated in controlled laboratory experiments involving irradiated mixtures of reactive gases in air or filtered ambient air [e.g., Husar et al. (7)]. In the atmosphere alone, a definition becomes difficult because of the complexities of air chemistry involving a series of processes that are both sunlight-induced and thermal in nature. A traditional operational definition of photochemical aerosols in the troposphere is as follows: those clouds of particles in air that (a) are identified with hazes where ozone is present, and (b) are formed and dissipated on a time scale of hours with maximum visibility degradation at midday.

Perhaps the best known example is the intense haze formation associated with photochemical smog over the Los Angeles area. For the details of aerosol photochemistry one must rely heavily on observations taken in Los Angeles, for they are far better documented here than elsewhere.

One semiquantitative measure of the concentration of haze is the part of the extinction coefficient, b_{scat}, for visible radiation resulting from the scattering of light from airborne particles. Although b_{scat} has been found not to correlate uniquely with oxidant behavior, an important correspondence in the relationship appears to exist for b_{scat} taken when ozone is measured at its maximum on a given day in the Los Angeles area. This relation is shown in Figure 1. Here the strong increase in light scattering (visibility reduction) with intensity of photochemical smog is demonstrated. The relationship is supported further bv observations taken by blimp traveling at midday with the winds on a north-south path across zones of increasingly high ozone concentration over central Los Angeles.

1. Physical Properties. The scattering coefficient, b_{scat}, depends on the particle size-number distribution of the haze and the particle index of refraction, as well as the wavelength of incident light. The number-size distribution and its moments can

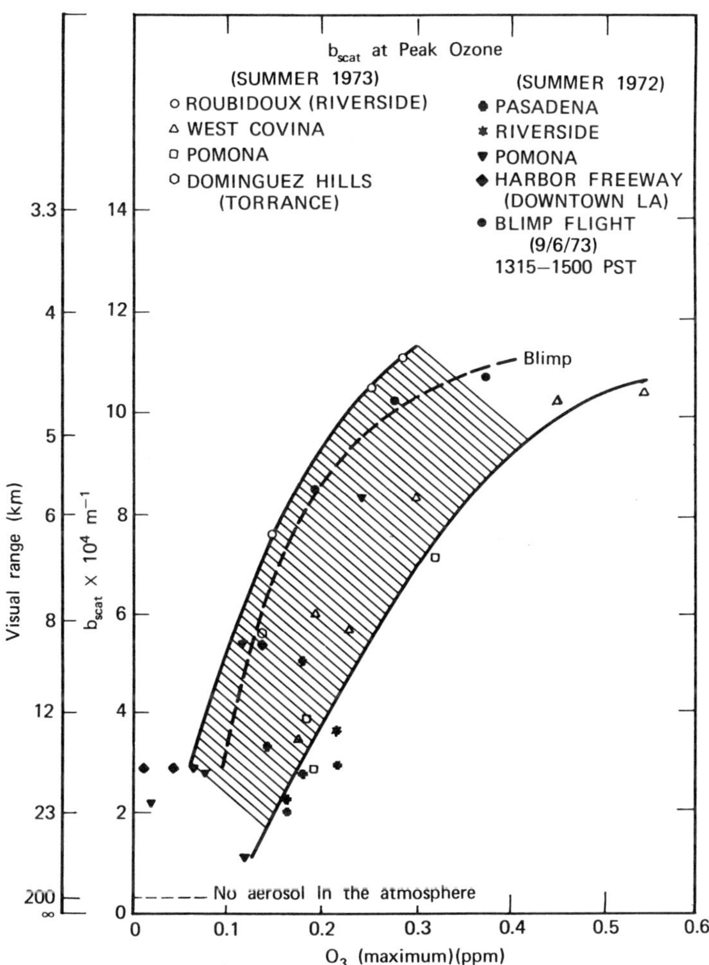

FIGURE 1. Correlation between b_{scat} and O_3) (maximum). Based on 2-hr averaged data taken in the Los Angeles area (6).

be considered the key physical properties of concern in characterizing atmospheric aerosol behavior. This distribution is often defined in terms of volume, v, or the distribution function, $n(v, t, \vec{x})\, dv$, refers to the number of particles per unit volume in the particle volume range v and v + dv at any time t and at any point in space given by $\vec{x}$.

The particle size distribution is difficult to measure, so that only moments of the distribution are monitored routinely. The zeroth moment of the distribution function corresponds to the total number concentration of particles, N, the two-thirds moment is proportional to the total surface area, S, per unit volume, and the first moment is proportional to the volume fraction, V. For particles of an average mass density $\bar{\rho}$, the product of $\bar{\rho}V$ is the total mass concentration; b_{scat} is a product of the Mie scattering function and S summed over all particle sizes. Evidence has accumulated from several experiments [e.g., Charlson et al. (8)] that b_{scat} is dominated by light scattering in the particle diameter range 0.1 to 1.0 μm. Indeed, observations taken in the ACHEX (6), as well as the 1969 Pasadena study, show that particle evolution in smog is strongly concentrated in this size range. This is illustrated readily by measurements of the change in volume - diameter distribution from early morning to midmorning in smog. An example of data taken in Pomona in 1973 is shown in Figure 2. These and other data also illustrate that few new

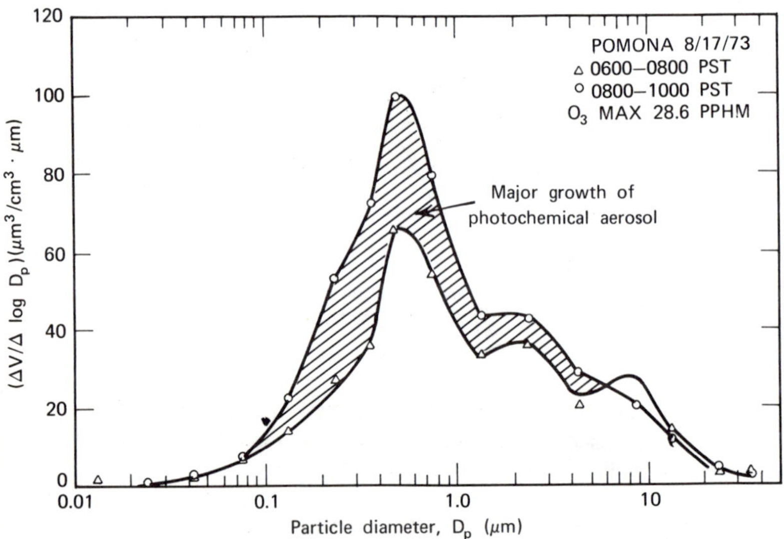

FIGURE 2. Evolution of the volume distribution of smog aerosol taken at Pomona, California, on August 17, 1973 (6).

nuclei smaller than 0.1 μm are produced in smog, in contrast to intense growth of existing particles during the evolution reactions of smog, which are believed to produce the condensable precursors.

Based on thermodynamic consideration of the Kelvin effect, there is reason to believe that the stability of aerosols is strongly influenced by the change in vapor pressure of materials with particle size. For most condensable species the surface energies dictate that the stable particle is approximately 0.1 μm in diameter, suppressing growth and new particle formation on ranges less than this size [see also Heisler et al. (9); Hidy (10)].

The fraction of particles smaller than 5 to 10 μm in diameter is maintained at a low concentration by several removal processes, including gravitational fallout [for details see Hidy (11)]. Haze evolution is spectacular over cities with polluted air like Los Angeles because of high concentration of precursors combined with intense sunlight. However, it is believed that similar reactions must take place almost universally in the troposphere at slower rates, modulated by repeated irradiation by sunlight and intermittent removal processes involving clouds.

2. Chemical Properties. Chemical analysis of airborne particles provides the most revealing information about the nature and significance of aerosol formation by atmospheric chemical processes. There is an accumulation of evidence from many studies, as reviewed, for example, by Junge (12) and from more recent reports such as those of the National Air Surveillance Network (13), that water-soluble sulfate, nitrate, and organics constitute a substantial fraction of the tropospheric aerosol. The importance of these constituents is illustrated in Table 1 for urban and nonurban aerosol samples in California. Examination of such samples also suggests that the sulfate and nitrate exist as ammonium salts, though observations in other geographical areas may indicate the presence of sulfuric acid. For example, Charlson et al. (14) have reported evidence of free sulfuric acid from nonurban air near St. Louis, and Cunningham (15) has identified by infrared analysis acid sulfate in urban samples of aerosol collected near Chicago.

The mass concentration of sulfate and nitrate cannot be accounted for by either natural or man-made primary sources [Robinson and Robbins (16); Hidy and Brock (17)]. Thus, the production of aerosols in the atmosphere is believed to involve mainly the highly complex chemistry of sulfur, nitrogen, and organic carbon. These constituents come from gaseous emissions of both natural and anthropogenic origin and are key ingredients in the "classical" hierarchy of gas reactions in photochemical smog. Most of the nitrate appears to be inorganic, with only a

TABLE 1. Chemical Composition of Filter-Collected Aerosol Samples

	Amount (wt %)			
	Nonurban		Urban	
Constituent	Pacific Coast[a] Offshore	Desert[b]	Fresno[b]	Pomona[b]
Al	2.8	2.4	2.4	0.99
Si	5.0	5.2	6.3	2.2
Na	3.6	0.71	0.40	1.1
Cl	9.9	0.32	0.19	0.51
K	1.3	0.39	0.33	0.24
Ca	1.9	2.1	0.47	1.0
Ti	0.15	0.13	0.09	0.11
V	ND[c]	0.004	ND	0.008
Cr	0.02	ND	0.0058	0.013
Mn	0.04	0.028	0.025	0.025
Fe	1.6	1.4	1.03	0.86
Cu	0.6	0.028	0.0058	0.014
Zn	0.2	0.035	0.055	0.13
Br	0.2	0.059	0.098	0.36
PB	0.4	0.017	0.42	1.6
I	-	-	-	-
$SO_4^{=}$	17.5	2.8	2.0	10.6
NO_3^-	5.7	3.5	3.8	20.2
NH_4^+	2.2	1.7	1.5	9.1
Noncarbonate C	3.0	ND	5.0[d]	24.0[d]
Total measured ($\mu g/m^3$)	30	42.4	207	180

[a]Average of 20 samples at San Nicolas Island [Hidy et al. (64)].

[b]Single 24-hr sample [Hidy et al. (6)].

[c]Not detected.

[d]Estimated as 3 x cyclohexane-extractable organics.

minor organonitrate contribution, evidently linked with photochemical processes [O'Brien et al. (18); Hidy (10)].

The organic fraction is characterized by more oxygenated material than is accounted for in the vapor. High-resolution mass spectroscopy and solvent extraction methods, combined with infrared analysis, suggest that more than half the organic aerosol is oxygenated, with a large carboxylic acid component [e.g., Schuetzle (19); Hidy et al. (6); Grosjean and Friedlander (20)].

The size distribution of "secondary" aerosol constituents is of interest for checking the evolutionary processes indicated by physical measurements. Observations of the mass concentration of sulfate over 2-hr intervals in smoggy air over Los Angeles and other cities indicates that at least half of this material is found in the fraction of aerosol less than 1.0 μm in diameter. Under conditions of heavy smog, most of the sulfate in Los Angeles air is concentrated in the small-particle fraction during the day, but at night and in the early morning the sulfate shifts to larger particles [Hidy et al. (6)]. In contrast, nitrate appears to be present in somewhat larger particles in the Los Angeles aerosol except in areas such as Riverside, which is far downwind of the city under the most common daytime conditions.

The distinction between the size distributions of sulfate and nitrate in smog aerosol is illustrated in the data shown in Figure 3. The case shown in Figure 3a corresponds to a 24-hr sample in West Covina, California, during which the maximum ozone concentration exceeded 0.4 ppm in the east central part of the Los Angeles basin. Another case from the Riverside area farther east is shown in Figure 3b in which the maximum ozone concentration was above 0.3 ppm. These samples illustrate that the mass distributions of nitrate and sulfate constituents in smog more aged than that at West Covina are almost identical, but much more nitrate is present downwind, farther to the east.

3. Diurnal Patterns of Evolution. An important description of the evolution of photochemical smog is obtained from examination of the changes during the day of the reactive gases in the air. The classical picture of smog formation is revealed in the set of data in Figure 4. Here the familiar pattern of early morning depletion in NO and hydrocarbons takes place with a midmorning maximum in NO_2 and a midday maximum in ozone concentration. Such behavior has been simulated successfully many times in smog chamber experiments. Accompanying the increase in ozone is an increase in b_{scat}, reflecting the aerosol growth and subsequent visibility reduction in smog through the midday. Changes in total number concentration are less well correlated with the reactive gas behavior, and are more closely related to local combustion sources.

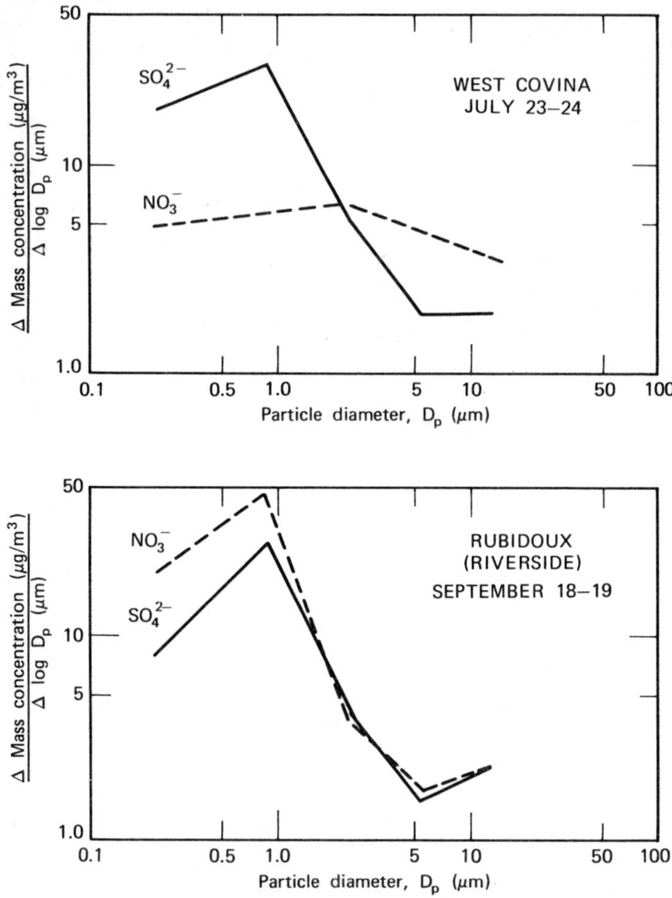

FIGURE 3. Mass distribution of $SO_4^=$ and NO_3^- with particle size at two locations in Los Angeles based on data taken in 1973 (6).

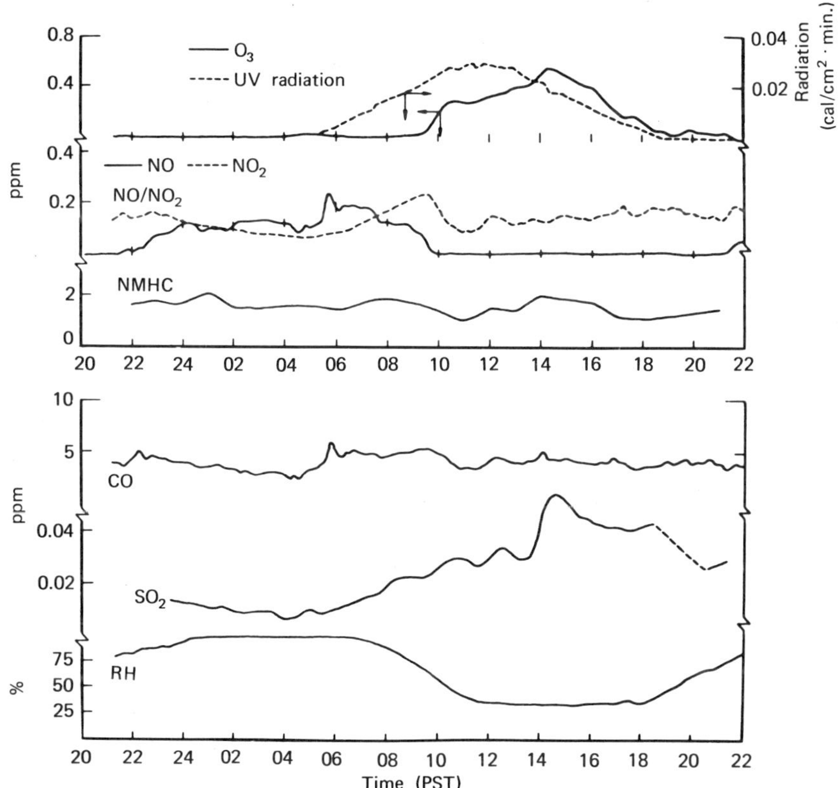

FIGURE 4. Diurnal patterns for West Covina, July 23-24, 1973. V_T = total particle volume, V (<1 μm) = particle volume less than 1 μm diameter, V ($0.01 > D_p > 0.1$) = particle volume in the range 0.01 to 0.1 μm diameter (D_p), and CNC = condensation nuclei count (6).

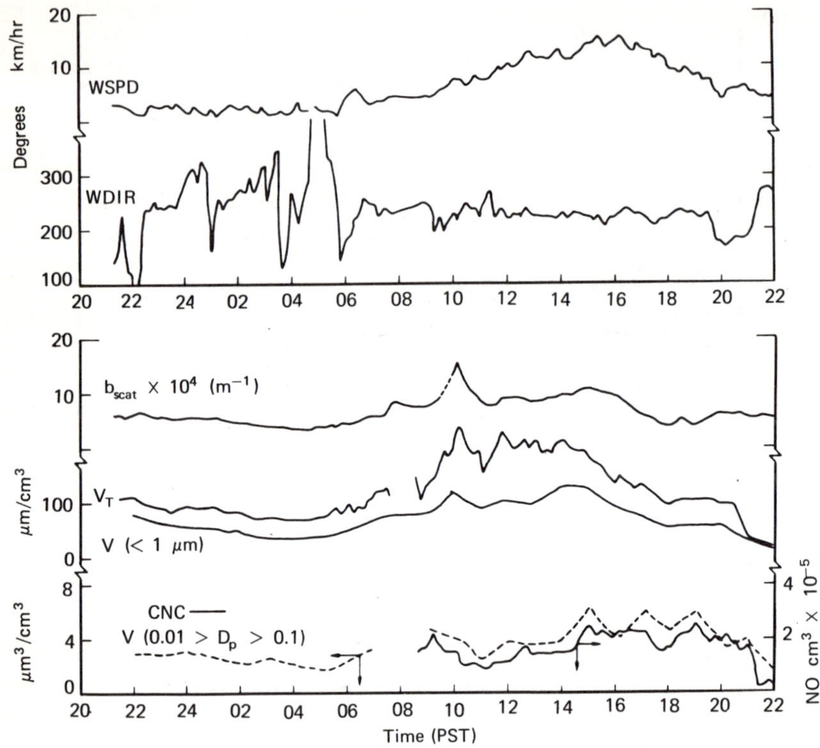

FIGURE 4 (continued).

The diurnal changes in sulfate, nitrate, and carbon constituents accompanying the gas chemistry are shown in Figure 5. In many (but not all) cases studied in the ACHEX [Hidy et al. (6)], observations showed that sulfate and noncarbonate carbon follow closely the ozone development with maxima in midafternoon. In addition, $SO_4^=$ follows SO_2 changes during the day. In contrast, nitrate was found to develop a maximum concentration by midmorning with the NO_2 pattern. Thus the components of the aerosol behave in different ways, and evidently sulfate and organics are generated by distinctly different chemical mechanisms from those for nitrate.

4. Conversion Ratios. A measure of the extent of conversion of an aerosol precursor with condensed material is the

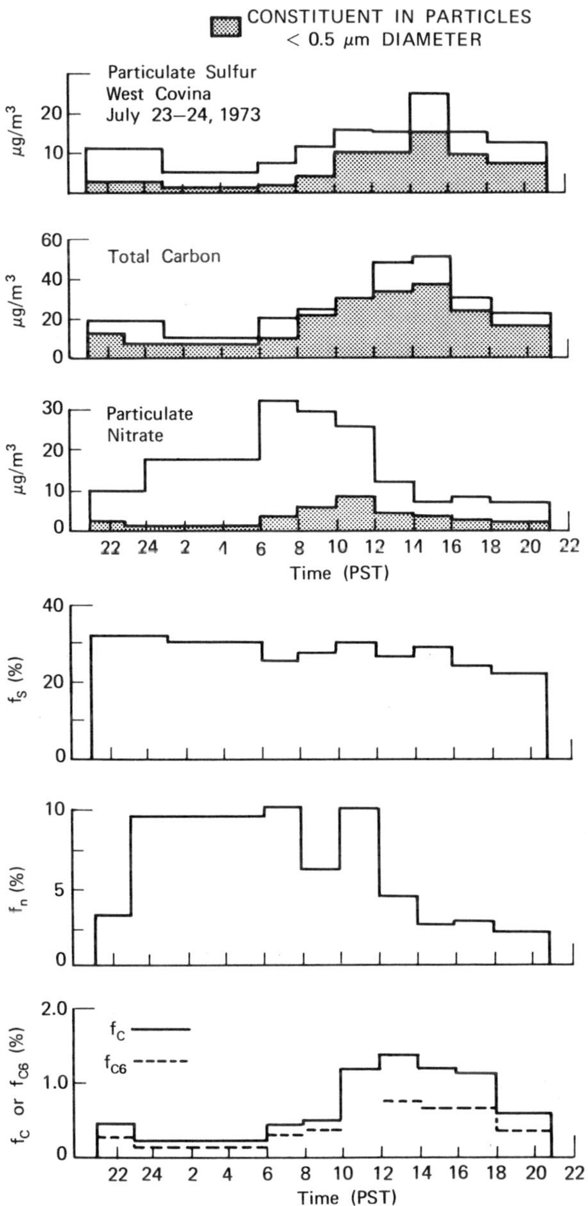

FIGURE 5. Aerosol diurnal pattern for West Covina, July 23-24, 1973 (6).

so-called conversion ratio. These ratios may be defined as follows:*

$$f_S = \frac{\text{mass of particulate S based on } SO_4^{=}}{\text{mass of particulate S + mass of gaseous S based on } SO_2}$$

$$f_N = \frac{\text{mass of particulate N based on } NO_3^{-}}{\text{mass of particulate N + mass of gaseous N based on } NO_2}$$

$${}^{*}f_C = \frac{\text{mass of particulate noncarbonate C - primary non carbonate C}}{\text{mass of particulate noncarbonate C - primary noncarbonate C + mass of reactive gaseous C from NMHC}}$$

In the troposphere the data on these ratios are limited, but a few values are given in Table 2 for comparison. Depending on the location of sampling, the values of the conversion ratios vary widely, but air masses containing aged aerosol or intense photochemical smog conditions appear to give the highest values.

The diurnal changes in f_S and f_N corresponding to the behavior of the reactants and products in West Covina are shown in Figure 5. Typically, f_S remaing roughly constant during the day with a value of approximately 0.2, while f_N is more variable with a maximum in the night through midmorning in this case. In the Los Angeles area $f_N \approx 0.06$ in central areas, but it reaches more than 0.30 farther downwind in Riverside.

The conversion ratio for carbon, f_C, is indicated according to the definition above in Figure 5. In this case it displays a maximum in midafternoon accompanying the ozone peak, with average values equal to or less than 0.01.

The conversion ratio f_C can be defined in a somewhat better way without assuming arbitrarily that the organic aerosol precursors are uniquely proportional to the nonmethane hydrocarbon concentration. In some cases the NMHC concentration has been measured through C_6 with chromatographic analysis (6). In view of certain later discussion, a better normalization for f_C might be in terms of the mass of noncarbonate carbon plus the mass of

*The secondary organic carbon is defined as the total noncarbonate carbon less the carbon identified with motor vehicle sources. For Los Angeles the latter is accounted for approximately as a factor, α, proportional to the airborne lead concentration; α is taken as unity for the purposes of this study, after Friedlander (62). NMHC = nonmethane hydrocarbon vapor concentration (in $\mu g/m^3$).

TABLE 2. Some Examples of Conversion Ratios from Atmospheric Observations (Numbers in parentheses are heights above the ground)

Investigator	Location	Ratio (%) f_S	f_N	f_C
Georgii (63)	Colorado	19 (5 km)	-	-
	Germany	17 (0.8 km)	-	-
		77 (2.8 km)	-	-
Rhode (67)	Sweden	6.7-13.4 (0.4-2.8 km)	-	-
Cuong et al. (66)	Antarctic	90	-	-
	South Pacific	86	-	-
	Mediterranean	72	-	-
Grosjean and Friedlander (20)	Pasadena	29	5.8	3.3[a]
ACHEX (6)	East/central Los Angeles basin	21	6.6	0.44
	Eastern Los Angeles basin	27	31	0.60

[a]Defined without correction for primary C contribution.

noncarbonate carbon plus the mass of total NMHC minus the NMHC $\leq C_6$, f_{C6}. On the basis of this definition, f_{C6} is shown for comparison in Figure 5. The relative diurnal change in f_{C6} is the same for f_C, but the former ratio is somewhat smaller because of the conversion factor used: parts per million HC to micrograms of carbon per cubic meter.*

Comparison between f_S and f_N for West Covina and Rubidoux (Riverside) much farther eastward (downwind) of the Los Angeles area shows that f_S is roughly the same in both locations. Yet f_N was found to be significantly higher at the downwind sites.

---

*For f_C, NMHC is determined using NMHC as propane, or the gas density is 2×10^{-3} g/cm^3. For hydrocarbons and vapors $\leq C_6$, a vapor density of 4×10^{-3} g/cm^3 has been assigned for purposes of the calculation.

Thus qualitatively the sulfate production process appears to generate a relatively uniform conversion over Los Angeles even though the principal sources of SO_2 are in the west and southwest portions of the basin. In contrast, there appear to be two distinct modes of nitrate production, one a rapid, localized process peaking with morning NO_x or NO_2, and one that is slower, causing daily accumulation of NO_3^- far downwind of major NO_x sources in Los Angeles.

Comparison of the variation in conversion ratios with key changes in trace gases provides additional insight into the mechanisms of aerosol formation in the atmosphere. For example, analysis of the ACHEX data taken in 1973 shows that f_S is not significantly dependent on relative humidity. But f_S for particles less than 0.5 μm in diameter is negatively correlated with relative humidity. Furthermore, a strong apparent influence of photochemistry is demonstrated for submicron sulfate formation by the system increase in f_S ($\leq$0.5 μm) with ozone concentration. The consistency of this trend is shown in Figure 6.

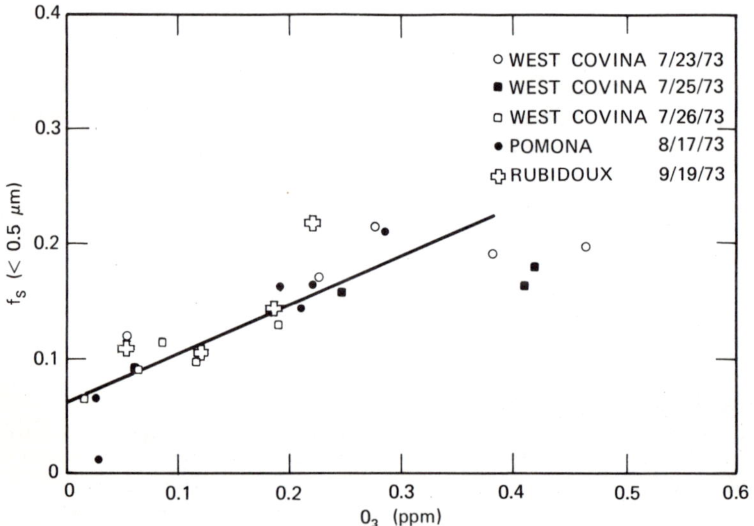

FIGURE 6. Scatter diagram of the conversion ratio, f_S, based on particles less than 0.5 μm versus 2-hr averaged ozone concentration for data in summer 1973 (6).

Further clues regarding the sulfate formation mechanism come from comparison of sulfate conversion and the behavior of particulate noncarbonate carbon. The conversion ratio f_S ($\leq$0.5 μm) is correlated with changes in particulate carbon, as indicated in Figure 7. Thus the formation of aerosols involving sulfate and organic carbon is closely coupled in Los Angeles smog.

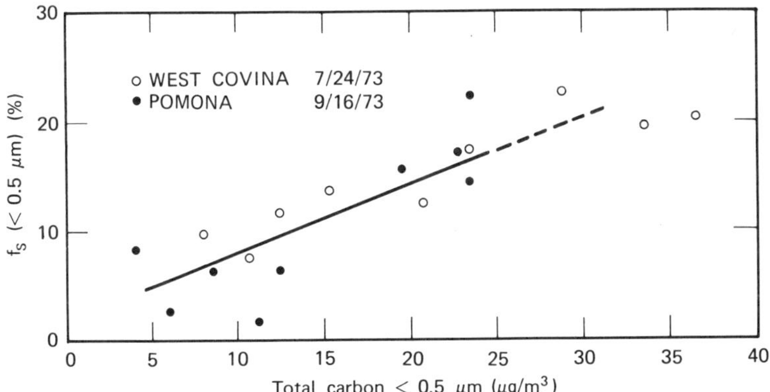

FIGURE 7. Scatter diagram of the conversion ratio, f_S, based on particles less than 0.5 μm versus total carbon less than 0.5 μm in diameter (based on 2-hr averaged filter data taken in summer 1973 (6).

The behavior of particulate nitrate is difficult to explain phenomenologically. It is puzzling, for example, that f_N for either the total particle mass concentration or that confined to 0.5 μm is poorly correlated with humidity, as well as NO_2, NO_x, and ozone concentration. The apparent independence of f_N or NO_2 is illustrated by the scatter diagram shown in Figure 8. Such results suggest that high nitrate content of the aerosol cannot be the result of a spurious absorption of gaseous NO_2 at the moist filter medium, for if this were the case, one would expect a proportionality between f_N and NO_2 concentration. By the same argument NO_3^- does not appear to form by a gas-diffusion-limited absorption of NO_2 in wet particles. On the other hand, the data suggest that the conversion process to nitrate may be dominated by other nitrogen oxide intermediates such as NO_3^-, N_2O_5, or HNO_3 which build up in the morning with the photochemical activity. However, these intermediates are more likely to be in high

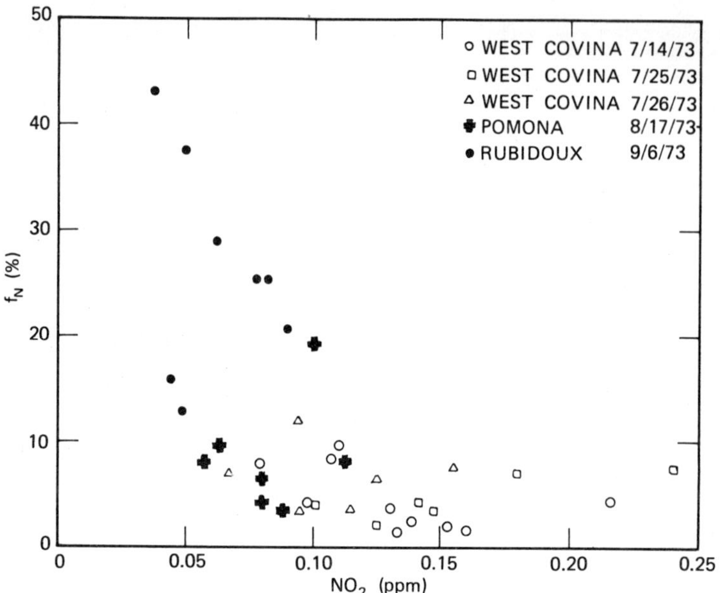

FIGURE 8. Scatter diagram of the nitrogen oxide conversion ratio, f_N, versus 2-hr averaged NO_2 concentration (6).

concentration at midday than in midmorning in smog. The rapid transient increase in nitrate in the midmorning does not seem to be related closely to the systematic increase in nitrate eastward across the Los Angeles basin.

5. Estimated Rates of Conversion. The information presently available on the rates of conversion of aerosol precursor gases in the atmosphere is very limited. There are reported in the literature that tropospheric residence times of SO_2 and NO_x are on the order of a few days, while reactive hydrocarbon vapors of high molecular weight evidently will not survive more than a day.

Some estimates of the quasi-first order rates of SO_2 oxidation that have been reported range from 0.1 to approximately 10%/hr, depending on many factors, including relative humidity [e.g., Calvert (21)]. Recent work, for example, of Husar et al. (22) near St. Louis suggests an oxidation rate of less than 1% in the absence of photochemical reactions. However, in the presence of photochemical activity SO_2 oxidation rates varying from 2 to 13%/hr have been estimated from Los Angeles data from the ACHEX (6) indicate similar rates of oxidation of SO_2 in the Los Angeles basin.

Virtually no quantitative information is available on the conversion rates of NO_x to nitrate. However, the Los Angeles experience previously discussed suggests that two extremes may be present. A rapid transient conversion may exist as part of the morning photochemical nitrogen oxide cycle, giving at high humidity a formation of some nitrate at a local rate exceeding that of SO_2 oxidation. In contrast, there appears to be a second nitrate-forming process that may be similar in rate to the sulfate conversion, or slower, depending on the location of the principal sources of NO_x. This conclusion comes from the marked differences downwind in nitrate and sulfate distribution in the Los Angeles basin.

The rate of organic aerosol production in the troposphere is essentially unknown. On the basis of the Los Angeles data, however, the conversion rate should be about the same as the sulfate formation rate because of the close diurnal correspondence between the two. The conversion rate of certain hydrocarbons may be even faster than that of sulfate since the hydrocarbon vapor sources are more evenly distributed over the city than are the sulfur oxide sources. Yet the conversion ratio for organics as presently defined is substantially less than that for sulfate or nitrate.

6. Summary. The existing knowledge of the phenomenology of aerosol formation in the troposphere relies heavily on the photochemical smog observations in Los Angeles. A summarizing list of key observations is given in Table 3.

B. The Stratosphere

The principal evidence for aerosol formation in the stratosphere comes from the discovery in the early 1960s and the subsequent confirmation of the dominance of sulfate in aerosol samples taken at high altitude. Observations have suggested the presence of a maximum concentration of particles greater than $\sim 0.1\ \mu m$ in diameter at altitudes of about 20 km. The "model" composition for the stratosphere aerosol is shown in Table 4. These data, based on the work of several investigators, including Junge et al (24) and Lazrus et al. (25), show that sulfate is by far the major component of aerosol at altitudes of ~ 20 km.

In contrast to the tropospheric material, both nitrate and organics, as presently reported, are essentially absent from the stratospheric particles. In the latter case this is nor surprising because it is expected that condensable organic vapors would have to be high in molecular weight and probably would not

TABLE 3. Some Properties of Tropospheric Aerosols Relevant to Their Formation Chemistry

Phenomenon	Urban	Nonurban
1. Production of new particles	Limited	Limited
2. Formation of nuclei in filtered, irradiated air	Copious nuclei generated	Uncertain
3. Haze formation and optical scattering	Strong diurnal changes accompany smog evolution (Los Angeles); correlation of scattering with maximum daily ozone	Widespread regional increase of haze over continents below 2 to 5 km altitude
4. Particle growth in 0.1 to 1.0 μm-diameter range	Predominant	Predominant
5. Role of liquid water in particle growth	Strong evidence supporting role in growth- urban aerosols generally hygroscopic	Suspected; not proved universally
6. Key chemical components		
Carbon-oxygenated organics	Major constituent, sometimes produced by reactions; confined to submicron particles	Highly variable constituent, but normally present
Sulfate	Universally present; large fraction in submicron range; submicron particle production correlated with organics; not correlated with humidity	Universally present

TABLE 3 (cont.)

Phenomenon	Urban	Nonurban
6. (cont.)		
Nitrate	Universally present; highest concentration in photochemical smog or in high humidity with NO_2 (not correlated with NO_2 or $SO_4^=$ concentration); sometimes in larger particles than sulfate; spatial distribution in Los Angeles differs from that of sulfate	Normally present
Ammonium	Normally present; often insufficient to "neutralize" acid anions	Normally present; usually insufficient to "neutralize" acid anions
7. Conversion ratios		
f_S: sulfate sulfur to total sulfur oxides	∿0.1 (within hours of SO_2 source); observed 2-hr average maximum ∿0.6 in aged air without new sources of SO_2; rate of formation 1 to 15%/hr	Sometimes approaches unity in upper troposphere
f_N: inorganic nitrate to total oxidized nitrogen	∿0.1 (hour from sources); observed 2-hr averaged maximum ∿0.4 in smog	Uncertain
f_C: organic carbon to carbon in nonmethane hydrocarbons	0.005-0.020 (based on nonmethane hydrocarbons) in Los Angeles	Unknown

TABLE 4. Model Chemical Composition of Stratospheric Particles at 20-km Altitude during the Period 1969-1973.

Substance	Concentration in air ($\mu g/m^3$) (ambient)	Observed concentration range ($\mu g/m^3$) (ambient)
Sulfate	0.6[a]	0.01-4 (Agung peak)[a]
Basalt[b] (calculated from Si)	0.05	0-0.7
NH_4^+	0.005	0-0.01
NO_3^-[c]	0	0
NO_2^-	0	0
Na	0.01	0.001-0.05
Cl	0.04	0.002-0.09
Br	0.002	0-0.003
Total	0.71	0.01-1

Source: Hidy et al. (28).

[a]For 8-year averaged data at 18 km from Castleman et al. (65).

[b]This includes all other components of the basalt, for example, Al, Ca, Mg.

[c]Particulate NO_3^- as contrasted with HNO_3 vapor.

survive substantial degradation above the tropopause. The absence of nitrate in the aerosol evidently is related to the high vapor pressure of HNO_3 relative to that of H_2SO_4 (see below); NH_4^+ or cations other than H^+ in large quantities are not expected to be present at high altitude. Apparently, the most acceptable steady-state theory for the presence and profile of sulfate comes from (photochemically related) oxidation of SO_2 _in situ_ at high altitudes [e.g., Friend et al. (3); Junge (26); Harrison and Larson (27)].

The details of the physical mechanism of sulfur oxide transition into particles in the stratosphere are unknown. The evidence from size distribution observations, combined with nuclei and optical data, suggests, however, that stable sulfate accumulation takes place mainly in existing particles rather than through formation of new nuclei. There is a general depletion in nuclei concentration with altitude, but the particle number in the $\sim$0.1 μm diameter range varies much less with height [e.g., Hidy et al. (28)]. On the basis of the few existing measurements, the number mean diameter for stratospheric aerosols is approximately 0.5 μm, and the total mass concentration is estimated to be 0.1 to 1.0 $\mu g/m^3$ at 20 km.

The rate of oxidation of SO_2 above the tropopause is unknown and very difficult to estimate because no data on SO_2 concentrations above 10 km are available. Rough estimates of the reaction time for SO_2, based on known SO_2 oxidation chemistry, now range from months [Harker (29)] to many years [Friend et al. (3)]. The former value is accepted by most chemists as the more likely one.

III. MECHANISMS FOR AEROSOL FORMATION

The chemical and physical processes that lead to aerosol formation in the atmosphere are probably very complex, with no single mechanism dominating the phenomenon. Unlike the chemical laboratory, the atmosphere is very "dirty" in that it contains substantial amounts of airborne particles which are continuously interacting with trace gases. Water as vapor and in the condensed phase probably plays a critical role in aerosol evolution, both in growth and in removal from the atmosphere. Thus, in elucidating atmospheric aerosol behavior, one must account for both homogeneous processes of precursor formation and heterogeneous interaction.

A. Physical Constraints

Regardless of the chemistry, certain physical constraints exist on aerosol-gas interactions. These restrictions relate mainly to (a) the thermodynamic stability of condensed particles of small diameter, (b) particle nucleation and condensation processes, and (c) gas-phase diffusion-limited rates combined with absorption or adsorption on particles.

Particles must be close to or at equilibrium with respect to the surrounding vapor to exist in air for any length of time. Thus the partial pressure of condensed species on particles essentially must be less than or equal to the saturation vapor pressure at atmospheric temperature for stability. This presents no great problem for most inorganic salts or for sulfuric acid, even at parts per billion concentration in the gas phase. However, it places a severe constraint on the ability of HNO_3 to exist as a pure compound or as an acid diluted in water.* The vapor pressure data reviewed by Toon and Pollack (30), for example, indicate that HNO_3 has a partial pressure over aqueous solutions more than 100 times higher than that of concentrated H_2SO_4 making nitric acid far more volatile in the atmosphere than sulfuric acid.

The requirements of low vapor pressure are particularly important to the stability of organic aerosols. The bulk of the organic vapors in the atmosphere that have been identified are in the range of carbon number less than four. Even if such materials reacted to form oxygenated materials, review of vapor pressure data [Hidy (10)] suggests that only material with a carbon number much greater than C_6 would be thermodynamically stable in the condensed phase at the concentrations of $\sim$1 ppb.

Accumulation of condensed material as aerosols in the atmosphere may take place by two basic processes: first, by condensation of supersaturated vapor or by chemical reaction leading to spontaneous formation of new particles; and second, by condensation, absorption, or reaction on existing particles. In the latter case the chemical reactions actually may take place on the surface of existing particles or within them.

For condensable precursors, the rate of particle formation may occur by homogeneous nucleation or by heterogeneous nucleation. It is generally accepted, but not proved, that heterogeneous processes are most likely in the atmosphere because of the large number of existing nuclei. One can readily see, however, that growth by condensation is limited by the rate of diffusion of vapor to the surfaces of nuclei. If conditions exist in which aerosol precursors evolve chemically at a rate exceeding the diffusional transfer, supersaturation could build up to high enough levels to permit heteromolecular (homogeneous) nucleation of new particles. It can be shown that H_2SO_4 can undergo heteromolecular nucleation at atmospheric concentrations in the absence of nuclei, but it is unlikely that HNO_3 can nucleate because of its relatively high vapor pressure. So little is known about the products of aerosol-forming organic reactions and their relevant physical properties that nothing can be said about the importance of homogeneous nucleation in this case.

---

*Vapor pressure data for HNO_2 or HNO_3 over solutions of ammonium sulfate or sulfuric acid are not readily available, so that the volatility of these acids over multicomponent aqueous media is unknown.

The fact that few new particles are observed in cities away from combustion sources or in rural areas, particularly in a highly reactive atmosphere like that of Los Angeles, makes it unlikely that heteromolecular nucleation is a widely important formation process in the troposphere. The general decrease in nuclei concentration with altitude, with more nearly constant large-particle concentration, suggests that new particle formation normally plays a small role at high altitude, too. Yet, this conclusion is by no means well established in the case of sulfuric acid behavior at altitudes above 10 km.

Growth of particles by accumulation on existing particles can be classified into two broad processes. If the precursor is supersaturated, growth will occur at a vapor-diffusion limited rate, which depends on the supersaturation, the temperature, the particle size, and the accommodation coefficient at the surface. The proportionality of particle size changes with the ratio of particle diameter and mean free path of the suspending gas. At one extreme the growth depends on volume to the 2/3 power; at the other, growth is proportional to volume to the 1/3 power. When the precursor is unsaturated, growth still may take place by irreversible absorption, or by chemical reactions in the particle. In this case the rate law should be proportional to the particle volume if the reaction is uniform throughout the particle. If the formation of material is limited by reactions in the particle, the conversion ratio should not be dependent on the gaseous precursor concentration.

There are insufficient data available to determine the rate law or physical mechanism most likely to predominate in atmospheric aerosol growth. However, there are clues to differences in the processes from the Los Angeles data. The shape of the particle volume-number distribution of tropospheric aerosol is such that the 1/3 (diameter) and 2/3 (surface) moments are concentrated in the submicron fraction, while the first moment (volume) is weighted toward larger particles. Thus the observed accumulation of sulfate and organic carbon on the small particles in smog suggests a surface or vapor-diffusion-controlled process. In contrast, the collection of NO_3^- in larger particles and the independence of NO_2 may indicate a volume-controlled reaction. The shift in NO_3^- to smaller particles between the east central and the eastern parts of the Los Angeles basin is of interest in this case and remains unexplained.

It is of interest also that thermodynamic equilibrium must influence the growth process of particles. If the radius of the particles is too small, the partial pressure condensable species can increase significantly by the influence of radius of curvature. From Kelvin's relation, examination of values of surface tension for a range of materials suggests that this effect will constrain growth to particles greater than 0.05 to 0.1 μm in diameter. This appears to be consistent with available observations of atmospheric growth and the distribution of secondary chemical components.

B. Chemical Reactions for Precursors

Physical processes of phase change, absorption, and stabilization restrict the maximum rate of formation and the classes of materials expected in aerosols generated in the atmosphere. Yet in many instances the rate-controlling process in the production mechanism may be the precursor formation. In the case of any of the three secondary constituents, a variety of possible chemical mechanisms may occur in the atmosphere. It is likely, in fact, that different mechanisms take place in parallel, or complement one another, so that no one process is dominant. One example of this is the difference in behavior of sulfate in Los Angeles air, as indicated by the changes in sulfate-particle size distributions.

1. Sulfate Reactions. The oxidation reactions of SO_2 have been reviewed by several investigators, including Bufalini (31) and Calvert (21). There are more than a dozen sulfate-forming reactions that may be relevant to atmospheric processes. These can be grouped in terms of homogeneous gas-phase reactions and heterogeneous reactions involving suspended particles. A summary of the two groups is given in Tables 5 and 6. The homogeneous reactions are broken down into subcategories whose end products are SO_3 or RO_2SO_2, and $ROSO_2$. Although the rates of reaction of these species with water or other species have not been reported, it is assumed that the reactions with water are fast to form H_2SO_4 and are not the rate-determining step in $SO_4^=$ production.

Listed in Table 5 are three classes of homogeneous reactions. The first consists of inorganic oxidation mechanisms to form SO_3, while the second involves organic radical oxidation agents generating SO_3 or RO_2SO_2. The third group of reactions forms $ROSO_2$ species. All of the reactions listed are exothermic and are favored thermodynamically. However, the first five reactions have been considered severely rate-limited on the basis of available rate data [e.g., Calvert (21)]. The remaining reactions listed appear to have rates that are sufficiently rapid to be of importance in the atmosphere, at least for polluted air with active photochemical processes. For comparison, estimates of the fractional conversion rate of SO_2 to $SO_4^=$ are given in the table. These results are based on Calvert's computer simulations. These, in turn, were based on an estimation of reactive intermediates after a 30-min period of sunlight irradiation at a zenith

TABLE 5. Estimated Rates of Theoretically Possible Homogeneous Removal Paths for SO_2 in a Simulated Polluted Atmosphere

Reaction	ΔH_{298} (kcal/mole)	Approximate rate (%/hr)
A. Inorganic reactions forming SO_3		
(1) $SO_2 + (1/2)O_2$ + sunlight $\rightarrow SO_3$	−24	<0.021
(2) $O(^3P) + SO_2 + \underline{M} \rightarrow SO_3 + \underline{M}$	−83	0.014
(3) $O_3 + SO_2 \rightarrow SO_3 + O_2$	−58	0.00
(4) $NO_2 + SO_2 \rightarrow SO_3 + NO$	−10	0.00
(5) $NO_3 + SO_2 \rightarrow SO_3 + NO_2$	−33	0.00
(6) $N_2O_5 + SO_2 \rightarrow SO_3 + N_2O_4$	−24	0.00
B. Organic reactions forming SO_3		
(7) $CH_2\text{—}CH_2$ (bridged by O_3) $+ SO_2 \rightarrow SO_3 + 2CH_2O$	−81	<0.4-3.0
(8) $\cdot CH_2OO\cdot + SO_2 \rightarrow SO_3 + CH_2O$	∿−117	<0.4-3.0
$CH_2{=}O \rightarrow O + SO_2 \rightarrow SO_3 + CH_2O$	∿−85	
(9) $HO_2 + SO_2 \rightarrow HO + SO_3$ (a)	−19	0.85
$\rightarrow HO_2SO_2^{\cdot}$ (b)	<−25	?
(10) $CH_3O_2 + SO_2 \rightarrow CH_3O + SO_3$ (a)	−30	∿0.16
$\rightarrow CH_3O_2SO_2^{\cdot}$ (b)	<−25	?
C. Reactions forming $HOSO_2$ or $ROSO_2$ radical		
(11) $HO + SO_2 \rightarrow HOSO_2^{\cdot}$	∿−82	∿0.23-1.4
(12) $CH_3O + SO_2 \rightarrow CH_3OSO_2^{\cdot}$	∿−73	∿0.48

Total potential rate of conversion of SO_2 to SO_3 (or sulfates) in moderate smog ≌ 1.7 to 5.5%/hr

Source: Calvert (21).

TABLE 6. Heterogeneous Reactions to Form Sulfate

A. Aqueous

(13) $SO_2 + H_2O(\ell) \rightleftarrows H_2SO_3$

$H_2SO_3 \rightleftarrows H^+ + HSO_3^-$

$HSO_3^- \rightleftarrows H^+ + SO_3^=$

(13A) $2SO_3^= + O_2(aq) \rightarrow 2SO_2^=$

(13B) $SO_3^= + O_3(aq) \rightarrow SO_4^= + O_2$

B. Nonaqueous

(14) $\left.\begin{array}{l} SO_2\ (ads) \\ H_2O\ (ads) \\ O_2\ (ads) \end{array}\right\} + \text{carbon(s)} \rightarrow H_2SO_4\ (ads)$

angle of 40°, with initial concentrations (in ppm) of NO_2 = 0.025, NO = 0.075, C_4H_8 = 0.10, CH_2O = 0.10, CH_3CHO = 0.06, CO = 10, CH_4 = 1.5, RH = 50% at 25°C, and SO_2 = 0.05 to 0.1.

Reactions 7 and 8 in Table 5 correspond to the interpretation of Cox and Penkett's (32) observations that SO_2 is oxidized at appreciable rates in the dark in ozone-olefin-air mixtures. The higher rate of 3%/hr was found for cis-2-pentene, and the lower rate of 0.4%/hr for propylene. Cox and Penkett suggested that either of two intermediates was involved in the SO_2 oxidation reaction: the ozonide reaction 7 or the zwitterion intermediates from the O_3-olefin reaction. The zwitterion has a diradical character and may be illustrated as reaction 8. Calvert's calculations of olefin-O_3 intermediates such as those in reactions 7 and 8 do not favor their importance as oxidizing agents. However, other radical species from the ozonide or zwitterion intermediates may be of interest, including those summarized by reactions 9 to 12. These classes of reactions may well account for SO_2 oxidation in the ozone-olefin mixtures.

Recently, the potential importance of radical addition reactions in the third class listed in Table 5 for SO_2 oxidation has become more fully appreciated. Such reactions are exemplified in the series 9 to 12.

The rate of the HO_2 reaction has been determined by Davis et al. (33). With such data the fractional rate of SO_2 disappearance may reach ∿1% in a moderate photochemical smog. Assuming that the rate of the CH_3O_2-SO_2 reaction is the same as that of the HO_2-SO_2 reaction, Calvert has estimated that the former will contribute a fractional oxidation rate of 0.2%/hr in moderate smog.

The OH radical-SO_2 reaction, reaction 11, appears to be of particular importance in the lower stratosphere, where OH concentrations are estimated to be high (27, 29). Calvert (21) has estimated the typical reaction rates for OH or CH_3 radical SO_2 reactions on the basis of analogies to reaction rates of CH_3, C_2H_5, and SO_2 reactions on the basis of analogies to reaction rates of CH_3, C_2H_5, and CFH_2CH_2. These rates are listed as 0.23 and 0.5%/hr, respectively, at the lower range.

Measurements of the rate constant for the $OH + SO_2 + M \rightarrow HOSO_2 + M$ reaction are emerging from recent fundamental studies. Hamilton (34) has summarized the preliminary values of the rate constant for this reaction. At ∿300°K the rate constant ranges from Wayne's value of $\sim 7 \times 10^9$/mol sec for 1 atm Ar to Payne et al.'s (35) value of 1.1×10^8/mol sec for 18 torr N_2 and 20 torr H_2O. Calvert (21) has used 1.1×10^9/mol sec to estimate the upper limit of 1.4%/hr listed in Table 4.

The radical addition products, such as $HOSO_2$, should react rapidly with other species to generate sulfuric acid, peroxysulfuric acid, alkylsulfates, and mixed intermediates such as $HOSO_2ONO_2$. Any of these ultimately should lead to sulfate in the presence of water.

Summing all of the known homogeneous reactions for SO_2 oxidation, it is possible to rationalize a theoretical rate of sulfate production in the range 1.7 to 5.5%/hr for moderate photochemical smog conditions. However, such rates are clearly highly dependent on the presence of unstable intermediates at relatively high concentrations. This is by no means a universal condition in nonurban air or in cities with minimal photochemical activity, as measured, for example, by ozone levels.

For conditions where photochemically induced homogeneous reactions cannot be important, the heterogeneous processes must be considered. The known reactions, which are listed in Table 6, have been categorized as aqueous and nonaqueous. The class of reactions that have been used most frequently to explain high SO_2 rates in the presence of liquid water containing aerosols is the system involving SO_2 absorption in water followed by oxidation by dissolved O_2 to form sulfate. Catalysis of the oxidation by heavy metal salts such as Mn^{++} ion can realize rates of oxidation in excess of 1%/hr in clean water solutions [e.g., Johnstone and Coughanowr (36); Matteson et al. (37)]. The absorption of SO_2 can be promoted by the buffering effect of simultaneous absorption of ammonia. Scott and Hobbs (38) have shown that the aqueous SO_2 oxidation process can be enhanced significantly by ammonium ion. Indeed, the estimates and experiments of Miller and dePena (39) and of Corn and Cheng (40) suggest that rates of SO_2 oxidation approaching 10%/hr can be achieved in fog.

It is well known that ozone is quite soluble in water. Therefore one expects that ozone absorption with SO_2 would contribute to significant oxidation of SO_2. Experiments of Penkett (41) have shown that oxidation of SO_2 in air at 7 ppb absorbed in water droplets with ozone, present in surrounding air at 5 pphm, can be as large as 13%/hr. Thus foggy or cloudy air mixed with photochemical smog, such as is sometimes observed along the Pacific Coast, could well be an important medium for $SO_4^=$ formation. Furthermore, such an aqueous mechanism could be significant at middle altitudes over continents even at background ozone levels.

The reported rates of SO_2 oxidation in clean water droplets must be considered maximum values. It is questionable whether they can ever be achieved in the atmosphere since such aqueous reactions have been shown to be suppressed significantly by organic contaminants. The work of Fuller and Christ (43) and later of Schroeter (44) has indicated that the aqueous absorption of SO_2 and its subsequent oxidation are reduced by as much as an order of magnitude by dissolved organic acids or alcohols. Since this type of material is known to be present in the atmospheric aerosol sampled at the ground, one can expect that the aqueous oxidation will probably be most efficient in relatively clean conditions of clouds well away from the earth's surface.

The heterogeneous mechanisms of SO_2 oxidation in the absence of liquid water are poorly understood. However, the recent work of Novakov et al. (42) has shown that $SO_4^=$ can be produced rapidly on the surface of carbon particles suspended in water vapor and air. These workers have observed that significant amounts of $SO_4^=$ can be found on carbon particles generated by combustion of hydrocarbons in ppm-level SO_2-enriched air.

It is difficult to assess the significance of carbon or organic particles for SO_2 oxidation in the free atmosphere. There is little doubt that absorption of SO_2 on carbon particles freshly generated by combustion can provide a surface-catalyzed oxidation medium. Indeed, experiments such as those of Yamamoto et al. (45) have shown that SO_2 oxidation can be as high as 30%/hr on activated charcoal particles $\sim$5 mm in diameter. Their work also indicates that this rate is strongly reduced by sulfuric acid collection in the micropores of the charcoal. The work of Yamamoto et al. further emphasizes that such a heterogeneous oxidation mechanism depends on a variety of factors, ranging from grain size of the carbon to temperature and to concentrations of SO_2, H_2O vapor, and oxygen, as well as to the micropore structure of the particle surface. It would seem that oily, gummy, wet particles collected from the atmosphere would be poorly suited for nonaqueous reactions to form sulfate since their micropore structure would be minimal. Yet such a mechanism cannot be ruled out from consideration at this time.

2. Nitrate-Forming Reactions. Like the production of sulfates, nitrates in atmospheric aerosols can be formed by a wide variety of homogeneous, as well as heterogeneous reactions. The pathways of nitrate generation are less well understood than those for sulfate, but it is likely that the two are interrelated at least in some circumstances.

Since both nitrous acid and nitric acid are much more volatile than sulfuric acid, it does not appear possible that particulate nitrogen oxide species can exist in the atmosphere in pure form as acids. Thus the production of nitrate must involve formation of a condensable species such as NH_4NO_3 in the gas phase, or the absorption of a nitrogen oxide constituent for particles followed by stabilization through chemical reaction. Such heterogeneous processes, of course, can take place in an aqueous or a nonaqueous medium.

The precursors for particulate nitrate formation are summarized in Table 7. These are classified in terms of (a) important nitrogen oxides, (b) volatile acids HONO and $HONO_2$, and (c) gaseous nitrates.

Because nitric oxide is relatively insoluble and nonreactive with water, the important nitrate-forming atmospheric oxides of nitrogen are believed to be NO_2, NO_3, and N_2O_5. These species are formed mainly in the atmosphere by the well-known "smog" reactions (16 to 20) and are not emitted primarily from material or anthropogenic sources.

With the nitrogen oxides coexisting with water vapor and sulfuric acid, the volatile nitrous and nitric acids can be formed via reactions 21 to 24. The mixed intermediates involving sulfuric acid are of interest as they link the NO_x and SO_x chemistry. Once the volatile acids are formed, they may react with ammonia in the gas phase to form, for example, NH_4NO_3 (reaction 25). The nitrogen oxides also react with radicals such as RO_2 to form organic nitrates and nitrites including peroxyacetylnitrate (PAN). Reaction 26 has been hypothesized by Calvert (21) on the basis of an analogy to the NH_3-HCl reaction. For concentrations of NH_3 and $HONO_2$ of $\sim$1 pphm, Calvert's (21) work suggests that the ammonia reaction may represent a significant removal path of $HONO_2$ to form aerosols.

TABLE 7. Reactions Potentially Involved in Nitrate Formation

Species	dNO_2/dt, or rate constant (ppm/min)[a]
A. Nitrogen oxides	
(15) $O_3 + NO \rightarrow NO_2 + O_2$	2.7×10^{-2}
(16) $O + M + NO \rightarrow NO_2 + M$	--
(17) $RO_2 + NO \rightarrow NO_2 + RO$	2.5×10^{-3}
(18) $O_3 + NO_2 \rightarrow NO_3 + O_2$	-4.0×10^{-4}
(19) $NO_3 + NO_2 \rightleftarrows N_2O_5$	-1.0 to -23×10^{-4}
B. Volatile acids	
(20) $N_2O_5 + H_2O \rightarrow 2HONO_2$	2×10^{-5}
(21) $HO + NO_2 + M \rightarrow HONO_2 + M$	-10^{-5} ($M = 1$ atm N_2)
(22) $NO + NO_2 + H_2O \rightarrow 2HONO$	--
(23) $HOSO_2O + NO \rightarrow HOSO_2\ ONO$ $+ H_2O \rightarrow H_2SO_4 + HONO$	--
(24) $HOSO_2O + NO_2 \rightarrow HOSO_2\ ONO_2$ $+ H_2O \rightarrow H_2SO_4 + HONO_2$	--
C. Gaseous nitrates	
(25) $NH_3 + HONO_2 \rightarrow NH_4NO_3$	$\sim 10^{-6}$
(26) $RO_2 + (N_2O_5, (NO_2 \rightarrow R^1C(=O)ONO_2 + R^1C(=O)ONO + \cdots$	-10^{-3}

[a]Typical for smog reactant concentrations in the first hour of reaction (e.g., Calvert and McQuigg, 68; Calvert, 21).

It is possible that condensable organic nitrates are formed via reaction 26 in the gas phase. Certainly the observation of such materials in smog chamber experiments [O'Brien et al. (46)] would provide some evidence for such cases. However, it is known that volatile organic nitrates such as PAN readily hydrolyze in an aqueous medium to form nitrite ion (reaction 31). Thus the presence of such compounds resulting from gas-phase reactions could lead to particulate nitrate after stabilization with ammonium ion or another cation.

Of the gas-phase reactions involved to form nitric acid vapor, 20 and 21 are believed to be most important. Using currently accepted rate constants for these reactions, Calvert has estimated that

$$R_{20} \simeq 0.5\text{-}2 \times 10^{-5} \text{ ppm/min}$$

$$R_{21} \simeq 2\text{-}6 \times 10^{-5} \text{ ppm/min}$$

in moderate photochemical smog. Thus it would appear that reaction 21 is of principal importance for $HONO_2$ formation in smog. The rate of conversion of NO_2 to $HONO_2$ by this reaction should be in the range of 2 to 8%/hr for the conditions used in the calculations in Table 5 (21). With absorption in wet aerosols and neutralization with ammonium ion, this value is not unreasonable for the upper limit of the estimated nitrate formation rate in Los Angeles smog in the morning "peak" condition. Once the nitrogen oxides are present or the acids begin to form, the interaction of these species with moist aerosols can take place. Some potentially important reactions are given in Table 8. Here the complexities of the aqueous reactions of NO_x become apparent. All of the nitrogen oxides contained in the atmosphere will react with liquid water to form traces of nitrate and nitrite. These reactions are generally reversible, however, so that the anions must be stabilized by a base such as NH_4^+. In the presence of dissolved oxygen or ozone, nitrite ion can be oxidized to nitrate in analogy to the aqueous sulfate formation reactions.

None of the nitrate reactions has been studied with atmospheric application in mind. However, there is a variety of information in the chemical literature dealing with NO_x absorption in water. There is ample evidence, for example, from studies reported by Nash (47) and Borok (48) that NO_2 at trace levels in air is readily absorbed in aqueous solutions. The efficiency of absorption varies widely, however, depending on the acid-base content of the solution.

If significant quantities of concentrated sulfuric acid are formed in atmospheric aerosols, nitrate formation via absorption of NO_2 to form nitrosylsulfuric acid, $HNOSO_4$, may be of interest.

TABLE 8. Aqueous Reactions of Nitrogen Oxides

(27)	$N_2O_5 + H_2O(\ell) \rightarrow 2H^+ + NO_3^-$
(28)	$NO + NO_2 + H_2O(\ell) \rightarrow H^+ + NO_2^-$
(29)	$2NO_2 + H_2O(\ell) \rightleftarrows H^+ + NO_3^- + HONO$
	$HONO + OH^- \rightarrow H_2O + NO_2^-$
(29a)	$2NO_2^- + O_2(aq) \rightarrow 2NO_3^-$
(29b)	$NO_2^- + O_3(aq) \rightarrow NO_3^- + O_2$
(30)	$2NO_2 + H_2SO_4 \rightleftarrows HNOSO_4 + HNO_3$
	$HNOSO_4 + H_2O(\ell) \rightleftarrows HNO_2 + H_2SO_4$
	$3HNO_2 \rightleftarrows HNO_3 + 2NO + H_2O$
(31)	$RONO_2 + H_2O(\ell) \rightarrow H^+ + NO_2^- + R^1OH$

The rate of absorption of NO_2 in sulfuric acid is fast, but the efficiency appears to be relatively low, according to Baranov et al. (49). Again, it appears that absorption of ammonia or another base has to be involved to drive the equilibria to nitrate production. Thus, in all of these aqueous reactions, nitrate formation may be limited by the concentration of ammonia rather than by any of the nitrogen oxide species.

3. Organic Particle Formation. Of the three major contributors to secondary aerosol production, the least is known about mechanisms for the organics. These processes have not yet been identified with any certainty. It is possible that organic materials can be polymerized in sulfuric acid solutions, for example. However, it is known that such reactions generally take place in very strong acid above 90% concentration. At equilibrium with water vapor in the lower atmosphere, H_2SO_4 should not exceed 40% concentration in water.

Another possible class of important organic aerosol reaction involves the attack of ozone on olefins. The ozone-olefin reaction mechanism remains a subject of controversy. One approach that can serve to illustrate the broad features of the reaction as related to aerosol formation is Story's model as shown in

Figure 9. The ozone adds to the olefinic bond forming a zwitterion, which can result in an epoxide or a molozonide. These species can follow one of three added paths to generate condensable peroxides or polymeric material. Addition of water in the system can provide added paths for decomposition of the peroxides to form acids or other oxygenated species, including alcohols and aldehydes. All of these materials have been identified in atmospheric aerosols.

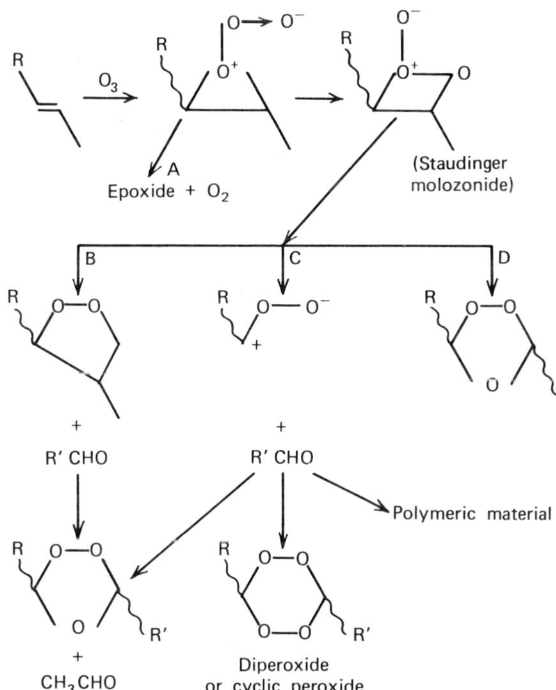

FIGURE 9. Story's model.

The intermediates identified by Cox and Penkett (32) appear in the early stages of the ozone-olefin reaction and offer a potential link between the sulfate chemistry and the organic aerosol chemistry.

The recent experiments of Franzblau et al. (50) have shown that the ozone-(terminal) olefin mechanism for aerosol production is initially first order in ozone and olefin. The constant relating the rate of mass concentration change increases with carbon number of the olefin. The significance of this effect is illustrated in the data shown in Figure 10. Here the experimental

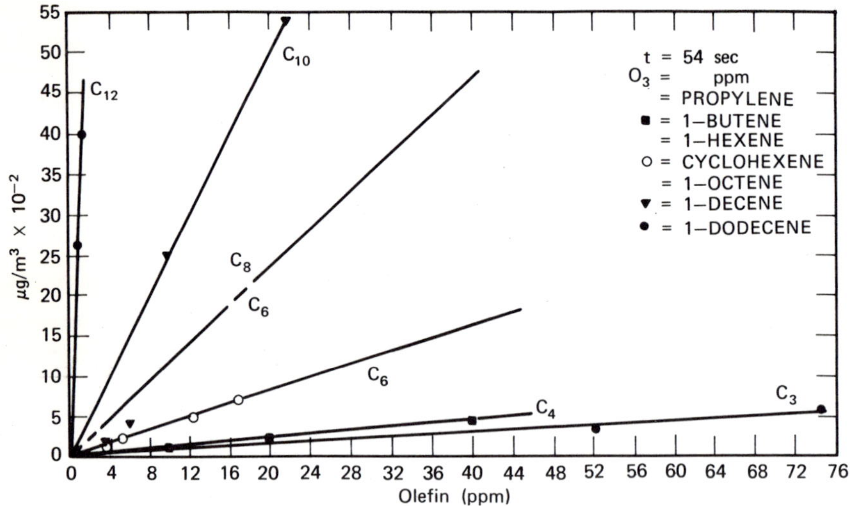

FIGURE 10. Mass concentration of aerosol formation from olefin-ozone reaction as a function olefin concentration.

variation in generation of condensable material from the reactions between ozone and several terminal olefins is shown for an ozone concentration of 12 ppm. Significant production of aerosol is observed over a reaction time (t) of 54 sec, mainly for the larger molecular weight species.

The empirical rates of production of aerosol from dry air ozone-olefin mixtures are shown in Table 9. Extrapolation of these data to olefins at the ppb level and ozone at 10 pphm

TABLE 9. Empirical Rate Relationships for Ozone-Olefin Reactions

Olefin	Rate ($\mu g/m^3 \cdot sec \cdot ppm^2$)
Propylene	1.0×10^{-2} ($\pm 0.3 \times 10^{-2}$)
1-Butene	1.7×10^{-2} ($\pm 0.5 \times 10^{-2}$)
1-Hexene	6.0×10^{-2} ($\pm 0.2 \times 10^{-2}$)
1-Octene	18.0×10^{-2} ($\pm 6.0 \times 10^{-2}$)
1-Decene	37×10^{-2} ($\pm 18 \times 10^{-2}$)
1-Dodecene	4.0 (± 1.0)
Cyclohexene	18.0×10^{-2} ($\pm 6.0 \times 10^{-2}$)

suggests a production rate for organic material of tenths of micrograms per cubic meter per hour, which is an order of magnitude lower than projected from atmospheric observations in Los Angeles. Thus either the estimate of total precursor concentration is too low, or the ozone-olefin mechanism cannot explain the atmospheric processes. It is also interesting that the experiments of Franzblau et al. (50) have demonstrated that aerosol formation in such mixtures, using 10 ppm O_3 and 10 ppm hexene in dry air, is heavily influenced by the addition of water vapor. Furthermore, traces of butyraldehyde strongly inhibit the formation of aerosols in this system.

This work suggests that the ozone-olefin reactions or oxygen atom-olefin reactions may be important. These classes of reactions involving high-molecular-weight olefins are known to produce polymerized oxygenated species that will condense out at ppb vapor concentrations. For carbon numbers greater than six, yields of polymerized material should exceed several percent of the vapor concentration. Such reactions are likely to be promoted on particle surfaces when the surfaces of large particles can act as preconcentrators for the olefin or air intermediate species.

C. Application of Laboratory Simulation

The review of potentially significant reaction mechanisms to form atmospheric aerosols could be an endless challenge for the chemist. Basic laboratory experiments to determine rate constants and important reaction steps play a key role in eliminating many candidates. However, the complexities of the chemical interactions in atmospheric processes cannot be elucidated by fundamental experiments alone. Predictions of the behavior of reactive mixtures should be consistent with atmospheric processes or with atmospheric simulation experiments. Simulation of atmospheric phenomena using controlled laboratory prototypes is a well-known technique, particularly as applied to fluid dynamic processes. However, simulation of atmospheric chemical phenomena is less well established. With the possible exception of the work of Friend et al. (3), only simulations of the lower atmosphere, more specifically urban air, have been attempted. The principal method that has been used by many investigators is the static reactor or smog chamber approach, where air mixtures containing reactive contaminants are studied over a period of several hours. Such studies are known to duplicate, qualitatively, at least, many of the features of urban photochemistry to form ozone and other smog products. The principal deficiencies as simulators include the low volume-to-surface ratio compared with the atmosphere, and the uncertainty regarding the role of wall reactions.

A limited number of smog chamber experiments have been undertaken to investigate aerosol formation in smog. Aside from the early studies reviewed by Leighton (1), these include the work of Groblicki and Nebel (51), Wilson and colleagues (52-54), O'Brien et al. (46), and the CALSPAN group (55). The results of these studies generally confirm that significant aerosol formation in simulated smog requires an induction time until ozone begins to build up, reflecting the time required for the buildup of reactive intermediates such as HO, HO_2, RO_2, and $HONO_2$ from chain reactions involving olefinic hydrocarbons. Aerosols found in chamber experiments are composed of oxygenated organic material, including carboxylic acids and organonitrates. If SO_2 is present, sulfate also is found. Much of the nitrate generated appears to be accumulated on the walls of the chambers.

The presence of SO_2 in irradiated NO_x-hydrocarbon-air mixtures enhances aerosol formation, but does not significantly influence the ultimate ozone level realized in the chamber. With a suitable choice of rate constants for reactions of HO and SO_2, and HSO_3 to form $SO_4^{=}$, Calvert (21) has indicated that computer modeling can duplicate many features of the Battelle chamber experiments [e.g., Wilson and Levy (52)].

The smog chamber studies have been useful in classifying the reactivity of hydrocarbons for photochemical aerosol production. With SO_2 the experiments generally indicate that the terpenoid compounds such as α-pinene are highly prolific aerosol formers, followed in rough order by diolefins, cyclic olefins, high-molecular-weight terminal olefins, and aromatics. Without SO_2 the work of O'Brien et al. (46) suggests that only diolefins and cyclic olefins of carbon number six or greater will generate significant quantities of aerosol at atmospheric pollutant concentrations for NO_x. Unfortunately such compounds have not been identified in air or in sources like motor vehicle exhaust. Their work also suggests that NO_2 tends to suppress aerosol formation. In another study using a smog chamber, Ripperton et al. (56) have obtained evidence of the importance of ozone attack on an olefinic bond in aerosol formation, particularly in "natural" production from terpenoid compounds.

To minimize the interactions between gases and the containing walls, another approach to simulation of aerosol-forming reactions has been attempted recently. Lipeles et al. (57) have reported a steady flow reactor technique in which the initial reactions of aerosol formation in irradiated gas mixtures are studied at reactant concentrations approaching those in the atmosphere. The initial experiments employing the flow method have demonstrated that nuclei can be generated in irradiated gas mixtures containing traces of NO_2 and olefins ($<C_6$) over 10-sec residence times. Their studies also have indicated a somewhat different reactivity scale than has the smog chamber work. Although α-pinene was the most prolific particle generator, high-molecular-weight terminal olefins and the conjugated hexadiene were superior to cyclohexene in aerosol production. The experiments of Lipeles et al. (58) further show that independent addition of SO_2 at ppb concentrations and of NH_3 at ppm concentrations to the reaction mixtures enhances aerosol production. Their work showed no consistent relation with NO_2 concentration; however, water vapor clearly suppressed aerosol formation in the flow reactor system. This observation is consistent with the smog chamber studies of Groblicki and Nebel (51), but raises questions about Wilson's conclusions (60) regarding the importance of stirring as the inhibiting influence of humidity.

To our knowledge only one of the simulation studies so far has attempted to duplicate the effect of existing particles in the reacting gas mixture. Only by the addition of particles under controlled conditions can the relative significance of heterogeneous reactions and homogeneous reactions be deduced. In the preliminary work of Heisler and colleagues (59) at Cal Tech, a study has been made of the nucleation and growth of aerosols in irradiated air mixtures doped with cyclohexene. These experiments have revealed that both nucleation and growth can take place in a medium containing $\sim 10^4$ particles/cm^3. In this case, the growth of particles appears to be consistent with a vapor-diffusion-controlled process influenced by the radius of curvature effect on vapor pressure.

IV. INTERPRETATION OF PRESENT KNOWLEDGE

The results of a large number of laboratory experiments and their interpretation provide strong evidence of the potential importance of several mechanisms for the production of atmospheric aerosols. The processes are clearly complicated, with many possible interacting avenues for formation of sulfate, nitrate, and organic material. However, the results of smog chamber simulations leave little doubt that photochemistry must play a role in aerosol evolution. At the same time heterogeneous processes involving liquid water droplets or moist aerosols certainly are key factors, at least in the troposphere. The evidence for nonaqueous heterogeneous reactions is less compelling at this time, but they cannot be ruled out as factors in the particle-gas interactions expected to take place in the atmosphere.

Until recently, the details of the changes in atmospheric aerosols over short time periods were unknown, so that meaningful intercomparisons between the atmospheric prototype and our mechanistic models could not be undertaken. With the results of new field experiments such as the ACHEX in hand, the significance of different mechanisms can now be tested, at least for urban air.

A. Sulfate Formation

In the case of photochemical smog in Los Angeles, comparison of the observations with the expectations of chemistry suggests qualitatively that several mechanisms probably play a role in sulfate and nitrate formation. The evidence is strong that submicron sulfate particles are produced photochemically as part of the smog mechanism. Such formation is coupled closely with the changes in organic aerosol during the day and is well correlated with the delay required to produce the high levels of oxidant.

The reactions that appear to be most significant for submicron sulfate are homogeneous in nature, involving either a dark reaction of ozone and olefins, or the groups of free radical reactions such as OH attack on SO_2. The accumulation of sulfate on the submicron-particle fraction during the day is consistent with a vapor-diffusion-limited process with a condensable or reactive precursor such as H_2SO_4 or HSO_3 radical. An alternate mechanism of catalytic SO_2 oxidation on carbon particles may be significant, but in our opinion it is a less likely mechanism than the photochemically related processes or reactions involving an aqueous medium.

The behavior of sulfate in the larger particles shows a poorer correlation with ozone and carbon in Los Angeles air, but the correlation with relative humidity change is also weak. It is likely that aqueous sulfate formation mechanisms play some role in total sulfate behavior in Los Angeles, particularly at night or in conditions of fog or haze with high liquid water content. Incidentally, the liquid water content of the Los Angeles aerosol varies with total mass concentration more strongly than with relative humidity (61). The inhibiting influence of organics in aqueous urban aerosols is uncertain, but may be a factor in the apparent variability of sulfate generation from day to day in Los Angeles.

Extrapolation of the experience in Los Angeles to other situations in the troposphere is difficult because of a lack of suitable observational data. However, it is likely that the photochemical formation of sulfate begins to be important whenever ozone exceeds 0.1 to 0.2 ppm in air and reactive hydrocarbons are present. At the ground the reactions in aqueous media are likely to be important in both urban and nonurban air if photochemically induced processes are absent. Away from the ground in the middle and upper troposphere, reactive hydrocarbon and free radical concentrations should be rather low under most circumstances. The presence of organics in aerosols at cloud level and above should be minimal, so that the aqueous oxidation processes globally should play a dominant role at altitudes between 2 and 10 km.

In the lower stratosphere the reactions of principal interest for sulfate formation appear to be photochemically induced. In particular, the homogeneous reaction of OH and SO_2 seems to be the leading candidate for $SO_4^=$ formation at this time.

B. Nitrate Formation

The knowledge of nitrate-forming processes applicable to the atmosphere provides a less convincing basis for explaining the behavior of this ion than that of sulfate. The key difference in mechanisms evidently is centered around the volatility of nitrous acid and nitric acid and their equilibria in aqueous solution. The Los Angeles experience emphasizes the distinct difference between nitrate and sulfate evolution. For nitrate there appear to be two extremes of behavior: first, a short-term transient production accompanying the NO_x peak in the morning, and second, a much slower, but systematic production such that high concentrations of NO_3^- appear after 2 to 3 hr of air mass transport across the Los Angeles basin. The overall NO_3^- production correlates poorly with ozone, NO_x, and NO_2, but the transient peak in the morning takes place at high relative humidity.

The evidence suggests that nitrate can be generated via homogeneous gas-phase reactions or by heterogeneous, aqueous processes. It seems that intermediate species other than NO_2 are involved, at least during the morning transient period. The independence of the nitrate conversion ratio of NO_2 or NO_x may suggest a droplet-rate-controlled NO_3^- formation after absorption. The complicated equilibria involved in nitrogen oxide ion solutions underscore the potential importance of basic substances such as NH_4^+ necessary to neutralize acidic particles. It is likely that NH_3 plays a critical role in stabilizing nitrate in aerosol particles. This conclusion can be drawn from the Los Angeles experience since there are relatively high ammonia concentrations in air over the eastern parts of the basin where the highest nitrate levels are observed.

Interpretation of nitrate formation in the troposphere is difficult at present because of the paucity of atmospheric aerosol data. However, it is likely that photochemical processes are important since the highest nitrates are generally observed in cities with high oxidant levels, particularly in Los Angeles. Again it would seem that NH_3 plays a crucial role in nitrate generation if acids are involved. Thus one would expect that less nitrate would be present at higher altitudes away from the principal sources of NH_3.

The absence of NO_3^- in stratospheric aerosols appears to be related to the high volatility of HNO_3 since this species has been observed as a gas at altitudes above 15 km.

C. Organic Carbon

Organic material is almost universally present in tropospheric aerosol samples but is probably absent from stratospheric material. The presence of organics in aerosols can be attributed to primary natural and anthropogenic sources, as well as to secondary processes in the atmosphere. The details of the secondary reactions are essentially unknown, but an important candidate, based on laboratory studies, is the ozone reaction with cyclic or linear olefins having carbon numbers larger than approximately six. The principal limitation in producing organic aerosols in the atmosphere appears to be the vapor pressure of condensable species.

The data taken in Los Angeles indicate that most of the noncarbonate carbon found in smog aerosol is secondary in nature. It is made up of a substantial amount of oxygenated material with some organic nitrate present. The organic reactions yield mainly submicron particles; production seems to correlate well with changes in ozone as well as with sulfate. In Los Angeles the organic fraction is normally the largest contributor by mass to filter-collected aerosol samples.

The concentration of organics in the submicron-particle fraction provides circumstantial evidence of a gas-phase formation reaction followed by vapor-diffusion-controlled growth on existing particles. The presence of significant quantities of oxygenated material and the correlation with ozone buildup are consistent with the expectation that the ozone-olefin reaction is important.

At this time there is virtually no information on the composition of olefins greater than C_6 in the ambient atmosphere, so that the organic aerosol precursors cannot be considered as known in urban air. However, taking the difference between the nonmethane hydrocarbon vapor concentration and the $<C_6$ hydrocarbon fraction identified by chromatography suggests that the $\geq C_6$ fraction may be about half of the NMHC concentration. Thus it is reasonable to expect that several high-molecular-weight olefins are present at ppb levels in urban air.

Near the ground, vegetation emanates significant quantities of hydrocarbon vapors. Thus organic aerosols can readily be formed in nonurban air from reactions between "background" ozone and such compounds as the terpenes from evergreens.

ACKNOWLEDGMENTS

This work was sponsored in part by the California Air Resources Board under Contract No. 358, by the Department of Transportation under an interagency transfer through Environmental Protection Agency Contract No. 68-02-1081, and by the Environmental Protection Agency under Grant No. R802160.

REFERENCES

1. Leighton, P. 1961. Photochemistry of Air Pollution. Academic Press, New York.

2. Went, F. 1960. Proc. Natl. Acad. Sci. U.S. 46, 212.

3. Friend, J. P., Leifer, R., and Trichan, M. 1973. J. Atmos. Sci. 30, 465.

4. Whitby, K. T. (ed.). 1971. "Aerosol Measurements in Los Angeles Smog," Air Pollution Control Office Publ. No. APTD-

0630, U. S. Environmental Protection Agency, Research Tri-Angle Park, NC.

5. Hidy, G. M. (ed.). 1972. Aerosols and Atmospheric Chemistry. Academic Press, New York.

6. Hidy, G. M. et al. 1974. "Characterization of Aerosols in California," SC524.25FR, Science Center, Rockwell International, Thousand Oaks, CA.

7. Husar, R. B. and Whitby, K. T. 1973. Environ. Sci. Technol. 7, 241.

8. Charlson, R. J., Ahlquist, N. C., Selvidge, H., and MacCready, P. B., Jr. 1969. J. Air Pollut. Control Assoc. 19, 937.

9. Heisler, S., Friedlander, S. K., and Husar, R. B. 1973. Atmos. Environ. 7, 633.

10. Hidy, G. M. 1973. "Theory of Formation and Properties of Photochemical Aerosols," presented at Fundamental Chemical Basis of Reactions in a Polluted Atmosphere, Battelle Summer Institute, Seattle, Washington, June 18.

11. Hidy, G. M. 1972. In Assessment of Airborne Particles (T. Mercer, ed.), Charles C. Thomas Co., p. 81.

12. Junge, C. E. 1963. Atmospheric Chemistry and Radioactivity. Academic Press, New York.

13. U.S. Environmental Protection Agency. 1972. "Air Quality Data from the National Air Surveillance Network, 1968," APTD-0978.

14. Charlson, R. J. et al. 1974. Science 184, 156.

15. Cunningham, P. and Johnson, S. A. 1976. Science 191, 77.

16. Robinson, G., and Robbins, R. 1970. J. Air Pollut. Control Assoc. 20, 233.

17. Hidy, G. M. and Brock, J. R. 1971. In Proceedings of the 2nd IUAPPA Clean Air Congress. Academic Press, New York, p. 1088.

18. O'Brien, R. et al. 1975. Environ. Sci. Technol. 9, 577.

19. Schuetzle, D. 1972. Ph.D. Thesis, University of Washington.

20. Grosjean, D. and Friedlander, S. K. J. Air Pollut. Control Assoc. 25, 1038.

21. Calvert, J. G. 1974. "Modes of Formation of the Salts of Sulfur and Nitrogen in an NO_x-SO_2-Hydrocarbon-Polluted Atmosphere," presented at Conference on Atmospheric Salts and Gases of Sulfur and Nitrogen in Association with Photochemical Oxidant, University of California, Irvine, Jan. 7-9; see also Proceedings of Conference on Health Effects of Pollutants, Assembly of Life Sciences, National Academy of Science, U.S. Government Printing Office No. 93-15, 1973, p. 19.

22. Husar, R. B. and Blumenthal, D. 1974. Unpublished report.

23. Roberts, P. and Friedlander, S. K. 1974. "Conversion of SO_2 to Particulate Sulfates in the Los Angeles Atmosphere," presented at Conference on the Health Consequences of Environmental Controls: Impact of Moble Emission Controls, U.S. Environmental Protection Agency, Durham, NC, Apr. 17-19.

24. Junge, C. E., Chagnon, C. W., and Manson, J. E. 1961. J. Meteorol. 18, 81.

25. Lazrus, A. L. et al. 1971. J. Geophys. Res. 76, 8083.

26. Junge, C. E. 1974. "Sulfur Budget of the Stratospheric Aerosol Layer," in Proceedings of the International Conference on the Structure, Composition, and General Circulation of the Upper and Lower Atmosphere, Vol. 1. IAMAP/IAPSO 1st Joint Assembly, Melbourne, Australia, Jan. 14-25, p. 85.

27. Harrison, H. and Larson, T. 1974. J. Geophys. Res. 79, 3095.

28. Hidy, G. M. et al. 1974. "Aerosols from Engine Effluents, Chap. 6 in Monograph III, Climatic Impact Assessment Program, Dept. of Transportation.

29. Harker, A. 1975. J. Geophysical Res. 24, 3399.

30. Toon, O. B. and Pollack, J. B. 1973. J. Geophys. Res. 78, 7051.

31. Bufalini, M. J. 1971. Environ. Sci. Technol. 5, 686.

32. Cox, R. A. and Penkett, S. A. 1972. J. Chem. Soc., Faraday Soc. 68, 1735.

33. Davis, D. D., Payne, W. A., and Stief, L. J. 1973. Science 179, 280.

34. Hamilton, E. J., Jr., private communication.

35. Payne, W. A., Stief, L. J., and Davis, D. D. 1973. J. Amer. Chem. Soc. 95, 7614.

36. Johnstone, H. F. and Coughanowr, D. R. 1968. Ind. Eng. Chem. 50, 1169.

37. Matteson, M., Stober, S., and Luther, H. 1969. Ind. Eng. Chem. Fundame. 8, 677.

38. Scott, W. D. and Hobbs, P. V. 1967. J. Atmos. Sci. 24, 54.

39. Miller, J. M. and dePena, R. G. 1971. Proceedings 2nd IUAPPA Congress. Academic Press, New York, p. 375.

40. Corn, M. and Cheng, R. T. 1972. J. Air Pollut. Control Assoc. 22, 871.

41. Penkett, S. A. 1972. Nature Phys. Sci. 240, 105.

42. Novakov, T., Chang, S. G., and Harker, A. B. 1974. Science 186, 259.

43. Fuller, E. C. and Christ, R. H. 1941. J. Amer. Chem. Soc. 63, 1644.

44. Schroeter, L. C. 1963. J. Pharm. Sci. 52, 559.

45. Yamamoto, K., Michiham, S., and Kunitaro, K. 1973. Nippon Kagaku Kaiski 7, 1268.

46. O'Brien, R. et al. 1975. Environ. Sci. Technol. 9, 568.

47. Nash, T. 1970. J. Chem. Soc. (A) 18, 3023.

48. Borok, M. T. 1960. Zh. Prikl. Khim. 33, 1761.

49. Baranov, A. V., Liberzu, E. A., and Popova, T. I. 1966. Tr. Sibiskogo Techologi. 38, 77.

50. Franzblau, E., Burton, C. S. and Hidy, G. M. "Aerosol Formation from Ozone-Olefin Reactions," in preparation.

51. Groblicki, P. and Nebel, G. J. 1971. "The Photochemical Formation of Aerosols in Urban Atmosphere," in Chemical Reactions in Urban Atmospheres, American-Elsevier, New York.

52. Wilson, W. E., Jr. and Levy, A. 1970. J. Air Pollut. Control Assoc. 20, 385.

53. Wilson, W. E., Jr. and Levy, A. 1972. Current Res. 6, 423.

54. Wilson, W. E., Jr., Levy, A., and Wimmer, D. B. 1972. J. Air Pollut. Control Assoc. 22, 27.

55. Kocmond, W. C. 1973. "Determination of the Formation Mechanism and Composition of Photochemical Aerosols," Final Report CAPA-8 Contract, CALSPAN Corp., Buffalo, NY, August.

56. Ripperton, L. A., Jeffries, H. E., and White, O. 1972. "Formation of Aerosols by Reaction of Ozone with Selected Hydrocarbons," in Photochemical Smog and Ozone Reactions, Advances in Chemistry, No. 113, p. 219.

57. Lipeles, M., Landis, D., and Hidy, G. M. 1977. "The Formation of Organic Aerosols in a Fast Flow Reactor", in Fate of Pollutants in the Air and Water, Part II. I. M. Suffet (ed.), J. Wiley & Sons, Inc., p. 69.

58. Lipeles, M. et al. 1973. "Mechanism of Formation and Composition of Photochemical Aerosols," Final Report, EPA-RS-73-036, July.

59. Heisler, S. 1975. Ph.D. Thesis, California Institute of Technology, Pasadena, CA.

60. Wilson, W. E., Jr. 1971. Discussion in Chemical Reactions in Urban Atmospheres (C. S. Tuesday, ed., American-Elsevier, New York, p. 264.

61. Ho, W. W., Hidy, G. M., and Govan, R. 1974. J. Appl. Meterol. 13, 871.

62. Friedlander, S. K. 1972. Environ. Sci. Technol. 7, 235.

63. Georgii, H. W. 1970. J. Geophys. Res. 75, 2365.

64. Hidy, G. M. et al. 1974. J. Appl. Meteorol. 13, 96.

65. Castleman, A. W., Jr. et al. 1974. Tellus 26, 250.

66. Cuong, N. B. et al. 1974. Tellus 26, 241.

67. Rhode, H. 1972. J. Geophys. Res. 77, 4494.

68. Calvert, J. and McQuigg, R. 1975. "The Computer Simulation of the Rates and Mechanisms of Photochemical Smog Formation" in Proc. Symposium on Chemical Kinetics Data for the Upper and Lower Atmosphere, Intl. J. Chem. Kinetics, Symposium No. 1, p. 113.

69. Spicer, C. W. and Schumacher, P. M. 1977. Atmos. Environ. 11, 873.

70. Husar, R. B. et al. 1978. Sulfur in the Atmosphere. Proc. of the International Symposium. Pergamon Press, N.Y.

71. Penkett, S. A. et al. 1979. Atmos. Environ. 13, 123.

ADDENDUM

Since the writing of this paper in 1974, considerable new work has been reported on secondary aerosol behavior in the atmosphere. One notable result has come from the work of Spicer et al. (69). They have shown that there is considerable nitrogen oxide gas (NO_x) absorption during particle collection on certain filter media, including the one used in the ACHEX. Thus, our earlier conclusions about the small significance of a nitrate artifact resulting from NO_x absorption is incorrect. However, more recent sampling in the Los Angeles area during 1977 still has yielded high concentrations of (particulate) nitrate approaching those observed in the ACHEX. Highest concentrations are observed east of the city, far downwind of the high density NO_x sources. Diurnal changes in Los Angeles air suggest that maxima in particulate nitrate are likely to be found at different times than sulfate maxima, with nitrate often highest at night.

Extensive research has been undertaken in atmospheric sulfate. Husar et al. (70) reported a comprehensive review of the state of accumulated knowledge and current research results. Notable among these are the findings that the aqueous reactions of SO_2 and H_2O_2 may be important in sulfate production (71). The aqueous oxidation process including O_2 and O_3 is highly pH dependent; the reactions are suppressed in an acidic medium. In contrast, the H_2O_2 reaction is much less acidity dependent, especially in the range below pH $\stackrel{\sim}{\sim}$ 4.5, where cloud and wet aerosol particle processes are likely to take place.

Formation of Organic Aerosols from Cyclic Olefins and Diolefins

DANIEL GROSJEAN[a] AND SHELDON K. FRIEDLANDER[b]
W. M. Keck Laboratories of Environmental Engineering
California Institute of Technology
Pasadena, California

---

[a]Author to whom correspondence should be addressed. Present address: Environmental Research and Technology, 2625 Townsgate Road, Westlake Village, California 91361.

[b]Present address: Department of Chemical Engineering, University of California, Los Angeles, California 90024.

Abstract

The chemical composition and rates of formation of organic aerosols formed upon sunlight irradiation, in a large outdoor smog chamber, of ppm levels of olefinic hydrocarbons and nitrogen oxides in air have been investigated.

Cyclohexene, cyclopentene, and 1,7-octadiene produced large amounts of organic aerosols (5 to 39% of the initial olefin concentration on a carbon basis) with aerosol production rates ranging from 1.5 to 40.6 μg C $m^{-3} \cdot min^{-1}$. The light-scattering efficiency of these aerosols was found to be about half that reported for southern California ambient aerosols.

Aerosol organic carbon (AOC) concentrations were measured as a function of irradiation time, and the rates of AOC production were found in some experiments to be proportional to the product of the olefin and ozone concentrations. The slopes of these relations, 7 to 680 μg C $m^{-3} \cdot min^{-1} \cdot ppm^{-2}$, are consistent with AOC production rates of $\sim$0.05 μg C $m^{-3} \cdot min^{-1}$ under typical ambient photochemical smog conditions.

Products identified by mass spectrometry included dicarboxylic acids and other α, ω-difunctional oxygenates bearing carboxylic, aldehyde, hydroxyl, or nitrate ester groups. Mechanisms involving initial cyclic olefin attack by ozone and by the hydroxyl radical to yield α, ω oxacids and α, ω-dials, respectively, followed by oxidation and other reactions in both gas and aerosol phases, are proposed to account for the observed products. The role of cyclic olefins and diolefins as hydrocarbon precursors of secondary organic aerosols in urban atmospheres is discussed.

I. INTRODUCTION

The formation of photochemical organic aerosols from hydrocarbon precursors, an area of research pioneered by Haagen-Smit more than two decades ago (1, 2), involves complex physical and chemical processes. Because particulate organic matter constitutes a significant fraction of urban aerosols (3) and contains carcinogenic (4, 5) as well as mutagenic (6) species, it is important to eludicate the chemical composition, to establish the rates of production, and to investigate the modes of formation of organic aerosols.

These objectives are addressed in the present study which involves chemical analysis and measurements of the rates of formation of "model" organic aerosols formed upon irradiation of a single olefinic hydrocarbon in the presence of oxides of nitrogen (NO_x). These experiments were performed in a large outdoor environmental chamber and involved sunlight irradiation of ppm levels of the olefin and NO_x in ambient air, that is, under conditions that closely resemble those prevailing in urban photochemical smog. Data on the physical mechanisms governing aerosol growth in these experiments have been reported elsewhere (7). Some of the chemical results reported here have been presented (8, 9) and have been included in a recent review on organic aerosols prepared by one of us for the National Academy of Sciences (10). These smog chamber studies also complement the field studied conducted as part of the California Aerosol Characterization Experiment (ACHEX) program, and are discussed throughout this chapter with reference to the ACHEX and other studies of ambient organic aerosols.

II. EXPERIMENTAL PROCEDURES

A. Outdoor Chamber Hydrocarbon-NO_x Irradiation Experiments

In a typical experiment, mixtures of NO, NO_2 (∿0.15 to 0.30 ppm), and hydrocarbon (∿0.5 to 2.0 ppm) were added to ambient Pasadena air and allowed to react in sunlight for up to 3 hr in a large outdoor chamber constructed from 2-mil FEP Teflon film and located on the roof of the Keck Laboratories Building, California Institute of Technology. The chamber surface was ∿117 m^2, and the initial chamber volume ranged from 53 to 134 m^3, with corresponding initial surface-to-volume ratios of 0.87 to 2.2 m^{-1}.

Hydrocarbon consumption, ozone formation, condensation nuclei concentration, aerosol light-scattering coefficient (b_{scat}) and solar radiation intensity were measured either continuously or at regular intervals throughout the runs. The outdoor chamber, monitoring instruments, instrument calibrations, and experimental procedures have been described in detail elsewhere (7,11).

No mechanical stirring device was employed during the runs. Internal mixing resulted from thermal convection and from wind action on the flexible chamber, which was supported by a wood frame raised 2 ft above the building's roof to allow circulation of air. Wilson et al. (12) have reported that mechanical stirring inhibits dramatically aerosol formation in smog chamber experiments similar to those reported here, that is, involving irradiation of HC-NO_x mixtures.

Experimental conditions and initial concentrations are listed in Table 1. Four olefinic hydrocarbons were studied: cyclohexene, cyclopentene, 1,7-octadiene (Aldrich Chemicals), and 2,3-dimethyl-

TABLE 1. Initial Experimental Conditions

Hydrocarbon	Run No.[a]	Initial concentration (ppm) HC	NO	NO_2	T (°C)	RH (%)	Nuclei (10^3 cm^{-3})	b_{scat} (10^{-4} m^{-1})
Cyclohexene	A07	1.0	0.33	0.17	40	∿10	28	0.4
	A91[b]	1.0	0.33	0.17		∿10	38	0.5
	A08	0.5	0.17	0.08			24	0.7
	A09	1.0	0.33	0.17	35	<15	8	2.2
	A92[b]	1.0	0.33	0.17	35	<15	15	3.1
	A10	1.0	0.33	0.17	35	<15	28.8	5.2
	A11	1.14	0.38	0.19	26	37	11	3.1
	A12	2.0	0.34	0.17	24	40	12	3.3
Cyclopentene	G01	2.10	0.30	0.30	31	24	26	1.5
	G03	0.52	0.30	0.15	33	∿10	15	0.2
2,3-Dimethyl-2-butene	E01	1.0	0.33	0.17			37	3.0
1,7-Octadiene	F04	1.0	0.33	0.17	45	∿10	14	3.2
	F05	1.06	0.38	0.38	41	21	20	2.2
	F91[b]	1.18	0.38	0.38	46	23	23	3.5
	F06	0.58	0.33	0.17			20	3.2

[a]Runs are coded as in Heisler and Friedlander (7).

[b]SO_2 (∿70 ppb) was also added in this run.

2-butene (Chemical Samples). All four hydrocarbons were $\geq$98% pure as verified by flame ionization gas chromatography. They were stored in the dark in a cold room and employed without further purification. Their concentrations during the irradiation experiments were measured using a Varian Model 1400 gas chromatograph equipped with a flame ionization detector. Columns employed were 1/8 in. x 10 ft 5% Carbowax 20 M and silver nitrate-ethylene glycol on Chromosorb W 60/80 mesh.

B. Chemical Analysis of Aerosol Samples

Two types of aerosol samples were collected for the determination of the total aerosol organic carbon concentration as a function of irradiation time and for identification of the organic products at the end of the runs, respectively. All aerosol samples were collected on Gelman Type A glass fiber filters that were extracted and fired before use in order to reduce organic impurities (3, 13).

Total organic carbon measurements were performed by collecting during each run 10 to 15 samples on 47-mm diameter filters (sampling flow rate 80 liters/min, sampling time 3 to 10 min, depending upon the aerosol concentration). Each sample was analyzed by direct combustion of a 5-mm diameter piece of the loaded filter, using a Dohrmann Model DC-50 organic carbon analyzer (13). Three to five pieces of filter were analyzed for each sample.

Samples were also collected at the end of each irradiation experiment on 8 x 10 in. filters (flow rate 60 cfm, sampling time 15 to 30 min) and subjected to mass spectrometry analysis. After removing a small portion of the filter for direct analysis by high-resolution mass spectrometry, the samples were subjected to organic solvent extraction (Soxhlet apparatus, ~6 hr with 1:1 v/v binary mixtures of Spectrograde quality methylene chloride and isopropyl alcohol or CH_2Cl_2-methanol), concentration by evaporation of the solvents at room temperature, and esterification (methyl esters) of the concentrated organic extracts with either diazomethane or BF_3 in methanol (40% w/v, Supelco Inc.) (14).

The esterified organic extracts of the aerosol samples were then analyzed by combined gas chromatography-mass spectrometry (GC-MS), using both electron impact (EI) and chemical ionization (CI), the latter with methane as the reagent gas. The EI-MS analyses were performed using an AEI Model 300 quadrupole instrument equipped with a SCC 4700 data system, a SCC 611 display unit, and a Hewlett-Packard Model 7620 A gas chromatograph. The CI-MS analyses were performed using a Hewlett-Packard Model HP 5930 A mass spectrometer equipped with a Model HP 5932 A data system and a Hewlett-Packard Model 5700 A gas chromatograph. Mass scans in the range m/e = 50 to m/e = 300 were generated every 4 sec. Monitoring of selected m/e values was also performed. Both gas chromatographs were dual-column, flame ionization instruments operated from 60 to 180°C at a rate of 6°C/min. Columns employed included 10 ft x 1/8 in. 20% OV-17 on Chromosorb W 100/120 mesh and 10 ft x 1/8 in. Silar 10C.

As mentioned earlier, small portions of the filter samples were subjected to direct (solid introduction probe) analysis by high-resolution mass spectrometry (HRMS), using a Du Pont Model 21-492B instrument operated at 75 eV. Comparison of the HRMS spectra with the data obtained by solvent extraction-esterification-GC-MS analysis makes it possible to establish whether any major products are not recovered by the latter method. High-resolution mass spectrometry also provides further information for interpreting the low-resolution mass spectral data. In addition, HRMS analysis of samples at several introduction probe temperatures provides a means of determining whether compounds separated and tentatively identified by GC-MS evolve from the filter according to their expected volatilities.

C. Authentic Samples

Authentic samples were either purchased or prepared, and their EI and CI mass spectra compared to those of the products found in the aerosol samples. Adipic acid, glutaric acid, and their dimethyl esters and glutaraldehyde were commercial samples (Aldrich). The nitrate ester 6-nitratohexanoic acid, $COOH-(CH_2)_4-CH_2ONO_2$, was prepared by reaction of silver nitrate with 6-bromohexanoic acid (Aldrich) in acetonitrile, according to the general method described by Ferris et al. (15) for organic nitrate esters. In the absence of literature data, its EI and CI (methane) mass spectra were established. Another sample of 6-nitratohexanoic acid was kindly provided by Dr. George Tsou of the California Air Resources Board Haagen-Smit Laboratories, El Monte, California.

III. SMOG CHAMBER PROFILES

Maximum ozone, condensation nuclei, and aerosol light-scattering coefficient, b_{scat}, are listed in Table 2 as well as the aerosol loadings at the end of the experiments. Typical changes in b_{scat} and in the hydrocarbon, ozone, and nuclei concentrations as a function of irradiation time are illustrated in Figures 1 and 2 and discussed in the following sections.

TABLE 2. Summary of Results

Hydrocarbon	Run no.	Maxima O_3 (pphm)[a]	Maxima Nuclei (10^3 cm^{-3})[b]	Maxima b_{scat} (10^{-4} m^{-1})[b]	Aerosol concentration (μg m^{-3})
Cyclohexene	A07	48.3	62	14.6	870
	A91[c]	53.0	120	10.5	480
	A08	33.7	58	4.4	220
	A09	66.2	19.1	10.7	550
	A92[c]	58.7	165	8.1	580
	A10	51.2	4.4	15.2	530
	A11	30.0	16	22.2	1300
	A12	18.9	8	45.7	3000
Cyclopentene	G01	2.1	210	71.1	4900
	G03	40.0	235	7.3	1400
2,3-Dimethyl-2-butene	E01	47.1	1260	<0.2	0
1,7-Octadiene	F04	89.1	72	21.8	
	F05	107	78	34.3	1350
	F91[c]	106	196	34.0	1750
	F06	74.2	31	5.8	440

[a]California Air Resources Board ultraviolet photometry method (52, 53).

[b]After substracting initial values listed in Table 1.

[c]Note that the addition of ∿70 ppb of SO_2 resulted in a significant increase in nuclei but no increase in b_{scat}.

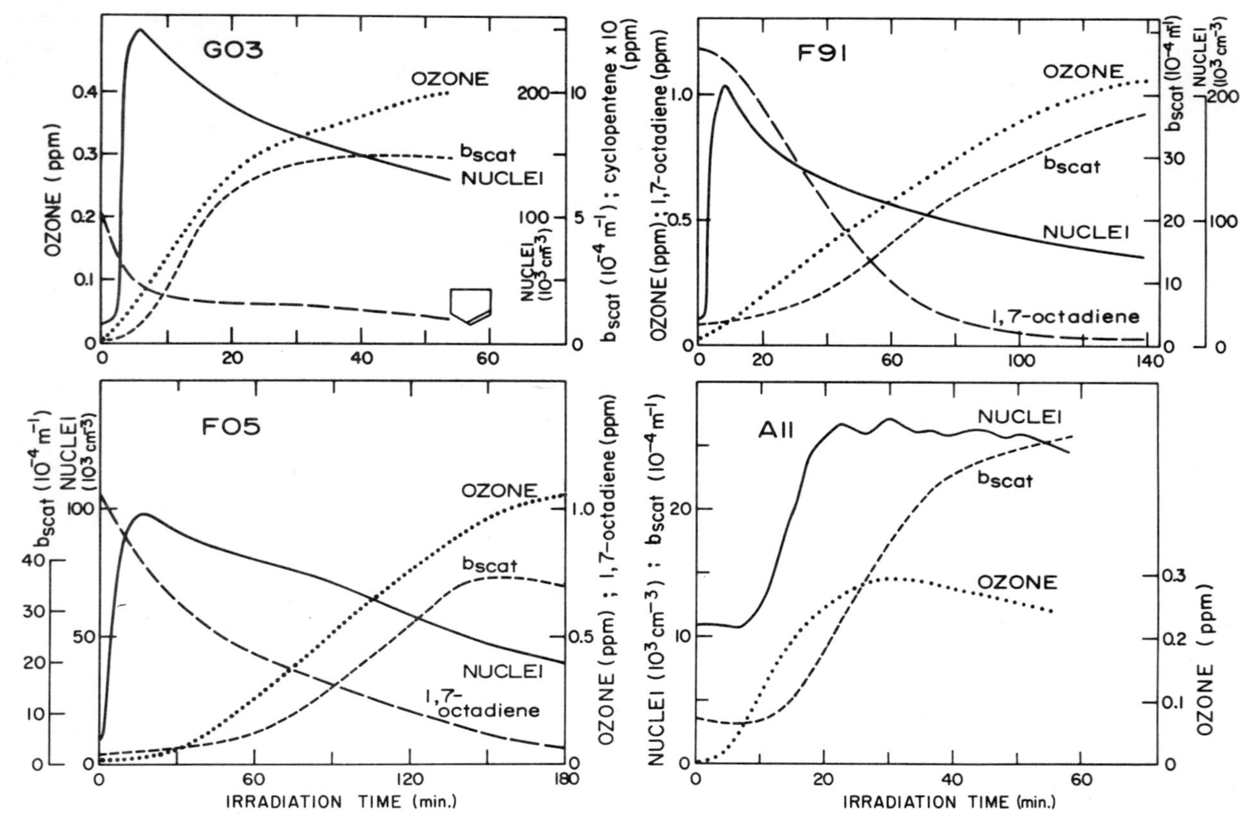

FIGURE 1. Ozone, hydrocarbon, condensation nuclei, and aerosol light-scattering coefficient, b_{scat}, as a function of irradiation time for outdoor smog chamber Runs G03, F91, F05, and A11. See Tables 1 and 2 for experimental conditions.

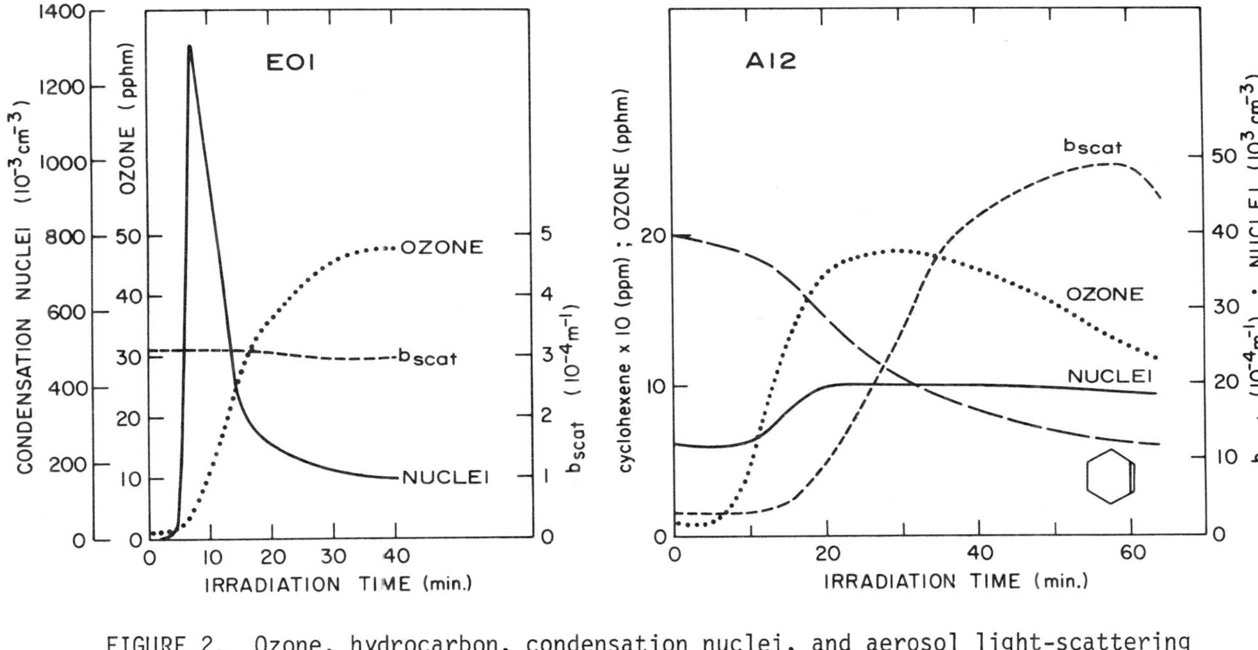

FIGURE 2. Ozone, hydrocarbon, condensation nuclei, and aerosol light-scattering coefficient, b_{scat}, as a function of irradiation time for outdoor smog chamber Runs E01 and A12. See Tables 1 and 2 for experimental conditions.

A. Ozone Production

Ozone was produced in all experiments after lag times ranging from less than 2 min for 2,3-dimethyl-2-butene to ∿20 min for 1,7-octadiene. The observed lag time sequence for ozone appearance (2,3-dimethyl-2-butene < cyclopentene < cyclohexene < 1,7-octadiene) parallels the NO photooxidation rates of these compounds (16), as is expected since ozone builds up only after the time required for the conversion of NO to NO_2. Maximum ozone concentrations obtained in the cyclohexene photooxidation experiments were in qualitative agreement with those predicted using Shen and Springer's (17) relation: $O_{3\ max} = 0.64\ (\text{cyclohexene})^{1/2}\ (NO_x)^{1/2}$. The three monoolefins produced similar amounts of ozone and somewhat less ozone than did 1,7-octadiene.

B. Nuclei Formation

Although this observation might be coincidental, it is of interest to note that, in all runs, new nuclei were formed after the onset of ozone production (Figures 1 and 2). The number of nuclei rapidly reaches a maximum (when the rates of nucleation and coagulation are equal) and then decreases as condensation predominates throughout the remainder of the run. The physical processes governing aerosol formation and growth in these experiments have been discussed in detail by Heisler and Friedlander (7). Of the four olefins studied, 2,3-dimethyl-2-butene was the most effective in generating new nuclei. Cyclohexene produced many fewer new nuclei than the other olefins studied. This behavior has been observed before for cyclohexene-NO_x mixtures irradiated in clean air (18-21), and is even more pronounced in our experiments conducted in ambient air, where existing particles provide ample surface for the diffusion of newly formed condensable species (7). Thus, increasing the initial ambient air light-scattering aerosol loadings resulted in a marked decrease in the production of new nuclei: maximum nuclei formed in the cyclohexene experiments were 62, 58, 19, 16, 8, and 4.4 $10^3\ cm^{-3}$ for initial b_{scat} values of 0.4, 0.7, 2.2, 3.1, 3.3, and 5.2 $10^4\ m^{-1}$, respectively.

C. Hydrocarbon Consumption

Hydrocarbon concentration as a function of time was measured during several of the cyclopentene, cyclohexene, and 1,7-octadiene runs. Ozone appearance in these systems always coincided with a marked increase in hydrocarbon consumption.

Hydrocarbon concentration as a function of time is given by

$$\frac{-d[HC]}{dt} = \sum_i k_i [X_i] [HC]$$

where k_i are the second-order rate constants of hydrocarbon attack by the species X_i, with X_i = O, OH, O_3, HO_2, etc. If it is assumed that only one species reacts with the olefin,

$$\frac{-d[HC]}{dt} = k_i [X_i] [HC]$$

or

$$\log [HC] = \log [HC]_0 + \frac{k_i}{2.303} \int_0^t [X_i] dt$$

where $[HC]_0$ is the initial hydrocarbon concentration.

Therefore, the relative importance of ozone in the overall hydrocarbon consumption can be tested further by plotting log [HC] as a function of $\int_0^t [O_3] dt$, as is shown in Figure 3 for cyclohexene (Run A12). With the exception of the first ∿10 min, during which an "excess rate" is evident, a good linearity is observed throughout the experiment, with a slope corresponding to a cyclohexene-ozone rate constant of 0.15 $ppm^{-1} min^{-1}$.

The reactivity of the four hydrocarbons (2,3-dimethyl-2-butene > cyclopentene > cyclohexene > 1,7-octadiene) is consistent with their known rates of reaction with ozone and with the hydroxyl radical OH (Table 3). It is difficult, however, to assess the relative importance of these two species in our experiments. The OH-cyclopentene reaction rate constant has not been measured, and the reported ozone-cyclohexene rate constants vary within a factor of 6. Thus, if one accepts the higher value reported by Japar et al. (22), cyclohexene disappareance in our system can be entirely accounted for by its reaction with ozone. On the other hand, the lower ozone-cyclohexene rate constants of Cadle and Schadt (23) and Vrbaski and Cvetanovic (24) are consistent with a significant participation of the OH-cyclohexene reaction in our experiments. In this case the linearity observed in Figure 3 suggests that the OH concentration should remain approximately constant after the initial ∿10 min of irradiation. Since some of the major products identified in our study are consistent with both OH- and O_3-cyclohexene reaction mechanisms (see below), additional kinetic data for these two reactions would be of great value.

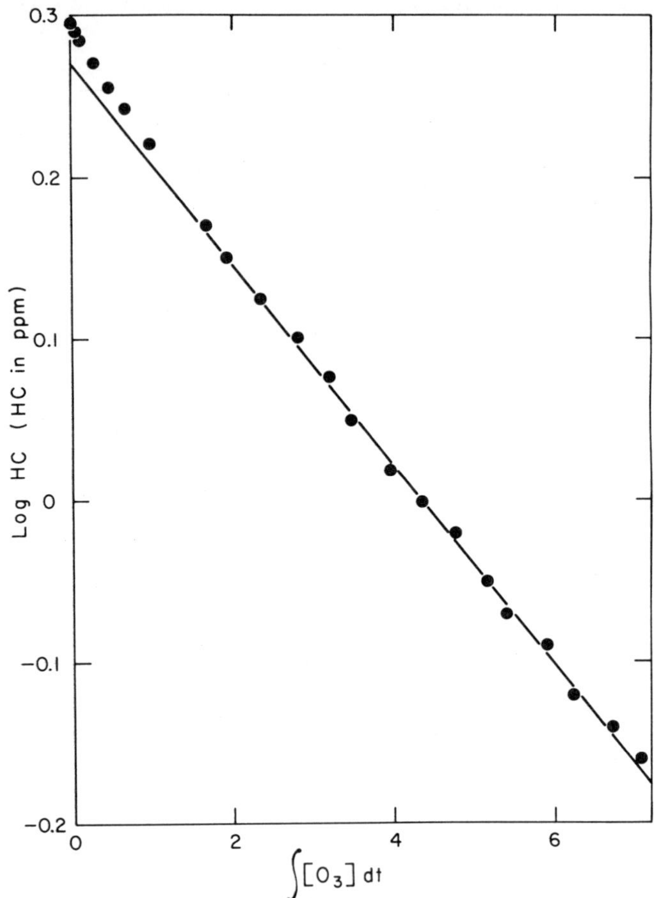

FIGURE 3. Cyclohexene concentration as a function of $\int [O_3]\ dt$, outdoor smog chamber Run A12.

D. Aerosol Light-Scattering Efficiency

As can be seen in Figures 1 and 2, a striking difference was observed between nuclei formation and production of light-scattering aerosols. The most reactive compound studied, 2,3-dimethyl-2-butene, also produced large numbers of nuclei but essentially no light-scattering particles. Cyclopentene, cyclohexene, and 1,7-octadiene generated large amounts of aerosols, in

TABLE 3. Room Temperature Reaction Rate Constants Relevant to Hydrocarbon Reactivity in our Smog Chamber Experiments

Hydrocarbon	O_3[a]	OH[b]	$O(^3P)$[c]	NO[d]
2,3-Dimethyl-2-butene	1510	110	5.11	59
Cyclopentene	813	No data (ND)	1.79	23
Cyclohexene	169 59[e] 30[f]	62.4	1.62	3.5
1,7-Octadiene	ND	ND	ND	ND
1-Hexene[g]	8.4	31.2	0.39	1.7
1,3-Butadiene[g]	11.1	68.5	1.45	4.3

[a] k_{O_3} in 10^{-18} $cm^3 \cdot molecule^{-1} \cdot sec^{-1}$. From reference 22 unless otherwise indicated.

[b] k_{OH} in 10^{-12} $cm^3 \cdot molecule^{-1} \cdot sec^{-1}$. From references 54-56.

[c] k_O in 10^{-11} $cm^3 \cdot molecule^{-1} \cdot sec^{-1}$. From relative rate data of Cvetanovic (57), placed on an absolute basis by using $k = 4.76 \pm 0.48$ 10^{-11} $cm^3 \cdot molecule^{-1} \cdot sec^{-1}$ for 2-methyl-2-butene (58).

[d] Nitric oxide photooxidation rate (ppb^{-1} min^{-1}). From reference 16.

[c] From reference 23.

[f] From reference 24.

[g] Included as indicative of the reactivity of 1,7-octadiene, for which no data are available.

agreement with previous studies (19, 25, 26). The final aerosol concentrations (Table 2) were approximately proportional to the initial hydrocarbon concentration in the range studied, 0.5 to 2 ppm. This has been observed before for aerosol formed from cyclohexene by reaction with ozone in the dark (27) or by photooxidation in the presence of NO_2 (21). The authors of these two studies (21, 27) also reported a marked leveling-off in aerosol production for initial cyclohexene concentrations greater than about 3 ppm.

The results in Table 2 also show that, for a given hydrocarbon, the final aerosol light-scattering efficiency did not change appreciably from run to run. Final aerosol concentrations and light-scattering coefficients correlated well for the eight cyclohexene runs:

$$b_{scat}\ (10^{-4}\ m^{-1}) = 2.67 + 1.45\ 10^{-2}\ \text{mass}\ (\mu g\ m^{-3})$$

with a correlation coefficient of 0.986. Including the results from the four 1,7-octadiene runs did not significantly modify the above relation:

$$b_{scat}\ (10^{-4}\ m^{-1}) = 2.27 + 1.58\ 10^{-2}\ \text{mass}\ (\mu g\ m^{-3}),\ R = 0.944$$

as expected since these two hydrocarbons yield aerosols of similar chemical composition (see below). Cyclopentene yielded particles of lesser light-scattering efficiency.

It is noteworthy that, although cyclohexene, cyclopentene, and 1,7-octadiene clearly produce large amounts of light-scattering particles, the aerosols formed in our smog chamber experiments are only half as efficient in scattering light, on a mass concentration basis, as ambient aerosols present in southern California air during photochemical smog episodes. Thus mass/b_{scat} ratios of $\sim$33 have been reported by the ACHEX and other investigators (3, 28–30) for ambient aerosols, compared to mass/b_{scat} ratios of $\sim$65 in our experiments with cyclohexene and 1,7-octadiene. This observation is consistent with the fact that the light-scattering efficiency of ambient aerosols is dominated by sulfate compounts (29, 30), while the particles formed in our chamber study consist mainly of organic compounds and inorganic nitrates.

IV. AEROSOL ORGANIC CARBON MEASUREMENTS

Aerosol samples collected during all runs except Run F04 (no aerosol was formed in Run E01) were analyzed for their organic carbon contents, which averaged only ∼35% by weight of the total aerosol. Inorganic nitrate was also produced in all runs and accounted for ∼10% by weight of the aerosol [from the nitrate ion concentration measured by the 2,4-xylenol colorimetric method and assuming all measured NO_3^- to be ammonium nitrate (3)]. Thus the low-carbon fraction indicates that the aerosol products formed are highly oxygenated species averaging ∼40 to 45% carbon by weight. This is indeed the case for the major organic products identified in these runs, for example, adipic acid and 6-nitratohexanoic acid (carbon - 41%) from cyclohexene.

Final aerosol organic carbon concentrations ranged from 5 to 39% of the initial hydrocarbon concentrations (Table 4). Thus a major fraction of the products resulting from photooxidation of the olefinic hydrocarbons studied remains in the gas phase. On the other hand, cyclopentene, cyclohexene, and 1,7-octadiene are more potent organic aerosol precursors than the hydrocarbon mix present in ambient southern California air, where carbon gas-particle distribution factors of only 1 to 3% are typically measured (3).

TABLE 4. Fraction of the Initial Hydrocarbon Converted to Aerosol Organic Carbon

Hydrocarbon	Run no.	A/G[a,b]	Hydrocarbon	Run no.	A/G[a,b]
Cyclohexene	A07	0.10	2,3-Dimethyl-		
	A91	0.06	2-butene	E01	0.0
	A08	0.05			
	A09	0.06	1,7-Octadiene	F04	
	A92	0.07		F05	0.11
	A10	0.06		F91	0.13
	A11	0.14		F06	0.07
	A12	0.17			
			Cyclopentene	G01	0.33
				G03	0.39

[a] A = aerosol organic carbon measured at the end of the run ($\mu g\ Cm^{-3}$); G = initial hydrocarbon concentration, converted from ppm to $\mu g\ Cm^{-3}$. (A/G is a dimensionless number).

[b] A/G values are presumably upper limits since some artifact aerosol is expected to form during sampling by adsorption of volatile acidic organics on the surface of the basic glass fiber filters.

Concentration-time profiles for aerosol organic carbon and hydrocarbon consumed are shown in Figure 4. It can be seen that after an induction period the aerosol fraction of the total products increases rapidly and then levels off.

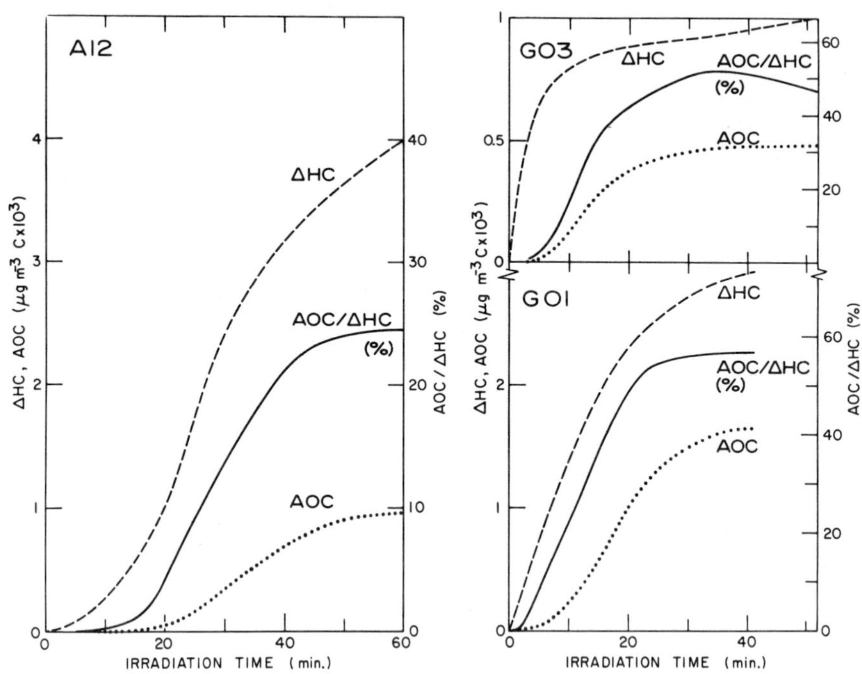

FIGURE 4. Hydrocarbon consumption (ΔHC), aerosol organic carbon (AOC) formation, and fraction of the hydrocarbon converted to aerosol (AOC/ΔHC,%) as a function of irradiation time, outdoor smog chamber Runs A12, G01, and G03.

Aerosol organic carbon formation rates, listed in Table 5, ranged from 1.5 to 41 μg C $m^{-3} \cdot min^{-1}$. In all runs these rates paralleled the rate of formation of ozone and the product of the hydrocarbon and ozone concentrations, that is, increased to the maximum values listed in Table 5 and then decreased upon further irradiation. Thus aerosol formation rates were found to correlate in an approximately linear fashion with the product of the olefin and ozone concentrations (Figure 5). This behavior is

TABLE 5. Aerosol Organic Carbon (AOC) Formation Rates

Hydrocarbon	Run no.	d(AOC)/dt (μg C $m^{-3} \cdot min^{-1}$) Average	Maximum
Cyclohexene	A12	16	39
Cyclopentene	G01	41	78
	G03	9.1	33
1,7-Octadiene	F05	2.5	3.2
	F91	4.3	8.0
	F06	1.5	2.5

consistent with the formation of condensable organic species in the gas phase followed by rapid diffusion to the condensed phase, that is, the rate of diffusion is fast relative to the rate-determining formation of the intermediate(s) in the gas phase. Note that an empirical relation of this general form (d (aerosol)/dt = $\alpha[O_3]$ [HC] as shown in Figure 5) does not necessarily imply that organic aerosol is produced only as a result of the ozone-olefin reaction. The relation may be composite, that is, may include terms representing the contributions of other species, including the hydroxyl radical. The kinetics of these systems has been discussed in detail elsewhere (9, 10).

The slopes of the above relations were 19, 7, 36, 153, and 680 μg C $m^{-3} \cdot min^{-1} \cdot ppm^{-2}$ for Runs F05, F06, F91, A12, and G03, respectively. Assuming that these linear relations hold for urban atmospheres (not necessarily a good assumption), one obtains, for typical ozone (0.1 ppm) and hydrocarbon (10 ppb) concentrations, organic aerosol production rates of $\sim$0.02, 0.15, and 0.6 μg C $m^{-3} \cdot min^{-1}$ for 1,7-octadiene, cyclohexene, and cyclopentene, respectively. For comparison purposes it should be noted that average rates of aerosol organic carbon production in southern California on smoggy days are on the order of 0.05 μg C $m^{-3} \cdot min^{-1}$ [estimated from the measured particulate organic carbon levels in the eastern part of the Los Angeles basin, 30 μg C m^{-3}, and assuming an air parcel transport time of 10 hr (31)].

V. AEROSOL CHEMICAL COMPOSITION

Products identified in the organic fraction of aerosols from cyclohexene (Table 6), cyclopentene (Table 7), and 1,7-octadiene

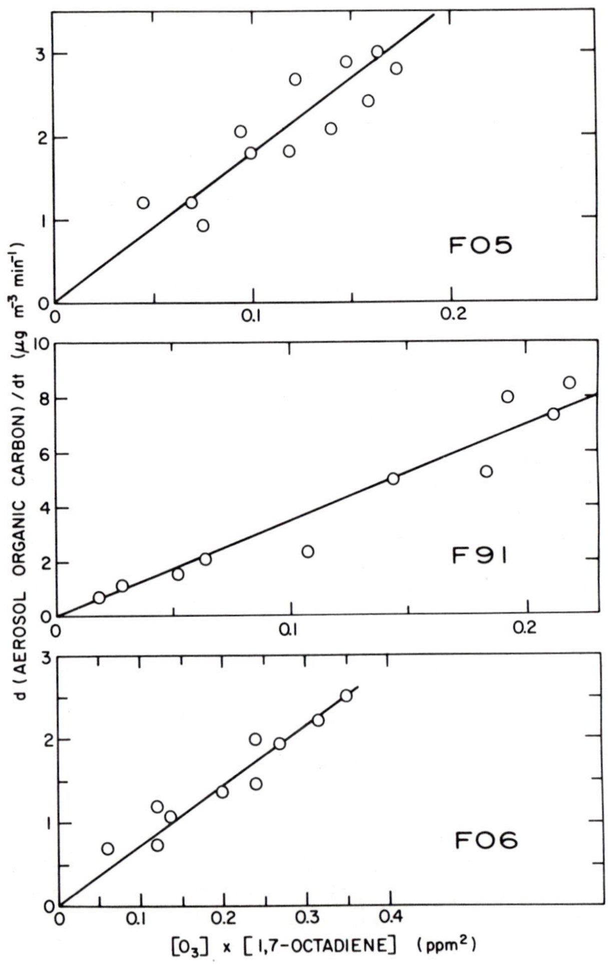

FIGURE 5. Aerosol organic carbon formation rate versus the product of ozone and hydrocarbon concentrations for Runs F05 (top), F91 (middle), and F06 (bottom).

TABLE 6. Cyclohexene Aerosol

Product[a]	Formula	Identification[b]
Adipic acid[c,d]	$COOH-(CH_2)_4COOH$	AC
6-Nitratohexanoic acid[c]	$COOH-(CH_2)_4CH_2ONO_2$	AS
6-Oxohexanoic acid	$COOH-(CH_2)_4CHO$	MS[e]
6-Hydroxyhexanoic acid	$COOH-(CH_2)_4CH_2OH$	MS
Glutaric acid	$COOH-(CH_2)_3COOH$	AC
5-Nitratopentanoic acid	$COOH-(CH_2)_3CH_2ONO_2$	MS[e,f]
5-Oxopentanoic acid	$COOH-(CH_2)_3-CHO$	MS[e]
5-Hydroxypentanoic acid	$COOH-(CH_2)_3-CH_2OH$	MS[e]
Glutaraldehyde	$CHO-(CH_2)_3CHO$	AC

[a] Acids and diacids were identified as methyl and dimethyl esters, respectively.

[b] By comparison with authentic samples, either commercial (AC) or synthesized (AS), or by interpretation of their mass spectra (MS).

[c] Major product.

[d] Also identified as a major product by direct (no extraction) high-resolution mass spectrometry-thermal analysis (HRMSTA) of a filter sample from Run A12. Glutaric acid was also identified. We thank Dr. R. L. Knights, University of Washington, Seattle, for analyzing the sample by HRMSTA.

[e] Also identified by Schwartz (35) in the acid, CH_2Cl_2-soluble, water-insoluble fraction of cyclohexene aerosol.

[f] The CI (methane) fragmentation pattern was identical to that of its homolog, 6-nitratohexanoic acid, synthesized in our laboratory. Schwartz (35) reported an identical CI (methane) spectrum for the same compound.

TABLE 7. Cyclopentene Aerosol

Product[a]	Formula	Identification[b]
Glutaraldehyde[c]	$CHO(CH_2)_3CHO$	AC
5-Oxopentanoic acid	$CHO(CH_2)_3COOH$	MS
Glutaric acid	$COOH(CH_2)_3COOH$	AC
5-Nitratopentanoic acid	$COOH(CH_2)_3CH_2ONO_2$	MS[d]
4-Hydroxybutanoic acid	$COOH(CH_2)_2CH_2OH$	MS
4-Oxobutanoic acid	$COOH(CH_2)_2CHO$	MS
1,4-Butanedial	$CHO(CH_2)_2CHO$	MS
4-Nitratobutan-1-al	$CHO(CH_2)_2CH_2ONO_2$	MS

[a,b,c,d]See footnotes a, b, c, and f in Table 6.

TABLE 8. 1,7-Octadiene Aerosol

Product[a]	Formula	Identification[b]
Adipic acid[c,d]	$COOH-(CH_2)_4COOH$	AC
6-Nitratohexanoic acid	$COOH-(CH_2)_4CH_2ONO_2$	AS
6-Oxohexanoic acid	$COOH-(CH_2)_4-CHO$	MS
7-Nitrato-1-heptene	$CH_2=CH-(CH_2)_4-CH_2ONO_2$	MS
6-Heptenoic acid	$COOH-(CH_2)_4-CH=CH_2$	MS
6-Hepten-1-al	$CH_2=CH-(CH_2)_4-CHO$	MS

[a,b,c]See footnotes a, b, and c in Table 6.

[d]Also identified by HRMSTA of a filter sample from Run F06. See footnote d in Table 6.

(Table 8) are all similar in structure. They are linear oxygenated compounds bearing carboxylic acid, carbonyl, hydroxyl, or nitrate ester functional groups. The two cyclic olefins yielded difunctional products having the same number of carbon atoms or one carbon less than the parent hydrocarbon. Products from the diolefin 1,7-octadiene included C_7 and C_6 difunctional compounds resulting from cleavage of one and two olefinic bonds, respectively.

The two major products identified in cyclohexene aerosol were 1,6-hexanedoic acid (adipic acid) and the nitrate ester 6-nitratohexanoic acid. These two compounds were also found in 1,7-octadiene aerosol, in agreement with O'Brien et al. (26). Glutaraldehyde was a major product in cyclopentene aerosol. Because further oxidation of the dialdehydes, acid-alcohols, and acid-aldehydes listed in Tables 6 to 8 would eventually lead to diacids, it is likely that longer irradiation times than those employed in our study would have yielded aerosols of simpler chemical composition, that is, aerosols consisting essentially of α, ω-diacids and α, ω-nitrato acids.

The products listed in Tables 6 to 8 were identified either positively, by comparison with authentic samples when available, or tentatively, on the basis of their EI and CI mass spectra. Literature data were available for both EI and CI of carboxylic acids and their methyl esters (32, 33). Boschan and Smith (34) reported EI spectra for several simple nitrate esters, but no CI data are available for these compounds. The major feature in our CI (methane) mass spectrum of 6-nitratohexanoic acid is the loss of nitric acid, yielding $(MH-HNO_3)^+$ as a major fragment (Table 9). The fragmentation pattern of the compound we tentatively identified as being its C_5 homolog, 5-nitratopentanoic acid, was identical. Schwartz (35) reported an identical CI (methane) spectrum that he also attributed to 5-nitratopentanoic acid. It is of interest that the CI spectrum of the tetranitrate ester $C(CH_2ONO_2)_4$ also contains a $(MH-HNO_3)^+$ fragment (36).

The major features of the CI spectra of the compounds tentatively identified (labeled MS in Tables 6 to 8) are summarized in Table 10, along with our interpretation of the fragmentation patterns. Fragments corresponding to the loss of methanol, $(MH-CH_3OH)^+$, originate from carboxylic acid methyl ester groups (33). Structural assignments for hydroxyl and carbonyl functional groups were made on the basis of CI data for simple alcohols (37) (no MH, some M-H, base peak is $M-H-H_2O$) and aldehydes (small MH, base peak is $MH-H_2O$). The CI spectra of the three unsaturated products of 1,7-octadiene (Table 8) contained alkenyl and alkyl fragments, as reported by Field (38) for alkenes. Although the proposed structures are consistent with both CI and EI mass spectra and with the mechanistic discussion

TABLE 9. CI (Methane) Mass Spectra of 6-Nitratohexanoic Acid and Its Methyl Ester

Compound	M	Base peak (BP)	Other major fragments[a,b]
COOH-$(CH_2)_4$-CH_2ONO_2	177	115 (MH-HNO_3)	160 (MH-H_2) 97 (BP-H_2O) 178 (MH)
$COOCH_3$-$(CH_2)_4$-CH_2ONO_2	191	129 (MH-HNO_3)	160 (MH-CH_3OH) 97 (BP-CH_3OH) 192 (MH)

[a]In order of decreasing abundance.

[b]Loss of H_2O and loss of CH_3OH are major fragments in the CI (methane) spectra of carboxylic acids and their methyl esters, respectively (33).

presented in the next section, all structural assignments based solely on mass spectral data must be regarded as tentative.

As was indicated in the experimental section, direct analyses of the filter samples were performed by electron impact high-resolution mass spectrometry (HRMS), using a solid introduction probe operated at 50, 180, 320, and 400°C. To a large extent the complex spectra obtained could be reconstituted from the individual EI spectra of the compounds identified by GC-MS in the corresponding extracts, thus indicating that extraction, concentration, esterification, or GC analysis did not result in the loss of major products. However, the HRMS spectra of cyclohexene and 1,7-octadiene aerosols contained small amounts of high-molecular-weight fragments, not seen by GC-MS, at m/e = 252, 280 (1,7-octadiene only), and 257 (both). These fragments were not present in control runs conducted with blank filters. Increasing the probe temperature from 50 to 400°C resulted in volatilization of the major products in the order expected from vapor pressure data (i.e., dicarboxylic acids last), and indicated that the high m/e fragments were associated with compounds of very low volatility. Thus small amounts of polymeric material may have been formed in our experiments or be present in the ambient air at the beginning of the runs. The HRMS analyses at four probe temperatures also showed that several of the products had appreciable volatilities and were eluted from the filter at ∿50°C.

TABLE 10. CI (Methane) Mass Spectral Data and Tentative Structural Assignments

Proposed structure	M	Base peak (BP)	Other Fragments
$CHO-(CH_2)_4-COOCH_3$	144	127 ($MH-H_2O$)	113 ($MH-CH_3OH$), 95 ($BP-CH_3OH$), small 145 (MH), no M-H
$CH_2OH-(CH_2)_4-COOCH_3$	146	127 ($M-H-H_2O$)	145 (M-H), 129 ($MH-H_2O$), no MH
$CH_2ONO_2-(CH_2)_3COOCH_3$	177	115 ($MH-HNO_3$)	146 ($MH-CH_3OH$), 83 ($BP-CH_3OH$), 178 (MH)
$CHO-(CH_2)_3COOCH_3$	130	113 ($MH-H_2O$)	101 ($MH-CH_3OH$), 81 ($BP-CH_3OH$), small 131 (MH), no M-H
$CH_2OH-(CH_2)_3COOCH_3$	132	113 ($M-H-H_2O$)	131 (M-H), 115 ($MH-H_2O$), no M-H
$CH_2OH-(CH_2)_2-COOCH_3$	118	99 ($M-H-H_2O$)	117 (M-H), 101 ($MH-H_2O$), no M-H
$CHO-(CH_2)_2COOCH_3$	116	99 ($MH-H_2O$)	87 ($MH-CH_3OH$), 67 ($BP-CH_3OH$), small 117 (MH), no M-H
$CH_2ONO_2-(CH_2)_2-COOCH_3$	133	71 ($MH-HNO_3$)	116 ($MH-H_2O$), 53 ($BP-H_2O$), small MH, no M-H
$CHO-(CH_2)_2-CHO$	86	69 ($MH-H_2O$)	51 ($BP-H_2O$), small 87 (MH), no M-H
$CH_2=CH-(CH_2)_4-CH_2ONO_2$	159	97 ($MH-HNO_3$)	95 ($M-H-HNO_3$), alkenyl fragments $(C_nH_{2n-2}CH_2ONO_2)^+$ at m/e = 158 (M-1, major), 144 (M-15), 130 (M-29), 116 (M-43, major), etc. Also alkyl fragments $(C_nH_{2n}CH_2ONO_2)^+$.

(continued)

TABLE 10 (cont.)

Proposed structure	M	Base peak (BP)	Other fragments
CH_2=CH=$(CH_2)_4COOCH_3$	142	111 (MH-CH_3OH)	109 (M-H-CH_3OH), alkenyl fragments $(C_nH_{2n-2}COOCH_3)^+$ at m/e = 141 (M-1, major), 127, 113, 99 (major), etc. Also alkyl fragments $(C_nH_{2n}COOCH_3)^+$
CH_2=CH-$(CH_2)_4$-CHO	112	95 (MH-H_2O)	93 (M-H-H_2O), alkenyl fragments $(C_nH_{2n-2}CHO)^+$ at m/e = 111 (M-1, major), 97, 83, 69 (major), etc. Also alkyl fragments $(C_nH_{2n}CHO)^+$

This may indicate that the basic surface of the glass fiber filters promotes additional retention, as aerosol, of acidic gaseous products during sampling.

VI. MECHANISMS OF FORMATION OF AEROSOL PRODUCTS

The formation of the aerosol organic products identified in our experiments can be accounted for in terms of olefin-ozone and olefin-hydroxyl radical reactions. Although these reactions have not been studied in the case of cyclic olefins, it is possible to propose mechanisms analogous to those established for the reactions of O_3 and OH with alkenes in HC-NO_x photooxidation systems (39-41).

A. Ozone-Cyclohexene Reaction Mechanism

The mechanism depicted in Figures 6 and 7 involves the formation of a peroxy biradical intermediate B from the initial molozonide M, and the subsequent formation of the dioxy radical D via a dioxirane-type intermediate (40, 41). The biradicals B and D, whose stability increases with chain length, are expected to be rather stable in the case of cyclohexene.

The only (but important) difference between the mechanism proposed for cyclic olefins (Figure 6) and that for alkenes lies in the fact that, after opening of the double bond, the initial number of carbon atoms is conserved and the chain carries both the carbonyl group and the biradical, whose further reactions lead to difunctional products (as opposed to monofunctional product from alkenes).

Reactions of biradical D include rearrangement to 6-oxohexanoic acid (found in our system), decomposition into CO_2, hydrogen atom, and the 5-oxoalkyl radical E, and possibly reaction with NO and NO_2 to yield 1,6-hexanedial. Since formation of the acid as a stable product is favored over decomposition as the chain length increases, formation of 6-oxohexanoic acid is presumably a major pathway in our system.

Further oxidation of 6-oxohexanoic acid to adipic acid presumably proceeds in the aerosol phase [oxidation of cyclohexene to adipic acid is a well-known reaction with industrial applications (42), and ozonolysis of cyclohexene in the vapor phase yields adipic acid as the major aerosol product (43)]. Nitric acid present in our system may also play a role in the heterogeneous oxidation of aldehydes to acids, this oxidation reaction by diluted HNO_3 being a standard preparation method (44). Free radical oxidation by OH and HO_2, as proposed by Horowitz et al. (45) for the oxidation of formaldehyde to formic acid, may also occur in the gas phase or in aerosol droplets.

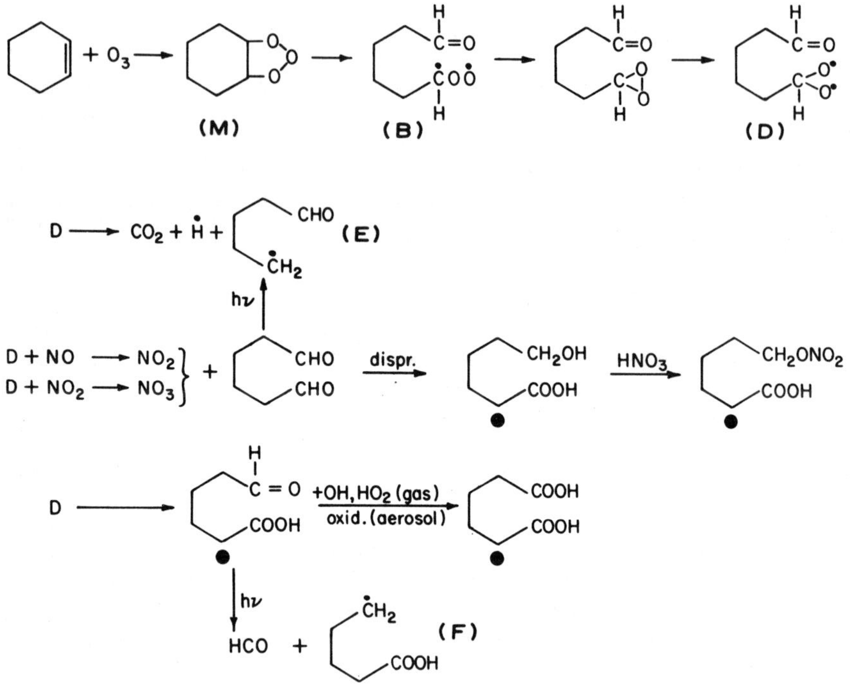

FIGURE 6. Proposed mechanisms for the reaction of cyclohexene with ozone and the formation of difunctional oxygenated products. Dots indicate products identified in cyclohexene aerosol. Further reactions of the alkyl radicals E and F are shown in Figure 7.

The formation of 6-nitratohexanoic acid, for which no gas-phase pathways could be identified, presumably proceeds in the aerosol phase and may possibly involve disproportionation or other reactions of 1,6-hexanedial, followed by esterification by HNO_3 or by reaction with NO_2 in acidic aerosol droplets.

Finally, further reactions of the C_5 alkyl radicals E and F (according to the well-known sequence $R\cdot + O_2 \rightarrow RO_2$, $RO_2 + NO \rightarrow RO + NO_2$, $RO + O_2 \rightarrow HO_2$ + aldehyde, and $RO + NO_2 \rightarrow RONO_2$) yield C_5 difunctional products bearing carboxylic acid, carbonyl, and

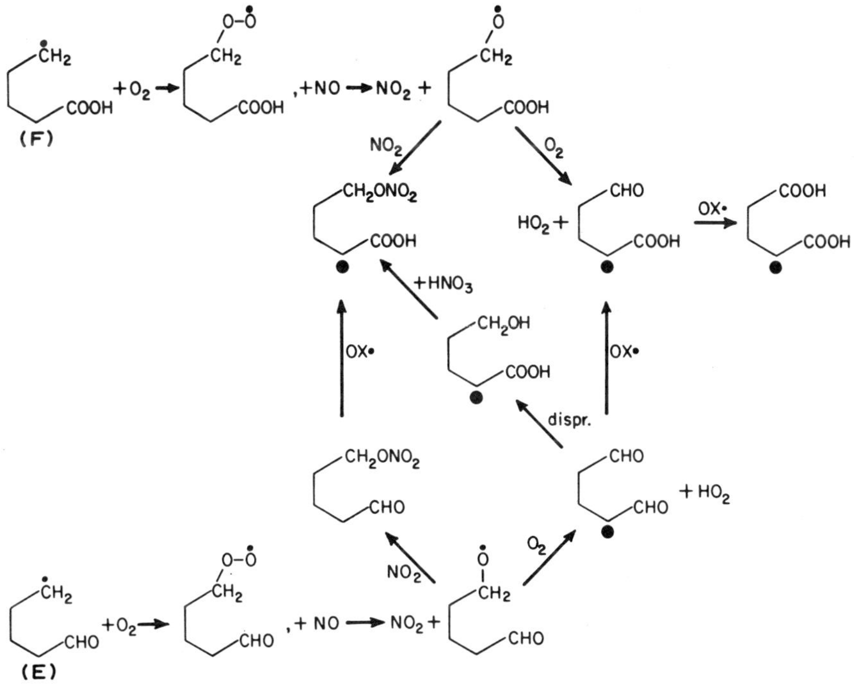

FIGURE 7. Proposed mechanisms for the formation of C_5 difunctional oxygenated products from the alkyl radicals E and F. Dots indicate products identified in cyclohexene aerosol.

nitrate ester groups (Figure 7). Further oxidation of aldehydes to acids may take place as described for the corresponding C_6 products.

B. Hydroxyl Radical-Cyclohexene Reaction Mechanism

Two reaction pathways are possible, namely, OH addition on the double bond and abstraction of an allylic hydrogen atom (Figure 8). Again by analogy with alkenes, we assume that abstraction, which would eventually lead to 1,5-pentanedial, is not important at room temperature in our conditions (47).

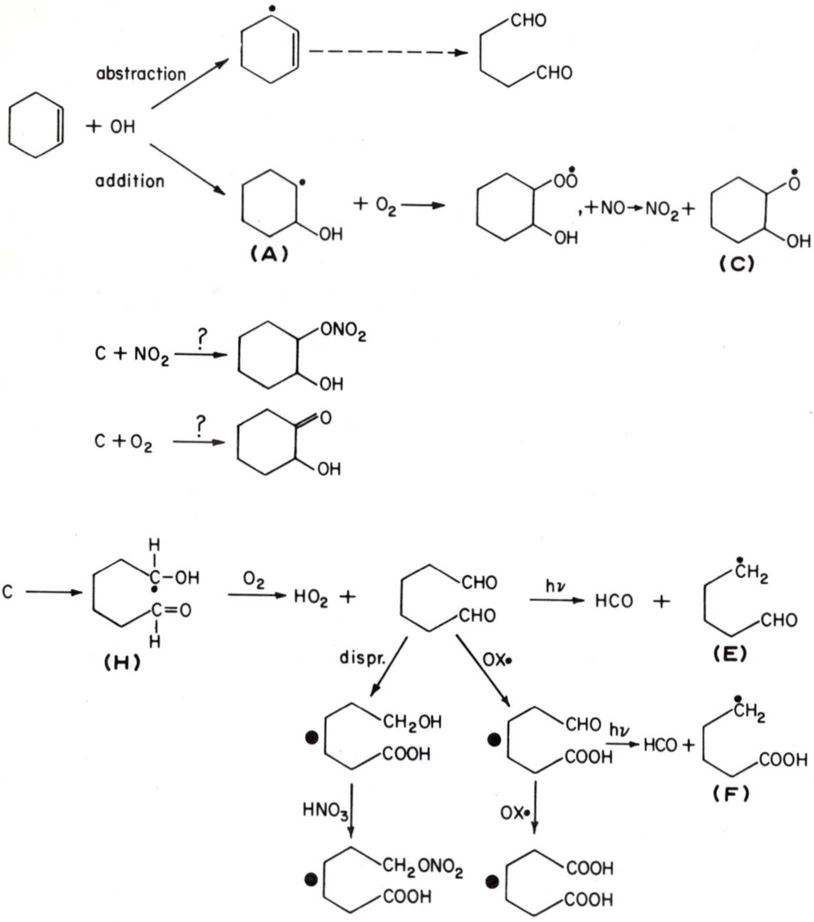

FIGURE 8. Proposed mechanism for the reaction of cyclohexene with the hydroxyl radical (OH) and the formation of difunctional oxygenated products. Dots indicate products identified in cyclohexene aerosol. Further reactions of the alkyl radicals E and F are shown in Figure 7.

The alkyl radical adduct A (2-hydroxy-1-cyclohexyl) yields the corresponding alkoxy radical C, which upon ring opening yields the hydroxyalkyl radical H. (Radical C may also possibly react with NO_2 and O_2 as shown in Figure 8.) Recent studies of the photooxidation of alcohols (47) have shown that hydrogen atom abstraction by molecular oxygen is the major reaction of radical H in our conditions, thus yielding 1,6-hexanedial as the major product. Further reactions of 1,6-hexanedial include oxidation to 6-oxohexanoic and adipic acids, photolysis to form the C_5 alkyl radicals E and F, and possibly disproportionation to give 6-hydroxyhexanoic acid, whose reaction with nitric acid yields 6-nitratohexanoic acid. These reactions and the fate of the radicals E and F were discussed in the preceding section.

C. Discussion

The mechanisms outlined above for cyclohexene apply to cyclopentene as well, and predict the formation of the α, ω-oxoacid and α, ω-dial as major products of cyclopentene reactions with O_3 and OH, respectively. As discussed above for cyclohexene, further reactions of these products would lead to other difunctional compounds, in agreement with the experimental results (Table 7). Octadiene products of the type $CH_2{=}CH{-}(CH_2)_4{-}X$ and $COOH{-}(CH_2)_4{-}X$ (X = COOH, CH_2ONO_2, or CHO; see Table 8) can also be rationalized in terms of ozone and OH reactions with one or two olefinic bonds, respectively. In all the above discussion it has been assumed that cyclic olefins behave like simple alkenes, and possible departure from the alkene-type mechanisms due to specific interactions between the two functional groups has not been considered. If the proposed mechanisms have some degree of validity, reactions of cyclic olefins with O_3 and with OH in our oxidative conditions yield essentially the same aerosol products.

VII. CYCLIC OLEFINS AND DIOLEFINS AS ORGANIC AEROSOL PRECURSORS IN THE ATMOSPHERE

Organic aerosol formation as a function of hydrocarbon type and structure has been reviewed and discussed in detail elsewhere (10), and only a brief and updated summary relevant to cyclic olefins and diolefins is given here. Grosjean and Friedlander (8) have compiled available vapor pressure data for mono- and difunctional oxygenated organic species (48-50) and have estimated for various olefins the minimal hydrocarbon precursor concentration required to achieve aerosol formation in the atmosphere (Table 11). The data in Table 11 clearly show that organic aerosol formation from most alkenes would require

TABLE 11. Lowest Ambient Olefin Concentration Required to Form the Corresponding Condensable Species in Excess of Its Saturation Concentration[a]

Olefinic precursor	Least volatile photooxidation product	Product vapor pressure (torr)	Minimal precursor concentration[b]
Propylene	Acetic acid	16	21,000 ppm
2,3-Dimethyl-2-butene	Acetic acid	16	
1-Butene	Propionic acid	4	5,200 ppm
1-Hexene	Pentanoic acid	0.25	327 ppm
1-Heptene	Hexanoic acid	0.02	26 ppm
1-Octene	Heptanoic acid	$\sim 9\ 10^{-3}$	11.8 ppm
1-Decene	Nonanoic acid	$\sim 6\ 10^{-4}$	0.8 ppm
1-Tridecene	Dodecanoic acid	10^{-5}	13 ppb
Cyclopentene	Glutaric acid	$2\ 10^{-7}$	∿0.26 ppb
Cyclohexene	Adipic acid	$6\ 10^{-8}$	0.08 ppb
1,7-Octadiene	Adipic acid	$6\ 10^{-8}$	0.08 ppb
3-Methylcyclohexene	Methyladipic acid	$\sim 2\ 10^{-8}$	∿0.03 ppb

[a]Adapted from Grosjean and Friedlander (8).

[b]These are lower limits, that is, were calculated assuming complete conversion of the precursor to the least volatile of the products and no vapor pressure lowering effect.

[c]Vapor-phase ozonolysis of 2,3-dimethyl-2-butene yields acetone (major product) and other products, including acetic acid (59).

unrealistic ambient concentrations (e.g., ∿50 ppm of 1-heptene to form hexanoic acid, Vp ∿2 10^{-2} torr), whereas cyclic olefins or diolefins, even when present as traces, can form aerosols (e.g., ∿0.1 ppb of cyclohexene to form adipic acid, Vp ∿6 10^{-8} torr). This is so because difunctional compounds are several orders of magnitude less volatile than the corresponding monofunctional species.

Diolefins having six or more carbon atoms have not been identified in gasolines, automobile exhaust, or ambient air (10), and their possible role as precursors of difunctional oxygenated aerosols is speculative at the present time.

Cyclic olefins are present in gasolines and automobile exhaust, and three of them have been found in the Los Angeles atmosphere at concentrations ranging from 2 to 10 ppb (Table 12). Therefore, on the assumption of 100% gas-phase reaction and 30% gas-to-aerosol conversion from our smog chamber data, 30 ppb of cyclic olefins would eventually form ∿30 μg m^{-3} of difunctional organic aerosols in ambient air. This estimate is consistent with atmospheric observations: aerosol organic carbon concentrations of ∿30 to 40 μg m^{-3} are typically measured on smoggy days in the Los Angeles basin, and 15 aliphatic dicarboxylic acids have recently been identified (51) in Riverside, California, air at concentrations ranging from 0.02 to 0.5 μg m^{-3}. Thus both smog chamber and field studies indicate that cyclic olefins, which are minor components of gasolines and engine exhaust and thus contribute little to oxidant formation, may constitute a major class of precursors for organic aerosol formation in photochemical smog.

ACKNOWLEDGMENTS

Dr. Steven L. Heisler participated in the smog chamber experiments. The Department of Chemistry, California Institute of Technology, and Dr. McPhee of the Southern California Air Quality Management District, Metropolitan Zone, provided access to their mass spectrometers. Mr. John Burke and Mr. John Wood assisted in the MS runs. We also thank Professor J. G. Calvert, Ohio State University, for a preprint of reference 45 before publication and Dr. E. E. Wigg, Exxon Research and Engineering Company for the detailed hydrocarbon composition of a gasoline and of the corresponding exhaust.

This work was supported in part by Environmental Protection Agency Grant No. R802160. The contents of this chapter do not necessarily reflect the view and policies of the Environmental Protection Agency, nor does mention of commercial products and trademarks constitute endorsement for use.

TABLE 12. Cyclic Olefins in Gasolines, Automobile Exhaust, and Ambient Air: Summary of Literature Data

Reference	Unit	Cyclic olefins[a]						
		C_5	3-MeC_5	1-MeC_5	C_6	3-MeC_6	4-MeC_6	1-MeC_6
Gasolines								
Sanders & Maynard (60)	wt %	0.13–0.18	0.04–0.08	0.28[b]	0.03			
Glasson & Tuesday (61)	mole %				0.14–0.55			
Neligan (62)	mole %	0.3						
Bureau of Mines (63)	-	+[c]	+	+	+			
Air Resources Board (64)	wt % carbon	0.12–0.14	0.09–0.10	0.38	0.03–0.04		0.01–0.03	
Wigg (65)	wt %	0.01–0.14				0.85		0.69
Automobile Exhaust								
Wigg (65)	vol %	0.14						
Dishart (66)	-				+			
Papa et al. (67)	-	+			+		+	
Neligan (62)	-	+						
Jacobs (68)	-						+	
Bureau of Mines (63)	-	+	+	+	+			
McEwen (69)	ppb	1			5	6[d]	6[d]	
Black & High (70)	-		+[e]	+[e]	+			

(continued)

TABLE 12 (cont.)

Reference	Unit	Cyclic olefins[a]						
		C_5	3-MeC_5	1-MeC_5	C_6	3-MeC_6	4-MeC_6	1-MeC_6
Ambient Air, Southern California								
Neligan (62) (Los Angeles)	ppb	2-9						
Stephens & Burleson (71, 72) (Riverside)	ppb	4.4-8						
LARRP Study (73) (Los Angeles)	ppb				10.7			4.7

[a] C_5 = cyclopentene, C_6 = cyclohexene, Me = methyl.

[b] Not resolved from *cis*-2-methyl-3-hexene.

[c] Identified but concentration not measured.

[d] Mixture of the two isomers 3-MeC_6 and 4-MeC_6.

[e] Isomer not specified.

REFERENCES

1. Haagen-Smit, A. J. 1952. Ind. Eng. Chem. 44, 1342.

2. Haagen-Smit, A. J., Bradley, C. E., and Fox. M. M. 1953. Ind. Eng. Chem. 45, 2086.

3. Grosjean, D. and Friedlander, S. K. 1975. J. Air Pollut. Control Assoc. 25, 1038.

4. Kotin, P., Falk, H. L., Mader, P., and Thomas M. 1954. Arch. Ind. Hyg. 9, 153.

5. Particulate Polycyclic Organic Matter, National Academy of Sciences, Washington, D.C., 1972.

6. Pitts, J. N, Jr., Grosjean, D., Mischke, T. M., Simmon, V. F., and Poole, D. 1977. Toxicol. Lett. 1, 65.

7. Heisler, S. L. and Friedlander, S. K. 1977. Atmos. Environ. 11, 157.

8. Grosjean, D. and Friedlander, S. K. 1975. "Kinetics and Mechanisms of Aerosol Formation from Cyclic Olefins and Diolefins," Paper No. 75, Joint American Chemical Society-Society for Applied Spectroscopy, Pacific Conference on Chemistry and Spectroscopy, North Hollywood, California, Oct. 28-30.

9. Grosjean, D. 1976. "Secondary Organic Aerosols and Their Gas Phase Precursors," Paper No. ENVT2, Symposium on Chemistry of Air Pollutants-1976, American Chemical Society Centennial Meeting, New York, Apr. 4-9; Div. Environ. Chem. Preprints 16, No. 1, pp. 4-7.

10. Grosjean, D. 1977. "Aerosols," Chap. 3 in Ozone and Other Photochemical Oxidants. National Academy of Sciences, Washington, D. C., pp. 45-125.

11. Heisler, S. L. 1975. "Gas-to-Particle Conversion in Photochemical Smog: Growth Laws and Mechanisms for Organics," Ph.D. Thesis, California Institute of Technology, Pasadena, California.

12. Wilson, W. E., Jr., Merryman, E. L., Levy, A., and Taliaferro, H. R. 1971. J. Air Pollut. Control Assoc. 21, 128.

13. Grosjean, D. 1975. Anal. Chem. 47, 797.

14. Metcalfe, L. D. and Schmitz, A. A. 1961. Anal. Chem. 33, 363.

15. Ferris, A. F., McLean, K. W., Marks, I. G., and Emmons, W. D. 1953. J. Amer. Chem. Soc. 75, 4078.

16. Glasson, W. A. and Tuesday, C. S. 1970. Environ. Sci. Technol. 4, 916.

17. Shen, C. H., Springer, G. S., and Stedman, D. H. 1977. Environ. Sci. Technol. 11, 151.

18. Stevenson, H. J. R., Sanderson, D. E., and Altshuller, A. P. 1965. Int. J. Air Water Pollut. 9, 367.

19. Groblicki, P. and Nebel, G. J. 1971. "The Photochemical Formation of Aerosols in Urban Atmospheres," in Chemical Reactions in Urban Atmospheres (C. S. Tuesday, Ed.), American Elsevier, New York, pp. 241-263.

20. Lipeles, M., Burton, C. S., Wang, H. H., Parry, E. P., and Hidy, G. M. 1973. "Mechanisms of Formation and Composition of Photochemical Aerosols," U. S. Environmental Protection Agency Rep. No. EPA-R3-73-036.

21. Shen, C. H. and Springer, G. S. 1977. Atmos. Environ. 11, 683.

22. Japar, S. M., Wu, C. H., and Niki, H. 1974. J. Phys. Chem. 78, 2318.

23. Cadle, R. D. and Schadt, C. 1952. J. Amer. Chem. Soc. 74, 6002.

24. Vrbaski, T. and Cvetanovic, R. J. 1960. Can. J. Chem. 38, 1053.

25. Prager, M. J., Stephens, E. R., and Scott, W. E. 1960. Ind. Eng. Chem. 52, 521.

26. O'Brien, R. J., Holmes, J. R., and Bockian, A. H. 1975. Environ. Sci. Technol. 9, 568.

27. Ripperton, L. A., Jeffries, H. E., and White, O. 1972. Adv. Chem. Ser. 113, 219.

28. Gartrell, G., Jr. and Friedlander, S. K. 1975. Atmos. Environ. 9, 279.

29. White, W. H. and Roberts, P. T. 1977. Atmos. Environ. 11, 803.

30. Grosjean, D., Doyle, G. J., Mischke, T. M., Poe, M. P., Fitz, D. R., Smith, J. P., and Pitts, J. N., Jr. 1976. Paper No. 76-20.3, 69th Air Pollution Control Association Annual Meeting, Portland Oregon, June 27-July 1.

31. Pitts, J. N., Jr., Grosjean, D., Shortridge, B., Doyle, G. J., Smith, J. P., Mischke, T. M., and Fitz, D. R. 1976. "The Nature, Concentration, and Size Distribution of Organic Particulates in the Eastern Part of the Southern California Air Basin," Paper No. ENVT1, Symposium on Chemistry of Air Pollutants-1976, American Chemical Society Centennial Meeting, New York, Apr. 4-9; Div. Environ. Chem. Preprints 16, No. 1, pp. 1-3.

32. Holmes, J. L. and St. Jean, T. 1976. Org. Mass Spectrom. 3, 1505.

33. Weinkam, R. J. and Gal, J. 1976. Org. Mass Spectrom. 11, 197.

34. Boschan, R. and Smith, S. R. 1957. "Mass Spectra of Nitrate Esters, Nitrite Esters, Nitro Compounds and Several Other Nitrogen Compounds," U.S. Naval Ordnance Test Station Research Dept., China Lake, California, July 23.

35. Schwartz, W. 1974. "Chemical Characterization of Model Aerosols," U.S. Environmental Protection Agency Rep. No. EPA-650/3-74-011.

36. Saferstein, R., Chao, J. M., and Manura, J. J. 1975. J. Assoc. Offic. Anal. Chem. 58, 734.

37. Munson, M. S. B. and Field, F. H. 1966. J. Amer. Chem. Soc. 88, 2621.

38. Field, F. H. 1968. J. Amer. Chem. Soc. 90, 5649.

39. Demerjian, K. L., Kerr, J. A., and Calvert, J. G. 1974. Advan. Environ. Sci. Technol. 4, 1.

40. Wadt, W. R. and Goddard, W. A., III. 1975. J. Amer. Chem. Soc. 97, 3004.

41. Martinez, R. I., Huie, R. E., and Herron, J. T. 1977. Chem. Phys. Lett. 51, 457.

42. Bailey, P. S. 1958. Ind. Eng. Chem. 50, 993.

43. Eastman, R. H. and Silverstein, R. M. 1953. J. Amer. Chem. Soc. 75, 1493.

44. Pryde, E. H. and Cowan, J. C. 1972. "Aliphatic Dibasic Acids," in High Polymers, Vol. 27: Condensation Monomers (J. K. Stille and T. W. Campbell, Eds.). Wiley-Interscience, New York, pp. 1-153.

45. Horowitz, A., Su, F., and Calvert, J. G. 1978. "An Unusual H_2-Forming Chain Reaction in the 3130 A Photolysis of Formaldehyde-Oxygen Mixtures at 25°C" (preprint).

46. Coffey, S., Ed. 1965. Rodd's Chemistry of Carbon Compounds, 2nd ed., Vol. I-D. Elsevier, Amsterdam.

47. Atkinson, R., Darnall, K. R., Lloyd, A. C., Winer, A. M., and Pitts, J. N., Jr. "Kinetics and Mechanisms of the Reactions of the Hydroxyl Radical with Organic Compounds in the Gas Phase," in Advances in Photochemistry, Vol. 11 (J. N. Pitts, Jr., G. S. Hammond, K. Gollnick, and D. Grosjean, Eds.). Wiley-Interscience, New York, 1979.

48. Davies, M. and Thomas, G. M. 1960. Trans. Faraday. Soc. 56, 185.

49. Hugues, E. E. and Lias, S. G. 1960. "Vapor Pressures of Organic Compounds in the Range below 1 mm of Mercury," National Bureau of Standards Tech. Note No. 70, Washington, D. C.

50. Jordan, T. E. 1954. Vapor Pressure of Organic Compounds. Interscience, New York.

51. Grosjean, D., Van Cauwenberghe, K., Schmid, J. P., Kelley, P. E., and Pitts, J. N., Jr. 1978. Environ. Sci. Technol. 12, 313.

52. De More, W. B. and Patapoff, M. 1976. Environ. Sci. Technol. 10, 897.

53. Wimmons, W. 1975. Filing sheet amending Section 70200, "Table of Standards Applicable Statewide," of Title 17 of the California Administrative Code, Sacramento, California, May 15.

54. Wu, C. H., Japar, S. M., and Niki, H. 1976. J. Environ. Sci. Health A11, 191.

55. Atkinson, R., Perry, R. A., and Pitts, J. N., Jr. 1977. J. Chem. Phys. 67, 3170.

56. Darnall, K. R., Winer, A. M., Lloyd, A. C., and Pitts, J. N., Jr. 1976. Chem. Phys. Lett. 44, 415.

57. Cvetanovic, R. J. 1960. Can. J. Chem. 38, 1678.

58. Atkinson, R. and Pitts, J. N., Jr. 1978. J. Chem. Phys. 68, 2992.

59. Smith, H. E. and Eastman, R. H. 1961. J. Amer. Chem. Soc. 83, 4274.

60. Sanders, W. N. and Maynard, J. B. 1968. Anal. Chem. 40, 527.

61. Glasson, W. A. and Tuesday, C. S. 1970. J. Air Pollut. Control Assoc. 20, 239.

62. Neligan, R. E. 1962. Arch. Environ. Health 5, 581.

63. Dimitriades, B., Raible, C. J., and Wilson, C. A. 1972. "Interpretation of Gas Chromatographic Spectra in Routine Analysis of Exhaust Hydrocarbons," Rep. No. RI 7700, U.S. Dept. of Interior, Bureau of Mines Energy Research Center, Bartlesville, Oklahoma.

64. Mayrsohn, H. and Crabtree, J. H. 1975. "Source Reconciliation of Atmospheric Hydrocarbons," California Air Resources Board, Div. of Technical Services, El Monte, California, March, pp, 8-10.

65. Wigg, E. E. 1975. Personal communication, Exxon Research and Engineering Company, December.

66. Dishart, K. T. 1970. Proc. Amer. Petrol. Inst. 50, 514.

67. Papa, L. J., Dinsel, D. L., and Harris, W. C. 1968. J. Gas Chromatogr. 6, 270.

68. Jacobs, E. S. 1966. Anal. Chem. 38, 43.

69. McEwen, D. J. 1966. Anal. Chem. 38, 1047.

70. Black, F. and High, L. 1977. "Automotive Hydrocarbon Emission Patterns and the Measurement of Nonmethane Hydrocarbon Emission Rates," Paper No. 770144, Society of Automotive Engineers International Automobile Engineering Congress, Detroit, Michigan, Feb. 28-Mar. 4.

71. Stephens, E. R. and Burleson, F. R. 1967. J. Air Pollut. Control Assoc. 17, 147.

72. Stephens, E. R. and Burleson, F. R. 1969. J. Air Pollut. Control Assoc. 19, 929.

73. Calvert, J. G. 1976. Environ. Sci. Technol. 10, 256.

PART III

PHYSICAL PROPERTIES

California Aerosols: Their Physical and Chemical Characteristics

KENNETH T. WHITBY AND GEORGE M. SVERDRUP*
Particle Technology Laboratory
Mechanical Engineering Department
University of Minnesota
Minneapolis, Minnetota

*Present address: Battelle Memorial Institute, 505 King Avenue, Columbus, Ohio 43085

I. INTRODUCTION

This paper, written just before publication of this book and after the writing of most of the other papers by the University of Minnesota group, is intended to serve several purposes. First, it is an introduction to the University of Minnesota's participation in the Aerosol Characterization Experiment (ACHEX). Second, it summarizes our most important findings about the size distributions of aerosols and their concentrations in California. Third, it provides a place to mention a number of minor but interesting details that may be pertinent for future study. And fourth, this paper describes the results of applying recently developed multimodal aerosol size distribution models to the California data, and a comparison of the California data with other data bases obtained by our group since the ACHEX program in 1972 and 1973. We have also made a limited application of the trimodal models to the characterization of chemical size distributions. Preliminary results are also included here.

A. University of Minnesota Participation in the Aerosol Characterization Experiment

At the same time of the planning of the ACHEX program, the University of Minnesota had participated in the Pasadena smog experiment [Whitby et al. (22, 24)]. Although results of the Pasadena experiment were only partly analyzed at the time the ACHEX was planned, the results showed that it was indeed practical to operate complex real-time gas- and aerosol-analyzing equipment in the field. Furthermore, there were suggestions of the general multimodal nature of aerosols, and that these modes might be useful tracers for determining the relative contributions of anthropogenic aerosol sources. It was also apparent that further measurements were going to have to be made in a variety of locations, and that this would require some sort of mobile laboratory. Therefore, when the project finally came into being in late 1971, the major task of the University of Minnesota group was to design, construct, and operate a large, movable laboratory.

The basic design of the mobile laboratory was done at the University of Minnesota; the detailed design and construction of the vehicle and its systems was performed at Thermo-Systems, Inc. (Sem et al., this book). The aerosol equipment was installed by the University of Minnesota Particle Technology Laboratory, the PDP 8 minicomputer data acquisition system was provided by Rockwell International, and the gas instrumentation was supplied by the Air and Industrial Hygiene Laboratory California State Department of Health.

The design of the vehicle and its systems are described by Sem et al. (this book), the list of instruments is given in the introduction to this book by Hidy, and further details of the placement of instruments and the operational characteristics relevant to particular measurements are provided in the various papers in this book that describe particular results.

II. DATA REDUCTION AND ANALYSIS

A. Data Base

About 4000 *in situ* aerosol measurements were made at nine sites during the operational period in 1972 when the University of Minnesota was operating the laboratory. More data were obtained by the staff of the Rockwell Science Center in 1973. After validation, 3341 10-min runs were selected at Minnesota for classification into episodes and for further averaging. The sites, dates of measurement, average run numbers, and type of average are shown in Table 1 for the 1972 data. The 2-, 6-, and 8-hr averages were prepared to match filter chemistry sampling periods. Episode averages were prepared for time periods with similar aerosol or meteorological characteristics. Each grand average is a simple unweighted average of all of the valid data from a given site.

Ideally, the data from each site should have been broken down into episodes, and a detailed correlation of aerosol with chemical and meteorological parameters made. This was done for the Harbor Freeway site [Whitby et al. (21); Nastrom and Whitby (10)] and for the Goldstone site [Sverdrup et al. (13)]. However,

TABLE 1. Summary of ACHEX Data Averages

Site	Dates	Run Nos.	Remarks
Richmond	8/10-8/16	1	Grand average
San Pablo	8/10-8/11	271	24-hr intensive average
	8/10-8/11	272-285	2-hr averages
	8/15-8/16	286	24-hr intensive average
	8/15-8/16	287-298	2-hr averages
San Francisco Airport	8/21-8/25	2	Grand average
	8/21-8/22	299	24-hr intensive average
	8/21-8/22	300-312	2-hr averages
	8/24-8/25	313-324	2-hr averages
Fresno	8/31-9/7	3	Grand average
	8/31-9/1	9	24-hr intensive average
	8/31-9/1	10-21	2-hr filter chemistry averages
	9/1 -9/7	230	24-hr intensive average
	9/6	227-229	2-hr averages
	9/6-9/7	22-33	2-hr filter chemistry averages
Hunter-Liggett	9/13-9/14	4	Grand average
	9/13-9/14	40-43	6-hr filter chemistry averages
	9/13-9/14	231-244	2-hr averages
Harbor Freeway	9/19-9/29	5	Grand average
	9/19-9/20	245	24-hr intensive average
	9/19	249	2-hr average
	9/19-9/20	49-61	2-hr filter chemistry averages
	9/20-9/21	250-256	2-hr averages
	9/19-9/20	247	8-hr filter chemistry averages
	9/20	248	10-hr filter chemistry averages

(continued)

TABLE 1 (cont.)

Site	Dates	Run Nos.	Remarks
Harbor Freeway (cont.)	9/27-9/28	246	24-hr intensive average
	9/27-9/28	62-73	2-hr filter chemistry averages
	9/28-9/29	263-270	2-hr averages
	9/27	256-262	2-hr averages
Pomona	10/5-10/26	6	Grand average
	10/5-10/6	425	24-hr intensive average
	10/5-10/6	86-96	2-hr filter chemistry averages
	10/5	325-326	1-1/2-hr filter chemistry averages
	10/5-10/6	327-335	2-hr averages
	10/11-10/13	336-366	2-hr averages
	10/16-10/17	367-380	2-hr averages
	10/18-10/20	381-402	2-hr averages
	10/23-10/24	403-410	2-hr averages
	10/24-10/25	74-85	2-hr filter chemistry averages
	10/25-10/26	411-424	2-hr averages
Goldstone	10/31-11/4	7	Grand average
	10/31-11/4	97-172, 182	1-hr average
	10/31	173	Episode A average
	10/31-11/1	174	Episode B average
	11/1	175	Episode C average
	11/1-11/2	176	Episode D average
	11/2	177	Episode E average
	11/2-11/3	178	Episode F average
	11/3	179	Episode G average
	11/3-11/4	180	Episode H average

(continued)

TABLE 1 (cont.)

Site	Dates	Run Nos.	Remarks
Goldstone (cont.)	11/4	181	Episode I average
	11/1-11/2	44-48	Filter chemistry averages
Point Arguello	11/8-11/11	8	Grand average
	11/8-11/11	186-218	2-hr averages
	11/9-11/10	34-39	Filter chemistry averages
	11/9-11/10	182	24-hr intensive average
	11/10	183-185	Filter chemistry averages
	11/18-11/21	219-226	Episode averages

for the other sites, data interpretation has been only partially completed. This paper will present some of these additional results, as well as summarize the already published results which are included for completeness in succeeding chapters.

B. Trimodal Size Distribution Model

To discuss aerosols quantitatively, it is necessary to adopt some interpretive framework. For the purposes of discussion and interpretation, the basic trimodal model and the nomenclature shown in Figure 1 will be used. The rationale behind this model

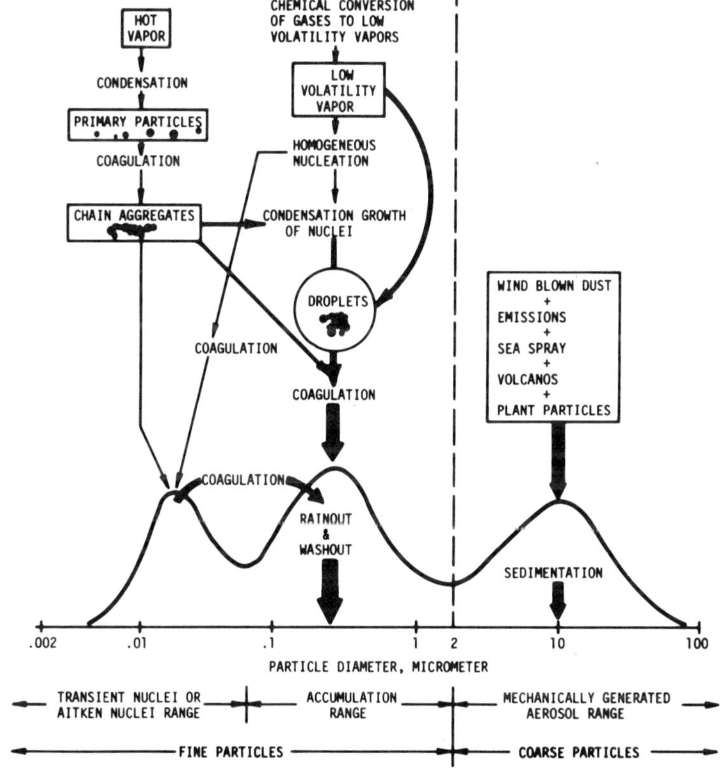

FIGURE 1. Schematic of an atmospheric aerosol size distribution showing the three modes, the main source of mass for each mode, and the principal processes involved in inserting mass into, and removing mass from, each mode.

has been previously discussed [Whitby et al. (22), Whitby (15, 16, 18); Willeke and Whitby (25)].

The distinction between "fine particles" (D_p < 2 μm) and "coarse particles" (D_p > 2 μm) is a fundamental one. There is now an overwhelming amount of evidence that not only are two modes in the mass or volume distribution usually observed [Whitby et al. (22); Lundgren (8); Hidy (5); Graedel and Graney (4); Kadowski (6); Mainwaring and Harsha (9)], but also these fine and coarse modes are usually chemically quite different [Appel et al. (1); Hidy (5); Dzubay and Stevens (3)]. The physical separation of the fine and coarse modes originates because condensation produces fine particles while mechanical processes produce mostly coarse particles. The dynamics of fine-particle growth ordinarily operates to prevent the fine particles from becoming larger than about 1 μm. Thus, as a first approximation, the fine and coarse modes originate separately, are transformed separately, are removed separately, and are usually chemically different.

Since the multimodal nature of atmospheric aerosols is now reasonably well established, the basic trimodal distribution, described in terms of three additive log-normal distributions, will be used to interpret both physical and chemical size distributions. For most atmospheric aerosols no more than two distinct modes will be seen in any one size distribution weighting. This is illustrated in Figure 2, where a typical urban aerosol size distribution model is presented in five different ways, and Figures 3, 4, and 5 where average distributions for location categories that will be discussed later have been plotted. From Figure 4 it is seen that background volume distributions typically show only the accumulation and coarse-particle modes. Generally, the number distributions show only the nuclei mode, although at background locations where the accumulation mode concentration is large relative to the nuclei mode both modes are sometimes visible.

C. Trimodal Fitting Procedures

We have found that the three modes of atmospheric aerosols can be modeled well by three additive log-normal functions [Whitby (16, 17, 18)]. As a measure of quality of fit we have used a normalized chi square, χ_N^2:

$$\chi_N^2 = \frac{1}{T} \sum \frac{(f_i - f_e)^2}{f_e} \tag{1}$$

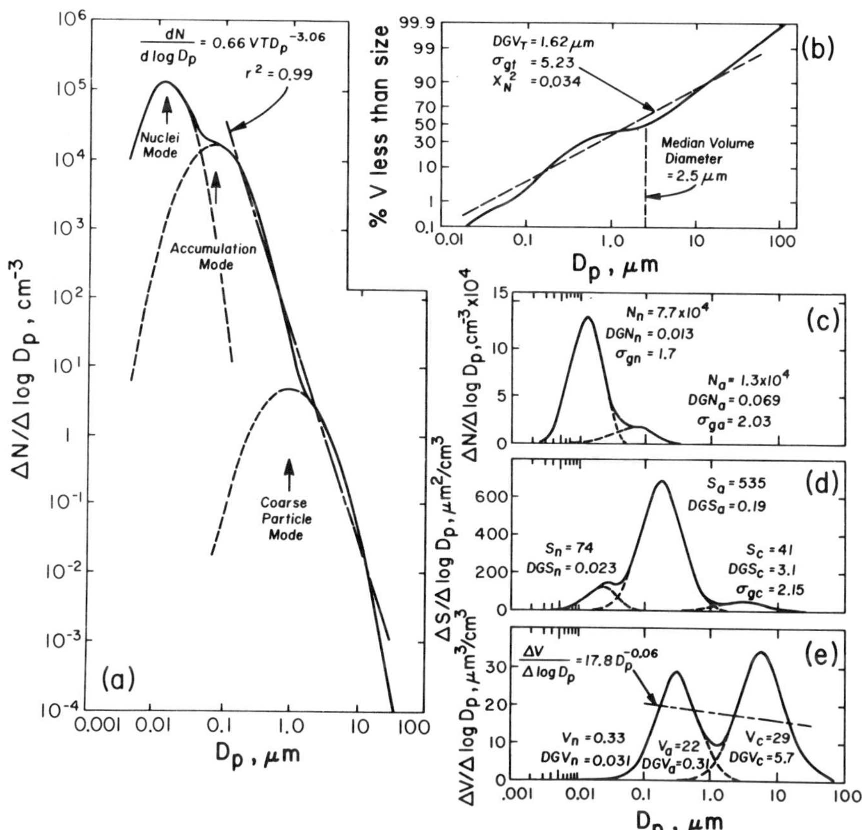

FIGURE 2. Average urban model aerosol distribution plotted in five different ways. In (a) a power function has been fitted to the number distribution over the size range 0.0 to 32 μm. In (b) a log-normal distribution has been fitted to the range 0.1 to 32 μm. It is evident that the modal nature of the aerosol is shown best by the plots of (c), (d), and (e).

where f_i is the experimental frequency in the size range, f_e is the expected frequency calculated from the added log-normal functions, and T is the total number (N), surface (S), or volume (V) in the mode being fitted or in the total distribution.

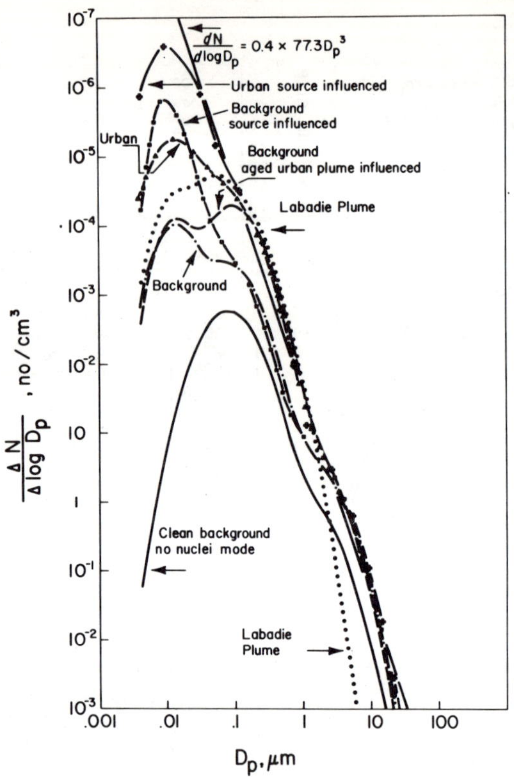

FIGURE 3. Number distribution corresponding to the last seven distributions in Table 4. Note that, although the large variations in the nuclei mode are clearly seen, the variations in the accumulation and coarse-particle modes are masked by this distribution plot.

If $f_e \ll f_i$, the χ_N^2 defined by Eq. 1 tends to blow up at the lower and upper ends of the distribution. To overcome this problem during the fitting of the impactor chemical distributions, which have only five size ranges, Eq. 1 has been modified:

$$\chi_N^2 = \frac{1}{T} \sum \frac{(f_i - f_e)^2}{(f_i + f_e)/2} \tag{2}$$

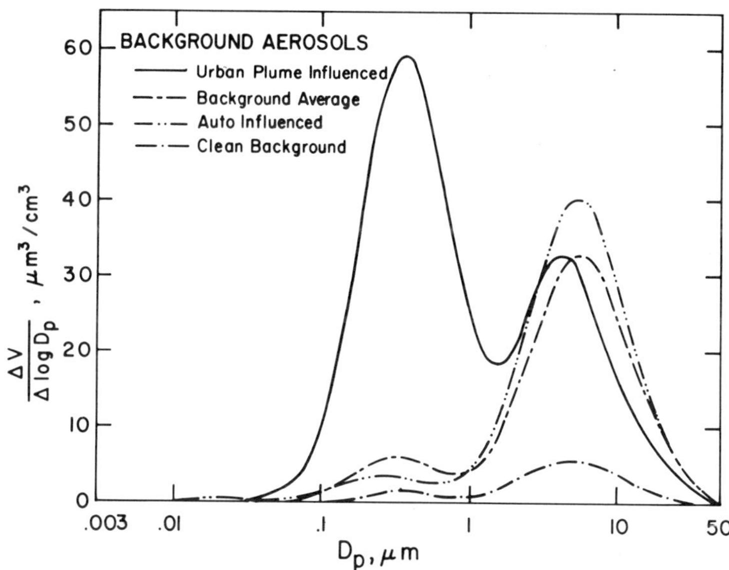

FIGURE 4. Linear-log plot of the volume distributions for the four background distributions shown in Figure 3 and Table 4. Notice how much the urban plume adds to the accumulation mode of the background.

By taking the average of the measured and expected frequencies, the $\Delta\chi^2$ at the ends of the distribution are prevented from dominating the χ_N^2 of the whole distribution. The rationale for using the average of f_i and f_e is that there is no *a priori* reason for considering one or the other to be more accurate.

Also, to calculate χ_N^2 for the impactor distributions, f_e has been calculated as the integral mass on each impactor stage by integrating the three added log-normal functions from the lower to upper cutoff size of the stage. For the filter mass, this was taken from 0.015 μm to the lower cutoff size of the last stage, for example, 0.5 μm for the Lundgren impactors used in the ACHEX. Thus, for the impactor fitting, $f_i = \Delta m_i$ and $f_e = \Delta m_e$ = the expected mass in the same size range.

Using χ_N^2 as the measure of goodness of fit, we obtained the trimodal parameters for the physical distributions of the various episodes, chemistry sampling periods, and other averages.

However, because the procedures for trimodal modeling of impactor chemical distributions have been developed only

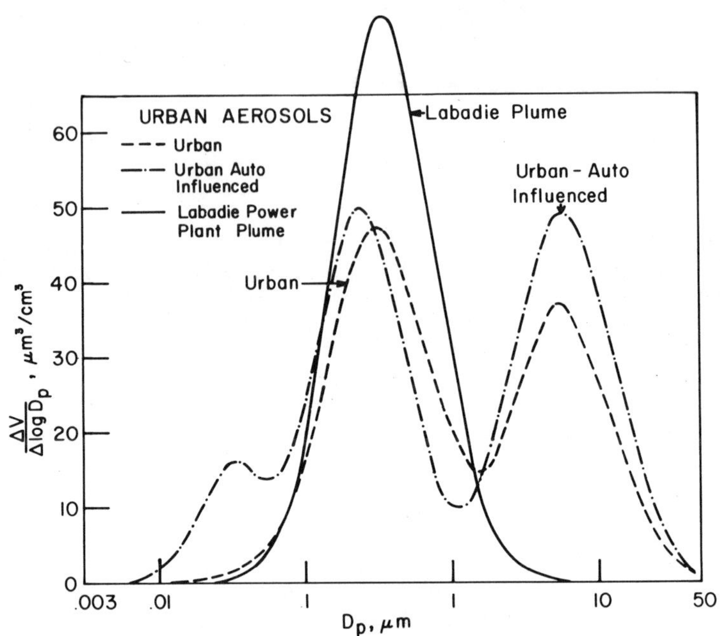

FIGURE 5. Linear-log plot of the volume distributions for two urban aerosols and a typical distribution measured in the Labadie coal-fired power plant plume near St. Louis [Whitby et al. (19)]. Size distributions measured above a few hundred meters above the ground generally have a rather small coarse-particle mode.

recently, only a few of the available chemical distributions have been modeled so far. Also, because the trimodal modeling procedure for impactor-collected chemical distributions is new, the procedure for determining the modal parameters will be described in more detail.

Since both submicron modes are in the size interval collected by the filter, the nuclei mode size is assumed equal to DGV_n, and the σ_g equal to that of the nuclei mode physical distribution as measured by the EAA. Only m_n, the mass in the nuclei mode, is determined by the fitting procedure.

For the accumulation mode, typically DGm_a's are in the 0.5 to 1 μm size range, so that a rough estimate of DGm_a can often be made by using the fitting procedure from the impactor mass distribution data. The σ_{ga} is assumed equal to that of the physical distribution as measured by the EAA and the optical counter.

For all of the impactor data, the upper cut size of the first stage was assumed somewhat arbitrarily to be 20 μm, since this was about the largest size that penetrated the sampling system. Even with the 20-μm cut size, not all of the coarse-particle mode was sampled by the impactor. Therefore a trial-and-error partial fitting procedure was used to obtain the best values of DGm_c, σ_{gc}, and m_c.

The question naturally arises as to whether the inclusion of the nuclei mode is really necessary to improve the fit over that obtained by fitting only the accumulation and coarse-particle modes. This was tested by obtaining the lowest χ_N^2 possible for only two modes on the Harbor Freeway bromine 0500 to 0700 PST sample (Table 7), a sample with a known dominant nuclei mode, and for the Goldstone bromine sample from 2100, November 1, 1972, to 2100, November 2, 1972, where the nuclei mode is quite small. For the freeway sample including the nuclei mode, χ_N^2 decreased from 0.10 to 0.0063, and for the Goldstone sample, from 0.34 to 0.005. Thus it is clear that, for both background and freeway-related impactor distributions, inclusion of the nuclei mode in the fitting procedure significantly improves the fit.

D. Aerosol Size Distribution and Concentration Categories

During the past 10 years, the University of Minnesota Particle Technology Laboratory has collected over 10,000 size distributions in a wide variety of locations both at and above the surface of the Earth. Perhaps 2000 of these have been characterized individually by the trimodal model, and most of the rest have been characterized at least in the form of fitting of the averages.

Examination of the modal parameters has shown that there are some distinct differences between background and source-related aerosol size distributions. Accordingly, we have categorized the aerosol observations to date on the basis of the aerosol concentration in the nuclei, accumulation, and coarse-particle modes. It is hypothesized that these categories are universally applicable for describing the physical properties of particle suspensions.

The seven categories adopted are as follows:

(1) Marine surface background. This represents typical background aerosol measured a few meters above the remote ocean. It is characterized by a low Aitken nuclei concentration (ANC)

of about 400 cm^{-3}, by low nuclei and accumulation mode concentrations, and by the steadiness of the concentrations. Here the term "background" is used because these steady concentrations are observed over large areas.

(2) Clean continental background. For these conditions the ANC is below 1000 cm^{-3}, the fine-particle volume is less than 2 μm^3 cm^{-3}, and the coarse-particle volume is on the order of 5 μm^3 cm^{-3}. Anthropogenic contributions are negligible.

(3) Average continental background. There are few places in the continental United States where anthropogenic activities will not make some contribution to the aerosol some of the time. Average background, therefore, is a mixture of clean background plus various kinds of anthropogenic aerosols. Typical ANCs are 4000 to 6000 cm^{-3}, fine-particle volume is on the order of 5 μm^3 cm^{-3}, and coarse-particle volume concentrations average around 25 μm^3 cm^{-3}.

(4) Background plus aged urban plumes. It is now well known that urban plumes can be transported in fairly high concentrations for several hundred kilometers. Because of the age of the urban plume, the nuclei mode is about the same as that for average background. However, the volume of aerosols in the accumulation mode is much higher, averaging 45 μm^3 cm^{-3}, almost 10 times the concentrations in average continental background. Because coarse particles would have settled out at these plume transport distances, coarse-particle concentrations are about the same as those for average continental background.

(5) Background plus local sources. The most common local source is the automobile. When this is added to clean background, it tends to raise the ANC into the 10,000 to 100,000 cm^{-3} range; however, the accumulation mode is little affected. Automobiles may stir up considerable quantities of dust, resulting in coarse-particle concentrations about one-third higher, on the average, than those for average background.

(6) Urban average. The average urban aerosol is characterized by an ANC of about 10^5 cm^{-3}, accumulation mode volumes of 30 to 40 μm^3 cm^{-3}, and coarse-particle concentrations about the same as those for average background.

(7) Urban plus freeway. Near a freeway or other large roadway the urban aerosol is modified by the presence of a large nuclei mode. Nuclei mode volume concentrations up to 30 μm^3 cm^{-3} have been observed on or immediately downwind of large freeways

in Los Angeles. Average nuclei mode volumes are on the order of 5 to 10 μm^3 cm^{-3}, compared to 0.6 for average urban aerosols and 0.037 for average background. Since automobiles are also efficient in resuspending coarse particles, coarse-particle concentrations near roadways are elevated somewhat relative to those for the average urban aerosol.

E. Comparison of California Aerosol Modal Parameters with Other Data

Using the procedures described earlier, we obtained the tri-modal parameters for the grand average aerosol size distributions at the different sites and for other distributions or averages of special interest. These parameters have been tabulated in categories along with other nonCalifornia data in Tables 2 and 3. Table 2 shows parameters for background aerosols; Table 3 shows them for urban aerosols. The average in each category was then computed and tabulated in Table 4. A number of comments need to be made about these tables.

The nuclei mode parameters are the least accurate of those presented because the EAA used in the ACHEX experiment could not measure below 0.01 μm, and the data in the 0.01 to 0.0178-μm range were subject to errors on the order of ±25%. Therefore, DGN_n and σ_{gn} have estimated errors of about ±40%.

Where $V_n = 0$ is shown, the volume in the nuclei mode was not significantly different from zero. More recent work [Whitby et al. (20)] with improved instruments has shown that a very small but ubiquitous nuclei mode is present in the atmosphere even though its mass concentration is extremely small. Because the nuclei mode disappears within a few hours of coagulation, this mode can exist only if there is continuous reinforcement by nucleation or by primary sources.

The volume, V_c, and size, DGV_c, of the coarse-particle mode is also subject to some uncertainty because the modal parameters are probably determined in some cases by the large-particle sampling efficiency of the Royco 245 inlet or the inlet to the Lundgren impactors. Although a partial fitting procedure was used for the coarse-particle mode in an attempt to minimize the effect of the upper end cutoff, it is probable that DGV_c is 10 to 15% lower than it should be.

The accumulation mode parameters are the most accurate because there were no instrumental cutoffs in this range, and the Royco 220 served to check the EAA data. It is believed that the accuracy of DGV_a, σ_{ga}, and V_a is about 10%.

TABLE 2. Modal Parameters for a Variety of Background Aerosols

Location	Comment	Nuclei		Accumulation		Coarse		Reference
		DGV (μm)	V ($\mu m^3\ cm^{-3}$)	DGV	V	DGV	V	
Fort Collins, CO	Grand average	0.029	0.1	0.35	3.8	20	230	Unpublished data
Goldstone, CA[a]	Grand average	0.038	0.063	0.33	4.77	5.3	5.7	This book
Goldstone, CA[a]	Very clean	--	∿0	0.31	1.15	6.8	0.81	This book
Point Arguello, CA[a]	Grand average	0.039	0.093	0.50	6.45	7.12	145	This book
Point Arguello, CA[a]	11/9/72, ocean air	--	∿0	0.35	1.1	6.8	53	This book
Milford, MI	10/21/75, WDIR = 350°, before test	0.027	0.026	0.23	1.98	4.9	2.9	Whitby et al. (23)
Glasgow, IL	1900–2000, 8/7/75	--	∿0	0.25	7.6	5.5	6.7	Unpublished data
Hunter-Liggett, CA[a]	Grand average	--	∿0	0.32	8.4	5.59	10.4	This book
	$\bar{x}$ =	0.033	0.035	0.33	4.41	7.8	56.8	
	σ =	0.006	0.044	0.082	2.89	5.0	85.4	
Background Site Influenced by Near Sources								
Milford, MI	10/21/75, WDIR = 350°, during test	0.020	1.2	0.21	2.7	5.4	4.4	Whitby et al. (23)
Richmond, CA[a]	No. 1217 refinery	0.044	0.33	0.23	13.9	3.8	11.2	This book
	$\bar{x}$ =	0.032	0.77	0.22	8.3	4.6	7.8	
	σ =	0.017	0.62	0.014	7.9	1.1	4.8	
Background Site Influenced by Urban Plumes								
Goldstone, CA[a]	Incursion from Los Angeles	--	∿0	0.42	34.1	3.66	9.02	This book
Glasgow, IL	2100–2200, 8/9/75, peak O_3 = 0.1 ppm	--	∿0	0.45	77.7	--	19.8	Unpublished data

(continued)

TABLE 2 (cont.)

Location	Comment	Nuclei		Accumulation		Coarse		Reference
		DGV (μm)	V ($\mu m^3\ cm^{-3}$)	DGV	V	DGV	V	
Milford, MI	10/23/75, WDIR = 180°, after test	0.032	0.14	0.24	14.8	4.8	13.5	Whitby et al. (23)
	$\bar{x}$ =		0.047	0.37	42.2	4.23	14.1	
	σ =		0.081	0.11	32.2	0.81	5.4	

[a] ACHEX mobile laboratory data.

TABLE 3. Modal Parameters for a Variety of Urban Aerosols

Location	Comment	Nuclei		Accumulation		Coarse		Reference
		DGV(μm)	V(μm^3 cm^{-3})	DGV	V	DGV	V	
Minneapolis, University of Minnesota	Grand average, 1966	---	0.8	0.33	37	8.0	50	Clark (2)
Pasadena, CA[a]	Grand average, 1969	0.039	0.8	0.36	37	6.0	30	Whitby et al. (22)
Pomona, CA[a]	1972 Grand average	0.041	0.98	0.36	44.4	6.0	29.8	This book
Pomona, CA[a]	1973 Grand average	0.034	0.42	0.35	39.5	5.0	36.9	This book
West Covina, CA[a]	1973 Grand average	0.040	0.86	0.38	52.8	4.1	39.3	This book
Rubidoux, CA[a]	1973 Grand average	0.049	0.43	0.45	67.6	26	2020	This book
Dominguez Hills, CA[a]	1973 Grand average	0.032	0.76	0.33	40.7	5.6	32.7	This book
Denver, CO	1971 Grand average, Welby	0.063	0.70	0.24	26	19	415	Willeke et al. (26)
Fresno, CA[a]	1972 Grand average	0.037	0.75	0.35	27.2	5.3	39.9	This book
	$\bar{x}$ =	0.042	0.722	0.35	41.4	5.71	36.9[b]	
	σ =	0.010	0.185	0.055	12.8	1.2	7.1	
Source-Influenced Aerosol								
Denver, CO	1971 Grand average, CMY rail yard	0.031	2.5	0.21	7.1	11.0	50	Willeke et al. (26)
Los Angeles, CA[a]	1972 Harbor Freeway, rush hour	0.044	25	0.25	61	4.0	40	This book
San Francisco Airport, CA[a]	1972 Grand average	0.031	2.25	0.29	12.8	5.0	34.8	This book
Milford, MI	Car, PG No. 12, 10/23/75, WDIR = 180°	0.026	22.9	0.30	26.6	---	---	Whitby et al. (23)
	$\bar{x}$ =	0.033	13.16	0.263	26.9	6.7	41.6	
	σ =	0.008	12.5	0.041	24.2	3.8	7.7	

[a] ACHEX mobile laboratory data.

[b] Average computed without Rubidoux and Denver.

TABLE 4. Modal and Integral Parameters for Eight Typical Atmospheric Size Distributions Derived from Averages of Measurements Made on Sites in Each Category

Aerosol Category	Nuclei DG (μm)	Nuclei σ_g	Nuclei V (μm^3 cm^{-3}) N (cm^{-3}x10^3)	Accumulation DG (μm)	Accumulation σ_g	Accumulation V (μm^3 cm^{-3}) N (cm^{-3}x10^3)	Coarse DG (μm)	Coarse σ_g	Coarse V (μm^3 cm^{-3}) N (cm^{-3})	Integral Parameters NT (cm^{-3}x10^3)	Integral Parameters ST (μm^2 cm^{-3})	Integral Parameters VT (μm^3 cm^{-3})
1. Marine, surface BK	0.019 0.01[a]	1.6[a]	0.0005 0.34[b]	0.3 0.071[b]	2[a]	0.10 0.06[c]	12 0.62	2.7	12 3.1	0.4	30	12.1
2. Clean continental BK	0.03 0.016	1.6	0.006 1.0	0.35 0.067	2.1	1.5 0.80	6 0.93	2.2	5 0.72	1.8	42	6.5
3. Average BK	0.034 0.015	1.7	0.037 6.4	0.32 0.076	2.0	4.45 2.3	6.04 1.02	2.16	25.9 3.2	8.6	148	30.4
4. BK and aged urban plume	0.028 0.014	1.6	0.029 6.6	0.36 0.12	1.84	44 9.6	4.51 0.83	2.12	27.4 7.2	16.2	938	71.4
5. BK and local sources	0.021 0.009	1.7	0.62 402	0.25 0.047	2.11	3.02 4.5	5.6 1.10	2.09	39.1 4.9	407	352	42.7
6. Urban average	0.038 0.014	1.8	0.63 106	0.32 0.054	2.16	38.4 32	5.7 0.86	2.21	30.8 5.4	138	1131	69.8
7. Urban and freeway	0.032 0.013	1.74	9.2 2120	0.25 0.062	1.98	37.5 37	6.0 1.08	2.13	42.7 4.9	2160	3201	89.4
8. Labadie plume (1976)[g]	0.015 0.0092	1.5	0.1 118	0.18 0.046	1.96	12 30	5.5 0.44	2.5	24 12	130	576	36
Averages V:	0.029±0.007	1.66±0.1	0.26±0.33[d]	0.29±0.06	2.02±0.1	21.5±20[e]	6.3±2.3	2.26±0.22	25.9±13	100±148[f]	460±440[f]	38.4±25[f]
N:	0.013±0.003		103±172[d]	0.068±.02		14.4±16	0.86±0.23		5.2±3.3			

[a]Assumed.

[b]Calculated from N_n = NT − N_a, NT = 400 cm^{-3}.

[c]Derived from data of Meszaros and Vissy (9a).

[d]Average of distributions 2 to 6 only.

[e]Average of distributions 2 to 8 only.

[f]Without No. 7.

[g]Typical distribution observed in the Labadie coal-fired power plant plume near St. Louis, Missouri.

F. Relationship between Modal Size and Amount of Aerosol in the Mode

The relationship between the mode size and the volume of aerosol in each mode has been investigated. Figure 6 shows a plot for the coarse particles of DGV_c versus V_c for the data in

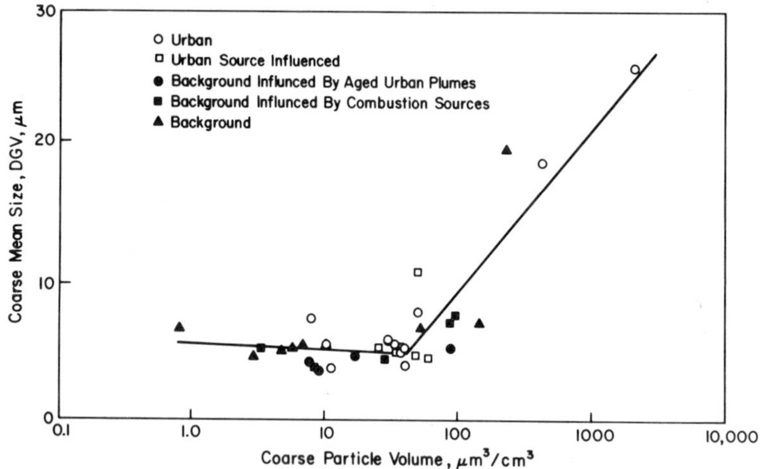

FIGURE 6. Relationship between the mean size of the coarse-particle mode and the volume in that mode for a variety of continental aerosols. The change in slope at V_c = 50 μm^3 cm^{-3} probably indicates a change in the particle suspension mechanisms at about that concentration.

Tables 2 and 3. From Figure 6 it is seen that DGV_c is rather independent of V_c up to $V_c \simeq 40$ μm^3 cm^{-3}, after which there is a linear correlation. The remarkable constancy of DG for $V_c <$ 40 $\mu m^3/cm^{-3}$ suggests that the factors determining coarse-particle size distribution below this size are also remarkably constant for a wide range of locations.

The correlation between DGV_c and V_c for $DGV_c > 40$ μm (Figure 6) for continental aerosols is $DGV_c = -15.7 + 5.53 \ln V_c$, $r^2 =$ 0.82. The correlation between DGV_c and V_c for the marine aerosols measured by Meszaros and Vissy (9a) is $DGV_c = -0.53 + 5.88 \ln V_c$,

$r^2 = 0.84$. The fact that the slopes are nearly the same for these greatly different locations suggests that, on the average, the same dynamic or distribution-shaping mechanisms are operative in suspending dusts over the continents as in suspending salt aerosols over the oceans.

In Figure 7, DGV_a has been plotted versus V_a for the accumulation mode for the same data. Except for the urban aerosols

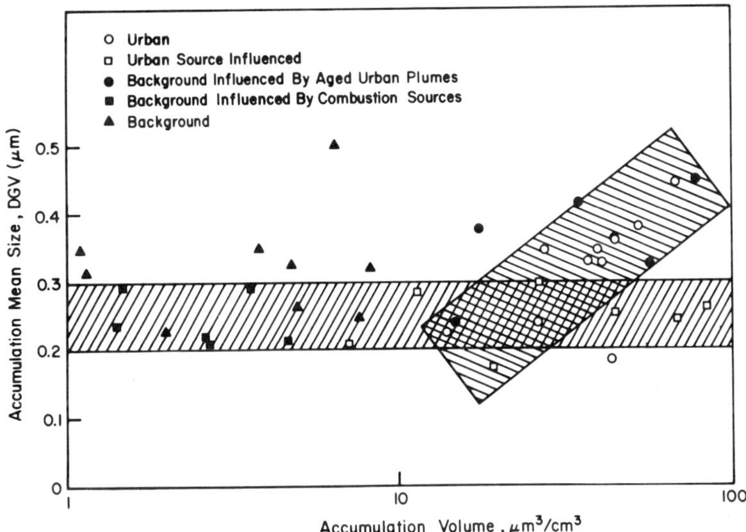

FIGURE 7. Relationship between the mean size of the accumulation mode and the volume concentration in the accumulation mode for the same data shown in Figure 6. Except for a rather weak correlation for the urban aerosol, the lack of correlation indicates that size is independent of concentration.

there is no clear correlation. Only for the high concentrations of urban aerosol, measured mostly in Los Angeles, is there any significant correlation between DGV_a and V_a. The constancy of the accumulation mode size at about 0.3 μm for so many different locations and for such a wide concentration range is a remarkable and useful fact. The constancy of DGV_a seems to derive from the opposing effects of nuclei growing into the accumulation mode by coagulation, and accumulation mode growth by condensation. Both mechanisms are governed by the same source of aerosol mass.

Although the ability of the EAA and the condensation nuclei counter (CNC) used during the ACHEX program to resolve the size of the nuclei mode was inadequate to systematically detect changes in size with concentration, Sverdrup (12) has analyzed about 900 size distributions obtained with improved instruments and calibration during the General Motors Sulfate Study [Whitby et al., (23)] conducted at a track near Detroit, Michigan. He found that DGN_n and N_n could be correlated by the following equation:

$$DGN_n = \left[\frac{4.7 \times 10^{-10}}{N_n}\right]^{0.13} \qquad (3)$$

These data included a wide range of background as well as automobile-related aerosols. The correlation described by Eq. 3 shows that the change in size observed over the concentration range from $N_n = 10^3$ to 10^7 cm^{-3} is only about a factor of 3.

Thus, because of balancing aerosol dynamic effects, both the nuclei and accumulation modes show a rather narrow range of sizes for many orders of magnitude change in mode concentration.

III. CALIFORNIA AEROSOLS

Using the interpretative framework presented above, we now discuss the size distribution of California aerosols.

A. Physical Size Distributions and Concentrations

An examination of the background aerosol data in Table 2 shows that clean continental background conditions, where the volume in the nuclei mode was negligible and the volume in the accumulation mode was below 2 μm^3 cm^{-3}, occurred for only a short period of time at Goldstone and at Point Arguello. Most of the time, at even the so-called background sites such as Goldstone, Point Arguello, and Hunter-Liggett, the accumulation mode volume ranged from 5 to 30 μm^3 cm^{-3}, indicating the presence of aged urban aerosol. The size distribution of clean continental aerosol is shown as the hatched area at the bottom of Figure 8.

The average background shows nuclei mode volumes up to 0.093 at Point Arguello. Much of the time at Point Arguello, spikes in the otherwise steady ANC were observed, indicating that plumes from automobiles or from a train were being sampled.

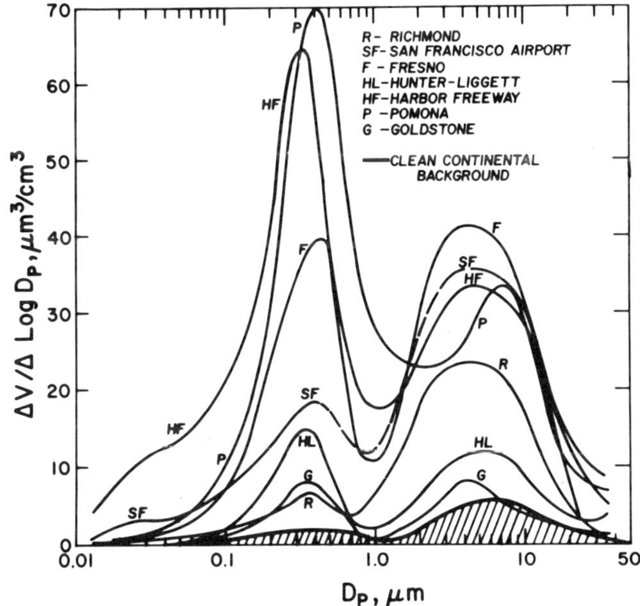

FIGURE 8. Grand average volume-size distributions for seven of the sites characterized by the mobile laboratory during the ACHEX experiment in 1972. A size distribution for clean continental aerosol is shown at the bottom for comparison. Note that even the background sites of Hunter Liggett and Goldstone are well above clean background.

The rarity of clean background data during the ACHEX study shows clearly that going out into remote or rural areas or to some location that is supposedly a background site does not guarantee that background air is being sampled. The term "background" must be qualified in terms of the origin of the air mass, and in terms of the steadiness of observed aerosol concentrations over large areas that have little anthropogenic contributions.

The urban site grand averages in Table 3 all show significant nuclei mode volumes ranging from 0.43 at Rubidoux to 0.98 at Pomona. No doubt this is caused by the ever-near automobiles and trucks in urban areas, since these are the largest and strongest source of nuclei mode aerosols.

Average accumulation mode aerosol volumes ranged from 27.2 at Fresno to 67.6 μm^3 cm^{-3} at Rubidoux. The relative uniformity of concentration of the accumulation mode in urban areas is no doubt due to the relative uniformity of source distribution and the stability of the accumulation mode.

The average coarse-particle concentration and size vary widely, all the way from 29.8 in Pomona to 2020 μm^3 cm^{-3} in Rubidoux. Familiarity with the site suggests that the high V_c observed at Rubidoux is due to a dusty site and high winds. Figure 6 shows that there is a strong relationship between concentration and size for the coarse-particle mode. Concentrations much above the average of 36.9 for the urban sites show a linear correlation between coarse-particle mode size and concentration, probably because any suspension mechanism that can suspend particles much larger than 6 μm in mean size can also sustain higher concentrations.

B. Grand Averages for the Eight Sites

The average modal parameters are presented in Tables 2 and 3 for the eight sites. Table 5 presents the corresponding meteorological and gas chemistry grand averages, and Table 6 gives the experimental grand average number distributions. Table 6 is included for those wishing to further analyze the data. The average size distributions for seven of the eight sites are shown in Figure 8. From this figure, it will be noted that all of the averages are well above the clean continental background.

C. Chemical Size Distributions

The chemical size distributions derived from samples taken on the Lundgren impactors were analyzed by, and are discussed by, personnel from the Air and Industrial Hygiene Laboratory of the State of California; see the papers by Appel et al. in this book. However, we were interested in the relationship of the physical to the chemical size distributions. As was mentioned earlier, we have developed procedures for making estimates of the elemental concentrations in the three modes from the impactor data. Results for lead (Pb), bromine (Br), and aluminum (Al) for several sites are shown in Table 7, and model and experimental size distributions for Br at the Harbor Freeway and at Point Arguello are shown in Figures 9 and 10.

Figure 9 represents a distribution for which 90% of the Br is in the nuclei mode, whereas Figure 10 from Point Arguello represents a distribution for which 79% of the Br is in the coarse mode and only 3.3% in the nuclei mode. From Table 7 it will be noted that most of the Pb (73%) and most of the Br (63%) are,

TABLE 5. Meteorological and Gas Chemistry Grand Averages for Eight Sites in 1972

Site	WDIR (deg)	WSPD (km/hr)	UVR (cal $cm^{-2} \cdot sec^{-1}$)	BBR (cal $cm^{-2} \cdot sec^{-1}$)	TO (°C)	TDP (°C)	RH (%)	O_3 (ppm)	SO_2 (ppm)	CO (ppm)	NO (ppm)	NO_2 (ppm)	CH_4 (ppm)	HCT (ppm)
Richmond	161	14.7	0.011	0.446	16.9	13.2	78	0.008	NA	1.08	0.013	0.017	3.68	24
San Francisco Airport	307	10.0	0.014	0.52	18.6	12.8	70	0.011	NA	2.11	0.093	0.032	1.78	1.63
Fresno	282	2.25	0.012	0.443	28.3	14.3	45	0.070	NA	1.9	0.012	0.039	1.96	3.26
Hunter-Liggett	215	1.2	0.014	0.34	19.4	2.5	35	0.084	NA	0.94	0.0027	0.0067	1.66	2.61
Harbor Freeway	233	3.2	0.015	0.35	21.9	17.7	81.8	0.010	NA	5.7	0.504	0.102	1.91	5.72
Pomona	191	1.04	0.006	0.20	17.9	13.5	73.4	0.028	0.009	2.98	0.047	0.062	2.38	3.47
Goldstone	180	5.21	0.009	0.347	11.1	1.69	55.8	0.0353	NA	0.531	0.024	0.023	1.58	1.88
Point Arguello	337	13.9	0.058	0.171	13.6	12.3	92.9	0.035	NA	0.48	0.016	0.053	1.57	1.92

NOTE: See the nomenclature for the meaning of the column mnemonics.

TABLE 6. Grand Average Number Distributions (number/cm^{-3}) for Eight Sites in 1972

Interval boundary, D_p (μm)	Richmond	San Francisco Airport	Fresno	Hunter-Liggett	Harbor Freeway	Pomona	Goldstone	Point Arguello
0.01								
	2.8 + 4	2.2 + 5	4.76 + 4	2.04 + 2	8.62 + 5	4.57 + 4	3.87 + 3	5.16 + 3
0.0178								
	9.89 + 4	1.04 + 5	2.99 + 4	4.93 + 2	3.56 + 5	2.89 + 4	1.98 + 3	2.57 + 3
0.0316								
	4.01 + 3	2.08 + 4	1.16 + 4	8.23 + 2	8.19 + 4	1.47 + 4	1.33 + 3	1.82 + 3
0.0502								
	1.66 + 3	6.06 + 3	5.13 + 3	1.12 + 3	1.91 + 4	7.39 + 3	8.84 + 2	9.01 + 2
0.1								
	5.56 + 2	1.89 + 3	2.43 + 3	7.71 + 2	5.47 + 3	3.23 + 3	4.13 + 2	3.14 + 2
0.178								
	1.51 + 2	5.07 + 2	1.03 + 3	3.87 + 2	1.90 + 3	1.50 + 3	2.01 + 2	1.28 + 2
0.316								
	3.21 + 1	9.06 + 1	1.78 + 2	7.24 + 1	3.1 + 2	2.99 + 2	3.86 + 1	2.53 + 1
0.422								
	9.94	a	8.19 + 1	1.91 + 1	7.58 + 1	1.40 + 2	1.22 + 1	1.43 + 1
0.562								
	4.43	a	1.37 + 1	3.12	2.33 + 1	4.12 + 1	2.13	9.90
1.0								
	2.19	a	3.13	6.76 - 1	3.73	4.89	3.22 - 1	4.93
1.78								
	0.778	a	1.26	2.86 - 1	9.76 - 1	8.11 - 1	1.54 - 1	1.71
3.16								
	8.34 - 2	a	2.63 - 1	6.36 - 2	1.25 - 1	1.70 - 1	5.1 - 2	1.83 - 1
5.62								
	3.35 - 2	3.93 - 2	4.32 - 2	1.29 - 2	3.37 - 2	3.81 - 2	5.22 - 3	1.74 - 1
10.0								
	2.20 - 3	4.40 - 3	4.62 - 3	1.19 - 3	5.63 - 3	3.65 - 3	6.38 - 4	2.74 - 2
17.8								
	9.21 - 5	1.08 - 4	3.66 - 4	1.01 - 4	3.33 - 4	2.69 - 4	4.44 - 5	1.75 - 3
31.6								
	7.36 - 6	4.29 - 6	3.67 - 5	1.40 - 5	2.31 - 5	2.78 - 5	5.16 - 6	8.70 - 5
38								

[a] Royco 220 data invalid.

TABLE 7. Modal Parameters for Pb, Br, and Al for Several ACHEX Mobile Laboratory Sites

Site and sampling period	Species	Measured mT ($\mu g\ m^{-3}$)	Nuclei: DG^a (μm)	Nuclei: σ_g^a	Nuclei: m_n ($\mu g\ m^{-3}$)	m/mT (%)	Accumulation: DG (μm)	Accumulation: σ_g^a	Accumulation: m_n ($\mu g\ m^{-3}$)	m/mT (%)	Coarse: DG (μm)	Coarse: σ_g^a	Coarse: m_c ($\mu g\ m^{-3}$)	m/mT (%)	χ_N^2	Calculated mT ($\mu g\ m^{-3}$)
Harbor Freeway 2100–2300, 9/19	Pb	2.41	0.05	1.6	1.44	57	1.1	2.0	0.63	25	10	2.4	0.45	18	0.042	2.52
2300, 9/19 to 0100, 9/20	Pb	3.25	0.05	1.6	2.95	86	0.5	2.0	0.1	2.9	11	2.3	0.4	12	0.009	3.45
0100–0300, 9/20	Pb	6.01	0.05	1.6	5.2	76 73 ± 15	0.8	2.0	1.15	17 15 ± 11	12	2.4	0.45	6.6 12.2 ± 6	0.016	6.80
Pomona 1200–1400, 10/24	Br	0.69	0.05	1.6	0.25	36	1.3	2.0	0.28	41	3.5	2.2	0.16	23	0.018	0.69
Hunter-Liggett 2100, 9/13 to 2100, 9/14	Br	0.0025	0.05	1.6	0.0184	74	0.7	2.0	0.0026	11	6	2.0	0.0038	15	0.02	0.0248
Harbor Freeway 2100, 9/19 to 0500, 9/20	Br	2.14	0.05	1.6	1.42	69	1.0	2.0	0.50	24	11	2.2	0.15	7	0.005	2.07
0500–0700, 9/20	Br	6.85	0.05	1.6	6.32	90	0.8	2.0	0.40	5.7	8	2.4	0.30	4.3	0.0063	7.02
Point Arguello 2100, 11/9 to 2100, 11/10	Br	0.120	0.05	1.6	0.0041	3.3	0.9	2.0	0.022	17.5	6	2.4	0.10	79.2	0.051	0.126
Goldstone 2100, 11/1 to 2100, 11/2	Br	0.026	0.05	1.6	0.012	46 63 ± 22[b]	0.9	2.0	0.0075	29 22 ± 14[b]	10	2.4	0.0068	25 15 ± 9[b]	0.005	0.0263
Hunter-Liggett 2100, 9/13 to 2100, 9/14	Al	1.45	0.04	1.6	0.052	2.6	0.8	2.0	0.65	32.4	13	2.1	1.3	65	0.10	2.00
Pomona 1200–1400, 10/24	Al	0.89	0.04	1.6	0.024	2.3	0.9	2.0	0.105	9.7	10	2.1	0.95	88	0.017	1.08

(continued)

TABLE 7 (cont.)

Site and sampling period	Species	Measured mT ($\mu g\ m^{-3}$)	Nuclei			m/mT (%)	Accumulation			m/mT (%)	Coarse			m/mT (%)	χ_N^2	Calculated mT ($\mu g\ m^{-3}$)
			DG[a] (μm)	σ_g[a]	m_n ($\mu g\ m^{-3}$)		DG (μm)	σ_g[a]	m_a ($\mu g\ m^{-3}$)		DG (μm)	σ_g[a]	m_c ($\mu g\ m^{-3}$)			
Harbor Freeway 2100, 9/19 to 0500, 9/20	Al	2.00	0.04	1.6	0.039	1.5	1.0	2.0	0.26	10.5	12	2.1	2.22	88	0.05	2.52
Goldstone 2100, 11/1 to 2100, 11/2	Al	0.88	0.04	1.6	0.027	2.8	0.9	2.3	0.10	10.2	6.0	2.1	0.85	87	0.076	0.98
						2.3 ± 0.6				15.7 ± 11				82 ± 11		
Overall average: 0.89 ± 0.19																

[a]Values assumed based on values from physical distributions.

[b]Point Arguello omitted from average.

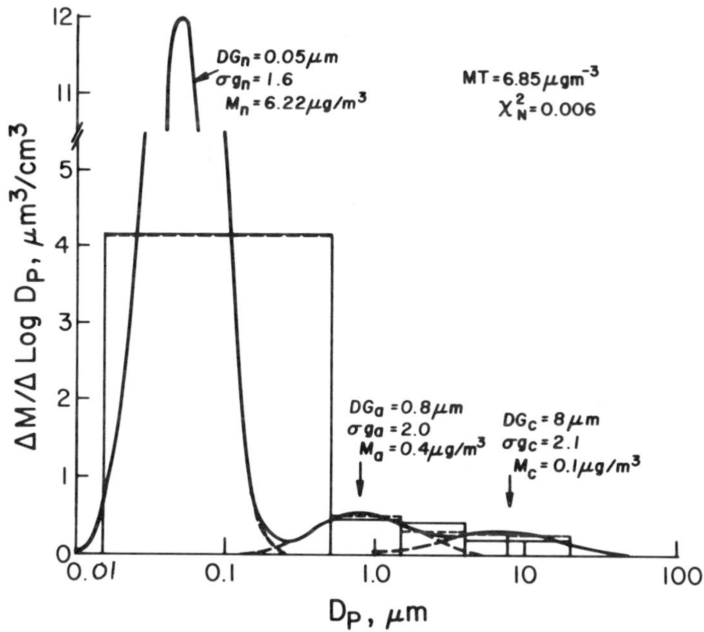

FIGURE 9. Bromine size distribution sampled from 0500 to 0700 at the Harbor Freeway on September 20, 1972, using the Lundgren impactor. Solid bars are the actual data, and dashed bars are the fitted trimodal model masses on each stage. The curves are the three modes. This represents a distribution dominated by the nuclei mode.

on the average, found in the nuclei mode. This is reasonable since the source for both is undoubtedly the automobile, which is known to emit a strong nuclei mode.

The fact that most of the Pb and most of the Br are in the nuclei mode, and not in the accumulation mode, for these samples means that the mean size of the Pb and Br is almost an order of magnitude smaller than has usually been assumed.

As expected, 82% of the Al, which is most likely to be associated with the coarse particles, is found in the coarse mode. It is also seen that the fraction of Al in the coarse mode is relatively constant for the four sites evaluated.

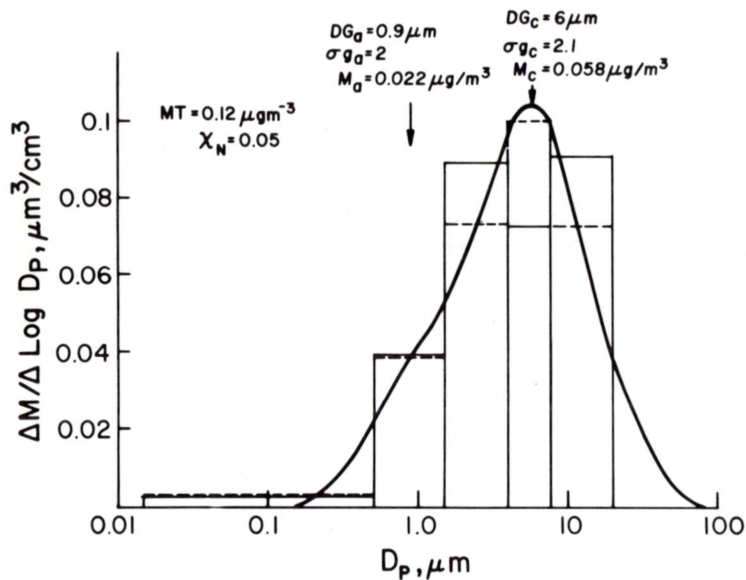

FIGURE 10. Bromine size distribution sampled from 2100, November 9, 1972, to 2100, November 10, at Point Arguello, using the Lundgren impactor. Solid bars are the actual mass on each stage, and the dashed bars are the fitted trimodal model masses on each stage. This represents a distribution dominated by the coarse-particle mode, with no detectable nuclei mode.

It is also interesting that the fractions of Pb, Br, and Al found in the accumulation mode are about the same, for example, 15 to 22%, but probably for different reasons. Probably the Pb and Br found in the accumulation mode arrive there by coagulation from the nuclei mode. On the other hand, the Al in the accumulation mode either is the lower tail of the coarse-particle mode or else comes from larger particles that have bounced off the upper stages of the impactor.

In Table 8 the mode volumes of the impactor samples are compared with the mass concentrations for Pb, Br, and Al. The ratio of $(m_n/V_n)_{Pb}$ for the Harbor Freeway averages 1.8 ± 0.37, suggesting that about one half of the nuclei mode volume there consists of Pb compounds.

TABLE 8. Comparison of Modal Volumes and Modal Masses of Pb, Br, and Al

Site and sampling period	Run No.	Species	VT ($\mu m^3 cm^{-3}$)	Nuclei DGV (μm)	Nuclei σ_g	Nuclei V_n ($\mu m^3 cm^{-3}$)	Nuclei m_n ($\mu g cm^{-3}$)	$\frac{m_n}{V_n}$ ($g cm^{-3}$)	Accumulation DGV (μm)	Accumulation σ_g	Accumulation V_n ($\mu m^3 cm^{-3}$)	Accumulation m_a ($\mu m^3 cm^{-3}$)	$\frac{m_n}{V}$ ($g cm^{-3}$)	DGV (μm)	σ_g	V_n ($\mu m^3 cm^{-3}$)	m_c ($\mu g cm^{-3}$)	$\frac{m_n}{V}$ ($g cm^{-3}$)
Harbor Freeway 2100-2300, 9/19	49	Pb	57.7	0.034	1.6	0.984	1.44	1.46	0.295	1.96	18.7	0.63	0.034	7.25	2.74	40.9	0.45	0.011
2300, 9/19 to 0100, 9/20	50 + 51	Pb	86.0	0.036	1.32	1.34	2.95	2.20	0.309	2.07	31.8	0.1	0.003	7.16	2.20	55.5	0.4	0.0072
0100-0300, 9/20	52	Pb	86.5	0.031	1.28	2.99	5.2	1.74	0.273	1.89	32.1	1.15	0.036	7.18	2.20	51.5	0.45	0.0087
								1.8 ± 0.37					0.024 ± 0.02					0.009 ± 0.002
Pomona 1200-1400, 10/24	76	Br	135	0.031	1.60	0.42	0.25	0.60	0.536	2.06	115	0.28	0.0024	3.58	2.44	24.4	0.16	0.0066
Hunter-Liggett 2100, 9/13 to 2100, 9/14	40 - 43	Br	19.4	0.035	1.6	0.0209	0.0184	0.88	0.318	1.61	8.39	0.0026	0.003	6.02	1.98	9.08	0.0038	0.0004
Harbor Freeway 2100, 9/19 to 0500, 9/20	247	Br	79.8	0.032	1.6	3.45	1.42	0.41	0.292	2.00	33.4	0.5	0.015	7.08	2.32	45.4	0.15	0.0033
0500-0700, 9/20	54	Br	131	0.029	1.60	11.5	6.32	0.55	0.287	2.06	82.6	0.4	0.0048	6.69	2.06	25.8	0.3	0.0116
Point Arguello 2100, 11/9 to 2100, 11/10	34-36 183-185	Br		0.027	1.60	0.074	0.0041	0.055	1.6	2.99	15.4	0.022	0.0014	10.2	2.10	245	0.1	0.0004
Goldstone 2100, 11/1 to 2100, 11/2	44 - 48	Br	13.6	0.031	1.6	0.032	0.012	0.6	0.301	1.65	4.75	0.0075	0.0016	5.26	2.33	9.22	0.0068	0.0007
								0.60 ± 0.17[a]					0.0048 ± 0.0052					0.0038 ± 0.0045
Hunter-Liggett 2100, 9/13 to 2100, 9/14	40 - 43	Al				0.0209	0.052	2.49			8.39	0.65	0.078			9.08	1.3	0.141
Pomona 1200-1400, 10/24	76	Al				0.42	0.024	0.057			115	0.105	0.0009			24.4	0.95	0.039
Harbor Freeway 2100, 9/19 to 0500, 9/20	247	Al				3.45	0.039	0.011			33.4	0.26	0.0078			45.5	2.22	0.049
Goldstone 2100, 11/1 to 2100, 11/2	44 - 48	Al				0.032	0.027	0.844			4.75	0.10	0.021			9.22	0.85	0.092
							0.036 ± 0.013		0.47 ± 0.43			0.279 ± 0.26					1.33 ± 0.62	
Hunter-Liggett and Goldstone averages								1.67 ± 1.16	0.33 ± 0.09[a]				0.05 ± 0.04					0.118 ± 0.036
Pomona and Harbor Freeway averages								0.034 ± 0.033					0.0044 ± 0.005					0.044 ± 0.007

[a]Without Point Arguello.

The ratio of $(m/V)_{Br}$ to $(m/V)_{Pb}$ is 0.33, 0.20, and 0.42 for the nuclei, accumulation, and coarse-particle modes, respectively. The theoretical ratio of Br/Pb for ethyl fluid used in gasoline is 0.39, although a ratio of 0.33 has usually been observed for particulate emissions from lead-fueled automobiles. Thus the observed ratios for all three modes are equal to the theoretical ratio within an experimental error of about 20%.

For the Hunter-Liggett and Goldstone samples $(m_c/V_c)_{Al}$ = 1.67 ± 1.1, suggesting that most of the nuclei mode volume probably consists of particles that bounced off the upper stages of the impactor. For the Pomona and Harbor Freeway samples, the fraction of nuclei mode volume that is Al is only a few percent.

D. Sulfate Size Distributions

Whitby (18) determined the accumulation mode parameters of the California sulfate distributions presented by Appel et al. (this book), along with four other data sets by Lee and Patterson (7), Wagman et al. (14), Patterson and Wagman (11), and Kadowaki (6). Using a single-mode fitting procedure, he concluded that about 96% of sulfate was in the accumulation mode and obtained accumulation mode parameters of $\overline{DGm_a}$ = 0.48 ± 0.1 μm and $\overline{\sigma_{ga}}$ = 2.0 ± 0.29.

The California data for 1973 has been reanalyzed using the multimodal impactor analysis procedure described earlier. Results are presented in Table 9. The reanalysis gave almost identical accumulation mode parameters, $\overline{DGm_a}$ = 0.58 ± 0.1 μm, compared to 0.55 ± 0.09 μm for the earlier analysis. However, a small but persistent amount of sulfate was found in the coarse-particle mode, for example, 1.77 ± 0.2 μg m^{-3} for all of the samples. The origin of this coarse-particle mode sulfur is not known, but its most likely source is soil dust since coarse-particle mode concentrations are also quite constant. Only for Dominguez Hills on October 10 and 11, 1973, was a significant amount found in the nuclei mode. It is noteworthy that this is the sample with the lowest total sulfate, 6.1 μg m^{-3}, compared to 20 to 34 for the others taken in Los Angeles in 1973. (Dominguez is near Stauffer Chemical, a source of H_2SO_4 emissions.) The absence of nuclei mode sulfate may be due either to coagulation removal by large amounts of volume in the accumulation mode or possibly to a difference in formation mechanisms. On October 10 and 11 the humidity was lower, average Pb, V, Ca, Mn, and Fe concentrations were higher, and a sea breeze was experienced for part of the day. It is possible that a chemical plant or refinery plume was sampled on this day, and this might explain the nuclei mode observed.

TABLE 9. Modal Parameters for Six California Sulfate Size Distributions[a]

Site and date	Expt'l m_T ($\mu g/m^3$)	Nuclei m_n ($\mu g/m^3$)	Accumulation DGm_a (μm)	Accumulation σ_{ga}	Accumulation m_a ($\mu g/m^3$)	Coarse DGm_c (μm)	Coarse σ_{gc}	Coarse m_c ($\mu g/m^3$)	χ^2_N
Rubidoux September 5,6	18.9	0	0.51	1.93	17.9	10	2.4	1.65	0.0029
Rubidoux September 18,19	23.0	0	0.70	1.75	22.0	13	2.4	1.6	0.006
Pomona August 16,17	27.5	0	0.61	1.95	26.3	10.1	2.4	1.75	0.002
West Covina July 23,24	32.7	0	0.55	1.95	31.6	9.7	2.4	1.70	0.0003
Dominguez Hills October 4,5	33.9	0	0.67	1.75	32.0	8.5	2.4	2.2	0.0022
Dominguez Hills October 10,11	6.1	1.05	0.43	2.5	3.7	8.5	2.4	1.70	0.0012
Averages			0.58 ± 0.1	1.97 ± 0.28		9.97 ± 1.6		1.77 ± 0.2	0.006 ± 0.008

*Data from Appel et al. (1).

It should be kept in mind that these fitting procedures cannot separate the nuclei mode from the accumulation mode unless the nuclei mode mass is greater than perhaps 20% of the accumulation mode mass.

IV. MISCELLANEOUS OBSERVATIONS

In this section a number of miscellaneous data are discussed.

A. Sodium and Aluminum Concentrations as a Function of Distance from the Ocean

In Figure 11 are plotted the Al/mT and Na/mT ratios as a function of the nominal distance from the ocean. The presumption

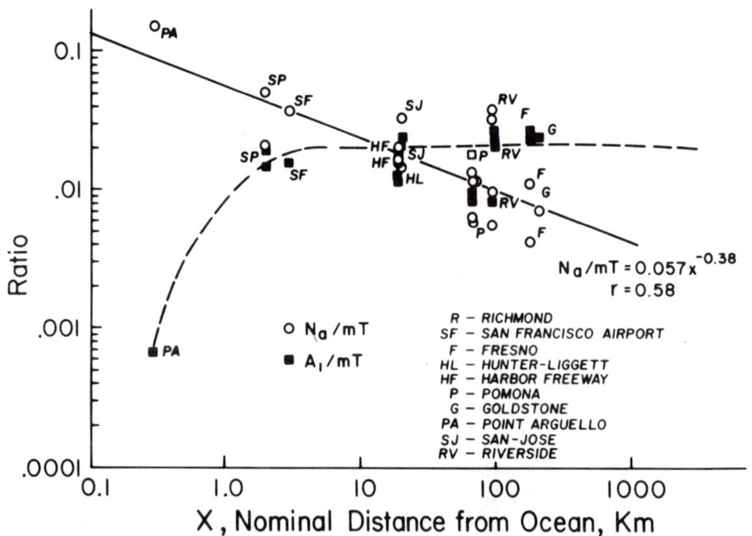

FIGURE 11. Na/mT and Al/mT ratios as a function of the nominal distance from the ocean. The Na/mT ratio decreases steadily as the distance from the ocean increases. However, aluminum appears to increase to a more or less constant value within about 1 km of the seashore.

here is that, if most Na is coming from the ocean, there should be a decrease in the Na/mT ratio with distance from the ocean. Conversely, if most of the Al comes from soil dust, there should be an increase in Al/mT with distance from the ocean. Figure 11 shows that indeed this is what happens along the coast, but while the decrease in Na/mT is gradual, the Al/mT ratio (which is soil dust-related) rises quickly to a more or less equilibrium value within 1 km of the seashore.

Earlier it was found that the coarse-particle mode contained a relatively constant concentration of sulfate equal to 1.77 $\mu g\ m^{-3}$. To answer the question of whether this coarse mode sulfate could have come from the sea, the Na/mT ratio from Figure 11 and the Na/sulfate ratio for seawater were used to calculate the expected sulfate concentration. If mT = 100 $\mu g\ m^{-3}$, then the expected coarse-particle sulfate concentration is 0.09 $\mu g\ m^{-3}$, a factor of 2000 smaller than that observed. It is very unlikely that the sea contributes significantly to airborne sulfate in California. The most likely source of the observed coarse-particle sulfate is soil dust. Except for a few rare periods at background sites, the concentration of coarse particles is quite constant, as is the observed coarse-particle sulfate. Possibly the coarse-particle sulfate comes from fine particles deposited on the ground, which coat the surface dust particles and then are resuspended with the dust.

B. Sea Salt Aerosol Generation at Point Arguello

The Point Arguello site was selected in the hope that it would provide information on background aerosol concentrations and size distributions. Actually, out of the 90 hr of data that were analyzed, true oceanic background (e.g., ANC $< 1000\ cm^{-3}$) was observed for only about 6 hr. For this period the accumulation mode volume was about 1.4 $\mu m^3\ cm^{-3}$. For the rest of the time there was evidence that some anthropogenic aerosol was being sampled, even though the wind was from the sea.

As expected for a site this close to the sea (150 m), large concentrations of sea-generated coarse particles were observed when the wind was off the ocean. For the 12 hr when the wind was from the land, the average coarse-particle concentration was 14 ± 6.5. For the 80 hr when it was off the ocean, the concentration was $217 \pm 249\ \mu m^3\ cm^{-3}$, even though the wind speeds were comparable, for example, 16 versus 20 km hr^{-1}. When the wind was from the sea sector, there was no correlation between high concentrations in the coarse-particle mode and concentrations in the accumulation mode. Thus the sea did not generate any detectable concentrations of accumulation mode-size particles.

From an examination of the 2-hr averages there appeared to be no obvious correlation between wind speed and coarse-particle aerosol volume, although one would be expected.

V. SUMMARY AND CONCLUSIONS ABOUT THE PHYSICAL CHARACTERISTICS OF CALIFORNIA AEROSOLS

A. Size Distribution and Concentration

California aerosols from all sites can be modeled well using three additive log-normal distributions.

The nuclei mode volume mean sizes are in the 0.02 to 0.05 μm range. Nuclei mode concentrations range from essentially 0 for aged background aerosols, through 0.7 $\mu m^3\ cm^{-3}$ for average urban aerosols, to as high as 25 $\mu m^3\ cm^{-3}$ near the Harbor Freeway during the morning rush hour.

Accumulation mode volume mean sizes range from 0.22 to 0.7 μm during very smoggy episodes in the Los Angeles basin. Accumulation mode volume concentrations range from a low of 1.1 $\mu m^3\ cm^{-3}$ during episodes of the cleanest continental background, through the 30 to 50 range during average conditions in urban areas, to a very high value of about 300 at Dominguez Hills in 1973. Since the accumulation mode-size particles are removed the most slowly from the atmosphere, they may be transported long distances. Such episodes were observed at Goldstone and Rubidoux, and evidence suggests that most of the time at all of the background sites some anthropogenic-generated accumulation mode aerosol was present.

Coarse-particle mode sizes ranged from 3 to 30, with a value of about 6 μm most of the time. Coarse-particle volume mean size was found to be unrelated to concentration up to a concentration of 40 $\mu m^3\ cm^{-3}$, but to be correlated with concentrations above that value.

Coarse-particle concentration appeared to be more unrelated to location and the size of the urban area than was found for accumulation mode. It is likely that the nearly uniform distribution of roads and people, coupled with hot, dry conditions much of the time in most places in California, causes coarse-particle suspension to be rather uniform.

B. Ratio of Submicron Aerosol Volume to Light-Scattering Coefficient

Sverdrup and Whitby (this book) found that the ratio of submicron aerosol volume, V3-, to light-scattering coefficient,

b_{scat}, varied from a low of 5 under clean background conditions to a high of 70 under source-influenced urban conditions. The large range of this ratio suggests that some caution is necessary in using light-scattering methods for the estimation of fine-particle aerosol volume.

C. Relative Humidity Effects on Size Distribution Measurement

A mismatch was observed by Sverdrup and Whitby (this book) between the size distribution data obtained with the Royco 220 optical counter (OPC), located inside the mobile laboratory, and the Royco 245 OPC, whose sensor was located outside the laboratory. The mismatch was shown to be caused by evaporation as the particles were transported from the lower outside ambient temperature to the higher inside temperature. An estimate of the mass fraction of water in the aerosol from the mismatch was in agreement with microwave, nephelometer, and filter measurements.

D. Size Distribution of Freeway Aerosols

Whitby et al. (21, and this book) have analyzed aerosol measurements taken alongside the Harbor Freeway in Los Angeles. By taking the difference between the size distribution when the wind was from, and away from, the freeway, it was determined that most of the freeway aerosol is in the nuclei mode. The morning rush hour contributed about 17.1 $\mu m^3\ cm^{-3}$ to the nuclei mode.

E. Aerosol Size Distributions and Concentrations in the Mojave Desert

Data from the Goldstone site in the Mojave Desert, analyzed by Sverdrup and Whitby (13, and this book), showed that periods of clean continental background ($V_a < 2\ \mu m^3\ cm^{-3}$) were rare. Most of the time there was evidence of local contributions from automobiles. An incursion of aged urban aerosol from the South Coast Air Basin was observed which raised the accumulation mode volume to about 30 $\mu m^3\ cm^{-3}$.

NOMENCLATURE

Roman Notation

ANC Aitken nuclei concentration
D_p particle diameter

DGN	geometric number mean size
DGS	geometric surface mean size
DGV	geometric volume mean size
DGm	geometric mass mean size
m	mass concentration
mT	total mass concentration in whole distribution
N_n	number concentration of particles in the nuclei mode
N_a	number concentration of particles in the accumulation mode
N_c	number concentration of particles in the coarse-particle mode
NT	total number concentration of particles in whole distribution
r	particle radius or correlation coefficient
T	total N, S, or V in Eqs. 1 and 2
V_n	volume concentration of particles in the nuclei mode
V_a	volume concentration of particles in the accumulation mode
V_c	volume concentration of particles in the coarse-particle mode
VT	total volume concentration in whole distribution

Subscripts

n, a, and c	designate the nuclei, accumulation, and coarse-particle modes, respectively
$\overline{N}_n$	the bar designates an average

Greek Notation

ρ_p	particle density
σ_g	geometric standard deviation
χ_N^2	normalized chi square, Eqs. 1 and 2

REFERENCES

1. Appel, B. R., Wesolowski, J. J., and Hidy, G. M. 1974. "Analysis of the Sulfate and Nitrate Data from the Aerosol Characterization Study," AIHL Rep. No. 166, Air and Industrial Hygiene Laboratory, California State Department of Health, Berkeley, California, pp. 1-16.

2. Clark, W. E. 1972. "Measurements of Aerosols Produced by the Photochemical Oxidation of SO_2 in Air," Ph.D. thesis, University of Minnesota, Minneapolis, Minnesota.

3. Dzubay, T. G., and Stevens, R. K. 1975. Ambient Air Analysis with Dichotomous Sampler and X-ray Fluorescence Spectrometer, Environ. Sci. Technol. 9, 663-668.

4. Graedel, T. E. and Graney, J. P. 1974. Atmospheric Aerosol Size Spectra: Rapid Concentration Fluctuations and Bimodality, J. Geophys. Res. 79, 5643-5645.

5. Hidy, G. M. 1975. Summary of the California Aerosol Characterization Experiment, J. Air Pollut. Control Assoc. 25, 1106.

6. Kadowski, S. 1976. Size Distribution of Atmospheric Total Aerosols, Sulfate, Ammonium, and Nitrate Particulates in the Nagoya Area, Atmos. Environ. 10, 39.

7. Lee, R. E. and Patterson, R. K. 1969. Size Determination of Atmospheric Phosphate, Nitrate, Chloride, and Ammonium Particulate in Several Urban Areas, Atmos. Environ. 3, 249-255.

8. Lundgren, D. A. 1973. "Mass Distribution of Large Atmospheric Particles," Ph.D. thesis, University of Minnesota, Minneapolis, Minnesota.

9. Mainwaring, S. J. and Harsha, S. 1975. Size distribution of Aerosols in Melbourne City Air, Atmos. Environ. 10, 57-60.

9a. Meszaros, A. and Vissy, K. 1974. Concentrations, Size Distribution, and Chemical Nature of Atmospheric Aerosol Particles in Remote Oceanic Areas, J. Aerosol Sci. 5, 101-109.

10. Nastrom, G. D. and Whitby, K. T. 1976. Influence of Air Parcel Trajectories on Aerosol Variability in the Los Angeles Basin: A Case Study for September 20, 1972, Atmos. Environ. 10, 195-197.

11. Patterson, R. K. and Wagman, J. 1977. Mass and Composition of an Urban Aerosol as a Function of Particle Size for Several Visibility Levels, J. Aerosol Sci. 8, 269-279.

12. Sverdrup, G. M. 1977. "Parametric Measurement of Submicron Atmospheric Aerosol Size Distributions," Ph.D. thesis, University of Minnesota, Minneapolis, Minnesota.

13. Sverdrup, G. M., Whitby, K. T., and Clark, W. E. 1975. Characterization of California Aerosols - II. Aerosol Size Distribution Measurements in the Mojave Desert, Atmos. Environ. 9, 482-494.

14. Wagman, J., Lee, R. E., and Axt, C. J. 1967. Influence of Some Atmospheric Variables on the Concentration and Particle Size Distribution of Sulfate in Urban Air, Atmos. Environ. 1, 479-489.

15. Whitby, K. T. 1973. "On the Multimodal Nature of Atmospheric Aerosol Size Distributions," presented at the VIII International Conference on Nucleation, Leningrad, U.S.S.R.

16. Whitby, K. T. 1974. "Modeling of Multimodal Aerosol Distribution," Proceedings of the 1974 Conference on Aerosols in Nature, Medicine, and Technology, GAF Bergweg 7, Vorderhinderlang 8973, West Germany.

17. Whitby, K. T. 1975. "Modeling of Atmospheric Aerosol Size Distributions," Report on Grant No. R-800971, submitted to Atmospheric Aerosol Research Sect., Div. of Chemistry and Physics, Air Pollution Control Office, Environmental Protection Agency.

18. Whitby, K. T. 1977. The Physical Characteristics of Sulfate Aerosols, Atmos. Environ. 12, 135-159.

19. Whitby, K. T., Cantrell, B. K., Husar, R. B., Gillani, N. V., Anderson, J. A., Blumenthal, D. L., and Wilson, W. E. 1977. Aerosol Formation in a Coal-Fired Power Plant Plume, submitted to Atmos. Environ.

20. Whitby, K. T., Cantrell, B. K., and Kittelson, D. B. 1977. Nuclei Formation Rates in a Coal-Fired Power Plant Plume, Atmos. Environ. 12, 313-321.

21. Whitby, K. T., Clark, W. E., Marple, V. A., Sverdrup, G. M., Sem, G. J., Willeke, K., Liu, B. Y. H., and Pui, D. Y. H. 1975. Characterization of California Aerosols - I. Size Distributions of Freeway Aerosol, Atmos. Environ. 9, 463-482.

22. Whitby, K. T., Husar, R. B., and Liu, B. Y. H. 1972. The Aerosol Size Distribution of Los Angeles Smog, J. Colloid Interface Sci. 39, 177-204.

23. Whitby, K. T., Kittelson, D. B., Cantrell, B. K., Barsic, N. J., and Dolan, D. F. 1977. Aerosol Size Distributions and Concentrations Measured during the General Motors Proving Ground Sulfate Study, to be published in *Proceedings of Symposium on Chemical Properties of Automotive Emissions from Catalyst-Equipped Cars and Biological Effects of Auto Emissions-Related Pollutants* (R. K. Stevens, ed.), Ann Arbor Sciences, Ann Arbor, Michigan.

24. Whitby, K. T., Liu, B. Y. H., Husar, R. B., and Barsic, N. J. 1972. The Minnesota Aerosol Analyzing System Used in the Los Angeles Smog Project, *J. Colloid Interface Sci.* *39*, 136-164.

25. Willeke, K. and Whitby, K. T. 1975. Atmospheric Aerosols: Size Distribution Interpretation, *J. Air Pollut. Control Assoc.* *25*, 529.

26. Willeke, K., Whitby, K. T., Clark, W. E., and Marple, V. A. 1974. Size Distributions of Denver Aerosols - A Comparison of Two Sites, *Atmos. Environ.* *8*, 609-633.

Appendix A: Size Distribution of Freeway Aerosol

Kenneth T. Whitby, William E. Clark, Virgil A. Marple, George M. Sverdrup, Gilmore J. Sem, Klaus Willeke, Benjamin Y. H. Liu, and David Y. H. Pui

Particle Technology Laboratory
Mechanical Engineering Department
University of Minnesota
Minneapolis, Minnesota

Abstract

Aerosol along the Harbor Freeway in Los Angeles, California, was sampled and measured with the Air Resources Board Mobile Air Pollution Laboratory. Simultaneously with the taking of filter and impactor samples for chemical analysis, the aerosol particle size distribution was measured with four continuous instruments over the particle size range from approximately 0.00340 μm.

The instruments used and their general arrangement within the mobile laboratory are shown in Figure A1. Data were gathered and presented every 10 min by the on-board PDP/8E minicomputer.

Site Description

The site next to the Harbor Freeway in Los Angeles was selected to allow measurement of the direct contribution of a large freeway to aerosol concentration.

The mobile laboratory was parked at the corner of 33rd and Hope Streets next to the freeway. This location was approximately 1.6 km south of the San Bernardino Freeway and 34 km south of downtown Los Angeles. At this location the eight-lane Harbor Freeway runs parallel to Hope Street in a north-south direction 25° from north. Hope Street had very little traffic, being used only by student drivers and an occasional truck. Between the freeway and mobile laboratory were 10 m of paved parking lot, Hope Street (about 10 m wide), and a 10-m wide embankment. Vegetation on the embankment extended about 5 m above the freeway. The freeway was about 6 m above the laboratory location so that the freeway surface was approximately in line with the upper end of the mobile laboratory sampling pipes.

From comparisons of measurements when the wind was directly from the freeway with measurements when the wind was blowing

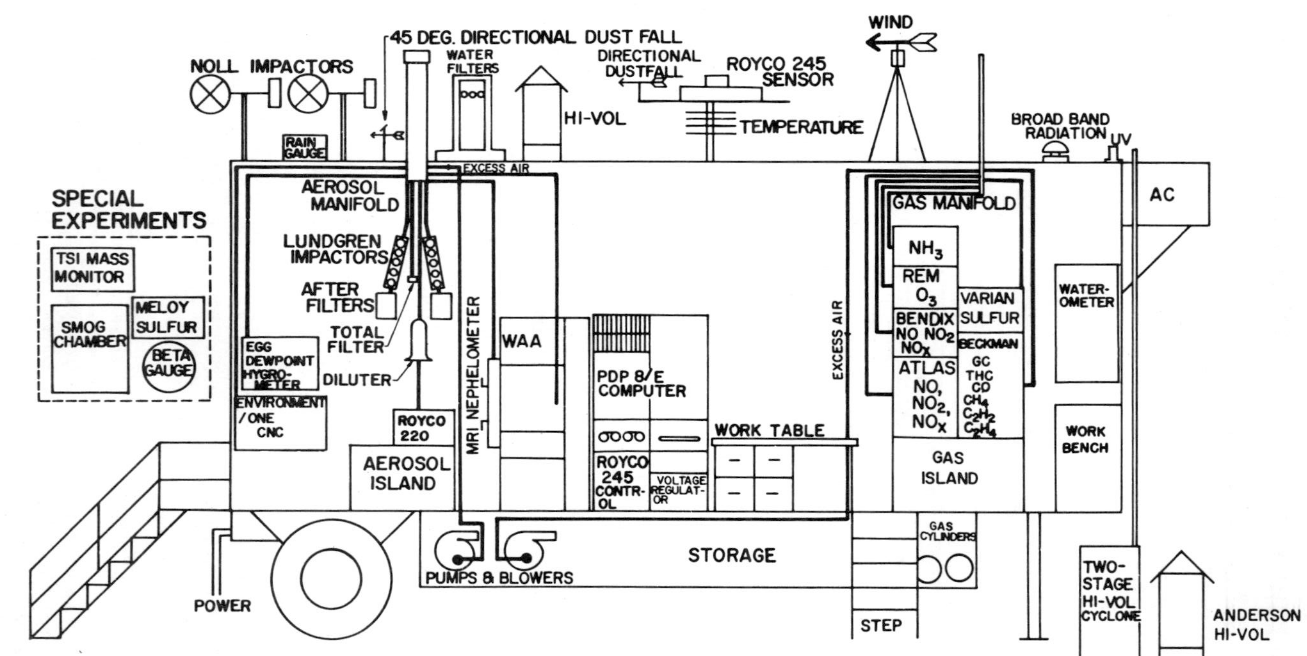

FIGURE A1. Schematic of the mobile laboratory, showing the general arrangement of gas analysis, aerosol sampling, and data acquisition equipment.

toward the freeway, it was possible to calculate by difference the direct contribution of the freeway traffic to the aerosol mixture.

From Figure A2 it can be seen that the freeway contributes most to the nuclei mode. For Run 54 minus Run 55, 17.1 μm^3 cm^{-3} of aerosol with a volume arithmetic mean size of 0.049 μm and a σ_g = 1.9 was produced by the traffic on the freeway. The fresh aerosol as measured at a distance of 30 m from the freeway is mostly smaller than 0.15 μm. The nuclei count was between 2 and 3 million particles cm^{-3}, and the surface area was about 3000 μm^2 cm^{-3}.

Comparisons of the size distributions measured alongside the freeway with size distributions measured in the dilution tunnel at Battelle-Columbus suggest that the dilutions are comparable.

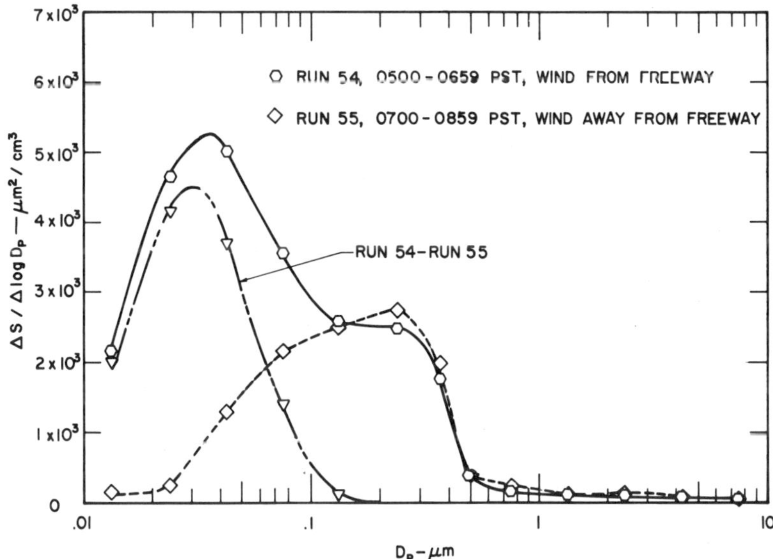

FIGURE A2. Plot of the surface size distributions $\Delta S/\Delta \log D_p$, for Run 54 when the wind was from the freeway and Run 55 when the wind was blowing toward the freeway, and the difference distribution, Run 54 minus Run 55, for D_p less than 0.15 μm.

Appendix B: Aerosol Size Distribution Measurements in the Mojave Desert

George M. Sverdrup, Kenneth T. Whitby, and William E. Clark

Particle Technology Laboratory
Mechanical Engineering Department
University of Minnesota
Minneapolis, Minnesota

Aerosols were sampled at the Goldstone tracking station in an attempt to characterize clean desert background air.

The mobile air pollution laboratory was located at the Pioneer Station of NASA's Goldstone Tracking Station. Pioneer Station is located approximately 200 km northeast of downtown Los Angeles and 58 km north of Barstow, California. Fort Irwin is located 19 km southeast of Pioneer Station. Five other Goldstone stations are located within 16 km of Pioneer to the south. One station is located 4.5 km to the northwest. All stations are connected by a lightly traveled paved road.

Pioneer Station is located in a basin approximately 3 x 1 km surrounded by 250-m hills to the northwest and 125-m hills to the east and south. The mobile laboratory was parked on a paved pad about 125 m west of the Pioneer buildings and the paved road leading to the basin. Directly north of the trailer was a dry lake approximately 0.4 km wide and 0.8 km long.

Results

Results fell into three groups, determined primarily by the origin of the air mass.

1. Clean Desert Aerosol. The period from 13:50 until 16:50 hr on October 31, 1972, was characterized by the lowest aerosol concentrations ever measured with the Minnesota Aerosol Analyzing System (MAAS). This episode occurred after a few days of unusual weather conditions. It rained the week before the laboratory arrived at Goldstone. While the laboratory was at Goldstone a portion of the dry lake directly north of the mobile laboratory was covered with water. The day the mobile laboratory arrived the wind was from the north with gusts up to 80 km/hr. By the time the laboratory was ready for operation at 13:30 hr on October 31, 1972, the wind was still from the north at 25.2 km/hr, blowing across the lake bed.

The total volume concentration of the aerosol was 1.85 $\mu m^3/cm^{-3}$ with 55% of that below 1 μm. Even though the average wind speed was 16.2 km/hr, the volume of particles greater than 1 μm was only 0.82 $\mu m^3/cm^{-3}$. The nuclei concentrations were below the lower sensitivity limit of the nuclei counter, for example, 500 cm^{-3}.

2. Average Background. The typical background desert condition measured during the sampling period was a total volume on the order of 8 to 13 $\mu m^3/cm^{-3}$, with the amount below 1 μm about 3 to 4 $\mu m^3/cm^{-3}$. The nuclei concentration was a few thousand. In contrast with other background data obtained at Fort Collins, Colorado, the volume of aerosol greater than 1 μm was much less, being only 60 to 80 % of the total volume. The occurrence of rain just before the sampling period makes it impossible to claim that typical background conditions over a long period of time were measured. Although large amounts of windblown dust had been expected at Goldstone, no evidence of such dust was recorded. The maximum volume greater than 1 μm was about 12 $\mu m^3/cm^{-3}$.

3. Incursion of Aged Aerosol. From 10:00 hr on November 3, 1972, until midnight the wind blew steadily from the south at about 15 km/hr. The wind speed increased at about 05:00 hr on November 4, 1972, and the scattering coefficient and submicron aerosol volume increased dramatically.

Figure B1 shows the buildup of the aged aerosol. In a period of 2 hr the total volume concentration and the light-scattering coefficient increased by a factor of 5. The 6-fold increase in volume between 0.1 and 1.0 μm is evidence of well-aged aerosol. There was essentially no increase in volume below 0.1 μm. The volume in the 1.0 to 10-μm range increased by a factor of 3.

The submicron aerosol volume observed at the peak of the episode, about 30 $\mu m^3/cm^{-3}$, was comparable to the average submicron volume observed in Pasadena during the 1969 study. This shows that the urban plume from Los Angeles can be observed at least 200 km away.

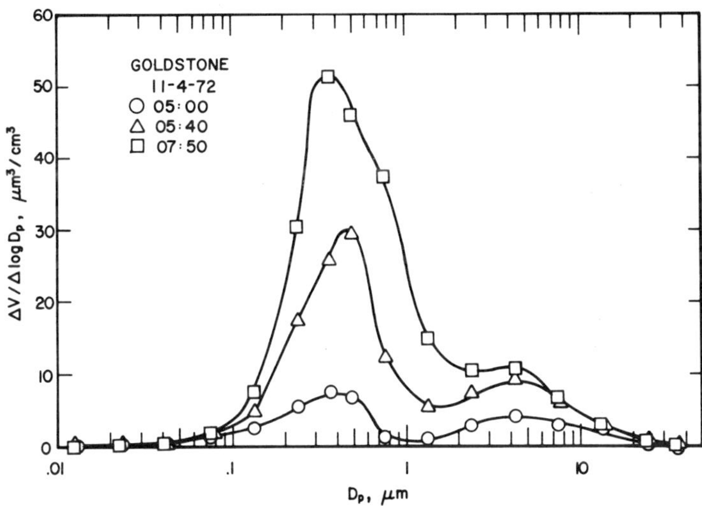

FIGURE B1. Incursion of aged aerosol from the South Coast Air Basin.

The Effect of Changing Relative Humidity on Aerosol Size Distribution Measurements

GEORGE M. SVERDRUP* AND KENNETH T. WHITBY
Particle Technology Laboratory
Mechanical Engineering Department
University of Minnesota
Minneapolis, Minnesota

Abstract

A mismatch between two aerosol size distribution segments obtained with optical particle counters was analyzed. One optical counter was located inside a mobile laboratory and covered the particle size range 0.42 to 5.6-μm diameter. The other optical counter, located on the roof of the laboratory, measured particles in the size range 5.6 to 38 μm. The mismatch at 5.6 μm was shown to be caused by the evaporation of liquid water from the particles as they traveled in the sampling line. An estimate of the mass fraction of liquid water in the particles as a function of relative humidity was found to be in reasonable agreement with microwave, nephelometer, and filter measurements.

I. INTRODUCTION

The amount of liquid water present in atmospheric aerosols and its subsequent effects have received considerable attention

*Present address: Battelle Memorial Institute, 505 King Avenue, Columbus, Ohio 43201.

during the past decade. Ho et al. (6) have shown that a significant fraction of the atmospheric aerosol is liquid water at relative humidities below saturation. Changing relative humidities result in changing aerosol size distributions as the particles equilibrate to their environment [Winkler and Junge (12); Winkler (11)]. This in turn leads to changes in aerosol light scattering [Covert et al. (3)] and hence visibility [Kasten (7); Hanel (4)].

This paper demonstrates the importance of changing relative humidity in aerosol sampling lines. From the aerosol size distribution measurements made during the California Aerosol Characterization Experiment (ACHEX), it is shown that the difference in relative humidity between the air outside the mobile laboratory and that inside the sampling tubes resulted in a change in the aerosol size distribution under certain circumstances. The change in the size distribution is quantified and compared with other measurements of the liquid mass fraction of the aerosol.

A. Sampling Aerosols with the Mobile Laboratory

Figure 1 shows three size distributions based on aerosol volume which were measured at the Harbor Freeway, using the

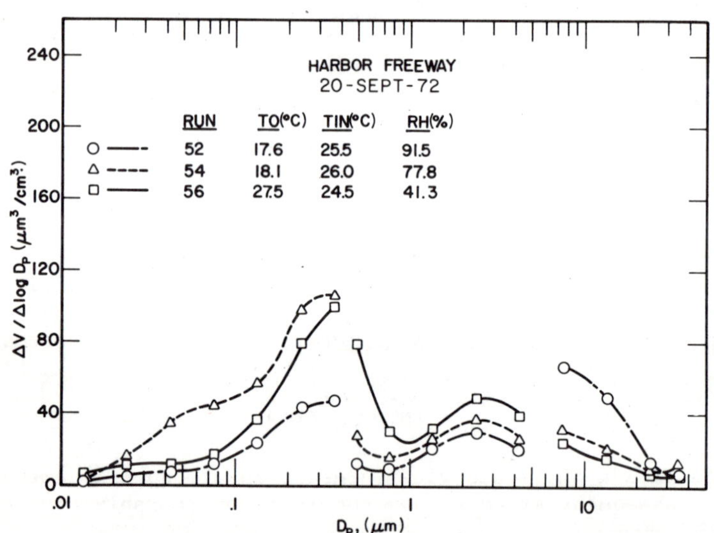

FIGURE 1. Effect of changing relative humidity on aerosol volume-size distributions from the Harbor Freeway.

mobile laboratory described by Whitby et al. (10). The distributions have been divided into three segments corresponding to data taken with the Whitby aerosol analyzer (WAA) in the particle size range between 0.01 and 0.422 μm, the Royco Model 220 optical single-particle counter (OPC) in the range 0.422 to 5.62 μm, and the Royco Model 245 OPC in the range 5.62 to 38 μm. For Run 56 the outside ambient temperature (TO) was greater than the temperature inside the mobile laboratory (TIN), and the outside relative humidity (RH) was relatively low (41.3%). It can be seen that the three segments of the distribution match very well. However, for Run 54 with TO less than TIN and a relative humidity of 77.8%, the segments of the distribution corresponding to the two OPCs do not match. At a higher relative humidity (Run 52), the mismatch is greater.

Examination of the paths that the aerosol followed to reach the instrument sensors suggests that the behavior of the three size distribution segments shown in Figure 1 can be explained in terms of different thermodynamic environments. The Royco 245 sensor was mounted on a stand 1.1 m above the roof of the mobile laboratory [Whitby et al. (10)]. The outside temperature sensor was located just below the Royco 245 sensor. Therefore, the segment of the distribution obtained with the Royco 245 was measured at the ambient temperature. These data were not affected by a difference between TO and TIN.

The Royco 220 sensor was located inside the mobile laboratory. The aerosol sample stream passed through a 0.25-in (0.63-cm) diameter by 50-in (1.27-m) long aluminum tube from the aerosol manifold near the ceiling to the Royco 220 sensor, and at most locations through a diluter.

The effect of the diluter on the Royco 220 performance was investigated at Point Arguello. From Figure 2 it can be seen that the mismatch was not due to the presence of the diluter.

For a flow rate of 0.1 cfm (47.2 cm^3/sec), the Reynolds number based on tube diameter was 122 and the transit time was 4.3 sec. For these conditions it is estimated that the bulk temperature of the sample stream increased from its value of TO to within about a degree of TIN by the time the air parcel reached the Royco 220 inlet.

The Royco 220 was modified with a sheath air inlet which removes about 85% of the total flow near the top of the inlet. This air passes through a filter and is returned to the bottom of the inlet to act as a clean air sheath around the sample stream. The sheath air line runs close to the Royco 220 light source. Thus the temperature of the sheath air can be expected to be higher when it returns to the inlet than when it first entered the instrument. To investigate this effect, a Royco 220 OPC similar to the one used in the ACHEX was operated with a

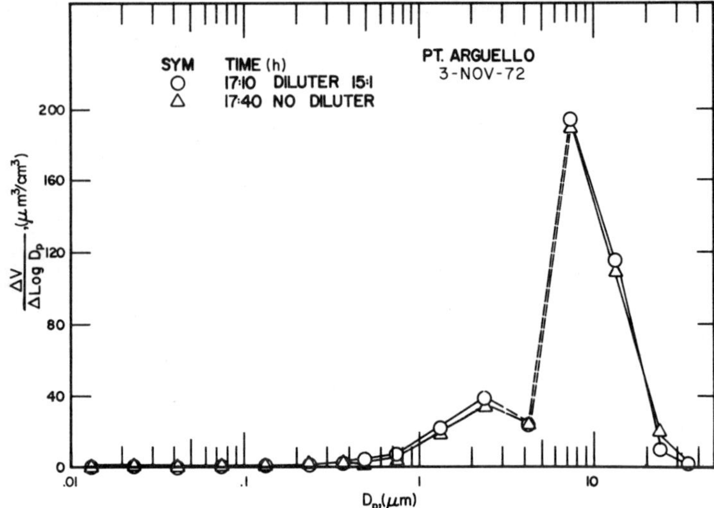

FIGURE 2. Effect of the diluter in the Royco 220 OPC sample line on aerosol size distribution.

thermocouple placed in the sheath air line just before the sheath air returned to the inlet and with a thermocouple placed in the Royco 220 box near the inlet. The returning sheath air temperature was found to be about 1.5 to 2°C higher than room temperature. Since the sheath air flow rate was 41.2 cm^3/sec, and the sample flow rate was 5.99 cm^3/sec, some heat was transferred from the sheath airstream to the sample airstream. Based on the above considerations, it is reasonable to assume that the aerosol stream was at temperature TIN when it passed through the viewing volume of the Royco 220 OPC.

The mismatch between the Royco 220 and Royco 245 is complicated by the fact that the refractive index of aerosol particles changes as their liquid water content changes. The responses of the two OPCs to changes in particle refractive index are quite different, since the Royco 245 uses a forward light-scattering system and the Royco 220 uses a 90° system. To determine the magnitude of this effect, the results of Berglund's (1) work were utilized. Figure 3 shows the results of a change in refractive index for Run 49 at the Harbor Freeway. The measured data (circles) were taken assuming a particle refractive index, m, of 1.49 based upon the dioctyl phthalate calibration aerosol. The data for m = 1.40 are based on Berglund's experimental work.

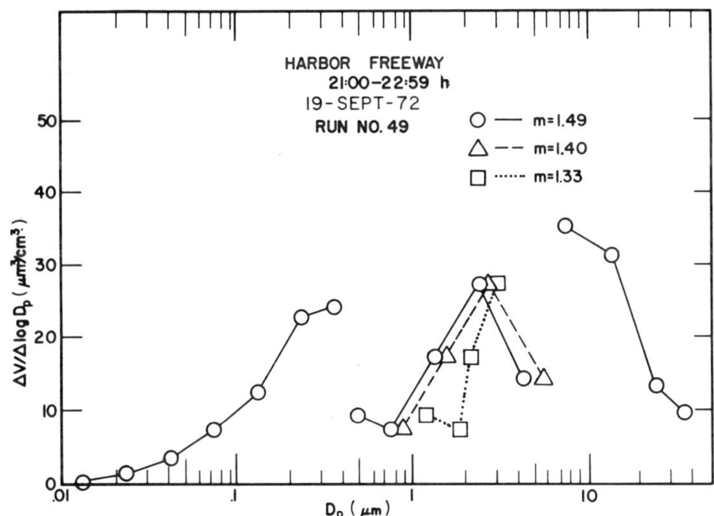

FIGURE 3. Effect of particle refractive index on optical particle counter aerosol data.

The data for m = 1.33 are presented as a limiting case and were calculated from Berglund's thesis using Quenzel's (9) theory. A change in refractive index did not affect the Royco 245 data. The data show that a change in refractive index alone is not enough to account for the mismatch.

The segment of the size distribution from 0.01 to 0.422 μm was obtained with the WAA. The aerosol traveled through a rubber hose between the aerosol manifold and the WAA. The aerosol was charged in the diffusion charger with a flow rate of 9.1 liters/min and a dry compressed air flow rate of 5.1 liters/min. Therefore, in the aerosol charging section much of the liquid in the aerosol was probably evaporated. In the mobility-analyzing tube the aerosol was introduced as a thin sheath around a clean air flow. The ratio of clean air to aerosol flow was 14:1. The temperature of the mobility tube was monitored and was usually within 1°C of TIN. The history of evaporation and recondensation of water on the aerosol is unknown; however, it is believed that the representative relative humidity at which the WAA measured the aerosol was probably close to that calculated using the inside laboratory temperature.

From Figure 1 it is evident that, if there was a mismatch between the WAA and the Royco 220 data, it is not apparent. To detect a mismatch between these two segments of data it is necessary to examine data from an aerosol with a large liquid water content, at a high ambient relative humidity, and low outside temperature. Figure 4 shows such a surface and volume distribution obtained at Point Arguello, California. The surface size distribution in Figure 4a clearly shows a mismatch between the WAA and Royco 220 data at 0.42 μm and between the Royco 220 and Royco 245 data at 5.6 μm. The volume distribution in Figure 4b shows only data from the WAA and the Royco 220. The Royco 245 are off scale.

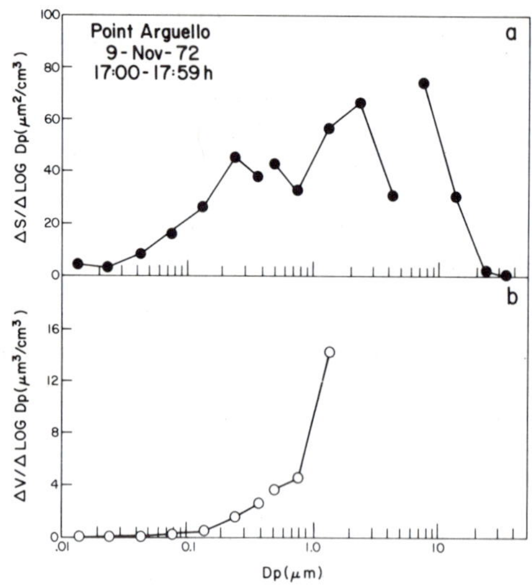

FIGURE 4. Aerosol size distributions from Point Arguello, showing the effects of extreme changes in liquid water content.

The magnitude of the mismatch between the WAA and Royco 220 data is much less than that between the two OPCs. The thermodynamic environments experienced by the aerosol en route to and in the WAA and Royco 220 were much more similar than the environments en route to the two OPCs.

B. Analysis and Discussion

An analysis of the data was made in order to quantify the mismatch of the three size distribution segments. No attempt was made to quantify the mismatch between the WAA data and the Royco 220 data. The Royco 220 data at 4.2 μm can be compared to the Royco 245 data at 7.5 μm, but there is no such standard with which to compare the WAA data.

A number of methods for quantifying the mismatch between the data from the two OPCs were tried. The method presented here, using the volume-size distribution, is simple enough that it can be applied to large quantities of data. Figure 5 shows a typical volume distribution and the method for determining the mismatch parameter, W. The Royco 245 data were extrapolated to find point A at 4.2 μm because the size distributions with no mismatch usually exhibited a volume mode at this particle size. The difference between points A and B is then a measure of the amount of water evaporated from the particles at a size of 4.2 μm, neglecting a change in particle radius with evaporation. This difference was then normalized by division by the value of point A. The value of W thus lies between 0 and 1 inclusive.

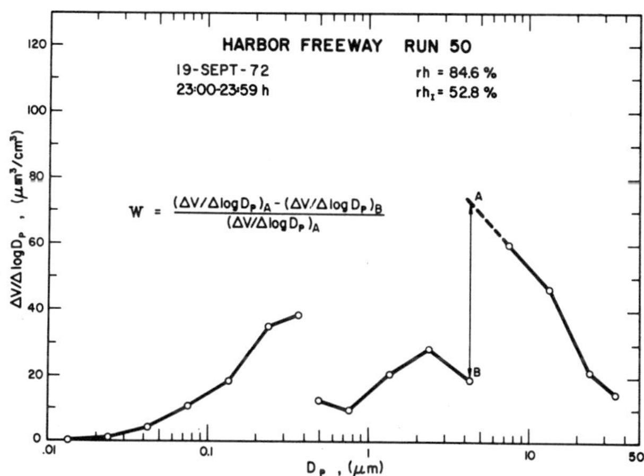

FIGURE 5. Method for quantifying the effects of changing relative humidity on aerosol size distributions.

A linear regression analysis was made of the variation of W with the difference (ΔRH) between the outside relative humidity and a relative humidity based upon the outside dew point temperature and the inside temperature, TIN. The nature of the data did not justify using a higher degree polynomial in the analysis. Generally a higher degree of correlation was obtained for data from a specific location, indicating more chemically uniform aerosols. Figure 6 shows the results of the calculations for all locations. The regression resulted in the equation

$$W = 0.213 + 0.0108(\Delta RH) \qquad (1)$$

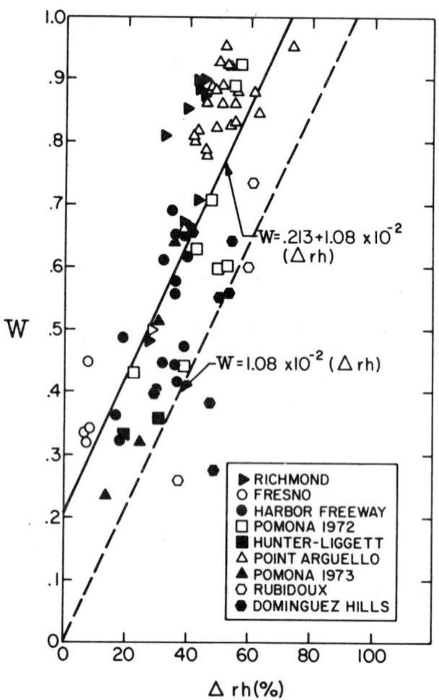

FIGURE 6. Mismatch parameter for the data from the two optical particle counters versus changing relative humidity for nine sites.

with a correlation coefficient of .72. The data obtained at Goldstone in the Mojave Desert gave negative values of W for all cases, indicating essentially no liquid water content in the aerosols, in agreement with the results of microwave measurements [Ho et al. (6)] and special filter measurements [Meyer et al. (8)].

It is of interest to compare the parameter W with measurements of the mass fraction of liquid water in the aerosol. For no change in relative humidity as the aerosol was being sampled, W should be equal to zero. Therefore, the regression line shown in Figure 6 was shifted to intersect the origin as shown by the dashed line. If the abscissa in Figure 6 is changed from ΔRH to an absolute relative humidity to yield W', W' then becomes a measure of the liquid water content. The value of W' should always be less than or equal to the true liquid water content of the aerosol, since all of the water might not have been evaporated from the particles.

Figure 7 is adapted from the work of Ho et al. (6). The data points shown are estimates of the mass fraction of liquid water, M_w/M_t, based upon microwave measurements of the liquid water and β-attentuation measurements of the total mass concentration. The dashed curves are estimates of the water fraction based upon light-scattering measurements conducted by Charlson (2). These data show that below a relative humidity of about 35% there is essentially no unbound water in the aerosol.

Therefore, the dashed curve of Figure 6 was changed to the following equation:

$$W' = 0.0108(RH - 35) \tag{2}$$

where W' is a measure of the volume fraction of liquid water in the aerosol.

The mass fraction of liquid water can be expressed in terms of particle density and volume as follows:

$$\frac{M_w}{M_t} = \frac{V_w}{V_w + \rho_d V_d} = \frac{1}{1 + \rho_d V_d/V_w} \tag{3}$$

where V_w is the volume of water, V_d is the dry aerosol volume, and ρ_d is the dry aerosol density. Substituting $W' = V_w/(V_d + V_w)$ into Eq. 3 gives the following equation for M_w/M_t:

$$\frac{M_w}{M_t} = \frac{1}{(\rho_d/W') + 1 - \rho_d} \tag{4}$$

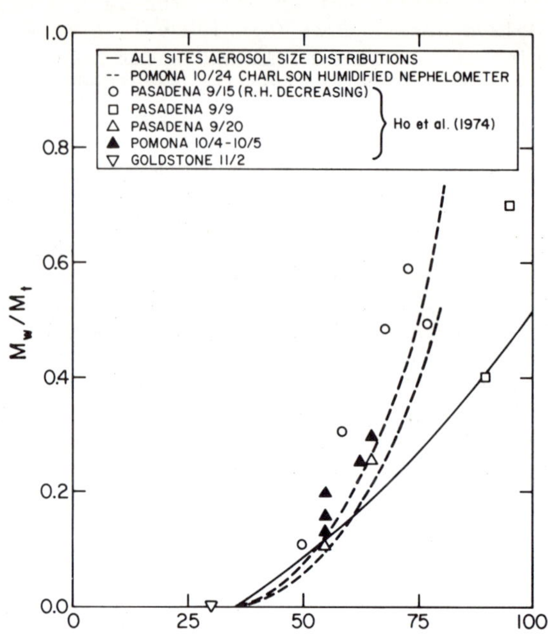

FIGURE 7. Estimates of the liquid water mass fraction of aerosols as a function of relative humidity based upon three experimental techniques.

Sverdrup and Whitby in the final ACHEX report [Hidy et al. (5)] computed aerosol density as a function of relative humidity for six sites using mass data from the total filter sampler and volume concentrations from the aerosol measurement system. Particle densities were computed for periods when the outside temperature was greater than the temperature inside the mobile laboratory. On the basis of their work, a value of 2.3 g/cm^3 was assigned to ρ_d. The resulting curve for M_w/M_t is shown as a solid line in Figure 7. This curve represents the minimum mass fraction of water in the aerosol because all of the water might not have been evaporated from the particles before their measurement.

The results shown in Figure 7 indicate that the observed mismatch in the aerosol size distribution was due to the evaporation of liquid water from the particles in the sampling lines and instruments. Therefore, the thermodynamic environment of the particles must be considered when sampling aerosols.

Equation 4, determined from the size distribution mismatch between two OPCs, cannot be considered to be equivalent to the other measurement techniques presented in Figure 7. The parameter W' in Eq. 4 was determined from a change in relative humidity when the latter was decreasing. Aerosol particles have been shown to be supersaturated under such conditions [Winkler and Junge (12)], exhibiting a hysteresis effect compared to increasing humidity. Furthermore, the relative humidity differences have been arbitrarily converted to absolute relative humidity (Eq. 2) without regard to the initial and final relative humidities in the sample lines.

II. CONCLUSIONS

A mismatch between aerosol size distribution segments determined with optical counters has been evaluated. The mismatch was due to evaporation of liquid water from the aerosol particles in the sample lines and instruments. The change in relative humidity experienced by the particles ranged from about 5 to 70%. An estimate of the liquid mass fraction of free water in aerosol was made and found to be consistent with microwave, nephelometer, and filter measurements. The liquid nature of atmospheric aerosols demonstrates that their thermodynamic environment must be considered while the particles are being sampled.

REFERENCES

1. Berglund, R. N. 1972. "Basic Aerosol Standards and Optical Measurement of Aerosol Particles," Ph.D. thesis, Dept. of Mechanical Engineering, University of Minnesota, Minneapolis.

2. Charlson, R. J. 1975. Private communication.

3. Covert, D. S., Charlson, R. J., and Ahlquist, N. C. 1972. A Study of the Relationship of Chemical Composition and Humidity to Light Scattering by Aerosols, J. Appl. Meteorol. 11, 968-976.

4. Hänel, G. 1971. New Results Concerning the Dependence of Visibility on Relative Humidity and Their Significance in a Model for Visibility Forecast, Beitr. Phys. Atmos. 44, 137-167.

5. Hidy, G. M. 1974. "Characterization of Aerosols in California," Vol. IV: "Analysis and Interpretation of Data," Final Report to California Air Resources Board, Contract No. 358, pp. 3-21 to 3-23.

6. Ho, W., Hidy, G. M., and Govan, R. M. 1974. Microwave Measurements of the Liquid Water Content of Atmospheric Aerosols, J. Appl. Meteorol. 13, 871-879.

7. Kasten, R. 1969. Visibility Forecast in the Phase of Precondensation, Tellus XXI, 631-635.

8. Meyer, R. A., Hidy, G. M., and Davis, J. H. 1973. Determination of Water and Volatile Organics in Filter Collected Aerosols, Environ. Lett. 4, 9-20.

9. Quenzel, H. 1969. Influence of Refractive Index on the Accuracy of Size Determination of Aerosol Particles with Light-Scattering Aerosol Counters, Appl. Opt. 8, 165-169.

10. Whitby, K. T., Clark, W. E., Marple, V. A., Sverdrup, G. M., Sem, G. J., Willeke, K., Liu, B. Y. H., and Pui, D. Y. H. 1975. Characterization of California Aerosols - I. Size Distributions of Freeway Aerosol, Atmos. Environ. 9, 463-482.

11. Winkler, P. 1973. The Growth of Atmospheric Aerosol Particles as a Function of the Relative Humidity - II. An Improved Concept of Mixed Nuclei, Aerosol Sci. 4, 373-387.

12. Winkler, P., and Junge, C. 1972. The Growth of Atmospheric Aerosol Particles as a Function of the Relative Humidity - I. Method and Measurements at Different Locations, J. Rech. Atmos. 6, 617-638.

The Variation of the Aerosol Volume to Light-Scattering Coefficient

GEORGE M. SVERDRUP* AND KENNETH T. WHITBY
Particle Technology Laboratory
Mechanical Engineering Department
University of Minnesota
Minneapolis, Minnesota

Abstract

Values of the ratios of submicron aerosol volume to aerosol light-scattering coefficient, K_p, for aerosols sampled in California in 1972 and 1973 are reported. The data show a variation in K_p of from 5 to 80. The variation of K_p is shown to be primarily a function of the aerosol size distribution. When K_p is plotted against the total aerosol number concentration, the data fall into groups that can be identified with general location.

---

*Present address: Battelle Memorial Institute, 505 King Avenue, Columbus, Ohio 43201.

I. INTRODUCTION

Current California ambient air quality standards are based upon total aerosol mass concentration and the prevailing visibility. For this reason, studies have been conducted to determine the relationship between these two parameters [Noll et al. (19); Air Resources Board (1)]. Determination of these quantities is not a fully automated process, but requires personnel to handle filters and to make visibility sightings. Integrating nephelometers have been proposed as instruments whose output of the light-scattering coefficient could be used in place of trained observers making visibility sightings. Middleton (18) and Charlson et al. (5) have discussed the relationship between visibility and the light-scattering coefficient.

A number of studies were reported in the 1960s and early 1970s that examined the relationship between the aerosol mass concentration and the total light-scattering coefficient, b_s, or the aerosol light-scattering coefficient, b_{sp} [Charlson et al. (3-5); Ettinger and Royer (10)]. The relationship between b_s and b_{sp} is shown in Eq. 1:

$$b_s = b_{Rg} + b_{sp} \tag{1}$$

where b_{Rg} is the Rayleigh scattering due to air molecules. In urban environments b_{sp} is normally much greater than b_{Rg}; in nonurban environments b_{Rg} and b_{sp} can be of the same order of magnitude. In a review paper Charlson (2) showed that for the self-preserving size distribution the ratio of the aerosol mass concentration to b_{sp} could be approximately a constant value. The experimental data up to that time gave a ratio of

$$\frac{MT\ (\mu g/m^3)}{b_{sp}\ (m^{-1})} = 4.5^{+4.5}_{-2.2} \times 10^5\ \mu g/m^2 \tag{2}$$

where MT is the total aerosol mass concentration. The superscripts are the 90% confidence limits. Charlson (2) and Griggs (12) suggested that this ratio was constant enough for use in air pollution work.

To better understand the dynamics of atmospheric aerosols, automated aerosol size distribution measurement systems were used to monitor the aerosols *in situ* [Whitby (28); Whitby (24)]. As a result of the 1969 Los Angeles Smog Experiment [Hidy (15)] the concept of the self-preserving size distribution over the entire particle size range was replaced by the multimodal concept [Whitby (23); Willeke and Whitby (29)]. When this concept is used, particles with diameters, D_p, less than about 2 μm are labeled fine particles. The fine-particle size distribution can consist of one or two modes. The first mode has been called the transient nuclei mode and is mainly the result of combustion processes. Its volume mean size is typically about 0.02 μm, which is below the light-scattering range where particles cause most of the visibility reduction. The accumulation mode has a volume mean size of from 0.2 to 0.4 μm. This mode has been shown to account for most of the visibility reduction and aerosol light scattering. Particles greater than 2 μm in diameter have been labeled coarse particles. For most atmospheric aerosols the coarse-particle mode is produced from ocean spray or suspended dust.

As a result of the California air quality standards and the increased knowledge of aerosol size distributions, the Tri-City Sampling Project was conducted from January 1970 through March 1971. This long-term project was conducted by the California Air Resources Board with the assistance of the Air and Industrial Hygiene Laboratory of the California State Department of Public Health. The study showed that in the three cities of Los Angeles, Sacramento, and Oakland b_s and prevailing visibility correlated very well. The total aerosol mass concentration, MT, was found to be a very poor measure of light scattering. Estimates of b_s from MT could be expected to vary by a factor of 2. The use of the refined mass concentration ($D_p < 3$ μm) improved the estimate of b_s only marginally. Using the refined mass was not statistically different from using MT.

Data obtained during the California Aerosol Characterization Experiment (ACHEX) [Hidy et al (17)] were used to correlate b_s with the total aerosol volume concentration, VT, the submicron volume concentration, V3- ($D_p < 1$ μm), and the volume concentration in the accumulation range, V3 ($0.1 < D_p < 1$ μm) ["Characterization of Aerosols in California," ACHEX), Final Report, 1974]. The results of the correlations are shown in Table 1 from which it is seen that the optical subrange V3 correlated the best with b_s, as expected. The high correlation coefficients which were obtained tend to indicate that b_s might be used as an indicator of fine-particle aerosol volume concentration; however, such simple correlations do not show the wide variations in the ratio of aerosol volume to light-scattering coefficient. These two studies conducted in California have suggested that the ratio of aerosol mass or volume to scattering coefficient cannot be used as a constant for air pollution purposes.

The purpose of this paper is to examine the variation in the ratio $K_p = V3-/b_{sp}$ obtained at various sites during the ACHEX.

TABLE 1. Correlation of Aerosol Volume Concentration and Total Light-Scattering Data from ACHEX Measurements

Variables	Correlation coefficient	Standard error of estimate
VT vs. b_s	.865	29.0
V3- vs. b_s	.937	10.8
V3 vs. b_s	.948	9.83

Other attempts have been made to explain such variations. Charlson et al. (3) theoretically studied the problem using a set of power law size distributions and an assumed particle refractive index and wavelength of light. They concluded that the variation of size distribution could explain (emphasis by Charlson et al.) the scatter in their data. Pilat and Ensor (20), using a log-normal distribution, theoretically studied the variation of the ratio VT/b_{sp} by varying the mean size and geometric standard deviation. Using the curves of Pilat and Ensor, and modeling the submicron volume mode as a single log-normal distribution, Greacen et al. (11) attributed the variations in the ratio V3-/b_{sp} to changes in the aerosol size distribution in the ACHEX Final Report. The variation of the ratio K = V3-/b_s with the total number concentration was also studied in the ACHEX Final Report (Vol. IV, Appendix C) in preliminary work by Sverdrup and Whitby.

This paper is a final version of the earlier work on the correlation of K with the number concentration, CNC. The variation of K_p is discussed here in terms of the integral size distribution parameters, CNC, V3-, and b_{sp}. Integral parameters were chosen as the basis of the discussion because they can be measured and because their significance is easily understood.

II. VARIATION OF PARAMETER K_p - EXPERIMENTAL

Figure 1 is a plot of the total aerosol number concentration, CNC, on the ordinate and the parameter, K_p, on the abscissa. The CNC values were obtained with an Environment/One Model Rich 100 condensation nuclei counter. Values of the submicron aerosol volume, V3-, were obtained using the automatic aerosol sizing instruments. The b_{sp} values were obtained with an MRI Model 1550

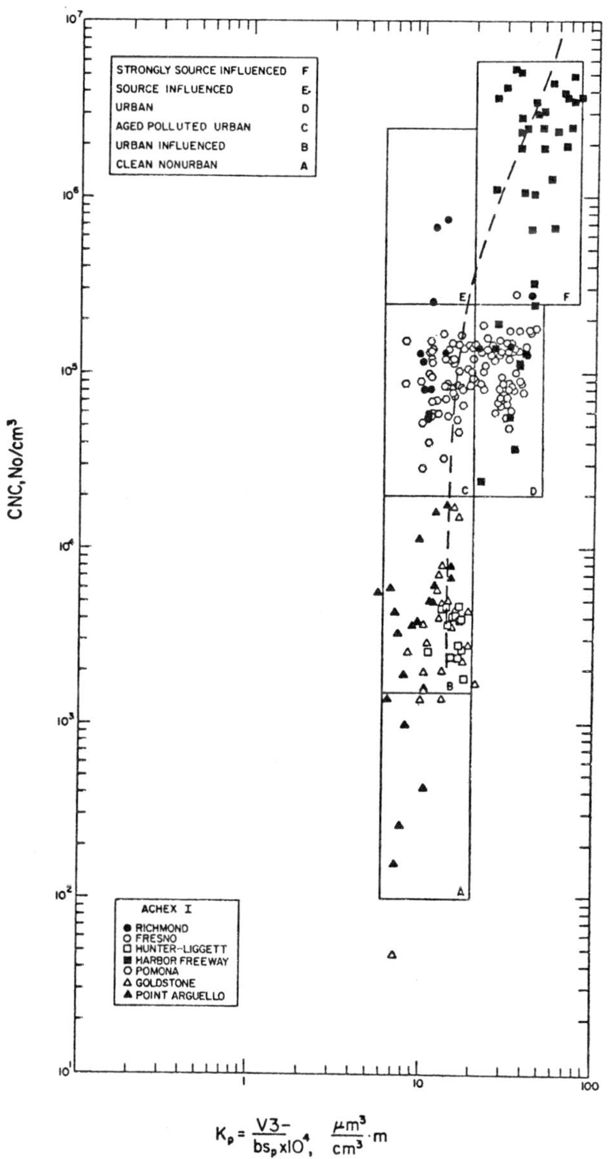

FIGURE 1. Variation of the aerosol submicron volume to aerosol light-scattering coefficient with the total aerosol number concentration from ACHEX I.

integrating nephelometer. The calibration, performance, and data-gathering procedures have been reported by Whitby (24) and in the ACHEX Final Report (Vol. II). Each datum point is an average of 10-min data over a 2-hr interval. The minimum number of 10-min runs allowed in each average for this paper was seven. Averaging the 10-min data reduced the quantity of data reported, allowing the 2-hr data to be discussed not merely on a statistical basis, but also in terms of the location at which the data were taken and the local atmospheric conditions.

Figure 1 contains data obtained during the 1972 sampling program (ACHEX I). Data from 1973 (ACHEX II) are plotted in Figure 2, using the same scales for the purpose of clarity. The CNC values have a range of five decades. The lowest concentration was obtained in the Mojave Desert [Sverdrup et al. (22)]; the highest concentration was obtained at the Harbor Freeway [Whitby et al. (26)]. Values of K_p range from 5.5 to 81.

To estimate how well the present values of K_p compare to Charlson's value of MT/b_{sp} in Eq. 1, the following steps were performed. The average density of 57 2-hr averages from ACHEX I was computed to be 1.5 g/cm^3 in the ACHEX Final Report. Converting Charlson's units to the present units and dividing MT by the particle density to obtain VT*, we obtain

$$\frac{VT^*}{b_{sp}} = 30^{+30}_{-15} \times 10^4 \ \mu m^3 \ m/cm^3 \qquad (3)$$

where the asterisk indicates an approximate value. The average submicron-volume-to-total-volume ratio for the 2734 10-min data points represented in Figure 1 is 0.42. Multiplying both sides of Eq. 3 by 0.42, we obtain

$$\frac{V3\text{-}^*}{b_{sp}} = 13^{+13}_{-7} \times 10^4 \ \mu m^3 \ m/cm^3 \qquad (4)$$

Values of $V3\text{-}^*/b_{sp}$ range from 6 to 26, indicating that the ACHEX data and the data reported by Charlson are consistent. Charlson et al. (6) have recently reported similar values for aerosols measured at Mauna Loa Observatory in Hawaii.

There appears to be a minimum in the experimental data for K_p of about 6. Charlson et al. (3), using power law distributions, and Pilat and Ensor (20), using log-normal models, showed theoretically that a minimum should exist. They reported values of about 22 and 10 (for a refractive index of 1.50 - 0.0 i), respectively.

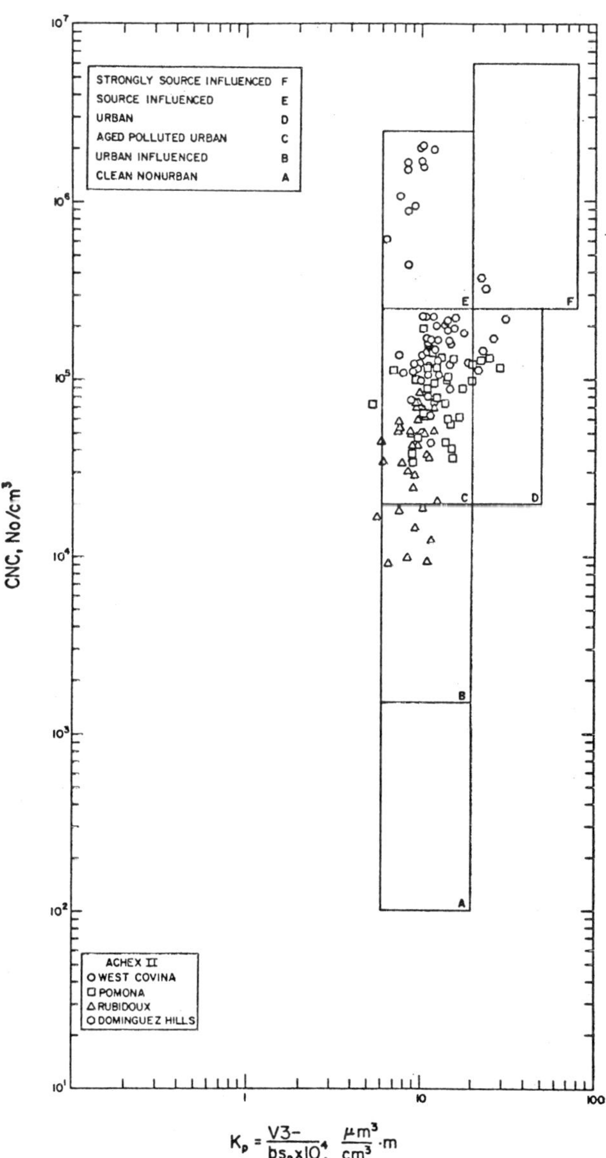

FIGURE 2. Variation of the aerosol submicron volume to aerosol light-scattering coefficient with the total aerosol number concentration from ACHEX II.

Charlson (2) has discussed the several factors that influence the value of K_p. By using aerosol volume concentrations rather than mass concentration, the effect of aerosol density is removed assuming that the instruments are insensitive to such changes. Covert (7) has shown that aerosol light scattering is a strong function of relative humidity for deliquescent aerosols. To determine what effect the relative humidity might have played, the data were divided into three categories according to the ambient relative humdiity (RH):

$$RH < 50\%, \quad 50\% < RH < 75\%, \quad 75\% < RH$$

No systematic variation was observed, probably because it has been shown that the instrumentation inside the Air Resources Board mobile laboratory sampled the aerosols at a temperature close to the laboratory temperature, thus minimizing wide variations in the RH (ACHEX Final Report, Vol. IV, Appendix A). Hänel (13) has shown that the ratio of extinction to mass concentration should be constant within a factor of 50% for RH less than about 97%.

The large variation in the values of K_p is believed to be primarily due to changes in the aerosol size distribution. Near sources of fresh aerosol (of which the CNC value is an indicator) aerosol volume is added to the nuclei mode and to the smaller sizes of the accumulation mode, increasing V3- without a similar increase in b_{sp} because of the relatively small light-scattering cross sections of these sized particles. This factor would increase the value of K_p.

A. Aerosol Categories

In an attempt to classify the data for analysis, they have been grouped into categories that can be identified with general location. The categories are listed in Table 2, along with the appropriate values of CNC and K_p. The number of data sets, the mean, the percent standard deviation, and the range of the data for each category are shown in Table 3.

1. Clean Nonurban. Clean nonurban conditions were rare during the ACHEX. From Figure 1 it is seen that such conditions were encountered at Goldstone and Point Arguello. The lowest average values for CNC, b_{sp}, and V3- were found for this category. The average value for b_{sp} of $0.347 \times 10^{-4}\ m^{-1}$ indicates that the atmospheric Rayleigh scattering was of the same order of magnitude as the aerosol light scattering.

TABLE 2. Aerosol Categories and the Range of K_p and CNC for Each Category

Category	$K_p = V3-/b_{sp}$ ($\mu m^3\ m/cm^3 \times 10^{-4}$)	CNC (number/cm^3)
Clean nonurban	6-20	100-1500
Urban influenced	6-20	1500-20,000
Aged polluted urban	6-20	20,000-250,000
Urban	20-50	20,000-250,000
Source influenced	6-20	250,000-2,500,000
Strongly source influenced	20-80	250,000-6,000,000

The lower bound for CNC of 100 cm^{-3} was set arbitrarily to the nominal detection limit of the condensation nuclei counter. The upper CNC bound of 1500 cm^{-3} was set to be below the data from Hunter-Liggett and Goldstone data influenced by local sources. Also, the CNC boundaries for each category were set to provide a range of about one decade.

2. Urban Influenced. This category is significant because it appears that much of the background aerosol in southern California is influenced by urban sources. This category includes data collected at Goldstone, Point Arguello, Hunter-Liggett, and Rubidoux (Figure 2).

The data from Goldstone include afternoon hours when the local traffic acted as an aerosol source, and a period when there was an incursion of aged aerosol from the South Coast Air Basin [Sverdrup et al. (22)].

The data from Point Arguello include a period when the wind was off the ocean, but all aerosol parameters of the fine-particle modes indicated aged polluted aerosol. It is believed that during this period aged aerosol from land which had blown out to sea was blown back over the coast. This category also contains data taken at Point Arguello during the evening and early morning, when trains passed near the mobile laboratory. Figure 3 is a plot of CNC versus K. The parameter K is defined by using the total light-scattering coefficient, given in Eq. 1. The Rayleigh scattering coefficient was assumed to be 0.23 x

TABLE 3. Aerosol Categories, Number of data Sets in Each Category, and Range, Average, and Standard Deviation of Four Aerosol Parameters for Each Category for ACHEX Data

		Parameter															
		CNC (number/cm^3)				b_{sp} ($m^{-1} \times 10^4$)				V3- ($\mu m^3/cm^3$)				K_p (μm^3 m/cm^3 x 10^{-4})			
Category	Number of data points	Min.	Max.	Avg.	SD (%)	Min.	Max.	Avg.	SD (%)	Min.	Max.	Avg.	SD (%)	Min.	Max.	Avg.	SD (%)
Clean nonurban	8	47.3	1400	759	78.5	0.123	0.951	0.347	81.3	0.863	5.94	2.87	66.2	6.25	13.7	8.68	27.9
Urban influenced	63	0.157×10^4	1.87×10^4	6140	78.3	0.075	12.4	1.18	208	1.27	79.9	11.0	150	5.51	20.5	12.5	29.5
Aged polluted urban	7[a]	5.83×10^4	1.34×10^5	9.56×10^4	33.5	0.385	0.563	0.472	14.0	4.17	7.58	5.15	22.5	6.46	9.56	7.29	14.3
	160	2.09×10^4	2.27×10^5	9.92×10^4	49.5	0.530	20.8	4.55	71.0	8.63	159	50.1	53.2	6.11	19.8	12.4	26.3
Urban	76	2.50×10^4	2.16×10^5	1.16×10^5	35.2	0.235	2.62	0.864	57.5	7.19	72.8	25.1	52.7	21.2	45.9	30.2	18.6
Source influenced	3[a]	2.68×10^5	7.65×10^5	5.73×10^5	46.6	0.495	0.600	0.534	108	5.76	7.08	6.62	11.2	11.4	14.3	12.5	12.7
	12	4.42×10^5	2.06×10^6	1.38×10^6	40.9	2.94	5.51	3.92	19.0	24.6	47.8	35.0	19.9	6.31	11.7	9.01	16.1
Strongly source influenced	34	2.53×10^5	5.45×10^6	2.42×10^6	65.7	0.16	3.56	1.28	60.4	6.84[b]	109	53.3	43.9	22.2	81.0	47.1	32.6

[a]These data were obtained at Richmond downwind from an oil refinery.

[b]There are only two runs with V3- less than 30 $\mu m^3/cm^3$.

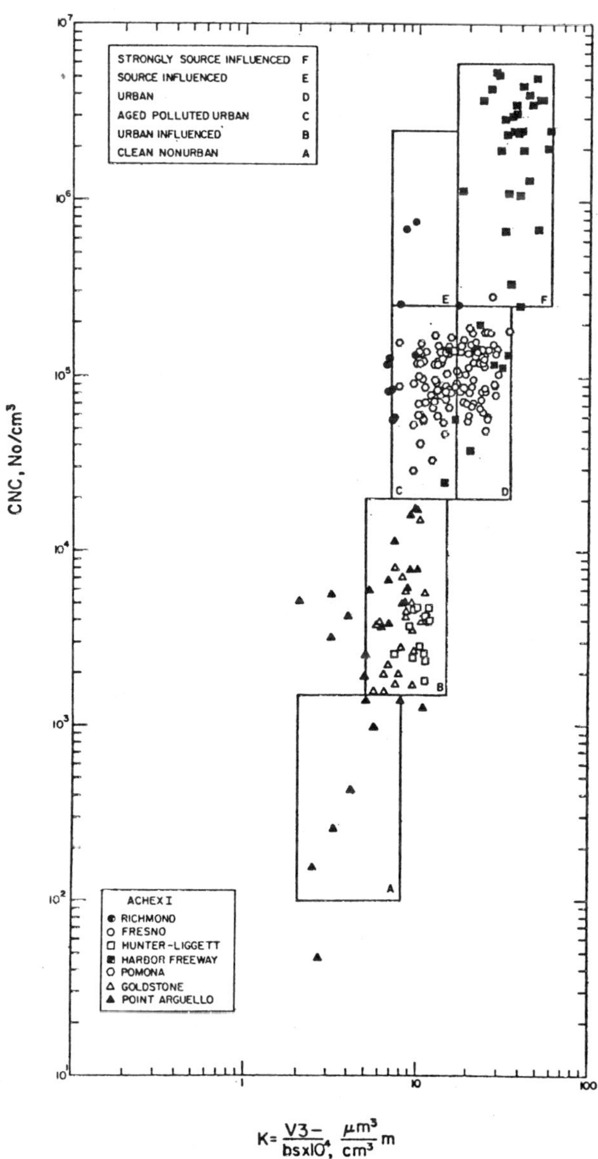

FIGURE 3. Variation of the aerosol submicron volume to total light-scattering coefficient with the total aerosol number concentration from ACHEX I.

10^{-4} m^{-1}. The parameter K was used in the preliminary work described in the ACHEX Final Report. It is apparent that, if K is used instead of K_p at lower values of CNC, the values of K become significantly smaller than those of K_p, and new boundaries for the categories must be drawn. This is so because at the low concentrations the atmospheric Rayleigh scattering is of the same order of magnitude as the aerosol light scattering. The boundaries shown in Figure 3 were drawn based upon the boundaries for K_p. Use of K rather than K_p gives a greater range to the values, resulting in more resolution on the abscissa. Figure 3 also shows four data points to the left of the urban-influenced category. These data were taken as trains passed by. Ten-minute values of CNC as high as 31,600 cm^{-3} were recorded. Without the influence of the trains the data would have fallen into the clean nonurban category.

The urban-influenced category contains data taken at Rubidoux a few kilometers west of Riverside. These data were taken during a period marked by winds from the north and low aerosol volume concentrations.

It is interesting that data reported by Hidy et al. (16), taken at San Nicolas Island located about 130 km west-southwest of Los Angeles, fall into this category. The authors reported a 2-month period labeled marine-enriched aerosol, with the wind direction generally from 180 to 300°, a CNC value of 2500 cm^{-3}, $b_s = 0.5 \times 10^{-4}$ m^{-1}, and a refined mass concentration of 12.1 $\mu g/m^3$. This results in a value of K = 25 based on mass.

3. Aged Polluted Urban, and Urban. Two categories for urban aerosols not affected by local sources are the aged polluted urban and the urban. The lower CNC limit was established as being above the nonurban values for Point Arguello or Goldstone. The upper CNC limit was set above values obtained for general urban locations. This value of 250,000 cm^{-3}, established by coagulation, is in agreement with values previously reported for Los Angeles [Whitby et al. (27)].

General urban aerosols have been divided into two categories, with the dividing line of $K_p = 20$ being the same value as the upper limit for the clean nonurban and urban-influenced categories. The fact that the aged polluted urban category has the same K_p boundaries as the two categories below it is reasonable, since aged aerosols would also be expected to make up those two categories.

From Figure 2 and Table 3 it is seen that seven data points from Richmond lie in the aged polluted urban category, even though their b_{sp} and V3- values are considerably lower than those for the other 160 points. The Richmond data represent conditions when relatively clean aerosol was being blown off San Francisco

Bay with a source of nuclei from an oil refinery superimposed upon it. There was so little aerosol in the accumulation mode that a high nuclei concentration was possible despite some coagulation of the aerosol. This indicates that values of CNC and K_p alone are not enough to classify the aerosol; however, when CNC, K_p, and b_{sp} or V3- values are used, the Richmond data can be separated from the other aged polluted urban data.

The other 160 data points represent aged polluted urban aerosol. The highest mean values of b_{sp} and V3- were obtained for this category. Data taken at West Covina during periods of heavy pollution are in this category. All Pomona data obtained on October 24, 1972 and August 16, 1973, fall into this category. Data from Pomona on October 5, 1972, a day during which Santa Ana conditions prevailed, are in the transition zone between the categories.

The urban category has K_p boundaries of 20 and 50. This category represents urban locations with less pollution based on b_{sp} and V3- values, or a fresher aerosol having more particles with sizes below the optical range. All Fresno data fall into this category. Data from the Harbor Freeway taken during periods when the wind was not from the freeway or when there was little traffic also are in this category. No data from Rubidoux or West Covina fall into this category.

4. Source Influenced. The source-influenced category is representative of sites with a local source of nuclei, but not in sufficient quantities to substantially increase V3- as compared to b_{sp}. This category has the same K_p boundaries as the other aged aerosol categories. Data from Richmond and West Covina fall into this category. Inspection of Figure 2 reveals two groups of data from West Covina. The West Covina data in the source-influenced category were obtained when aerosol from a local incinerator was sampled.

5. Strongly Source Influenced. This category represents data taken during periods when local sources were contributing significant aerosol volume concentrations of particles with sizes below the optical range. The highest mean values for CNC and K_p were obtained for this category. Table 3 indicates that b_{sp} was only slightly greater than that for the urban category and much less than for the aged polluted urban and source-dominated categories. The category contains 1 datum point from Fresno and 2 from Dominguez Hills with automobiles as the source, 1 point from Richmond with an oil refinery as the source, and 28 points from the Harbor Freeway.

B. Denver Aerosols

Data obtained by Willeke et al. (30) at two sites in Denver were analyzed in the ACHEX Final Report. The data fell into the two urban and two source categories. Their position on a plot such as Figure 3 was explained by the local conditions experienced while the aerosols were sampled.

III. VARIATION OF PARAMETER K_p - THEORETICAL

An aerosol size distribution model was employed to determine whether the variation of K_p with CNC could be explained by different aerosol size distributions. The model used was the trimodal log-normal distribution model proposed by Whitby (25). This model uses three log-normal frequency functions fitted to the nuclei, accumulation, and coarse-particle size distribution modes previously described.

The log-normal distribution is given by Eq. 5:

$$f(D_p)\, dD_p = \frac{A}{D_p \sqrt{2\pi}\, \ln \sigma_g} \exp \frac{-(\ln D_p - \ln \bar{D}_{pg})^2}{2(\ln \sigma_g)^2}\, d(D_p) \qquad (5)$$

where $\bar{D}_{pg}$ is the geometric mean size, σ_g is the geometric standard deviation, and A is the integral of the function over the particle size, D_p. The distributions that were used for the theoretical analysis of the variation of K_p describe the grand average size distribution (425 10-min runs) obtained at Goldstone in the Mojave Desert. The parameters used in these distributions are listed in Table 4. Integration of the size distribution model using the parameters in Table 4 for Goldstone yielded a value of 8750 for number concentration, as compared to a measured value of 8730 particles/cm^3. The model gave a value of 4.44 $\mu m^3/cm^3$ for V3-, compared to a measured value of 4.38 $\mu m^3/cm^3$, and gave a value of $0.278 \times 10^{-4}\ m^{-1}$, compared to $0.317 \times 10^{-4}\ m^{-1}$ obtained with the nephelometer.

The model was used to vary the total number concentration in the nuclei mode by simulating a source of fresh aerosols. For each value of the number concentration, K_p was calculated. The number concentration of particles in the nuclei mode was varied by varying the value of CNC and calculating the number concentration in the nuclei mode as CNC minus the particle concentrations in the other two modes.

TABLE 4. Parameters of the Log-Normal Functions Used to Fit the Experimental Goldstone Data

Parameter	Nuclei mode	Accumulation mode	Coarse mode
A (number/cm^{-3})	1.88×10^4	2.42×10^3	1.10
σ_g	2.38	2.07	2.32
$\overline{D}_{pg}$ (μm)	0.00695	0.0673	0.696

The value of K_p was computed from Eq. 6:

$$K_p = \frac{V3-}{b_{sp}} = \frac{\pi/6 \int_{0.01}^{1} D_p^{\,3} \sum_{i=1}^{3} f_i d(D_p)}{\pi/4 \int_{0.01}^{38} D_p^{\,2}\, \overline{Q}_s(D_p) \sum_{i=1}^{3} f_i d(D_p)} \tag{6}$$

The f_i represent the three log-normal functions. The integration interval was changed from decade to decade, varying from 0.005 μm in the range $0.01 < D_p \leq 0.1$ μm up to 5.0 μm in the range $10 < D_p \leq 38$ μm. Smaller integration steps did not affect the accuracy. The integration was performed using the trapezoidal rule with a minimum of four intervals together with Romberg quadrature.

The value of b_{sp} was calculated using a simple two-dimensional model of the integrating nephelometer. The effects of three dimensions [Rabinoff and Herman (21)] and the angular truncation [Ensor and Waggoner (9)] were not considered to be of importance for the present work. The wavelength response of the nephelometer was taken into account by numerically integrating values of the Mie scattering efficiencies for spheres, Q_s, over the nominal wavelength response of an MRI Model 1550 integrating nephelometer to obtain averaged values, $\overline{Q}_s$. The wavelength response was determined from manufacturer's data for the light source, optical filters, and photomultiplier. Values of Q_s were calculated using a modification of the downward recurrence routine reported by Dave (8).

Tables of $\overline{Q}_s$ as a function of D_p (36 geometric intervals per decade of particle size) were calculated for several refractive indices. For this work a value of 1.50 - 0.02 i was taken as representative for the complex refractive index, as was done by Heintzenberg and Quenzel (14). For the integration of Eq. 6, values of $\overline{Q}_s$ were determined by linear interpolation in the table.

The theoretical curve showing the variation of K_p with CNC is shown in Figure 1. The curve is evidence that changes in the aerosol size distribution alone are enough to explain the variation in K_p. From the curve it is seen that, within the urban and nonurban categories, the theoretical K_p is constant to within 10% for a given size distribution if only the number concentration in the nuclei mode is varied. The scatter of data about the curve is due to the particular size distribution, each represented by a datum point, having different characteristics from those of the Goldstone data, and also different refractive indices.

IV. CONCLUSIONS

From an experimental and theoretical study of the relationship between aerosol size distributions and light scattering, the following conclusions can be drawn:

(1) Submicron-aerosol-volume-to-light-scattering ratios, K_p, have been computed and found to be in agreement with values reported by others.

(2) The large variations in K_p (5 to 80) preclude its use as a good indicator of submicron aerosol volume or mass concentration.

(3) The variation of K_p has been examined and related to integral size distribution parameters. Using the particle number concentration from a condensation nuclei counter and the ratio K_p, it is found that the data fall into groupings which can be classified according to general location.

(4) The variation of K_p has been explained using a trimodal log-normal size distribution model.

Although particle refractive index can be significant, differences in the aerosol size distributions in different locations are sufficient to explain the variation of K_p.

REFERENCES

1. Air Resources Board, State of California. 1973. "Visibility, Light Scattering, and Mass Concentration of Particulate Matter," Report of the California Tri-City Aerosol Sampling Project.

2. Charlson, R. J. 1969. Atmospheric Visibility Related to Aerosol Mass Concentration: A Review, Environ. Sci. Technol. 3, 913-918.

3. Charlson, R. J., Ahlquist, N. C., and Horvath, H. 1968. On the Generality of Correlation of Atmospheric Aerosol Mass Concentration and Light Scatter, Atmos. Environ. 2, 455-464.

4. Charlson, R. J., Ahlquist, N. C., Selvidge, H., and MacCready, P. B., Jr. 1969. J. Air Pollut. Control Assoc. 19, 937-942.

5. Charlson, R. J., Horvath, H., and Pueschel, R. F. 1967. The Direct Measurement of Atmospheric Light Scattering Coefficient for Studies of Visibility and Pollution, Atmos. Environ. 1, 469-478.

6. Charlson, R. J., Porch, W. M., Waggoner, A. P., and Ahlquist, N. C. 1974. Background Aerosol Light Scattering Characteristics: Nephelometric Observations at Mauna Loa Observatory Compared with Results at Other Remote Locations, Tellus XXVI, 346-360.

7. Covert, D. S. 1974. "Light Scattering, Humidity, and Chemical Composition of Aerosols," Ph.D. thesis, Dept. of Civil Engineering, University of Washington.

8. Dave, J. V. 1968. "Subroutines for Computing the Parameters of the Electromagnetic Radiation Scattered by a Sphere," IBM Rep. No. 320-3237, IBM Scientific Center, Palo Alto, California.

9. Ensor, D. S. and Waggoner, A. P. 1970. Angular Truncation Error in the Integrating Nephelometer, Atmos. Environ. 4, 481-487.

10. Ettinger, H. J. and Royer, G. W. 1972. Visibility and Mass Concentration in a Nonurban Environment, J. Air Pollut. Control Assoc. 22, 108-111.

11. Greacen, E. S., Sverdrup, G. M., and Whitby, K. T. 1974. "A Preliminary Report on Correlation of b_{scat} vs. Submicron Volume," submitted to G. Hidy, Rockwell International Science Center.

12. Griggs, M. 1972. Relationship of Optical Observations to Aerosol Mass Loading, J. Air Pollut. Control Assoc. 22, 356-358.

13. Hänel, G. 1972. The Ratio of the Extinction Coefficient to the Mass of Atmospheric Aerosol Particles as a Function of the Relative Humidity, J. Aerosol Sci. 3, 455-460.

14. Heintzenberg, J. and Quenzel, H. 1973. Calculations on the Determination of the Scattering Coefficient of Turbid Air with Integrating Nephelometers, Atmos. Environ. 7, 509-519.

15. Hidy, G. M. 1972. Aerosols and Atmospheric Chemistry. Academic Press, New York.

16. Hidy, G. M., Mueller, P. K., Wang, H. H., Karney, J., Twiss, S., Imada, M., and Alcocer, A. 1974. Observations of Aerosols over Southern California Coastal Waters, J. Appl. Meteorol. 13, 96-107.

17. Hidy, G. M., et al. 1974. "Characterization of Aerosols in California (ACHEX)," Final Rep. SC524.25FR, submitted to the Air Resources Board, State of California.

18. Middleton, W. E. K. 1958. Vision Through the Atmosphere. University of Toronto Press, Toronto, Canada. Reprinted 1968.

19. Noll, K. E., Mueller, P., and Imada, M. 1968. Visibility and Aerosol Concentration in Urban Air, Atmos. Environ. 2, 465-468.

20. Pilat, M. J., and Ensor, D. S. 1970. Plume Opacity and Particulate Mass Concentration, Atmos. Environ. 4, 163-173.

21. Rabinoff, R. A. and Herman, B. M. 1973. Effect of Aerosol Size Distribution on the Accuracy of the Integrating Nephelometer, J. Appl. Meteorol. 12, 184-186.

22. Sverdrup, G. M., Whitby, K. T., and Clark, W. E. 1975. Characterization of California Aerosols - II. Aerosol Size Distribution Measurements in the Mojave Desert, Atmos. Environ. 9, 483-494.

23. Whitby, K. T. 1973. "On the Multimodal Nature of Atmospheric Aerosol Size Distributions," presented at the VIII International Conference on Nucleation, Leningrad, U.S.S.R., Sept. 28.

24. Whitby, K. T. 1974. "Mobile Laboratory Design, Preparation, Operation, and Data," ARB Aerosol Characterization Study, submitted to G. Hidy, Rockwell International Science Center, under Subcontract No. 262-1948.

25. Whitby, K. T. 1974. "Modeling of Multimodal Aerosol Size Distributions, Proceedings of the 1974 Conference on Aerosols in Nature, Medicine, and Technology, GAF Bergweg 7, Vorderhinderlang 8973, West Germany.

26. Whitby, K. T., Clark, W. E., Marple, V. A., Sverdrup, G. M., Sem, G. J., Willeke, K., Liu, B. Y. H., and Pui, D. Y. H. 1975. Characterization of California Aerosols - I. Size Distributions of Freeway Aerosol, Atmos. Environ. 9, 463-482.

27. Whitby, K. T., Husar, R. B., and Lui, B. Y. H. 1972. The Aerosol Size Distribution of Los Angeles Smog, J. Colloid Interface Sci. 39, 177-204.

28. Whitby, K. T., Lui, B. Y. H., Husar, R. B., and Barsic, N. J. 1972. The Minnesota Aerosol-Analyzing System in the Los Angeles Smog Project, J. Colloid Interface Sci. 39, 136-164.

29. Willeke, K. and Whitby, K. T. 1975. Atmospheric Aerosols: Size Distribution Interpretation, J. Air Pollut. Control Assoc. 25, 529-534.

30. Willeke, K., Whitby, K. T., Clark, W. E., and Marple, V. A. 1974. Size Distributions of Denver aerosols - A Comparison of Two Sites, Atmos. Environ. 8, 609-633.

Atmospheric Aerosols, Humidity, and Visibility

DAVID S. COVERT, ALAN P. WAGGONER, RAY E. WEISS,
NORMAN C. AHLQUIST, AND ROBERT J. CHARLSON
Departments of Environmental Health and Civil Engineering
University of Washington
Seattle, Washington

I. INTRODUCTION

Reduction of visibility and the presence of a visual haze are the effects of air pollution that are most often perceived by the general population. Visual range, V, is often 100 km or more in the absence of air pollution. In polluted atmospheres visual range is often reduced to 10 to 20 km and even as low as 1 km during severe smog episodes in the South Coast Air basin in southern California and elsewhere. Visibility reduction may be limited spatially to a local airshed or, in some places, to an area surrounding a specific source. More often, over relatively flat terrain it is a regional effect covering spatial scales of 100 to 1000 km and resulting from the combined emissions of many sources that cannot be distinguished. The latter situation prevails over the eastern third of the United States and western Europe.

It is the aerosol particles resulting from air pollution that affect the optical properties of the air and visibility through the air. Particles interact with visible radiation primarily by scattering and to a lesser extent by absorption to obscure distant objects. Visual range is often reduced by the interaction of water with aerosol particles at high relative humidities. Even low concentrations of particles may result in visibilities of 10 km or less if the humidity approaches 100%. This combination of causal factors (high particle mass and high relative humidity) has historically confounded the study of visibility reduction.

Over the last 10 years significant advances have been made in the measurement and understanding of the integral and differential aerosol parameters that control visibility in a polluted atmosphere. Initially point measurements of the particle light-scattering coefficient, b_{sp}, and particle mass concentration, [m], led to an observed relationship between these two integral aerosol parameters [Charlson et al. (2)]. Similarly, observations led to relationships between V and b_{sp}, and also [m] and V [Horvath and Noll (16); Middleton (22)]. These relationships relied on assumptions concerning the spatial and temporal uniformity of the mass concentration, size distribution, and optical properties of the aerosol.

More recently, measurements of differential aerosol properties such as the particle size distribution, particle mass, and chemical composition in the optical size range (0.1- to 2.0 μm diameter) where they interact most effectively with visible light became available [Willeke and Whitby (29); Hidy (13)]. These allowed considerable improvement in the accuracy of the b_{sp}, V, [m] relationships, and in the understanding of the aerosol properties which control them. At this point visibility reduction is probably the best understood aspect of air pollution. The physical and quantitative basis for this understanding will be outlined in the following sections. Special attention will be given to the effect that humidity exerts in modifying the optical properties of particles.

II. ATMOSPHERIC OPTICS AND AEROSOL PROPERTIES

The optical properties of the atmosphere are determined by the fixed components (nitrogen, oxygen, etc.) and the variable components, primarily the suspended particles. The parameters by which the extinction of light in the atmosphere is commonly described are formulated below.

The extinction coefficient, b_{ext}, of the atmosphere determines the change in intensity, ΔI, of light traversing a path length Δx according to the Beer-Lambert law:

$$\frac{\Delta I}{I} = -b_{ext} \Delta x \tag{1}$$

where b_{ext} is the sum of two terms, b_{scat} and b_{abs}, the scattering and absorption components, respectively. These terms can also be expanded into particulate and gas components such that

$$\begin{aligned} b_{ext} = b_{scat} + b_{abs} &= (b_{sp} + b_{Rg}) + (b_{ap} + b_{ag}) \\ &= (b_{sp} + b_{ap}) + (b_{Rg} + b_{ag}) \\ &= b_{ext,p} + b_{ext,g} \end{aligned} \tag{2}$$

where b_{Rg} denotes the Rayleigh scattering coefficient of air, and subscripts s, a, p, and g refer to scattering, absorption, particulate, and gas components, respectively. Thus b_{Rg} Δx is the fraction of incident light scattered in all directions by gas molecules in Δx, b_{ag} Δx is the fraction of incident light absorbed by gas molecules in Δx, b_{ap} Δx is the fraction of incident light absorbed by particles in Δx, and b_{sp} Δx is the fraction of incident light scattered in all directions by particles in Δx. For air at standard temperature and pressure, the value of $b_{ext,g} = b_{Rg}$ is about 2×10^{-5} m^{-1} for green light [Middleton (22)]. Except in situations where there is appreciable NO_2, b_{ag} is zero [Charlson et al. (4)].

Visibility, in the general sense of the word, refers to the clarity of the atmosphere and ability to see distant objects distinctly. There are several more rigorous definitions which apply to this and related terms, however. Daytime visibility, in weather observing practice, is the greatest distance in a given direction at which it is just possible to see and identify a prominent dark object against the horizon with the naked eye. Visual range, V, is the distance during daylight at which the apparent contrast between a specified target and its background becomes equal to the threshold contrast of the observer. The formula

$$V = \frac{1}{b_{ext} \ (\ell n\ C/C_o)} \tag{3}$$

defines visual range in terms of extinction along the visual path where C_o and C are the contrasts at distances of zero and V, respectively. For black objects against the horizon sky, C_o = -1. The threshold contrast, although it varies somewhat from observer to observer, has been determined to be about 0.02 for objects 1/2° or greater in size. Equation 3 becomes

$$V = \frac{3.9}{b_{ext}} \tag{4}$$

This formula also describes meteorological range with the distinction that meteorological range is defined by the extinction coefficient at one point and not integrated over visual path length. These definitions become identical in an atmosphere that is homogeneous along the sight path.

The particle extinction coefficient, $b_{ext,p}$, is dependent on a number of physical and chemical properties of the aerosol particles. The particle concentration and size distribution are the physical parameters of primary importance. Particle shape and refractive index are of secondary importance to integral optical effects. The chemical composition determines the complex refractive index and hygroscopic/deliquescent nature of the particles. An external parameter, ambient relative humidity (RH), has a strong influence on visibility through its effect on the size and refractive index of hygroscopic particles.

The relationship between particle properties and extinction coefficient is potentially complex. However, with the recent advances in aerosol measurement technology certain regularities in the size distribution, chemical composition, and composition-size distribution of atmospheric aerosols have become apparent. These have led to the current good understanding of aerosol effects on visibility.

Within the last 10 years particle size measurement technology has advanced sufficiently to provide details of the size distribution within the size range of 0.1 to 2.0 μm. Willeke and Whitby (29) and others [Lundgren and Paulus (21); Jaenicke et al. (17)] have observed atmosphere aerosol in areas affected by anthropogenic activity, but away from local sources, to be consistently bimodal in its volume-size distribution. The modes are centered at diameters of about 0.3 and 10 μm (termed accumulation and mechanical, respectively) with a distinct minimum near 1.0 to 2.0 μm. The volumes of aerosol in each mode are of comparable magnitude, generally within a factor of 3, but vary independently. Figure 1 illustrates a typical atmospheric size distribution plotted in terms of particle volume, d[v]/d(log D), as a function of particle diameter.

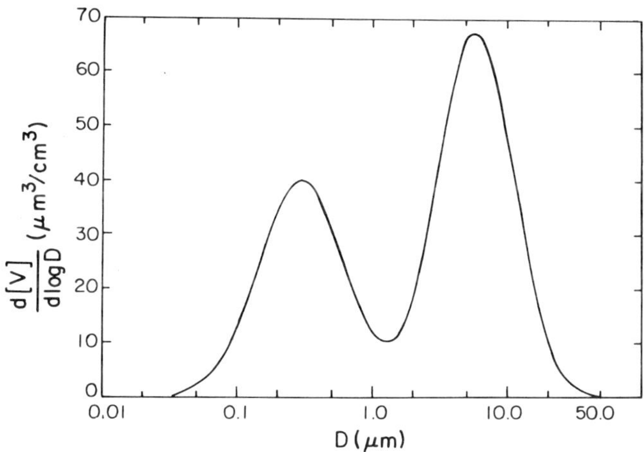

FIGURE 1. Typical bimodal atmospheric aerosol volume-size distribution. Fine mode: Dg - 0.3 μm, σg = 2. Coarse mode: Dg = 6 μm, σg = 2.

Refractive index, m, is composed of a real and an imaginary part, m = (m_{real} - i·m_{imag}). An often quoted range for m_{real} for atmospheric aerosol particles is 1.3 to 1.7; for m_{imag} the range is 0.001 to 0.05. Chemical compounds commonly found in atmospheric aerosols have an m_{real} that varies from 1.33 (water) to 1.5 (various inorganic salts). Thus for dry (RH < 60%) atmospheric aerosol, the m_{real} is likely to be between 1.4 and 1.6. The values of m_{imag} varies from 0 for pure salts and many organic compounds to 0.6 for graphitic carbon. A typical value of m for a dry atmospheric aerosol is (1.5 - 0.02 i). This, like chemical composition, is undoubtedly a function of particle size. Most of the imaginary part is due to particles less than 2.0 μm in diameter.

The integral optical effects of an aerosol of given size distribution and refractive index can be calculated according to the formula

$$b_i = \int_o^D \frac{\sigma_i(D)}{v} \cdot \frac{d[v]}{d(\log D)} \, d(\log D) \qquad (5)$$

Here b_i is the extinction, scattering, or absorption coefficient, and $d[v]/d(\log D)$, where [v] is particle volume concentration (in m^3/cm^3) and D is particle diameter (in μm), describes the volume-size distribution of the aerosol. Also, $\sigma_i(D)/v$ is the extinction scattering or absorption cross-section per unit of particle volume. This parameter, which is a function of particle diameter, can be computed from Mie theory [Mie (23)] for spherical particles. Calculated extinction, scattering, and absorption cross-sections per unit of particle volume for light of wavelength $\lambda = 0.550$ μm and a particle complex refractive index, m - (1.5 - 0.02 i) are illustrated in Figure 2. It can be seen that on a unit volume basis particles in the size range 0.1- to 2.0-μm diameter (often referred to as the optical subrange) scatter light most efficiently, and that scattering is about 90% of total extinction. The usefulness of this method of introducing and plotting optical cross-section as σ_i/v will become obvious in consideration of particle size distributions.

By multiplying the extinction cross-section function by the volume-size distribution, the distribution function of extinction for a typical atmospheric aerosol, $db_{ext,p}/d(\log D)$ is obtained as illustrated in Figure 3. The integral of this curve (Eq. 5) gives $b_{ext,p}$. This function graphically demonstrates that the fraction of particles in the accumulation mode controls the optical effects of atmospheric aerosols. Similarly, $db_{sp}/d(\log D)$ is plotted on the same graph for comparison.

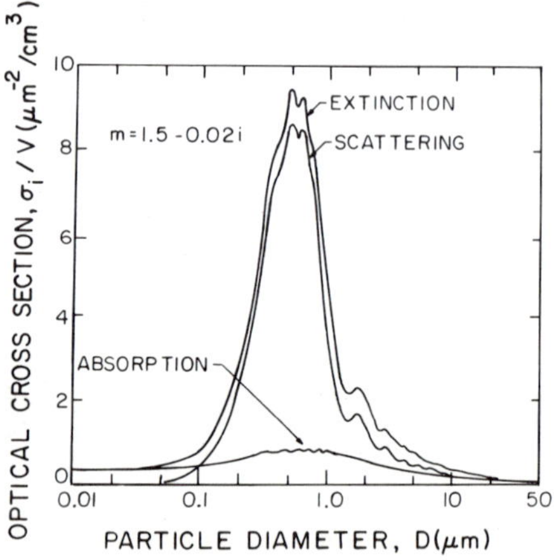

FIGURE 2. Particle extinction, scattering, and absorption cross-sections per unit of aerosol volume, as calculated from Mie theory.

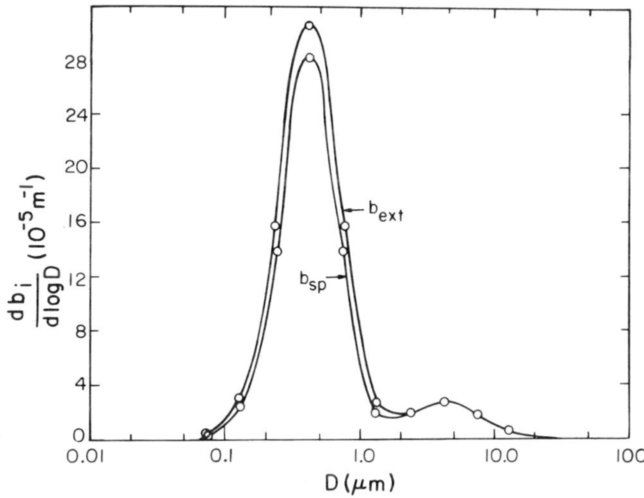

FIGURE 3. Particle extinction and scattering-size distributions as calculated from cross-sections and size distributions in Figures 1 and 2. The integral of db_{ext} = 1.9 x 10^{-4}. The integral of db_{sp} = 1.7 x 10^{-4} m^{-1}. Measured b_{sp} = 1.7 x 10^{-4} m^{-1} on the occasion that the above distribution was made [Willeke and Brockman (30)]. In the calculation of b_{ext}, it is assumed that the absorption coefficient of the coarse mode is zero. It can be seen that about 90% of the extinction is due to particles $<$ 2 μm in diameter.

For typical atmospheric aerosol size distributions 80 to 90% of extinction is due to particles in the accumulation mode. (This will not be true, however, in situations where the particle size distribution deviates radically from that shown in Figure 2, i.e., in dust storms or clouds.) For the rest of this discussion the emphasis will be on the accumulation mode and its effects on atmospheric optics and visibility.

The fact that the accumulation mode is relatively uniform (spatially and temporally with respect to mean size and standard deviation) and corresponds roughly to the mode of the optical cross-section functions means that submicrometer aerosol volume (or mass) should be well correlated with aerosol extinction or scattering coefficient. Data that will be presented later bear this out.

It has long been known that atmospheric aerosols are in some degree hygroscopic, that is, even at humidities below saturation they take up or release water in response to changes in ambient RH to maintain equilibrium water vapor pressure with the surrounding air. The hygroscopic properties of aerosol particles are determined primarily by their chemical compositions and to a much lesser extent by particle size (Kelvin effect) and shape (capillary absorption) [Winkler (31)]. For the particle sizes (0.1 to 2 μm) and humidity ranges (less than 95% RH) of interest to this discussion, chemical composition is the controlling factor. Recent measurements of aerosol chemical composition as a function of particle size have shown that most hygroscopic compounds are found in the accumulation mode [Charlson et al. (6)]. Measurements of aerosol properties as a function of RH have shown that an increase in RH from 30 to 80% will generally result in an increase in particle radius by a factor of 1.2 and a mass increase by a factor of 1.7 due to uptake of water [Winkler and Junge (32); Ho et al. (14)]. Thus hygroscopic growth can be expected to have a large effect on aerosol optical properties as it affects particle size and refractive index. The value of $b_{ext,p}$ can be expected to increase in nearly the same ratio as mass and has been observed to do so. It is important to note that because of this relatively small diameter change the increase in scattering is not due to smaller particles growing into the 0.1- to 2.0-μm optical subrange. Rather it is the larger increase in submicrometer mass due to water uptake that causes this increase. The decrease in refractive index that accompanies hygroscopic growth has the effect of lessening the increase in extinction cross-section, (σ_i/v), but this effect is small compared to that of increasing size.

Considerable hysteresis is associated with the uptake and release of water by hygroscopic aerosols [Junge (18)]. Although there have been no *in situ* measurements, data for salt aerosols and impactor-sampled ambient aerosol indicate that the hysteresis loop is closed at relative humidities <40% and >70 to 80% [Winkler and Junge (32)]. An exception to this generality may occur when significant amounts of nitrate compounds are present in the aerosol. Nitrate salt particles with and without ionic and/or solid impurities seem to exist as supersaturated solution dorplets (i.e., they retain liquid water) at very low humidities [Winkler et al. (33)].

III. MEASUREMENT OF AEROSOL OPTICAL PROPERTIES AND ATMOSPHERIC DATA

Instrumentation has been developed to measure the integral optical parameters, $b_{ext,p}$, b_{sp}, b_{ap}, and $b_{sp}(RH)$, in the atmosphere.

Long path extinction is usually used to measure $b_{ext,p}$. A large number of vertical extinction or turbidity data have been obtained by Volz turbidity instruments [Flowers and McCormick (11)] or astronomical telescopes [Hodge (15)], but these are not very useful except for astronomical purposes. The aerosol mixing or scale heights are usually unknown, and it is difficult to relate measurement of vertical path extinction to any other effects. Horizontal extinction is a more difficult measurement but much more useful since it measures an aerosol property near the surface and the aerosol is more nearly uniform in a horizontal path. A comparison of b_{ext} and b_{scat} at a remote location in the southwestern United States showed good agreement between these two parameters over visible wavelength [Waggoner and Charlson (26)].

The particle light-scattering coefficient, b_{sp}, can be measured with an integrating nephelometer [Charlson et al. (3)]. It is used throughout the world in air monitoring and aerosol research to measure b_{sp} and to estimate visibility and particle mass concentration.

A typical value for b_{sp} in urbanized regions is 10^{-4} m^{-1}. Maximum and minimum values in the lower troposphere are around 10^{-3} m^{-1} and 10^{-7} m^{-1}, respectively [Samuels et al. (24); Bodhaine and Mendonca (1)]. Measurements made at a number of locations throughout the continental United States indicate that temporal variability at any given site is greater than the range of average values between clean and polluted sites [Waggoner and Charlson (26)]. Data for b_{sp} are summarized in Table 1.

Light scattering by aerosol particles and aerosol volume or mass for particles less than 2 μm in diameter have been observed to be well correlated, as can be expected from the preceding theoretical discussion. Ninety percent of the data obtained in the ACHEX [Hidy (13)] fit the relationship

$$\frac{[v]\ (\mu m^3/cm^3)}{b_{sp}\ (m^{-1})} = 0.2^{+0.2}_{-0.1}\left(\frac{cm^3}{m^2}\right) \qquad (6)$$

where [v] is particle volume concentration in the size range less than 1.0-μm diameter. The variation in this ratio may be due to variations in the size distribution from location to location,

TABLE 1. Mean and Variation for Scattering Parameters in Eleven Locations[a]

Location	Average b_{sp} (530 nm)	Low b_{sp}	High b_{sp}	b_{ap}/b_{ext} (%)
Richmond, California	$0.4 \times 10^{-4}\ m^{-1}$	$0.2 \times 10^{-4}\ m^{-1}$	$1.4 \times 10^{-4}\ m^{-1}$	---
Point Reyes, California	0.12	0.04	0.4	---
Fresno, California	1.0	0.3	1.9	18 ± 5
Hunter Liggett, California	0.4	0.2	0.8	---
Pasadena, California	1.5	0.8	3.0	20 ± 8
Pomona, California	1.8	0.6	6.0	16 ± 8
Washington University, Missouri	1.58	1.12	2.24	11 ± 1
Tyson, Missouri	0.63	0.28	1.41	14 ± 4
St. Louis University, Missouri	0.71	0.40	1.25	14 ± 2
Henderson, Colorado	0.31	0.08	1.25	17 ± 8
Denver, Colorado	0.56	0.22	1.58	18 ± 5

[a]In all locations the sample air was heated to 5 to 20°C above ambient temperature. For each measurement parameter the range of that parameter containing 63% of the data is specified.

but is more likely due to instrument error. Measurements near combustion sources have indicated that particles less than 0.1 μm in diameter can contribute to the total volume, but contribute little to b_{sp}. The variation may also be due to instrumental or sampling errors. The correlation at a given site is usually better, often having a correlation coefficient greater than .9 and as illustrated by the data shown in Figure 4a. Measurements of submicrometer particle mass concentration, [m], and b_{sp} at locations in southern California, Missouri, and the Pacific Northwest have yielded the relationship

$$\frac{[m]}{b_{sp}} \frac{(\mu g/m^3)}{(m^{-1})} = 0.27 \pm 0.5 \ (g/m^2) \tag{7}$$

[Samuels et al. (24); Waggoner et al. (27)]. Recent data for $[m]/b_{sp}$ for one rural and three urban sites in the Pacific Northwest which yielded a ratio of 0.31 ± 0.06 (g/m^2) are illustrated in Figures 4b and 4c. Although these relationships are consistent with theoretical considerations, it must be remembered that they are in fact empirical and will vary slightly with location as the physical and optical parameters of the aerosol change. The more recent data were obtained with more accurate methods, instrumentation, and at sites away from local sources. The reduction of the deviation is due to these improvements in technique.

There are a number of ways of determining b_{ap}, none of which is entirely satisfactory. Long path extinction cannot be used because b_{ap} is generally 10^{-5} m^{-1} or less. Values determined by inversion of angular scattering data [Eiden (9); Grams et al. (12)] are subject to large uncertainty without precise knowledge of aerosol size distribution and the effects of irregularly shaped particles on angular scattering. The value of b_{ap} can be estimated from collected aerosol samples by transmission measurements using a dispersal in KBr pellets [Volz (25)] or by reflectance using a dispersal in white powder [Lindberg and Laude (20)]; accuracy is probably no better than a factor of 2. Collection of aerosol on an impactor or Nuclepore filter surface and measurement of refractive index in an integrating sphere [Fischer (10)] or absorption in an integrating plate apparatus [Lin et al. (19)] yields somewhat better results. Values of m_{imag} are 0.001 to 0.05 [Fischer (10), Lindburg and Laude (20)]. Values of b_{ap}, expressed as the ratio b_{ap}/b_{ext} are in the range 0 to 0.5 [Waggoner and Charlson (26)].

The most extensive measurement of b_{ap} have been made with the integrating plate technique of Lin et al. (19). The data show that b_{ap}/b_{ext} values in urban-industrial areas such as St. Louis, Denver, and Phoenix tend to be in the range 0.3 to 0.5;

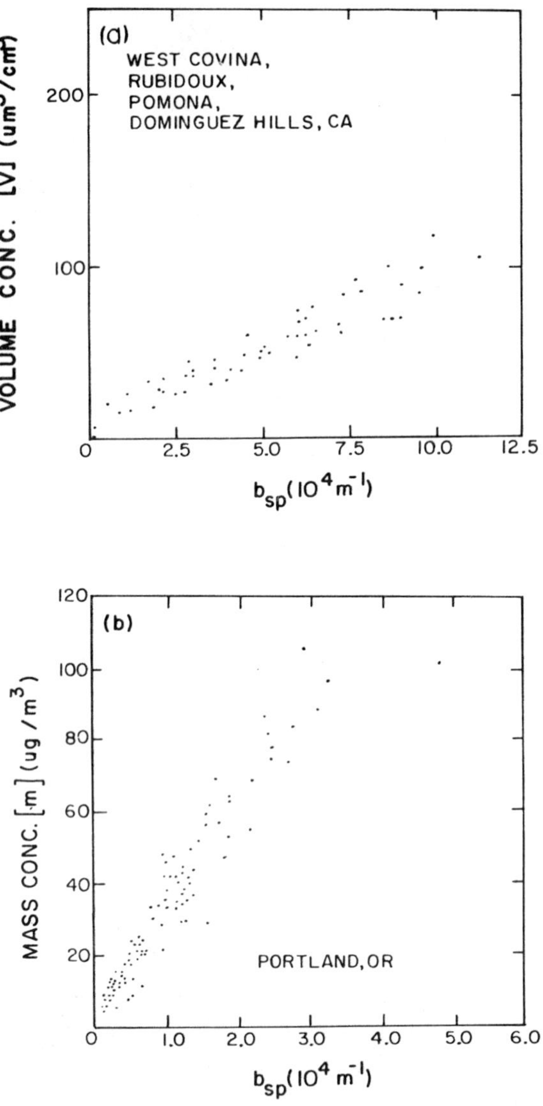

FIGURES 4a, 4b, 4c, 4d, 4e. Scatterplots of particle light scattering versus submicrometer particle volume or mass concentration show the high degree of correlation between these parameters.

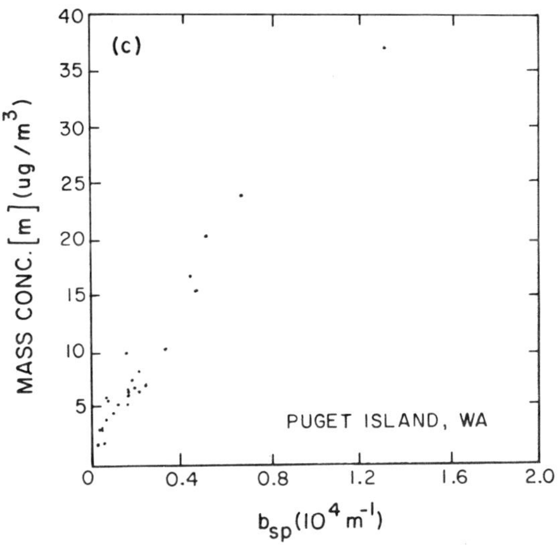

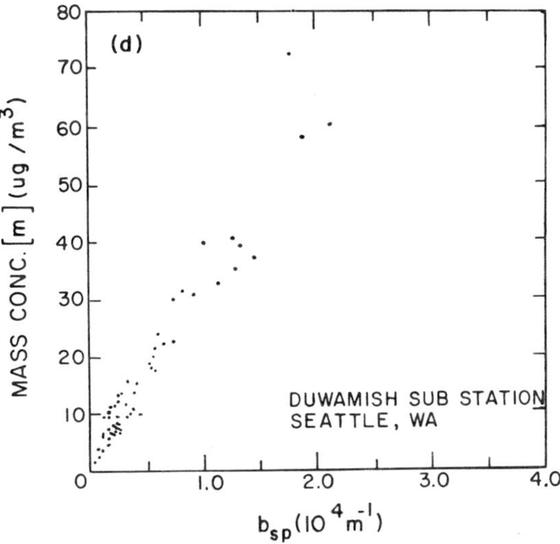

FIGURE 4 (cont.)

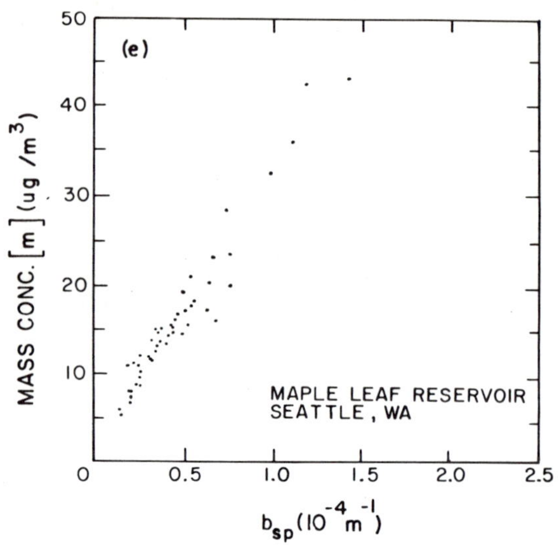

FIGURE 4 (cont.)

b_{ap}/b_{ext} in rural and suburban areas of the West and Midwest is 0 to 0.2. These values probably have an accuracy of better than ±20%.

It has often been assumed that b_{ap} is small and negligible compared to b_{sp} and thus that measured b_{sp} can be taken to be equal to $b_{ext,p}$ for purposes of calculating visibility and other optical properties of the atmosphere. Although this assumption may be valid throughout most of the troposphere, it is obviously in error in some locations, primarily urban.

The humidity dependence of the light-scattering coefficient, which will subsequently be expressed as the ratio $B = b_{sp}(RH)/b_{sp}(RH = 30\%)$, can be measured with a pair of nephelometers, one operated at constant RH (30%) and the other at variable RH through the addition of water vapor to the sample airstream. The ratio of the resulting outputs gives B, which is plotted as a function of measured RH in the variable humidity nephelometer as RH is scanned from 30 to 90%. An automatic system to make this measurement has been described by Covert (7) and used to chemically identify molecular sulfate species found as 0.1 to 1.0 μm particles in ambient air [Charlson et al (5); Weiss et al. (28)].

Data on the humidity dependence of light scattering have been collected for a wide range of aerosol types. Aerosols in urban, rural, and marine locations were observed to be generally hygroscopic in nature. They exhibited a measurable increase in light scattering due to hygroscopic growth at 60 to 70% RH and greater. At 80% RH the scattering increased by a factor of 1.6 to 1.9. Only under conditions where continental background aerosol (defined here as clean air from dry southwest desert or high plains areas) was present was the light-scattering increase at 80% RH less than a factor of 1.6. Typical humidograms (Figures 5a through 5h) for 12 locations which encompassed a wide range of aerosol types illustrate in more detail the effect of relative humidity on b_{sp}. Obvious differences in the data were related to the general chemical nature of the aerosols, which can be classified accordingly into four groups: marine, sulfate, urban/photochemical smog, and background continental. The aerosol in each of the first two of these groups is composed predominantly of one compound (or group of compounds): sea salt and the sulfate compounds [H_2SO_4, NH_4HSO_4, $(NH_4)_3H(SO_4)_2$, and $(NH_4)_2SO_4$], respectively. Their dominance determines the shape of the humidograms and is indicated by a deliquescence step [at $\sim$75% for sea salt and $\sim$80% for $(NH_4)_2SO_4$] or its appearance upon addition of NH_3 in the case of acid sulfate aerosols. Sulfate aerosol humidograms are illustrated by data obtained in the St. Louis area as shown in Figures 5a through 5d. Similar humidograms have been observed at two other sites in the midwestern United States: Milford, Michigan, and Hall Mountain, Arkansas [Weiss et al. (28)]. It can be concluded that over a large area of the Midwest the effect of humidity on light scattering and visibility will be dominated by sulfate compounds, and this effect can be predicted from those curves. Humidograms for most urban or photochemical aerosols are monotonic, indicating that strongly deliquescent compounds were lacking in the aerosol or at least they were not dominant. Inorganic nitrate compounds, although they are deliquescent in bulk form, have been observed to remain as supersaturated solution droplets at very low humidities, both alone and in combination with other inorganic salts. The presence of a significant amount of nitrates along with organic compounds in urban and photochemical aerosol is probably responsible for the monotonic increase in b_{sp} with humidity in spite of the presence of sea salt or sulfate compounds, as were indicated by independent chemical analyses or the proximity of sources (e.g., coastal site, Figure 5g).

Only in conditions where continental background aerosol was present was the light-scattering increase at 80% RH less than a factor of 1.6. During these periods the size distribution was presumably strongly dominated by coarse mode, nonhygroscopic soil

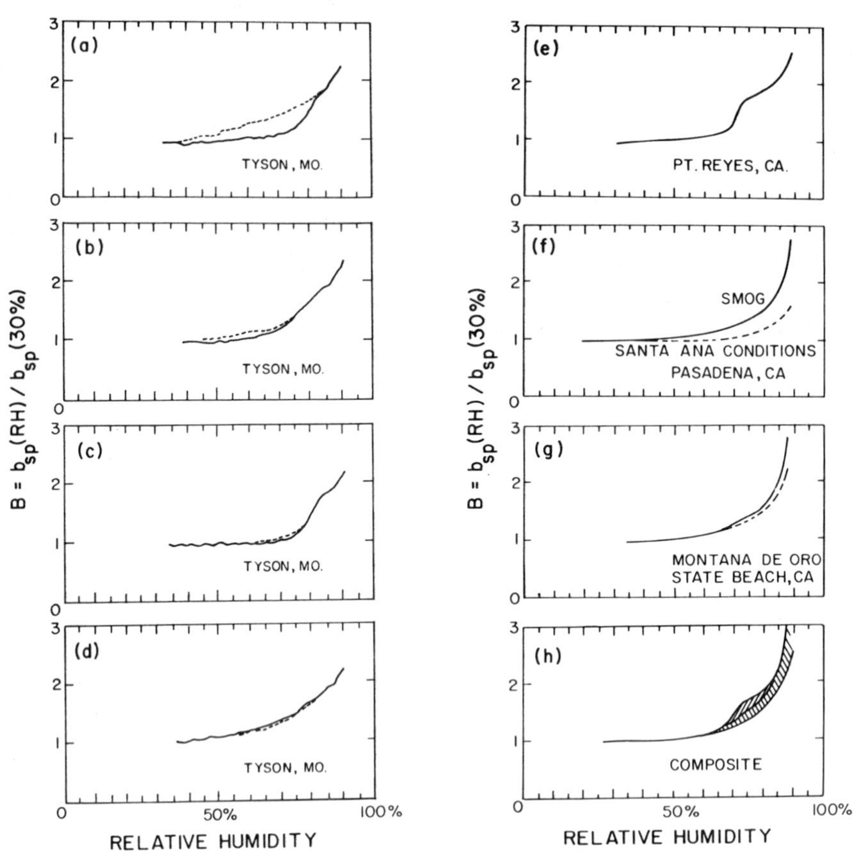

FIGURES 5a to 5h. Humidograms for a number of sites show the increase in b_{sp} which can be expected at elevated humidities for specific sites or aerosol types (marine, Point Reyes, CA; sulfate, Tyson, MO) and the range observed for a variety of urban and rural sites (composite).

dust and not by the accumulation mode. However, the increase in b_{sp} with relative humidity, if not the level of b_{sp} at low RH, was probably still controlled by the submicron fraction. Soil dust is notably hygrophobic and, according to the data of Winkler (31), would have provided an increase in b_{sp} less than a factor of 1.1 at 80% RH.

A composite of humidograms averaged by site (Figure 5h) indicates the range of hygroscopic behavior that may be expected for atmospheric aerosols. Another measure of the range of hygroscopic behavior of atmospheric aerosols is illustrated in Figure 6. Histograms of the frequency of occurrence of B values at several relative humidities were plotted after the data were sorted according to site and/or aerosol chemical type. Statistics for these data are presented in Table 2. Acid sulfate, marine, and urban or photochemical aerosols had a mean ratio of 1.1 to 1.2 at 60% RH, while continental background and sulfate salt aerosols had a ratio of 1.05 to 1.1. At 80% RH the ratios were 1.6 to 1.9 for all aerosol types except continental background, which had a ratio of 1.3 or less. Thus we see that (a) the effect of humidity is quite regular for each aerosol type, (b) the molecular composition affects the dependence of B on RH, and (c) in all cases and regardless of composition, the effect of 80% RH on the most hygroscopic aerosols is only modestly different from the effect on the least hygroscopic, so that forecasting the effect of RH on b_{sp} is possible.

IV. CONCLUSIONS

The major conclusions we wish to suggest are as follows:

(1) The degradation of visual range by aerosol particles is due to the accumulation mode (0.1 to 2.0 μm particle sizes) except in dust storms and fog.

(2) The correlation of accumulation mode mass concentration with optical effects is good.

(3) The optical properties controlling visual range can be increased with sufficient accuracy for most applications.

(4) Relative humidity has predictable and measurable effects on visual range, which are small if RH < 70%.

(5) As a result of the above, visibility degradation--even at RH up to 90%--is quite possibly the best understood effect of air pollution.

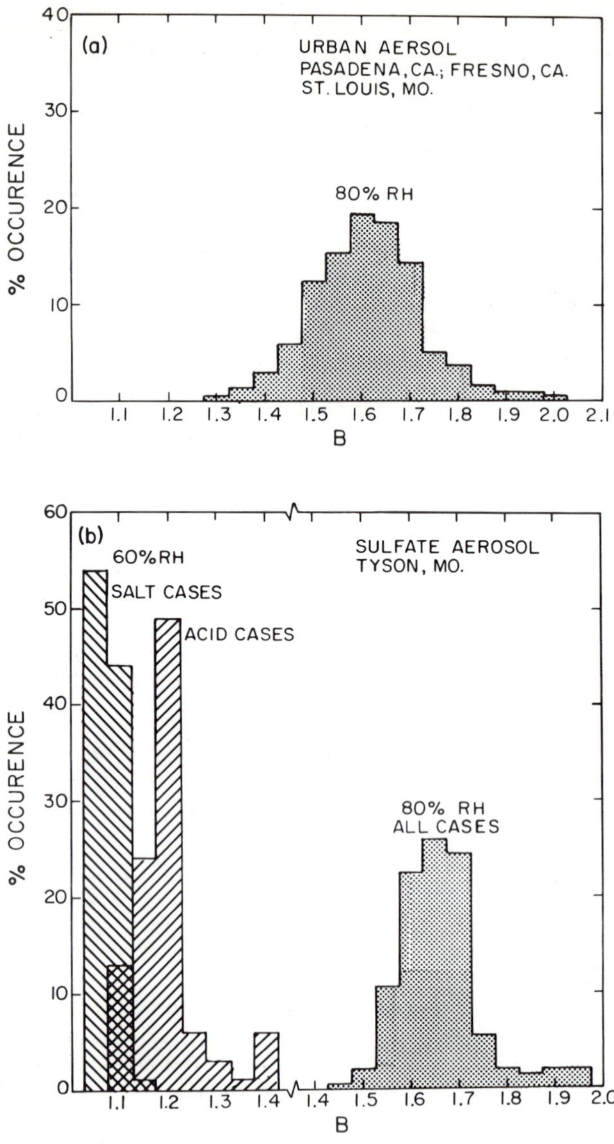

FIGURES 6a to 6d. Histograms of observed values of B at 80% RH and 60% RH.

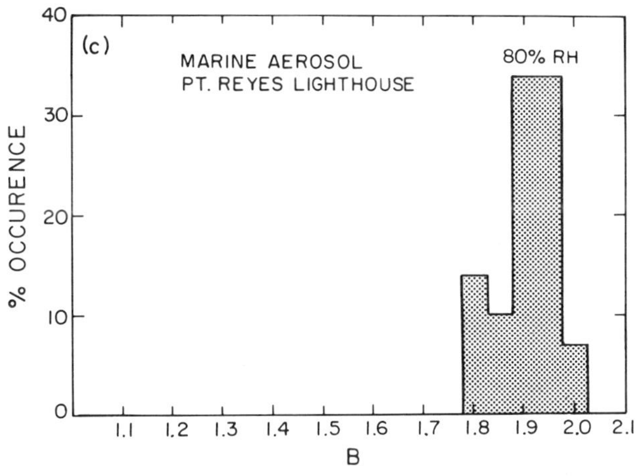

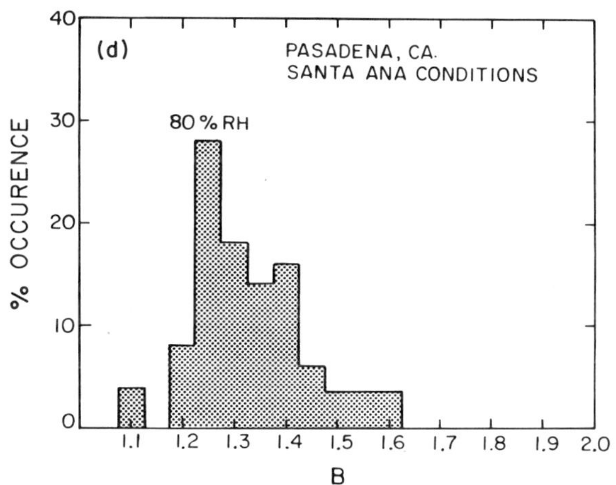

FIGURES 6a to 6d (cont.)

TABLE 2. Humidograph Statistics[a]

Site		Relative humidity (%) 60	70	80	85	Remarks
Point Reyes, CA	$\overline{B}$ =	1.13		0.89		
	s =	0.06		0.21		
	$\overline{B}$ =	(1.11)		(1.90)		(Sea salt)
	s =	(0.03)		(0.06)		
Tyson, MO		1.07	1.16	1.64	1.87	(Ammonium sulfate)
		0.03	0.04	0.07	0.09	
		(1.19)	(1.35)	(1.67)	(1.90)	(Acid sulfate)
		(0.07)	(0.09)	(0.10)	(0.12)	
Fresno, CA		1.09		1.55		
		0.05		0.12		
Hunter Liggett, CA		1.07		1.59		
		0.04		0.10		
Pasadena, CA		1.14		1.61		
		0.05		0.10		
		(1.04)		(1.34)		(Santa Ana)
		(0.04)		(0.10)		
Pomona, CA		1.19		1.95		
		0.05		0.21		

[a] $\overline{B}$ is the average; s is the standard deviation of $\overline{B}$ values observed at the indicated site and RH.

REFERENCES

1. Bodhaine, B. and Mendonca, B. 1974. Preliminary Four Wavelength Nephelometer Measurements at Mauna Loa Observatory, Geophys. Res. Lett. 1, 119.

2. Charlson, R. J., Ahlquist, N. C., and Horvath, H. 1968. On the Generality Correlation of Aerosol Mass Concentration and Light Scatter, Atmos. Environ. 2, 455.

3. Charlson, R. J., Ahlquist, N. C., Selvidge, A., and MacCready, P. B. 1969. Monitoring of Atmospheric Aerosol Parameters with the Integrating Nephelometer, JAPCA 19, 937.

4. Charlson, R. J., Covert, D. S., Tokiwa, Y., and Mueller, P. K. 1972. Multi-wavelength Nephelometer Measurements in Los Angeles Smog Aerosol-III, in Aerosols and Atmospheric Chemistry (G. M. Hidy, ed.), Academic Press, New York.

5. Charlson, R. J., Vanderpol, A. H., Covert, D. S., Waggoner, A. P., and Ahlquist, N. C. 1974. $H_2SO_4/(NH_4)_2SO_4$ Aerosol, Optical Detection in the St. Louis Region, Atmos. Env. 8, 1257.

6. Charlson, R. J., Covert, D. S., Larson, T. V., and Waggoner, A. P. Chemical Properties of Tropospheric Sulfur Aerosols, Atmos. Environ. 12, (in press).

7. Covert, D. S. 1974. Ph.D. thesis, University of Washington.

8. Dzubay, T. G. and Stevens, R. K. 1975. Ambient Air Analysis with Dichotomous Sampler and XRF Spectrometer, Environ. Sci. Technol. 9, 663.

9. Eiden, R. 1966. The Elliptical Polarization of Light Scattered by a Volume of Atmospheric Air, Appl. Opt. 5, 569.

10. Fischer, K. 1973. Mass Absorption Coefficient of Natural Particles in the 0.4 to 2.4 μm Wavelength Interval, Contrib. Atmos. Phys. 46, 89.

11. Flowers, E. C. and McCormick, R. A. 1969. Atmospheric Turbidity over the U. S., J. Appl. Meteorol. 8, 955.

12. Grams, G. W., Blifford, I. H., Gillette, D. A., and Russel, P. B. 1974. Complex Index of Refraction of Airborne Soil Particles, J. Appl. Meteorol. 13, 459-471.

13. Hidy, G., ed. 1974. Characterization of Aerosols in California, Final Report to Air Resources Board, State of California.

14. Ho, W., Hidy, G. M., and Govan, R. M. 1974. Microwave Measurements of the Liquid Water Content of Aerosols, J. Appl. Meteorol. 13, 871-875.

15. Hodge, P. W. 1971. A Large Decrease in Clean Air Transmission of the Air 1.7 km above L.A., Nature 129, 549.

16. Horvath, H., and Noll, K. E. 1969. The Relationships Between Atmospheric Light Scattering and Visibility, Atmos. Environ. 3, 543.

17. Jaenicke, R., Junge, C., and Kanter, H. 1971. Messungen der Aerosolgrossenverteilung uber dem Atlantik, Meteorol. Forsch. Erg. B7, 1-54.

18. Junge, C. 1963. Air Chemistry and Radioactivity, Academic Press, New York.

19. Lin, C. I., Baker, M. B., and Charlson, R. J. 1973. Absorption Coefficient of Atmospheric Aerosols: A Method for Measurement, Appl. Opt. 12, 1356-1363.

20. Lindberg, J. D. and Laude, L. S. 1974. Measurement of the Absorption Coefficient of Atmospheric Dust, Appl. Opt. 13, 1923-1924.

21. Lundgren, D. A. and Paulus, H. J. 1975. Mass Distribution of Large Atmospheric Particles, JAPCA 25, 1227.

22. Middleton, W. F. K. 1952. Vision Through the Atmosphere, University of Toronto Press, Toronto, Canada.

23. Mie, G. 1908. Beitrage zur Optik Truber Medien, Ann. Phys. 25, 377.

24. Samuels, H. J., Twiss, W., and Wong, E. W. 1973. Visibility, Light Scattering, and Mass Concentration of Particulate Matter, Report to California Air Resources Board, July.

25. Volz, F. 1972. Infrared Absorption by Atmospheric Aerosol Substance, J. Geophys. Res. 77, 1017-1031.

26. Waggoner, A. P. and Charlson, R. J. 1976. Measurement of Aerosol Optical Parameters, in Fine Particles (B. Y. H. Liu, ed.), Academic Press, New York.

27. Waggoner, A. P., Vanderpol, A. J., Charlson, R. J., Larsen, S., Lennart, G., and Tragdardh, C. 1976. Sulfate-Light Scattering Ratio as an Index of the Role of Sulfur in Tropospheric Optics, Nature 261, 120-122.

28. Weiss, R. E., Waggoner, A. P., Charlson, R. J., and Ahlquist, N. C. 1977. Sulfate Aerosol, its Geographical Extent in the Midwest and Southern U.S., Science 195, 979.

29. Willeke, K. and Whitby, K. T. 1975. Atmospheric Aerosols: Size Distribution and Interpretation, JAPCA 25, 529.

30. Willeke, K. and Brockman, J. E. 1977. Extinction Coefficients for Multimodal Atmospheric Particle Size Distributions, Atmos. Environ. 11, 975-999.

31. Winkler, P. 1969. Untersuchungen uber das Grossenwachstum naturlicher Aerosolteilchen mit der relativen Feuchte nach einer Wagemethode, Ann. Meteorol. 4, 134-137.

32. Winkler, P. and Junge, C. 1972. The Growth of Atmospheric Aerosol Particles as a Function of Relative Humidity, J. Rech. Atmos. (Henri Dessens Memorial), 617-638.

33. Winkler, P., Covert, D., and Heintzenberg, J. 1978. Unpublished results.

PART IV

AEROMETRIC AND SOURCE FACTORS

Summary of Meteorology in the Los Angeles Basin During ACHEX I and II

THEODORE B. SMITH, DONALD L. BLUMENTHAL, AND S. L. MARSH
Meteorology Research, Inc.
464 West Woodbury Road
Altadena, California

I. GENERAL METEOROLOGY

In the Los Angeles basin during the summer months, a late night and early morning stagnation condition generally occurs and pollutants accumulate near their sources. By midmorning a sea breeze starts to form, and the accumulated pollutants are transported across the basin. Several types of sea breeze flow patterns occur. During the early summer the afternoon surface streamlines frequently go north from the southern portion of Los Angeles County, then turn east as they approach the mountains. Figure 1 for July 25, 1973, is typical of this flow pattern. Later in the summer, as high pressure to the northeast and the resulting offshore pressure gradients become more prevalent, the streamlines tend to straighten out and the flow often becomes more westerly across the basin (see Figure 2 for September 20, 1972). The latter flow pattern tends to keep the northern portion of the basin cleaner. The impact of the early morning emissions in the southern Los Angeles County region is shifted away from the Azusa-Upland (AZU-CAB) area south to the Fullerton (FUL) area.

During ACHEX I (1972) most of the sampling in the Los Angeles basin was done in the late summer and early autumn, and the flow pattern in Figure 2 was typical. During ACHEX II (1973) the sampling started earlier, and the offshore pressure gradient conditions were not quite so prevalent, resulting in flow patterns closer to those in Figure 1.

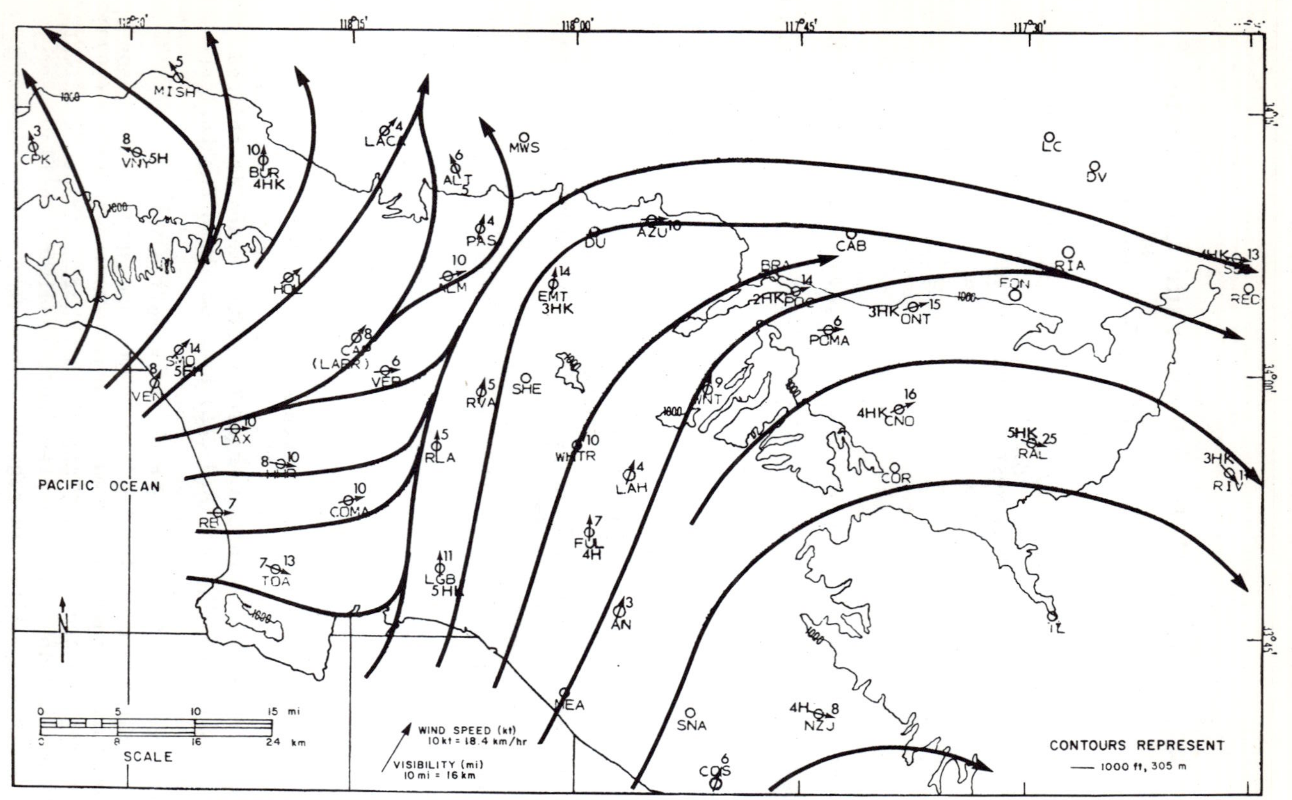

FIGURE 1. Surface wind streamlines: 1600 PDT, July 25, 1973.

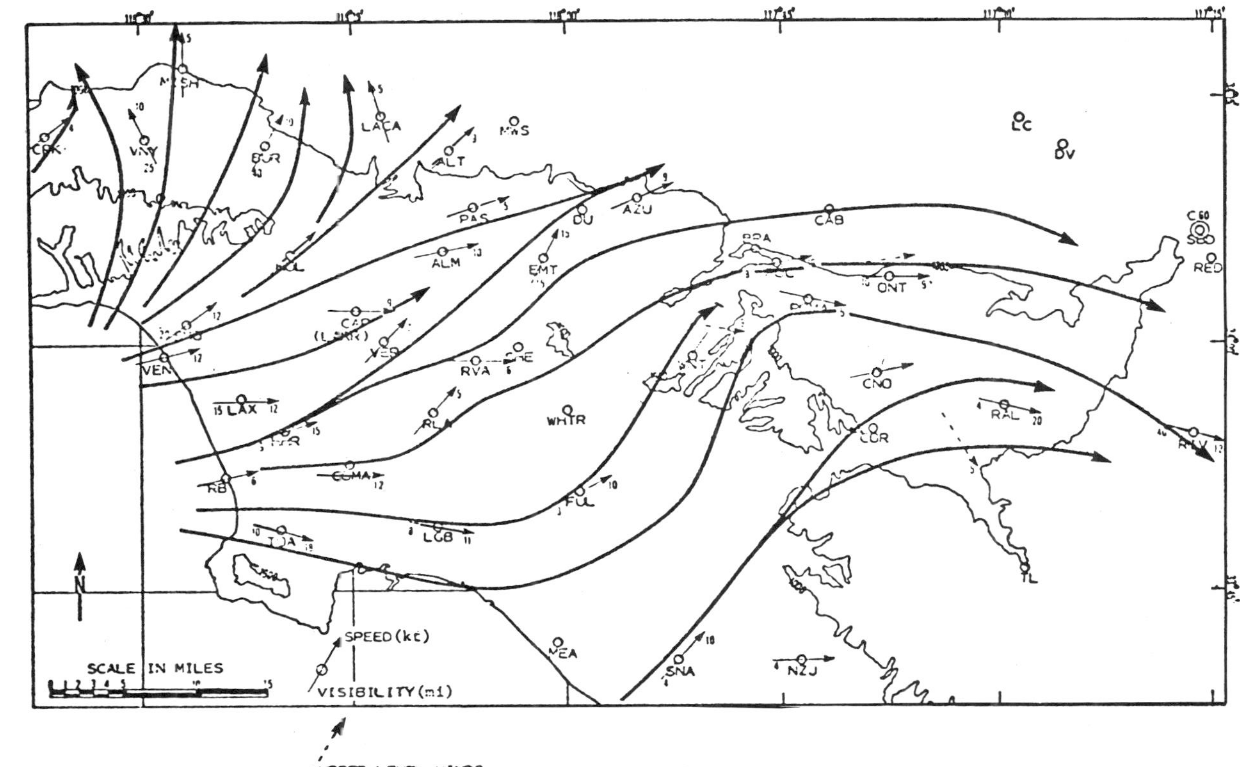

FIGURE 2. Surface wind streamlines: 1600 PDT, September 20, 1972.

II. METEOROLOGICAL INDEX OF POLLUTION POTENTIAL

The wind flow patterns in the basin determine the distribution and transport of pollutants across the basin, but the intensity of an episode is also determined by the strength and depth of the subsidence inversion over the basin and the associated trapping of pollutants.

The temperature aloft, for example, at 850 millibars (about 5000 ft msl), can be an excellent indicator of the relative importance of this semipermanent, low-level temperature inversion. To break through the low-level inversion and permit vertical dispersion of pollutants, the afternoon maximum surface temperature must equal or exceed the temperature aloft, with due regard for the adiabatic decrease in temperature with height. If one assumes a critical value for the temperature at 850 millibars, T_{850}, of 20°C for the warm months of July, August, and September, surface temperatures of 32°C (90°F) would be necessary to break any low-level inversions present. A typical situation on such days might be for the inversion to break inland, but not in the coastal sections where lower surface temperatures prevail. A similar critical value of 15°C has been selected for use in October. A surface temperature of 28°C (80°F) would then be necessary to break any low-lever inversions under these conditions.

To obtain an estimate of the pollution potential and representativeness of the sampling days, 850 millibar temperatures were plotted from daily radiosondes taken at Los Angeles International Airport (LAX), Santa Monica (SMO), and Vandenberg Air Force Base (VBG). Figure 3 is a graph of the 850-millibar temperatures, T_{850} (in °C), from the 0000 GMT (1600 PST) radiosonde at VBG for the months of July, August, September, and October, 1972 and 1973. Days selected for sampling by the ACHEX instrumented van are labeled, as are aircraft sampling and Metronics tracer study days. Peaks in the data signify warmer air aloft, resulting in stronger temperature inversions and an increased pollutant potential, while valleys imply cooler air aloft and weaker inversions.

Table 1 gives the frequency of occurrence of 850-millibar temperatures above the critical value for each month, along with the absolute maximum (in parentheses) for that month and year. The table indicates that September 1972 and September 1973 were unusually low in frequency of warm temperatures aloft, indicating reduced pollutant potentials. The months of July and August for both years, however, tend to be more representative of previous years with frequencies as well as peak values near normal.

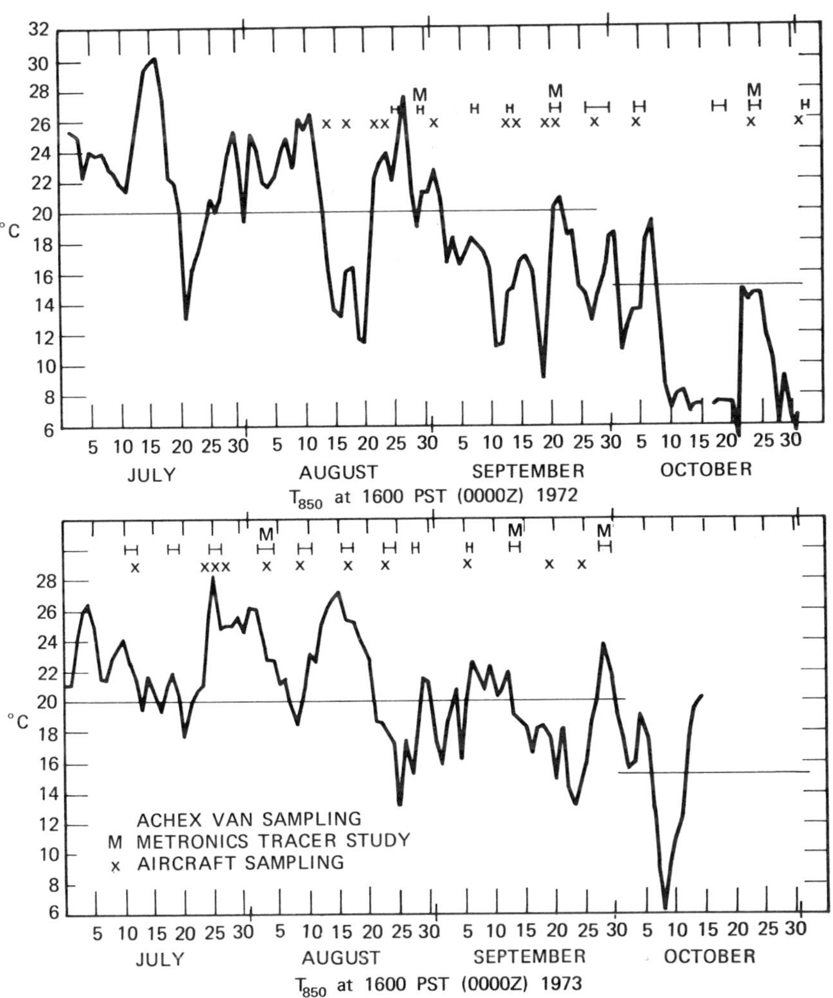

FIGURE 3. T_{850} at 1600 PST (0000Z), 1972 and 1973.

Since much of the 1972 experimental program was performed during late August and September, it is clear from Figure 3 and Table 1 that few high-pollution days were available for sampling. The program was continued into 1973 because of the lack of good sampling days in 1972. Sampling was started earlier to take advantage of the likelihood of a higher frequency of potential pollution days in the early summer. Many days with high-pollution potential were indeed sampled in 1973, and from Figure 3 it is clear that maximum advantage was taken of the few potential days that existed in late 1972.

TABLE 1. Frequency of Occurrence of 850-Millibar Temperatures Above Reference Levels[a,b] (Number of Days per Month)

Year	July (20°C)	August (20°C)	September (20°C)	October (15°C)
1972	26 (30.4)[c]	16 (27.5)[d]	2 (22.5)[d]	4 (18.0)[d]
1973	∿22 (26.0)	25 (25.6)	∿9 (25.8)	∿19 (20.0)
1959		23 (26.2)	13 (26.4)	22 (24.7)
1960	25 (30.6)	23 (26.3)	22 (25.6)	19 (19.6)
1961	26 (30.3)	22 (26.3)	9 (24.1)	17 (24.0)
1962	24 (25.7)	27 (28.0)	23 (25.7)	18 (21.6)
1963	24 (25.9)	25 (25.5)	20 (27.8)	17 (24.6)
1964	26 (28.4)	25 (26.0)	13 (23.6)	24 (24.5)
1966[c]		27 (28.3)	11 (22.2)	16 (21.1)
1967[c]		30 (26.4)	11 (23.5)	26 (22.9)
1968[c]	23 (24.7)	13 (25.2)	19 (25.0)	20 (21.9)
1969[c]	18 (24.5)	26 (27.7)	15 (24.6)	8 (21.7)
1970[c]	29 (26.4)	29 (28.3)	15 (22.7)	15 (24.1)

[a]All data from Santa Monica unless otherwise noted.

[b]Numbers in parentheses are the maximum 850 millibar (5000 ft) temperatures (°C) during month.

[c]Data from Vandenberg Air Force Base.

[d]Data from Los Angeles International Airport.

Three-Dimensional Distribution of Air Pollutants in the Los Angeles Basin*

RUDOLF B. HUSAR AND DAVID E. PATTERSON
Center for Air Pollution
Impact and Trend Analysis
Washington University
St. Louis, Missouri

DONALD L. BLUMENTHAL, WARREN H. WHITE,† AND THEODORE B. SMITH
Meteorology Research, Inc.
Altadena, California

*From Journal of Applied Meteorology, Vol. 16, No. 10, October 1977. Reprinted by permission of the American Meteorological Society.

†Present affiliation: 1180 North Chester, Pasadena, California, 91104.

Abstract

Data from a three-dimensional pollutant mapping program, conducted in the Los Angeles basin, were analyzed to obtain "grand average" vertical profiles sampled on 24 summer days in 1973. Morning and afternoon profiles at four locations show an erosion of the nighttime radiation inversion, increased temperatures, more intense mixing in the inland areas, and a semi-permanent subsidence inversion at higher levels. High values of primary pollutant parameters (NO_x and condensation nuclei) are seen in the western part of the basin at Hawthorne. Secondary pollutant parameters (O_3 and light scattering coefficient) were dominating at the inland receptor site, Riverside. Ozone concentrations in the morning were consistently higher aloft. The deficit near the surface is attributed to ozone scavenging by primary emissions.

I. INTRODUCTION

The experience of observing the smog layer from a descending airplane is probably the most convincing demonstration of the three-dimensional nature of the Los Angeles smog. In the summer months of 1972 and 1973, the Three-Dimensional Pollutant Gradient Study was conducted to map the vertical and horizontal pollutant structure, and to study the transport and transformation processes in the Los Angeles basin. Two aircraft made soundings, in the form of vertical spirals, from the top of the polluted layer to the surface at 17 locations with an average of about 40 spirals for each location. Three meteorological parameters, temperature, relative humidity, and turbulence (energy dissipation rate), as well as five air pollution parameters, O_3, NO_x, CO, light scattering coefficient, and condensational nuclei, were monitored continuously along with the necessary coding and position parameters. A list of the instruments used and their characteristics is given in Table 1.

The three-dimensional pollutant mapping program improved substantially our understanding of the pollutant transport and

TABLE 1. Aircraft Instrumentation

Instrument	Ranges available	Time response (90% for usual ranges)
1. MRI Integrating Nephelometer	10^a, 40, 100 x 10^{-4} m^{-1}	1 s
2. Environment One Condensation Nuclei Counter	1, 3, 10, 30, 100^a, 300^a, 10K x 10^3 CN cm^{-3}	5 s
3. REM 612 Ozone Monitor (chemiluminescent)	50^a, 200 pphm	5 s
4. REM 642 NO-NO_x Monitor (chemiluminescent)	0.5^a, 2^a, 10 ppm	5 s
5. Andros 7000 CO Monitor (dual isotope fluorescence)	20, 50^a, 100, 200 ppm	5 s
6. MRI Airborne Instrument Package		
Temperature	-5 to +45°C	5 s
Humidity	0 to 100 %	30 s
Turbulence	0 to 10 $cm^{2/3}$ s^{-1}	3 s (to 60%)
Altitude	0 to 10,000 ft	1 s
Indicated air speed	50 to 150 mph	≤ 1 s
7. Metrodata M/8 VOR Analog Converter (uses aircraft radio)		1 s
8. Metrodata 620 Data Logger (20 channels)		48 channels/second

[a] Range normally used.

transformation processes in the Los Angeles basin. The data indicate complex and varying patterns of these processes from one time or location to another. Detailed inspection of the data revealed, however, that certain phenomena are so characteristic for a given location or time period that they survive extensive averaging.

In this paper we present and discuss the "grand average" vertical profiles measured over the airports of Hawthorne, El Monte, Ontario, and Riverside. Their respective locations in the basin are shown in Figure 1. All 165 soundings, conducted on 24 days in 1973 at these locations, were used to calculate the mean and standard deviation of all measured parameters for each location, time period (a.m. or p.m. PDT), and hundred-foot altitude increment.

The average profiles presented in this section are calculated from a sample population which was biased toward photochemical smog conditions, as mapping was intentionally restricted to days with above average smog potential. Although they cannot be taken as representative on an overall basis, the mean profiles should be fairly typical of conditions during smoggy summer and early fall days since they are drawn from data on 24 days of this type. It is hoped that the findings presented here will have utility in the assessment of average transport and transformation processes and in providing information for the testing of three-dimensional diffusion-transformation models.

II. METEOROLOGICAL PARAMETERS

A. Temperature

A primary factor in the production of high pollutant concentrations in Los Angeles air is the frequent occurrence of temperature inversions which restrict the vertical dispersion of pollutants. This phenomenon is evident in Figure 2, which shows the mean morning and afternoon temperature profiles at Hawthorne, El Monte, Ontario, and Riverside. In the mean morning profiles at all four locations, the temperature of the air at 3000 ft MSL is higher than the temperature of the air near the surface, with a negative lapse rate from 3000 ft down almost to the surface. As the day progresses, the surfaces of inland areas of the basin are heated by the sun, warming the air near the ground and eroding the inversion from below [Edinger (5)]. By afternoon, the mean profiles for all three inland locations show a well-defined layer at the surface in which the lapse rate is adiabatic.

The effect of surface heating is seen from another perspective in Figure 3a, which displays the mean afternoon tempera-

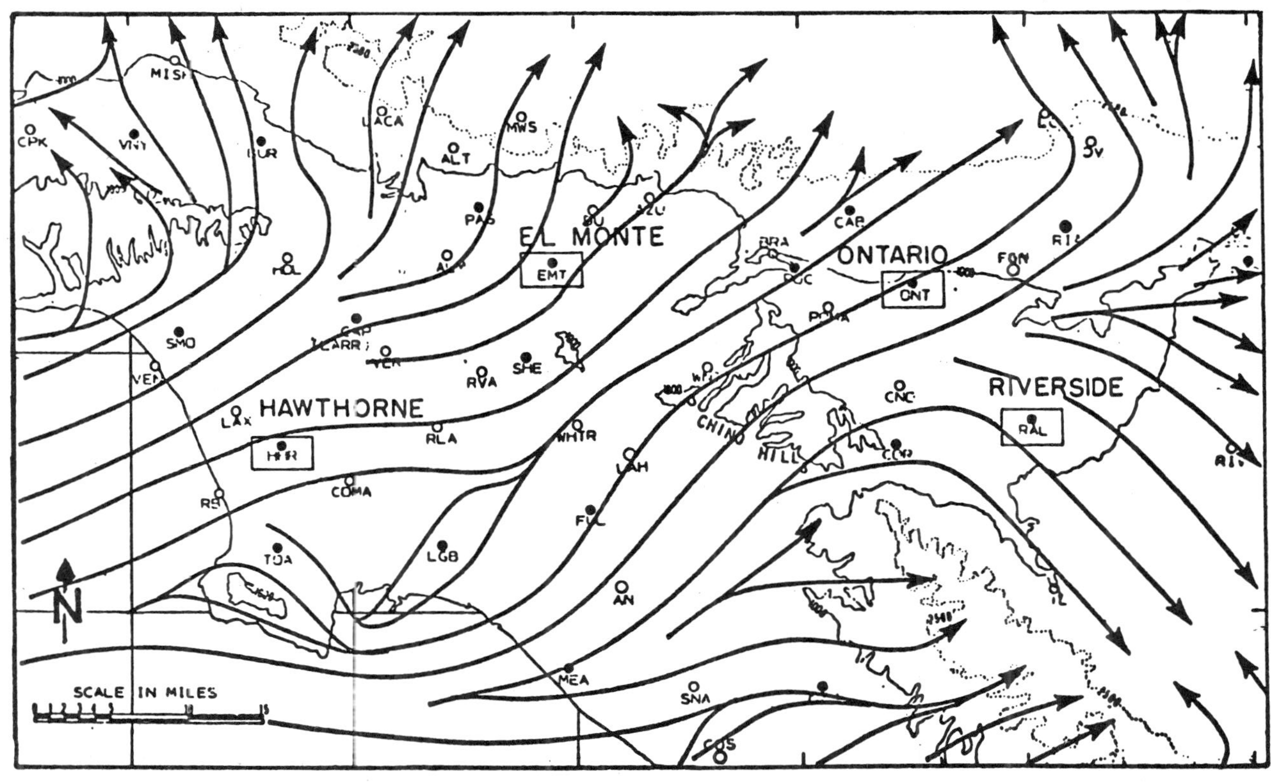

FIGURE 1. Map of the Los Angeles air basin indicating the locations where the "grand average" profiles were calculated. Streamlines show most frequent afternoon surface winds during July [after Blumenthal et al. (1)] .

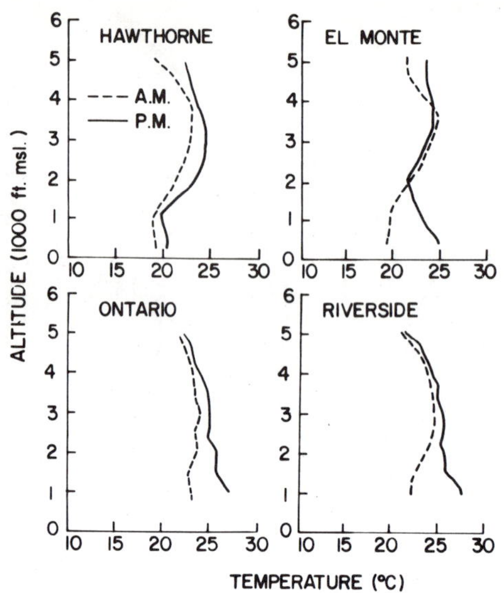

FIGURE 2. Mean morning and afternoon temperature profiles

ture profiles for the four sampling locations. At Hawthorne the afternoon profile is little changed from the morning profile due to the stabilizing influence of the nearby ocean. As air moves inland with the usual afternoon flow, however, temperatures near the surface increase, eroding the inversion. The unstable layer at the surface deepens until, at Ontario and Riverside, it may "break through" the inversion.

B. Turbulence

The mean profiles of small-scale turbulence intensity (Figure 4), as indicated by the energy dissipation coefficient, $\varepsilon^{1/2}$ [MacCready (14)], are compatible with the mean thermal structure. In the morning, when air is stable, turbulence is confined to a shallow layer at the surface where it is generated by mechanical effects and surface heating. In the afternoon, turbulence extends about 1000 ft above the morning mixing layer. In all profiles, the intensity of the turbulence decreases with height until it reaches a "background" value of about $\varepsilon = 0.5$ $cm^{2/3}$ s^{-1}, reflecting the decay of small-scale turbulence with increasing height.

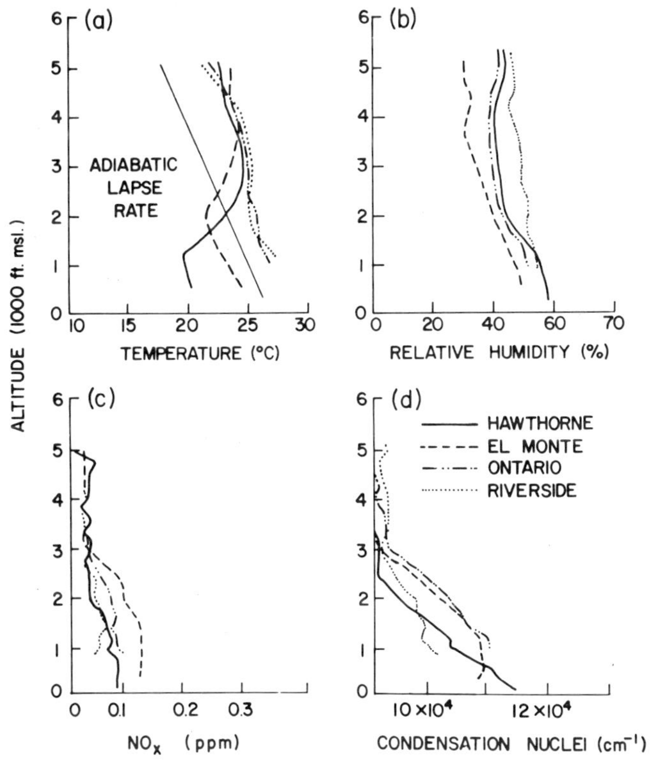

FIGURE 3. Mean afternoon profiles of temperature, humidity, NO_X, and condensation nuclei at the four sampling sites.

C. Relative Humidity

Mean afternoon profiles of relative humidity are shown in Figure 3b. Mean morning and afternoon relative humidities are generally in the range 30 to 40%, except in the mixing layers at Hawthorne and Riverside.

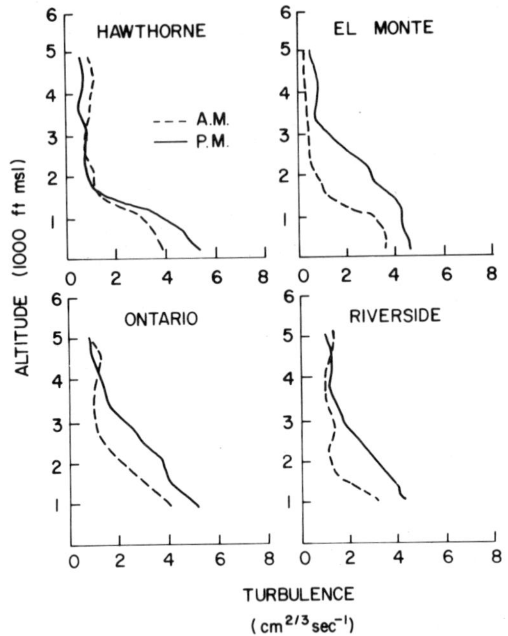

FIGURE 4. Mean morning and afternoon turbulence profiles.

III. REACTIVE GASES

A. Nitrogen Oxides

The first two oxides of nitrogen are important participants in the photochemistry of the Los Angeles atmosphere. Photodissociation of NO_2 ($NO_2 + h\nu \rightarrow NO + O$) starts the chain of reactions leading to the buildup of ozone. Oxidation of NO ($NO + O_3 \rightarrow NO_2 + O_2$) limits the rate of this buildup in its early stages. Although these two reactions affect the relative concentrations of NO and NO_2 in the air, they leave invariant their sum, the concentration of NO_x. NO_x is formed almost exclusively by combustion sources such as motor vehicles and power plants and is lost through reactions with surfaces and by the formation of nitrates.

Mean profiles of NO_x concentration are shown in Figures 5 and 3c. Substantial concentration gradients appear near the

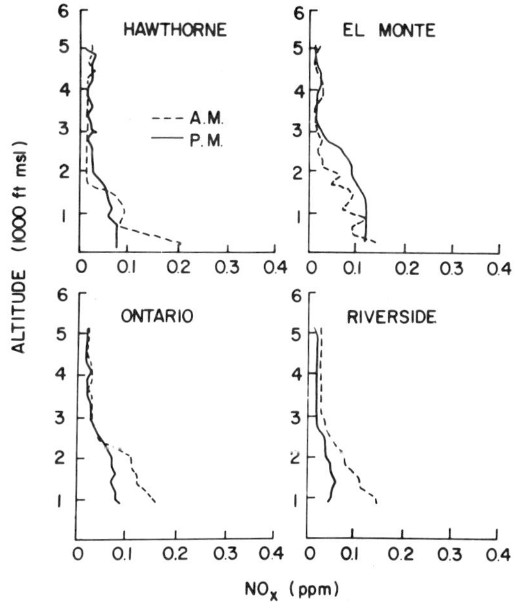

FIGURE 5. Mean morning and afternoon NO_x profiles.

surface in the morning profiles due to the generally poor mixing prevailing at this time. By afternoon, concentrations become fairly uniform through the unstable mixing layer. In both morning and afternoon profiles, mean concentrations drop to about 0.03 ppm above the mixing layer.

The smallest values for mean NO_x concentrations and the integral of these concentrations with respect to height are seen in the afternoon profile for Riverside. The latter half of a typical afternoon trajectory to Riverside passes over a predominantly rural area, where emissions of NO_x are apparently not sufficient to balance losses by aerosol formation and dry deposition. Peroxyacetyl nitrate (PAN) concentrations at Riverside can reach 0.05 ppm on smoggy days [Lundgren (13)]. Particulate nitrate concentrations > 100 $\mu g\ m^{-3}$, equivalent to over 0.04 ppm of NO_x, were measured there during the 1973 Aerosol Characterization Study [Hidy et al. (8)]. The measured high concentrations of PAN and particulate nitrates indicate that a substantial fraction of the NO emitted in the western portion of the air basin is converted to nitrates. It is also likely that a further, at present unknown, fraction of NO_x is lost to vegetation and ground [Hill (9)] enroute to Riverside.

B. Ozone

Unlike NO_x, ozone is not emitted directly, but is formed in the atmosphere through the sequence of reactions initiated by the photodissociations of NO_2. It reacts very rapidly with NO, the principal constituent of NO_x emissions, so that fresh emissions of NO_x tend to lower, rather than raise, O_3 concentrations. These characteristics are clearly demonstrated in the mean profiles of O_3 (Figure 6).

Note that ozone concentrations do not drop to background levels (∿0.04 ppm) above the mixing layer. At Hawthorne, for

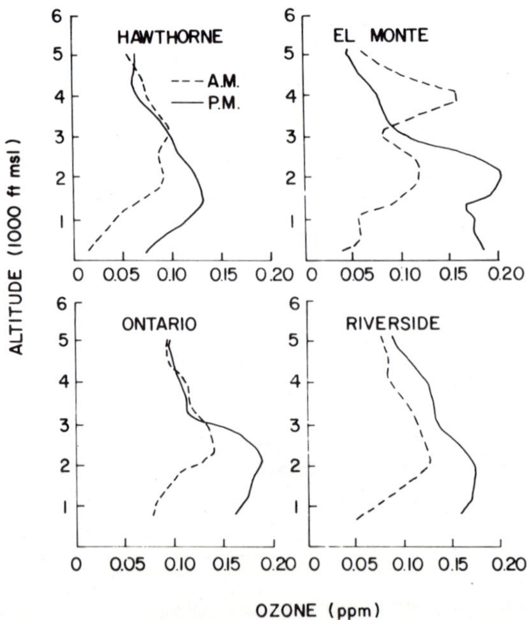

FIGURE 6. Mean morning and afternoon O_3 profiles.

example, the mean temperature, turbulence, and NO_x profiles all indicate a mixing layer strongly confined below 2000 ft MSL, yet the mean O_3 concentration at 3000 ft MSL is 0.1 ppm, greater than the federal standard of 0.08 ppm. Mechanisms by which polluted air occurs above the mixing layer are discussed elsewhere [Blumenthal et al. (1)]. What is important in the present context is that the O_3 at 3000 ft over Hawthorne is part of a polluted air mass which has aged above the mixing layer. In the absence of scavenging by fresh NO_x emissions, which are confined within the mixing layer, reactions have proceeded almost to completion, producing relatively large ozone concentrations from small concentrations of NO_x and hydrocarbons.

A second feature of the mean profiles which is unique to O_3 is the occurrence of deficits near the surface. The most striking example of this is the morning profile at Hawthorne, where mean O_3 concentrations near the surface are under 0.02 ppm, less than the 0.04-ppm background levels observed in clean air. In fact, all four morning profiles, as well as the afternoon profile at Hawthorne, show ozone concentrations which are lower within the mixing layer than they are immediately above it. These O_3 deficits are due principally to scavenging by fresh emissions of NO_x which are trapped within the shallow, well-defined mixing layers exhibited in these profiles.

IV. AEROSOL PARAMETERS

A. Condensation Nuclei Count

Since small particles are much more numerous than large particles, the condensation nuclei count (CN) is primarily an index of the small-particle fraction of the aerosol. Particle size distribution measurements [Whitby et al. (15)] have shown that particles under 0.1 μm in diameter account for nearly all of the CN. Particles in this size range are produced by combustion sources and (under conditions thought to be rare in ambient Los Angeles air) by homogeneous nucleation [Husar et al. (11), Whitby et al. (15)]. These particles coagulate rapidly with larger particles and with each other, so that the half-life of CN under typical Los Angeles smog conditions is on the order of an hour [Husar and Whitby (10), Husar et al. (11), Davies (4)].

Mean profiles of CN are shown in Figures 7 and 3d. Substantial CN gradients are observed near the surface in all four morning profiles and in the afternoon profile at Hawthorne. Above the mixing layer, CN counts drop to about 5×10^3 cm^{-3}, less than one-tenth of the counts within the mixing layer. Counts within the mixing layer are lowest at Riverside.

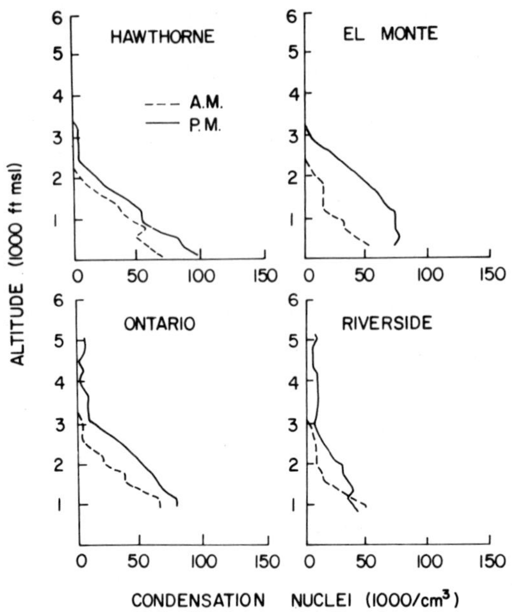

FIGURE 7. Mean morning and afternoon condensation nuclei profiles.

B. Light-Scattering Coefficient

The light-scattering coefficient (b_{scat}) of the atmosphere is determined largely by the concentration of particles with diameters in the range 0.1 to 1.0 μm [Ensor et al (6)], since these are the most efficient scatterers of visible radiation. Electron micrographs [Husar et al. (12)] and chemical analysis [White (16)] indicate that, under photochemical smog conditions, much of the aerosol in this size range is composed of secondary sulfates, nitrates, and organics [Hidy et al (8), Gartrell and Friedlander (7)] produced from the gas phase. Some of this material is hygroscopic, and b_{scat} is affected by high ambient relative humidities [Covert et al (3)]. Once formed, the light scattering fraction of the aerosol is a fairly stable component of the atmosphere, with low surface loss rates [Chamberlain (2)] and low coagulation rates [Husar et al (11), Davies (4)].

Of the four air quality parameters studied in this work, b_{scat} is the most stable and least sensitive to fresh emissions. These qualities are shown in the mean afternoon profiles of b_{scat} (Figure 8). In all four profiles b_{scat} is fairly uniform through the mixing layer and drops to low levels above the mixing layer.

V. POLLUTANT AGING

Figure 1 shows that the characteristic air masses sampled over Hawthorne and Riverside in the afternoon have different histories. Hawthorne is only about 30 min of air parcel travel time downwind from the ocean. During those 30 min over land en route to Hawthorne, air passes over the San Diego Freeway and State Highways 1 and 107, as well as the heavily industrialized region around El Segundo. The afternoon soundings at Hawthorne thus sample predominantly fresh emissions, superimposed on the marine background (which itself may contain well aged anthropogenic contaminants).

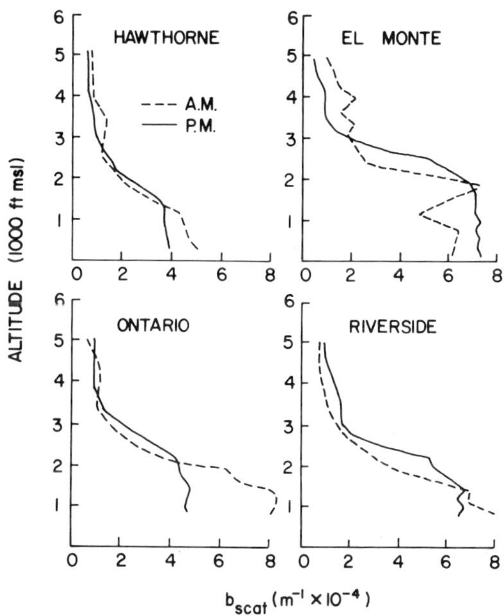

FIGURE 8. Mean morning and afternoon b_{scat} profiles.

Riverside, on the other hand, is from 3 to 6 hr downwind of the ocean. As marine air moves inland, it first passes over the urban and industrialized areas of Los Angeles and Orange counties then over the relatively rural region from the Chino Hills and Santa Ana mountains eastward to Riverside. As a result, most of the contaminants sampled in the afternoon soundings at Riverside have had 3 to 6 hr to age enroute.

The afternoon contaminant characteristics of the two locations are summarized and compared in Figure 9 which shows the afternoon means for NO_x, O_3, CN, and b_{scat} in the first 200 feet above ground level. It is apparent from Figure 9 that the afternoon contaminant mixtures at Hawthorne and Riverside reflect the different histories of the air sampled, with significantly more NO_x and CN at Hawthorne and significantly more O_3 and b_{scat} at Riverside. Direct emissions account for most NO_x and CN, at which concentration can only decay in the atmosphere. High values of these parameters are thus generally indices of fresh emissions. Ozone, and a large fraction of the light scattering aerosol, are not emitted directly, but are produced in the atmosphere as secondary pollutants. High values of these parameters are generally indices of pollutant aging.

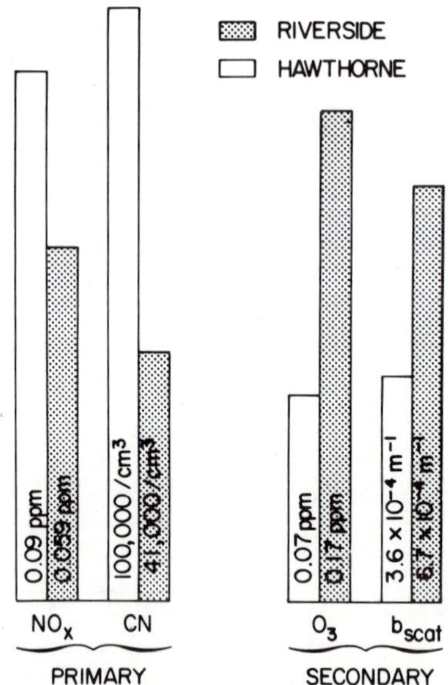

FIGURE 9. Mean afternoon contaminant characteristics in first 200 ft above ground at Hawthorne and Riverside.

VI. VARIABILITY OF THE PROFILES

The profiles presented in the previous sections are averages over 24 days of sampling. In some instances, the variability about the mean is also of interest. For illustration purposes, the mean and the standard deviation of temperature, turbulence, ozone, and b_{scat} for the afternoon soundings at El Monte are shown in Figure 10. The standard deviations typically range from 15% of the mean to over 40% of the mean; the pollutant parameters vary more than temperature and turbulence.

Since it is not convenient to display the profiles with their standard deviations for all sites for each parameter for both morning and afternoon, we choose to represent each profile with a single number--its vertical integral. The integrations were carried out from ground level to 5200 ft MSL, and the results are listed in Table 2. The vertical integral of the light scattering coefficient, i.e., the average optical depth of the summer smog layer, is between 0.25 at Hawthorne and 0.54 at El Monte. The integrals of condensation nuclei, ozone, and NO_x do not have a simple physical meaning, but they may be useful in matching photochemical smog models with observed data. Also, the standard deviations of the integrals are a measure of the variability of the profiles themselves.

VII. SUMMARY AND CONCLUSIONS

Morning and afternoon "grand average" vertical profiles for Hawthorne, El Monte, Ontario, and Riverside show an erosion of the nighttime radiation inversion during the day and increased temperatures and mixing due to surface heating in the inland areas. The semi-permanent subsidence inversion is also seen at higher levels over the four sampling stations.

Average contaminant profiles indicate high values of primary pollutants (NO_x and condensation nuclei) in the source area near the coast at Hawthorne. On the other hand, the receptor site, Riverside (several hours of transport time inland), is characterized by high values of secondary pollutants, O_3 and b_{scat}, and low NO_x and CN. Morning profiles of O_3 at all locations indicate higher concentrations aloft than at the surface. This deficit is attributed to the scavenging of ozone by primary emissions.

ACKNOWLEDGMENTS

This work was supported by the California Air Resources Branch. The authors appreciate the support and guidance provided

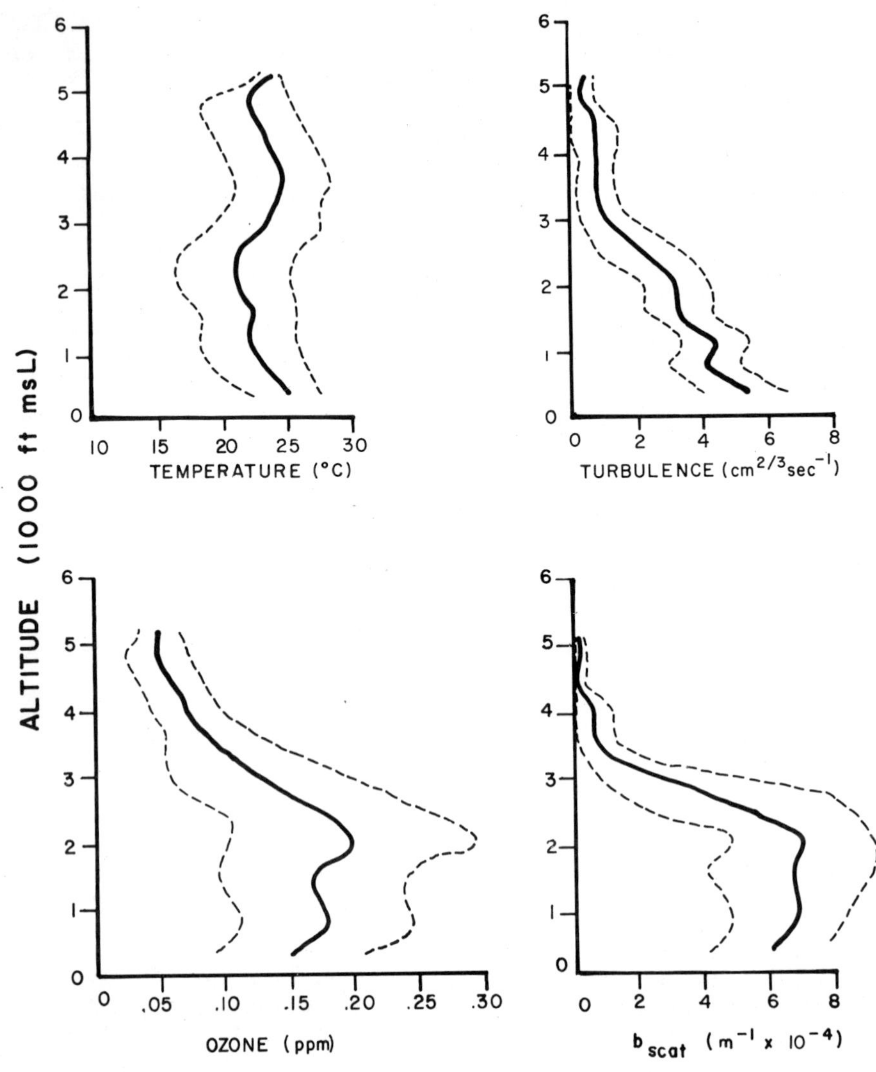

FIGURE 10. The mean and standard deviation of temperature, turbulence, ozone, and b_{scat} for the afternoon soundings at El Monte: solid line is the mean; dashed lines represent the mean ± 1 standard deviation.

TABLE 2. Vertical Integrals of the Profiles and Their Standard Deviations

Meteorological parameters and pollutants		Hawthorne		El Monte		Ontario		Riverside	
		Mean	$\pm\sigma$	Mean	$\pm\sigma$	Mean	$\pm\sigma$	Mean	$\pm\sigma$
Temperature	A.M.	30180	4816	30850	2621	32310	4572	32310	4389
(°C-m)	P.M.	32740	4511	33470	5242	36700	3901	37060	4206
Turbulence	A.M.	2170	1622	1658	902	2682	1341	2219	951
($cm^{2/3}$ s^{-1}-m)	P.M.	2268	1402	3097	1256	3999	1609	3901	1244
Condensation nuclei	A.M.	19.51	10.5	15.12	7.5	34.62	15.9	20.48	9.0
(10^{12} m^{-2})	P.M.	31.70	13.9	36.58	14.9	53.16	22.4	36.09	20.2
Ozone[a]	A.M.	100	40	142	61	141	63	135	51
(ppm-m)	P.M.	140	69	185	79	199	86	210	77
NO_x	A.M.	76.8	38.0	79.2	51.0	132.9	35.0	108.5	46.0
(ppm-m)	P.M.	64.8	38.0	101.2	27.0	74.4	41.0	53.6	24.0
b_{scat}	A.M.	0.295	0.166	0.492	0.239	0.671	0.269	0.507	0.254
(dimensionless)	P.M.	0.254	0.135	0.541	0.219	0.349	0.202	0.476	0.312

[a] The units of the vertical integrals were obtained as ppm-meters (rather than ppm-ft).

by the late Dr. D. G. Hutchinson, Mr. Harris Samuels, and Dr. Jack Suder.

REFERENCES

1. Blumenthal, D. L., Smith, T. B., White, W. H., Marsh, S. L., Ensor, D. S., Husar, R. B., McMurry, R. B., Heisler, S. L., and Owens, P. 1974. "Three-Dimensional Pollutant Gradient Study--1972-1973 Program," Tech. Rep. MR174FR-1261, prepared for the California Air Resources Board by Meteorology Research, Inc. [NTIS PB 241982/66].

2. Chamberlain, A. C. 1967. Radioactive Aerosols and Vapours, Contemp. Phys. 8, 561-581.

3. Covert, D. S., Charlson, R. J., and Ahlquist, N. C. 1972. A Study of the Relationship of Chemical Composition and Humidity to Light Scattering by Aerosols, J. Appl. Meteor. 11, 968-976.

4. Davies, C. N. 1974. Size Distribution of Atmospheric Particles, Aerosol Sci. 5, 239-300.

5. Edinger, J. G. 1973. Vertical Distribution of Photochemical Smog in the Los Angeles Basin, Environ. Sci. Tech. 7, 247-252.

6. Ensor, E. S., Charlson, R. J., Ahlquist, N. C., Whitby, K. T., Husar, R. B., and Liu, B. Y. H. 1972. Multiwavelength Nephelometer Measurements in Los Angeles Smog Aerosol, J. Colloid Interface Sci. 39, 242-251.

7. Gartrell, G. and Friedlander, S. K. 1975. Relating Particulate Pollution to Sources: The 1972 California Aerosol Characterization Study, Atmos. Environ. 9, 279-299.

8. Hidy, G. M., et al. 1974. "Characterization of Aerosols in California," Final Rep. to California Air Resources Board under Contract 358 [NTIS: Vol. 2, PB248799/66; Vol. 4, PB247947/66].

9. Hill, A. C. 1971. Vegetation: A Sink for Atmospheric Pollutants, J. Air Pollut. Control Assoc. 21, 341-346.

10. Husar, R. B. and Whitby, K. T. 1973. Growth Mechanisms and Size Spectra of Photochemical Aerosols, Environ. Sci. Tech. 7, 241-247.

11. Husar, R. B., Whitby, K. T., and Liu, B. Y. H. 1972. Physical Mechanisms governing the Dynamics of Los Angeles Smog Aerosol, J. Colloid Interface Sci. 39, 211-224.

12. Husar, R. B., White, W. H., and Blumenthal, D. L. 1976. Direct Evidence of Heterogeneous Aerosol Formation in the Los Angeles Smog, Environ. Sci. Tech. 10, 490-491.

13. Lundgren, D. A. 1970. Atmospheric Aerosol Composition and Concentration as a Function of Particle Size and of Time, J. Air Pollut. Control Assoc. 20, 603-608.

14. MacCready, P. B., Jr. 1964. Standardization of Gustiness Values from Aircraft, J. Appl. Meteor. 3, 439-449.

15. Whitby, K. T., Husar, R. B., and Liu, B. Y. H. 1972. The Aerosol Size Distribution of Los Angeles Smog, J. Colloid Interface Sci. 39, 177-204.

16. White, W. H. 1976. Reduction of Visibility by Sulphates in Photochemical Smog, Nature 264, 735-736.

The Evolution of the 3-D Distribution of Air Pollutants during a Los Angeles Smog Episode: July 24–26, 1973

DONALD L. BLUMENTHAL, THEODORE B. SMITH, AND WARREN H. WHITE
Meteorology Research, Inc.
464 West Woodbury Road
Altadena, California

I. INTRODUCTION

During the summers of 1972 and 1973 an independent study called the 3-D Pollutant Gradient Study was carried out in cooperation with the ACHEX program [Blumenthal et al. (1)]. The primary objective of the gradient study was to gain an understanding of the three-dimensional distribution and transport of pollutants in various California air basins.

The study utilized two instrumented light aircraft (a Cessna 205 and a Cessna 310). Observational systems were identical in the two aircraft and included instruments to measure the following parameters:

Air Quality	Meteorological
Light scattering, (b_{scat})	Temperature
Condensation nuclei	Humidity
CO	Turbulence
NO_x	Pressure (altitude)
O_3	

All data processing was computerized, so that data could be processed and checked between sampling missions; thus the sampling program was continually updated, taking into account data from previous flights. In addition to the data taken by the vehicles, extensive meteorological and pollution data were collected from the FAA, NOAA, local air pollution control agencies, and MRI Mechanical Weather Stations, so that the pollution data from any sampling mission could be interpretated in light of a firm understanding of the meteorology.

On any sampling day both aircraft operated in the same air basin. By flying two separate routes they could cover an entire basin in a few hours. The routes consisted of a series of spirals from the top of the polluted layer to ground level over about eight points per flight. The spirals provided a series of snapshots (vertical profiles) of the vertical distribution of pollutants and other parameters over fixed points on the ground. Each route was flown two or three times per day, giving a picture of both the spatial and temporal variations of pollutant concentrations.

One of the most interesting episodes studied during the program occurred from July 24 to July 26, 1973, in the Los Angeles basin. Although not the worst episode to occur in Los Angeles, this particular one was the first and only occasion on which federal offices have been closed because of smog. The episode was also particularly well documented by both aircraft and Air Resources Board (ARB) trailer measurements and was the only occasion when aircraft measurements were made on an around-the-clock basis.

The meteorological conditions during this period were typical of the early to midsummer regime during which the worst of the smog episodes in the Los Angeles basin have generally occurred in recent years. In particular, this episode was characterized by the "Upland anomaly" of very high ozone concentration in the northeastern part of the basin.

Because it was both representative and well documented, this episode was chosen for special study. This paper traces the buildup and decay of the episode and tries to put the three-dimensional distribution and transport of pollutants into perspective using the meteorology. Specific attention is paid to the eastern portion of the basin which includes Upland, Ontario, and Riverside (see map in Figure 1). The Upland anomaly, the mixing layer structure, the 3-D distribution of pollutants at night, and the effect of certain point sources in the eastern basin are all discussed. The sampling began on July 25, so that day is chosen for most of the discussion. Data from July 26 are used to illustrate the specific points.

II. SYNOPTIC METEOROLOGY OF THE EPISODE

The meteorology during the episode of July 24 to 26, 1973, was typical of that leading to early summer episode conditions in the inland valleys. The water offshore was 3 to 5°F colder than normal, causing fog conditions typical of early summer, while a thermal low existed in the desert areas. On July 23 a strong high-pressure ridge started developing near the coast, and by July 24 it was well developed.

During this episode subsidence associated with the high-pressure area lowered and strengthened the marine inversion and created a cap of hot air over the Los Angeles basin. In places a double inversion actually occurred, with a marine inversion below an independent subsidence inversion. In addition, on July 24 the ridge created higher pressure to the north and east of the basin and tended to retard the formation of the sea breeze. This high-pressure ridge and the inland thermal low are shown on the weather map for July 25 in Figure 2.

A crude indication of the occurrence of this subsidence condition, and thus of the likelihood of keeping pollutants trapped near the surface under an inversion, is the temperature at 850 millibars (approximately 1500 m msl). When this temperature is greater than 20°C (60°F), the surface temperature must reach about 35°C (95°F) before the inversion can be broken and pollutants ventilated. Table 1 was assembled using data from soundings at Los Angeles airport.

From this table it is easy to see that July 24 and 25 were likely to be bad smog days, but that by midday on July 26 the situation should have improved. This was exactly the case; on the 24th and 25th hourly average ozone values exceeding 60 pphm were recorded in the basin and air stagnation advisories were called, while on the 26th the highest value was 50 pphm and conditions improved rapidly in the afternoon. This high-pressure ridging along the coast and corresponding high temperatures aloft are typical of July weather. For 10 years between 1960 and 1973 the 850-millibar temperature in southern Caifornia exceeded 20°C in July for an average of 25 days.

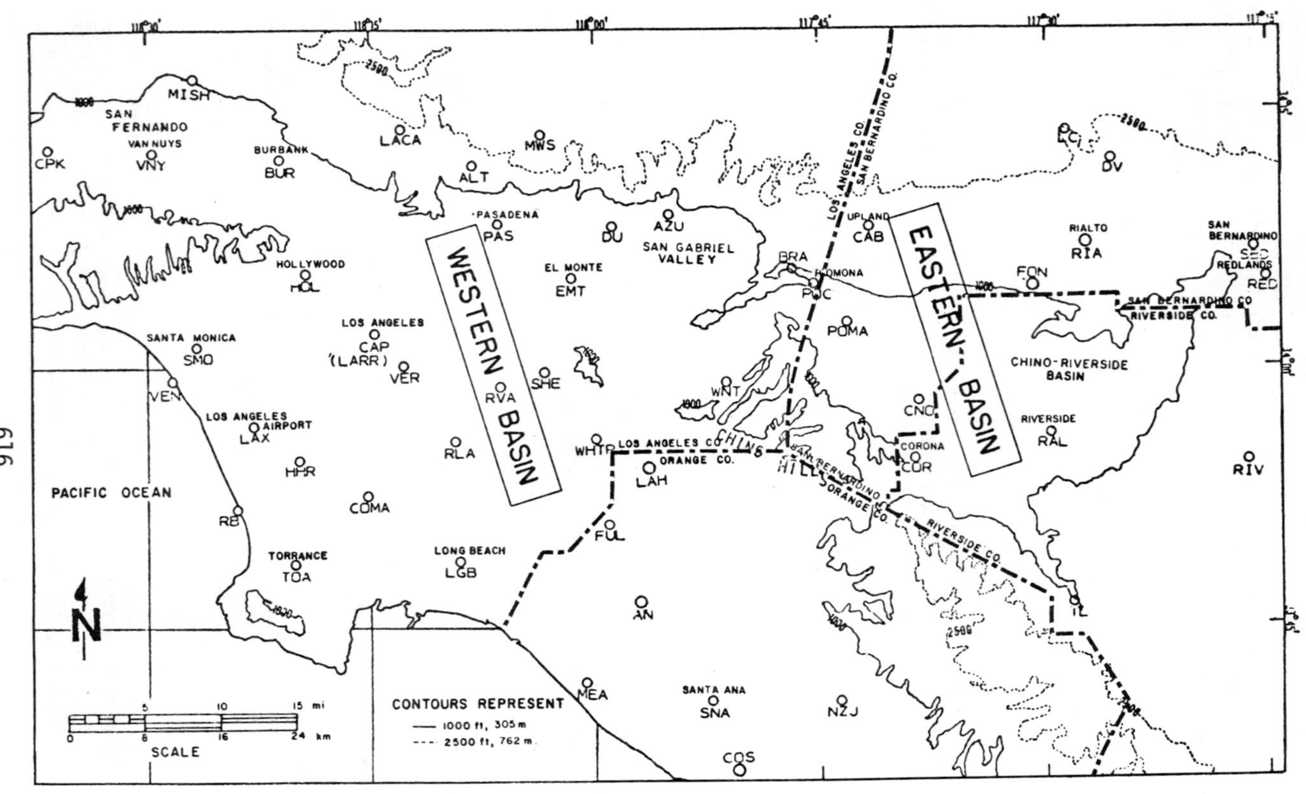

FIGURE 1. Map of the Los Angeles Air Basin showing east and west basins. [Circles are locations of weather or air pollution monitoring stations or airports (spiral points).]

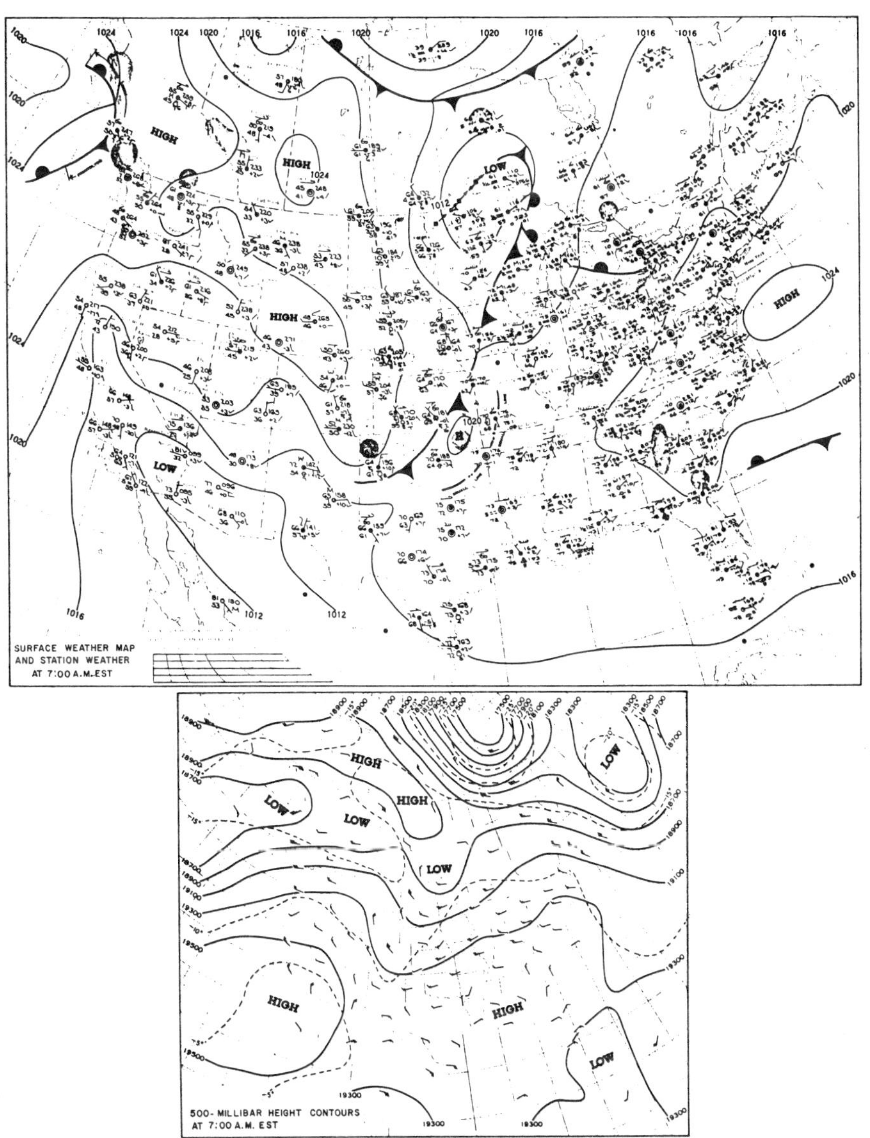

FIGURE 2. Weather map for July 25, 1973.

TABLE 1. Temperatures at 850 Millibars from July 23, 1973 to July 26, 1973

Date and Time		Temperature (°C)
July 23	5:30 a.m. PST	20
	11:30 a.m.	21
July 24	5:30	23
	11:30	25
July 25	5:30	27
	11:30	27
July 26	5:30	27
	11:30	23

The high-pressure ridge moved slightly north on July 25, allowing the sea breeze to build up, but maintaining the subsidence inversion. On the 26th the ridge weakened and moved farther north. Surface temperatures had risen to over 38°C (100°F) in the inland valleys during the subsidence condition; and when the temperature aloft decreased, the inversion was broken at inland points and pollutants were ventilated upward. On July 24 and 25 the winds aloft were generally light and variable, but by midday on the 26th the flow aloft had become organized and westerly, thus carrying off the pollutants ventilated through the inversion.

III. FLOW PATTERNS AND METEOROLOGY FOR JULY 25, 1973

This day began with widespread fog over most of the coastal plain in the early morning hours, extending inland as far as Pomona and Chino by 0700 PDT. Visibilities were down to "zero" in fog at some coastal localities, but increased to 4 to 5 km (2.5 to 3 miles) in haze and smoke in the San Gabriel and San Bernardino Valleys and as high as 24 km (15 miles) in the San Fernando Valley, where relatively clean drainage flows from the mountains to the north had cleansed the valley. Air movement during these early morning hours consisted of a weak, disorganized sea breeze in the coastal areas and a disorganized easterly flow inland. By 0900 PDT the coastal fog was scattered and lifting, and absent altogether in the inland valleys. Most inland areas were still reporting haze and smoke except Van Nuys, which reported 11 km (7 mile) visibility). A well-organized sea breeze did not start to develop until about 1100 PDT. By early afternoon the sea breeze had approached a steady-state situation, penetrating well inland to Redlands and Riverside. The streamlines changed little until after 1900 PDT, although the velocities varied slightly. The streamlines for 1600 PDT are shown in Figure 3. Note that the flow pattern transports air across the southern coastal plain northward into the San Gabriel Valley (EMT, AZU, etc.) and then eastward to Upland (CAB) and San Bernardino (SBD). At the same time air is transported across Orange County (AN, SNA) through Santa Ana Canyon (COR) and into the Chino-Riverside area (CNO, RAL). This is basically the flow pattern within the surface mixing layer. Above the mixing layer the winds were light and variable until mid-afternoon.

Winds aloft on July 25, 1973, were measured at four stationary sites. Pilot balloons (pibals) were released hourly at Chino (CNO), in the eastern basin, and at Elysian Park, midway between Hollywood (HOL) and the Los Angeles Railroad Yard (LARR) in the western basin. In addition, pibals were released at 1233 PDT from Los Angeles International Airport (LAX) and at 1335 and 1654 PDT from El Monte Airport (EMT).

Figures 4 and 5 show trajectories, computed from these wind observations, for the air arriving over Upland (CAB) at 1600 PDT and Redlands (RED) at 1800 PDT. The extreme northerly and southerly winds within the mixed layer (from pibal observations) were used to construct an envelope of possible trajectories, and the mean wind within the mixed layer was used to construct a characteristic trajectory.

On July 25 the California Air Resources Board station at Upland (CAB) reported an average ozone concentration of 63 pphm for the hour beginning 1600 PDT, the highest such concentration recorded in the Los Angeles basin that year. Figure 6 shows that surface concentrations were generally high in the northern part of the eastern basin at that time.* The trajectories shown in Figures 4 and 5 indicate that much of the air over Upland at the time of peak ozone at that location was onshore before the sea

---

*The method used by the Los Angeles County Air Pollution Control District (L.A. APCD) to calibrate its ozone monitors is not equivalent to the one used by the California Air Resources Board (ARB) and the other control districts in the South Coast Air Basin. To make the surface ozone maps internally consistent, and also consistent with the aircraft ozone data, the L.A. APCD ozone values in Figures 6 and 14 have been adjusted to equivalent ARB values as described by De More et al. (2). This does not reflect any judgment on the relative accuracy of the two calibration methods.

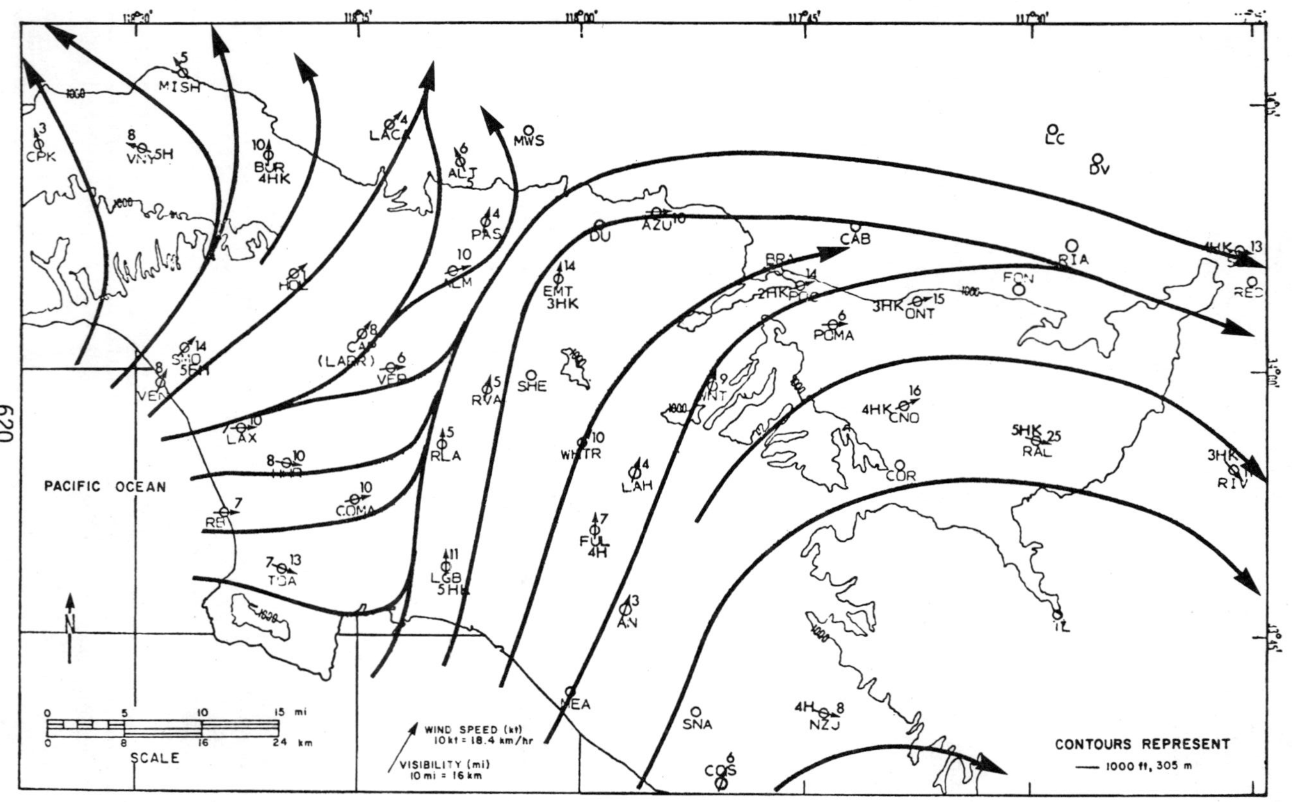

FIGURE 3. Surface wind streamlines: 1600 PDT, July 25, 1973.

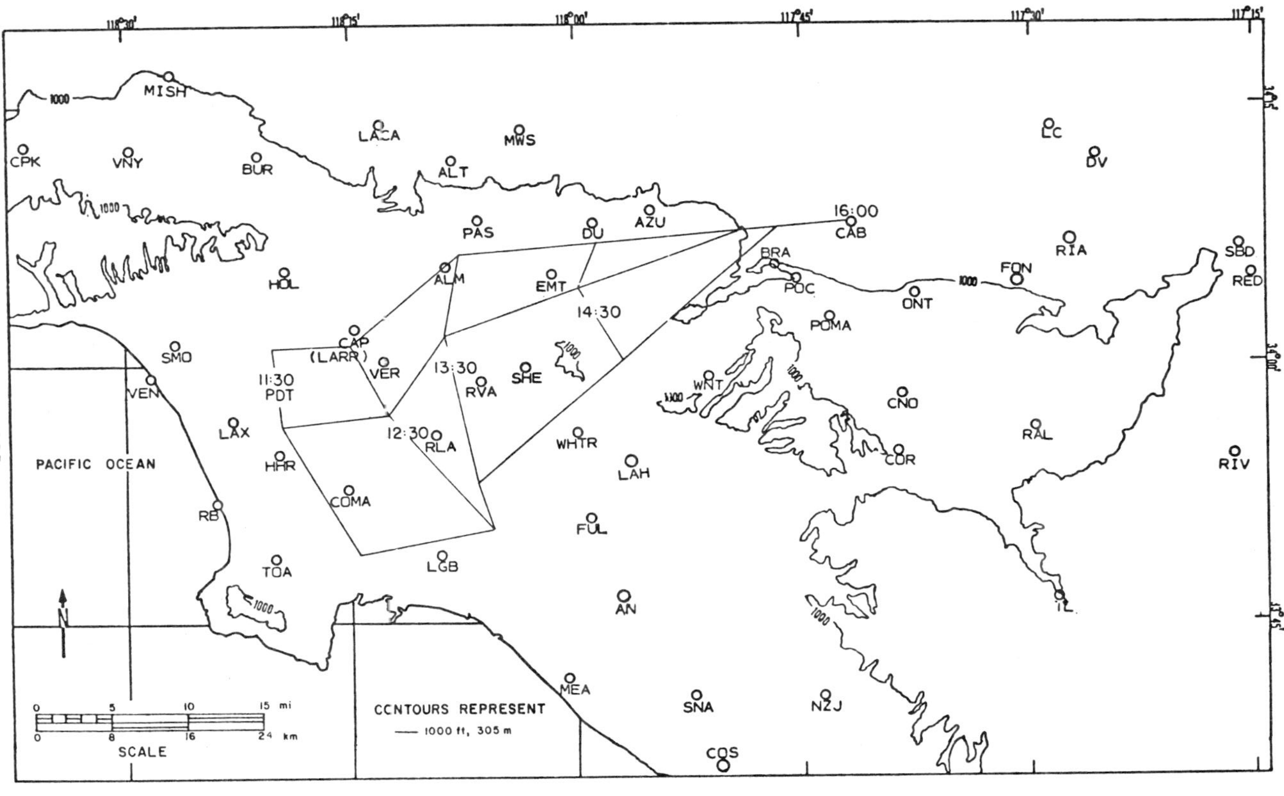

FIGURE 4. Trajectory envelope for air arriving at cable airport (Upland): 1600 PDT, July 25, 1973.

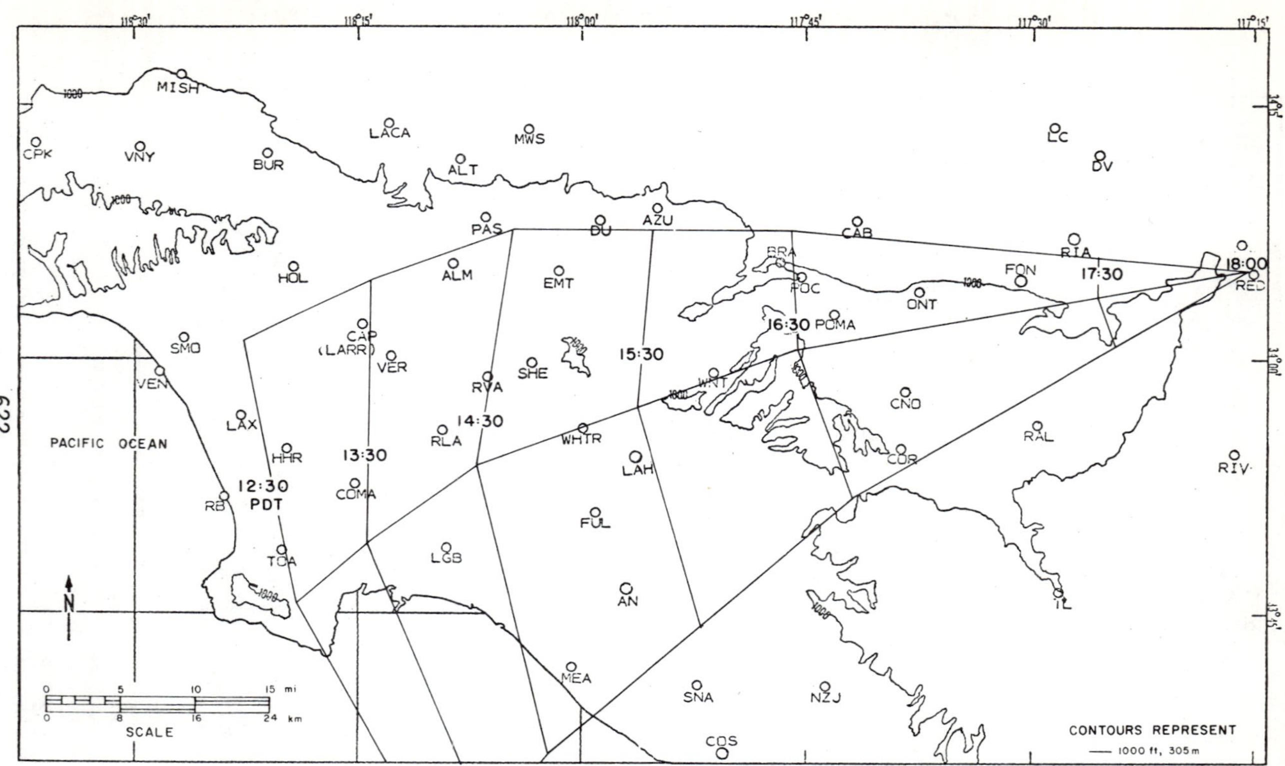

FIGURE 5. Trajectory envelope for air arriving over Redlands: 1800 PDT, July 25, 1973.

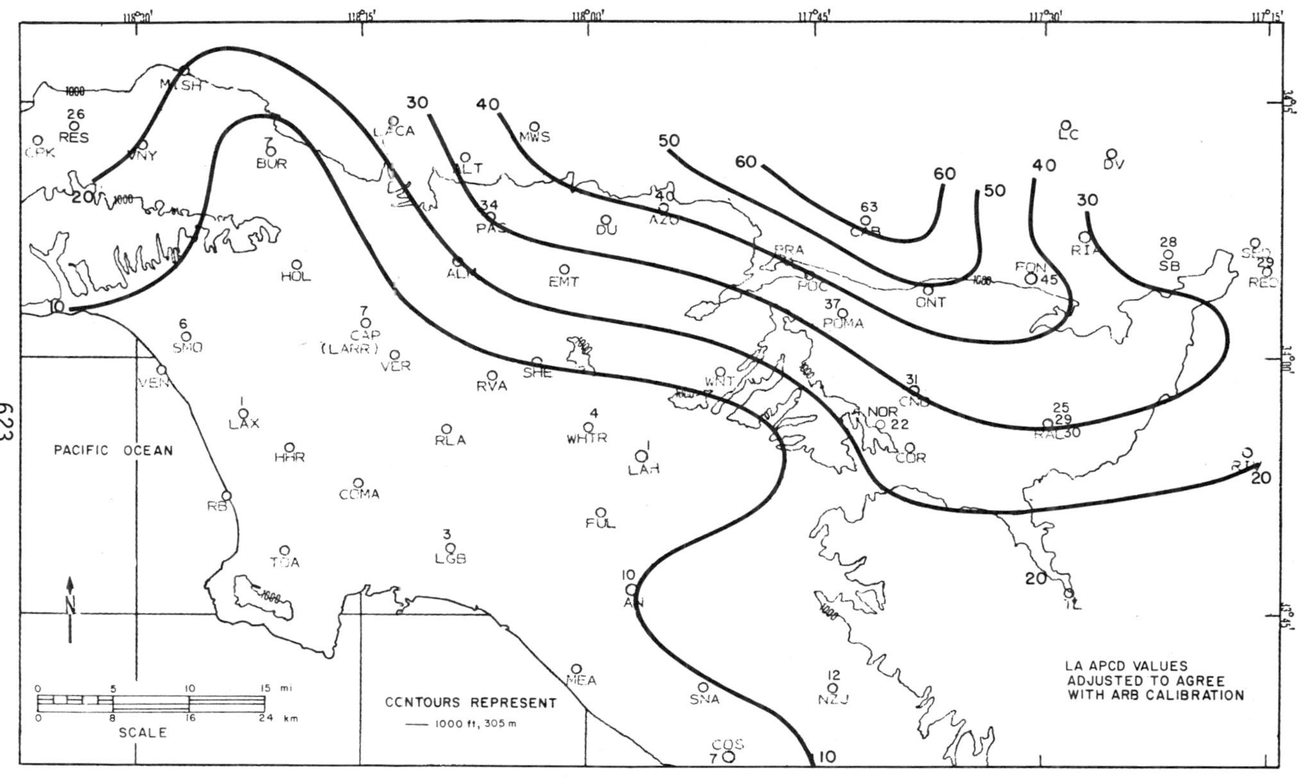

FIGURE 6. Surface ozone concentrations (pphm): 1600-1700, July 25, 1973.

breeze started, probably having accumulated pollutants since the night before, while much of the air over Redlands at the end of the afternoon had come onshore with the start of the sea breeze. The air arriving over Upland at 1600 PDT had the longest residence time in the western basin of any air arriving over Upland that day, a fact that may account for the timing of the ozone maximum there.

The streamlines and trajectories indicate that polluted western basin air was advected into the eastern basin during the afternoon of July 25. This conclusion is corroborated by the afternoon gradients in surface ozone concentration, which clearly shows the advance of cleaner air from the coast with the onset of the sea breeze and the displacement of the more ozone-laden air eastward. The hourly average surface ozone concentrations from 1600 to 1700 PDT (Figure 6) provide a snapshot of this effect, showing a strong gradient between La Habra (LAH) and downtown Los Angeles (CAP), which had already been flushed with marine air, and Azusa (AZU) and Upland (CAB), which had not.

Figure 7 shows the approximate times at which the sea breeze began to flush various locations in the basin during the afternoon. These times were determined from ozone data of the California Air Resources Board and local Air Pollution Control Districts (APCD) according to an objective criterion: the approximate time of arrival of the sea breeze is calculated as M:00 PDT, where M is the first integer, i, such that $\chi_{i+1} \leq 1/2\ \chi_{i-1}$, χ_i being the average ozone concentration during the hour beginning at i:00 PDT. Comparison of Figure 7 with the trajectories to Upland and Redlands in Figures 4 and 5 shows that these trajectories depict air moving just ahead of the advancing cleaner sea air. From 1600 to 1900 PDT the marine front in Figure 7 moved about 70 km (45 miles) inland at an average speed of about 24 km/hr (15 mph), in good agreement with observed wind speeds.

IV. MIXING LAYER STRUCTURE FOR JULY 25, 1973

In the morning before solar heating became important, the mixing layer over the basin was about 300 m (1000 ft) or less above ground level (AGL). Mixing was limited by a radiation inversion caused by the cooling of the ground during the night. Pollutants emitted near the surface were trapped within this layer. The subsidence inversion remained on top of the radiation inversion at about 900 m (3000 ft) above sea level (msl).

Figure 8 is a vertical profile taken over Brackett Airport (BRA) on July 25 at 8:42 PDT. The nighttime radiation inversion was still quite strong, and fresh pollutants were trapped within a mixing layer that extended up to about 500 m msl (1600 ft msl).

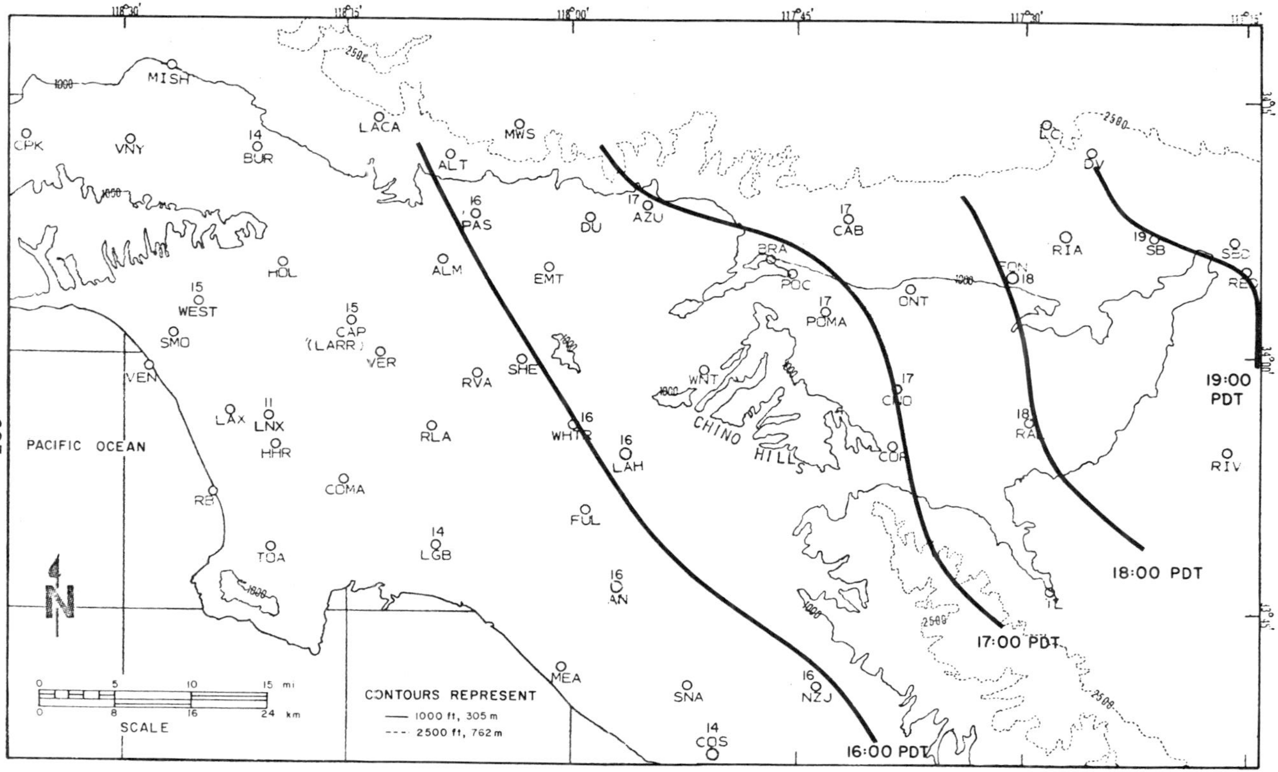

FIGURE 7. Approximate time (PDT) of arrival of sea breeze: July 25, 1973. (Objective criterion based on ground level ozone concentrations.)

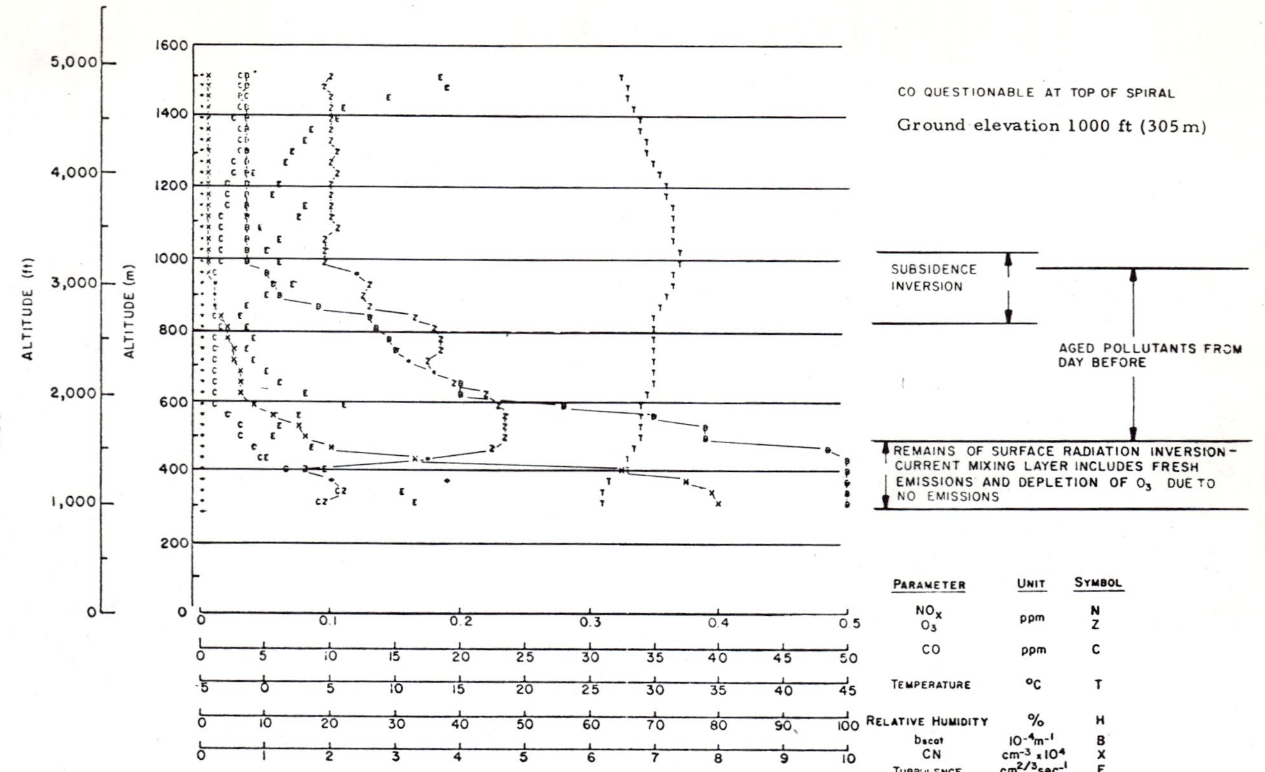

FIGURE 8. Vertical profile over Brackett (BRA): 0843 PDT, July 25, 1973.

Fresh emissions are indicated by the high concentrations of condensation nuclei (X) and CO (C) under the inversion. The high b_{scat} (B) values near the surface are indicative of the thick haze left in the basin as the fog burned off. The ozone values show a deficit under the radiation inversion, pointing out that ozone is destroyed by fresh emissions of NO.

Above the radiation inversion a sharp increase in ozone occurs. This high level of ozone is accompanied by high b_{scat} levels, but lower condensation nuclei and CO levels. This combination suggests an aged air mass left over from the day before and separated from the surface mixing layer. This layer aloft was trapped between the radiation inversion and the subsidence inversion, which starts at about 850 m msl (2800 ft msl), and pollutants were not carried off during the night because of light wind conditions aloft. By midday, surface heating had destroyed the radiation inversion in the eastern portion of the basin and the mixing layer extended up to the subsidence inversion, allowing the pollutants that had been trapped aloft to mix throughout the layer.

Figure 9 shows a midday vertical cross-sectional contour map of b_{scat} for the northern part of the basin from the coast at Santa Monica (SMO) to Redlands (RED). The contours (isopleths) are lines of equal b_{scat}. Superimposed on the contour plot are the winds-aloft vectors at points and times closest to the points shown on the map. The numbers listed over the various locations are actual b_{scat} values obtained from the spirals.

Throughout the day the subsidence inversion covered the whole basin at about 900 m msl (3000 ft msl), while either a marine or radiation inversion existed at a lower level. By midday a marine inversion sloped upward from about 300 m msl (1000 ft msl) near the coast until it merged with the subsidence inversion roughly 40 km (25 miles) inland. The structure of this double inversion is also shown in Figure 9.

During the night and morning hours the surface flow in the basin was relatively stagnant, allowing the buildup of pollutants within the surface layer. In Figure 9 the highest concentrations are seen within the source region in the western basin. Although some of the b_{scat} in this figure is due to humidity effects, much of it results from emissions accumulated in a relatively stagnant air mass. At the time represented in Figure 9, the sea breeze had just recently started and clean air had not yet come onshore to flush out the basin.

Figure 10 gives a slightly different perspective on the midday mixing layer structure. It is a mixing layer height contour map which shows that fresh emissions were confined to a relatively thin layer over the whole basin and that venting which normally occurs in the eastern half of the basin was prevented by the sub-

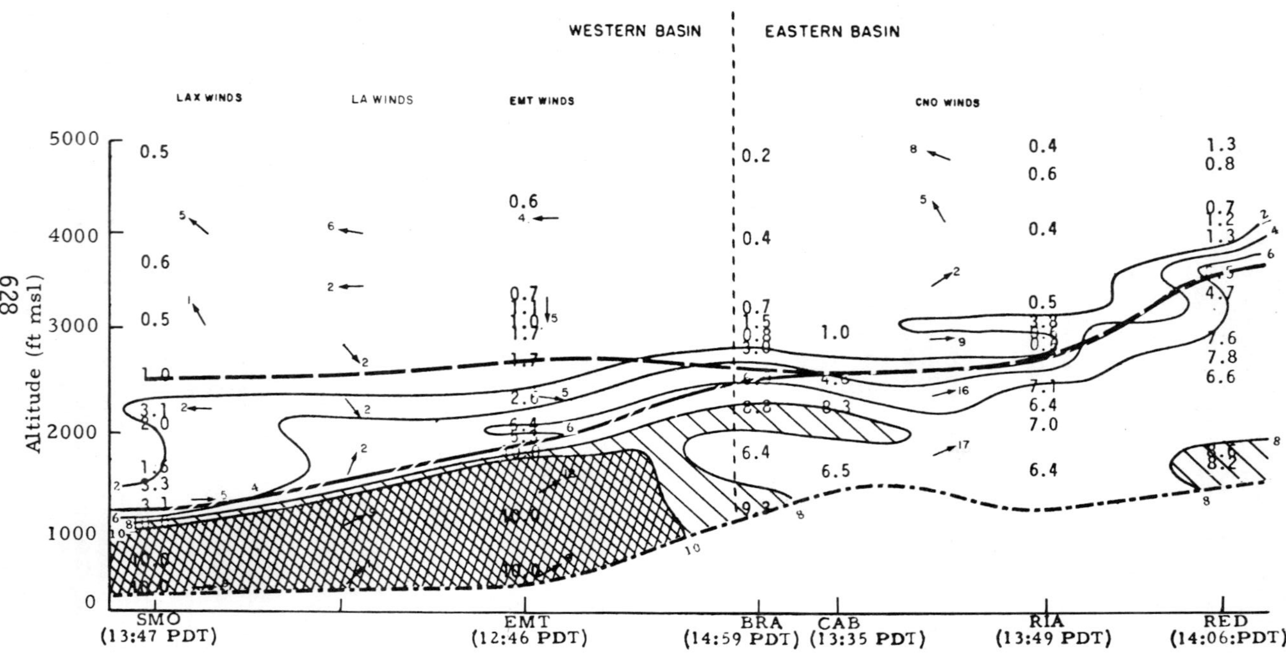

FIGURE 9. Vertical cross-section of b_{scat}: midday of July 25, 1973. (Units are 10^{-4} m^{-1}.)

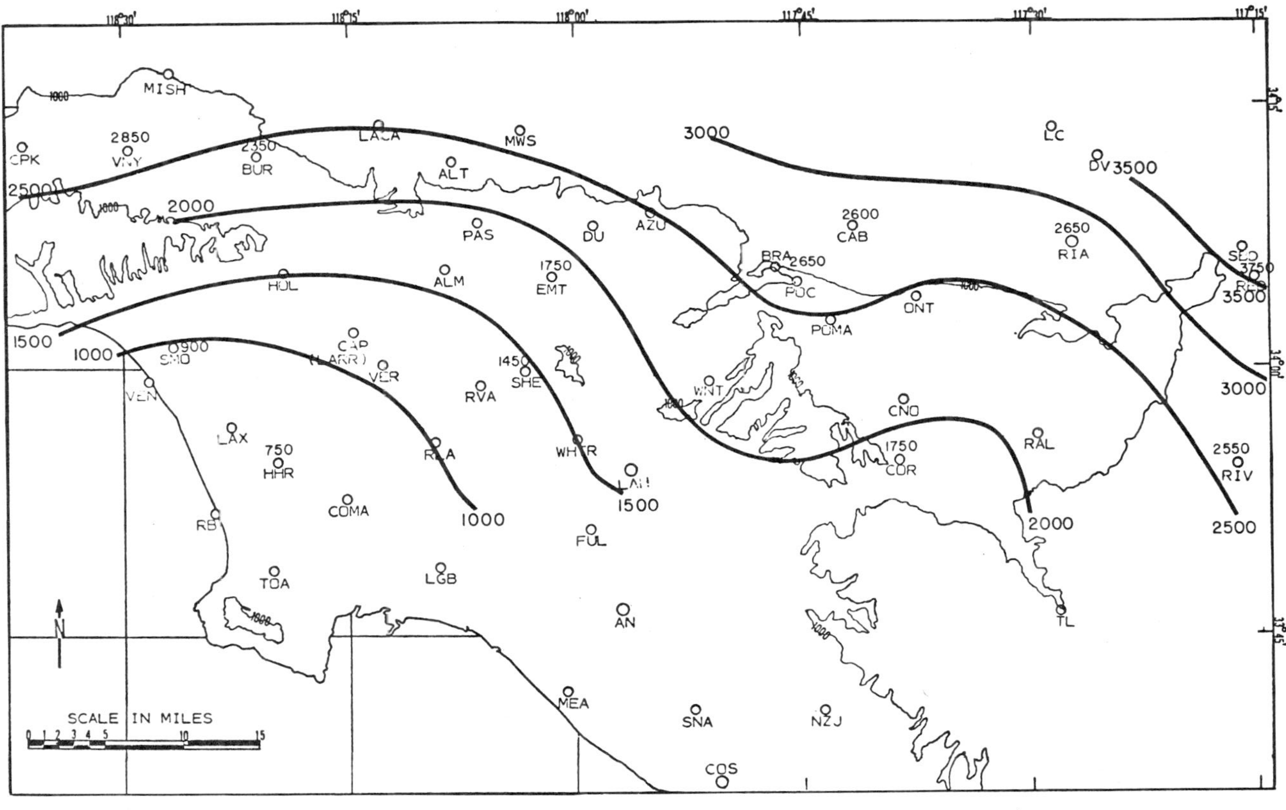

FIGURE 10. Mixing heights (ft msl): midday (11-1300 PST), July 25, 1973.

sidence. The mixing layer is higher over high terrain, but the thickness above ground is about the same over most of the basin. The tongue of lower mixing height in the Corona (COR) area is indicative of the fact that the terrain is lower in that area than farther north or east.

Later in the afternoon, as shown in Figure 11, the sea breeze had ventilated much of the surface layer in the western basin, replacing air that had had a long residence time over land with cleaner air with a shorter time onshore. The sea breeze "front" in Figure 11 is between El monte (EMT) and Brackett (BRA). The air just ahead of the "front" had accumulated emissions over the western basin earlier in the day and possibly during the night before, and had the longest residence time over strong source areas of any air in the basin. Note that the highest concentrations in Figure 11 are over Upland (CAB) at the time which corresponds to the peak ozone reading at Upland (and in the basin) for the day.

In both Figures 9 and 11 upper layers are seen between the two inversions in the western basin. Pollutant concentrations remain high in these layers all day, but the aircraft sounding and pibal data indicate that these layers are decoupled from the air below. The air in the upper layers is well aged and relatively stagnant compared to the surface air. These layers may be caused by upslope flow along some of the nearby hills, by lifting or undercutting of polluted air by the sea breeze, or by other means.

The difference between the air in front of and that behind the sea breeze front is seen in Figures 12 and 13. Figure 12 is a vertical profile over El Monte after the passage of the front. A surface mixing layer is well defined by the temperature, turbulence, and pollutant profiles extending up to about 425 m msl (1400 ft msl). Within this layer the b_{scat} and ozone values have dropped down from their peaks for the day. The condensation nuclei values are still high, however, indicating continuing fresh emissions. The air between about 425 m (1400 ft) and 600 m msl (2000 ft msl) is fairly well aged, but still has a moderate condensation nuclei population, possibly indicating that this air was previously part of the surface mixing layer but was recently undercut by the advancing sea breeze. The remainder of the layer aloft could be a remnant of the layer existing earlier in the midday soundings.

Figure 13 is a vertical profile at Upland (CAB) before the passage of the front. The mixing layer is well defined, and pollutant concentrations are high within the layer. The ozone concentration exceeds 0.5 ppm. The air within the mixing layer at Upland is similar to that just above the sea breeze in Figure 12. This profile represents the period of peak ozone concentration at Upland.

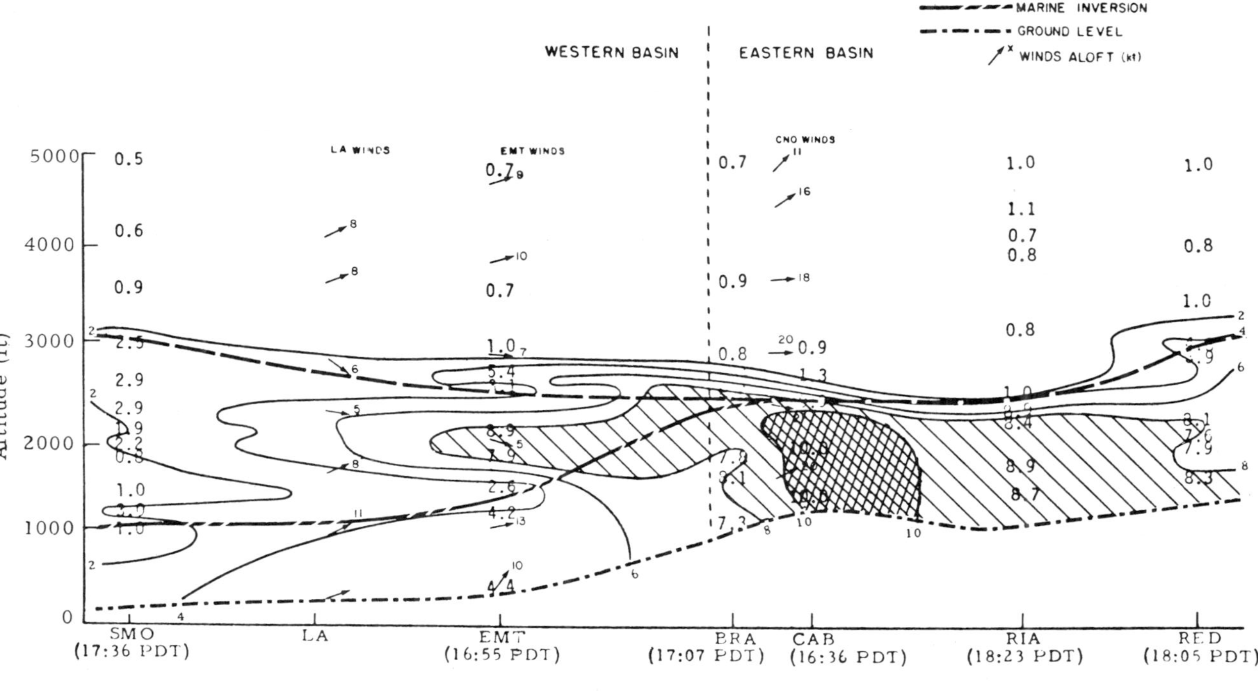

FIGURE 11. Vertical cross-section of b_{scat}: afternoon of July 25, 1973. (Units are 10^{-4} m^{-1}.

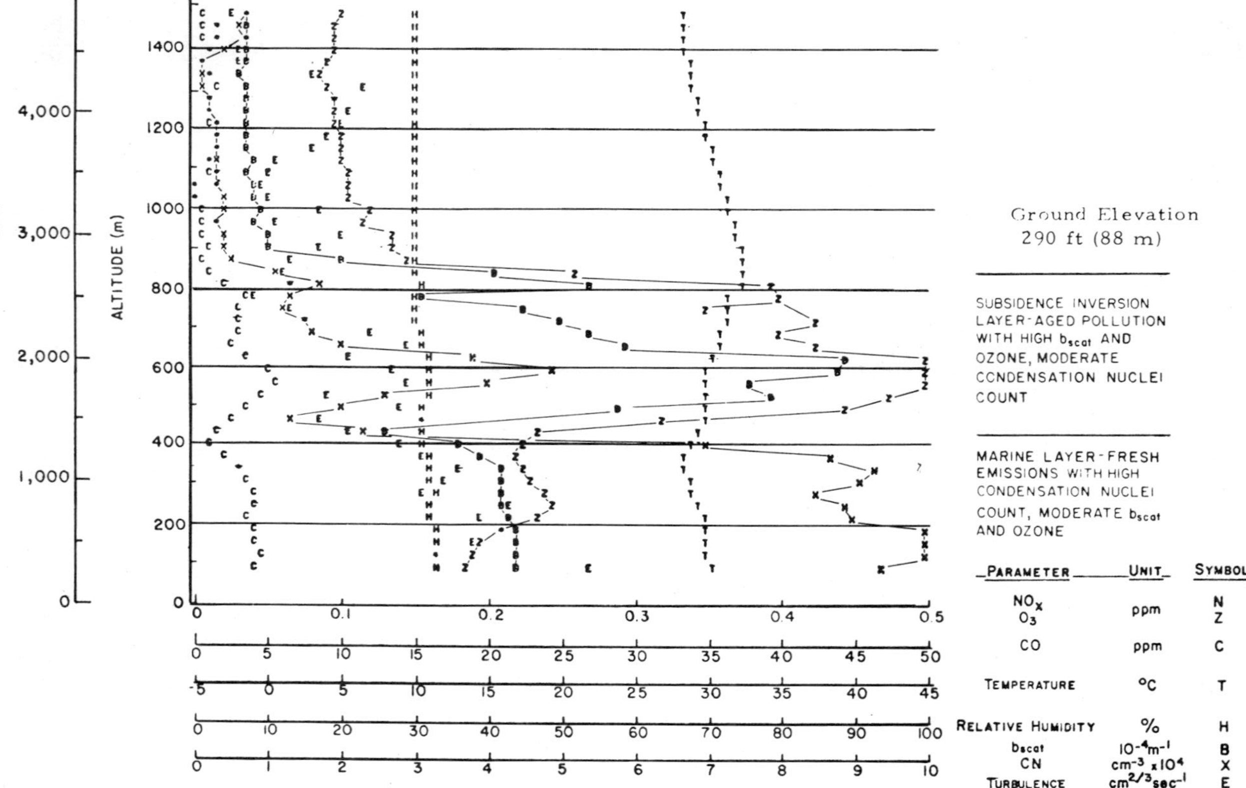

FIGURE 12. Vertical profile over El Monte (EMT): 1655 PDT, July 25, 1973, after passage of sea breeze front.

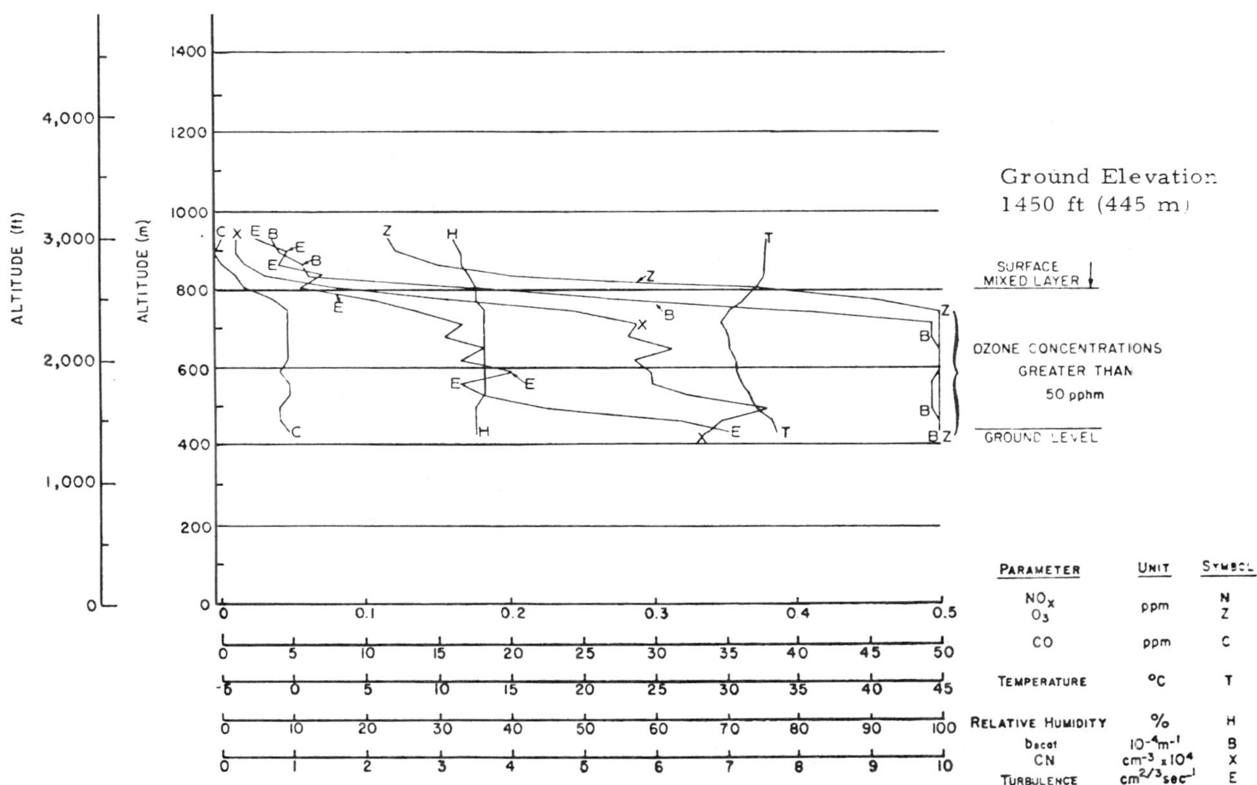

FIGURE 13. Vertical profile over cable (CAB): 1636 PDT, July 25, 1973.

V. OZONE DISTRIBUTION AND THE UPLAND ANOMALY

It is apparent from the preceding discussion that this episode was characterized by very high ozone levels in the northeastern part of the basin. The ozone concentrations well exceeded 50 pphm at Upland (CAB) on both July 24 and July 25. Figure 14 is an isopleth map of the surface level ozone concentration for 1300 PDT on the 25th. The highest concentrations of ozone occur in the foothill areas in the northern portion of the basin. These areas share several features in common which may account for their high concentrations:

1. At midday the air they have received has had a long residence time over heavy source areas and has not yet been influenced by the fresh onshore flow of marine air.

2. The air masses reaching them have been irradiated for several hours and were not recently shielded by fog.

3. The mixing layer in these areas has entrained ozone-containing air from the aged layers aloft, thus increasing the pollutant burden in the mixing layer.

4. The foothill areas generally have a lower source strength of NO than the more heavily industrialized upwind areas from which they are receiving their air; thus ozone in the surface layer is not scavenged as effectively as at upwind locations.

This combination of features points out that the areas of highest ozone concentration are those receiving heavily polluted air which is most aged in a photochemical sense. Other areas in the basin appear to be receiving air that is "younger" (less sunlight or more recent NO emissions) or less polluted to start with.

The importance of "aging" and lack of fresh NO emissions for the generation of ozone is indicated again in Figure 15, which presents an isopleth map for the midday peak ozone value above the mixing layer. The data were taken from the vertical profiles. The area of highest concentration in this case is the area above the coastal plain where pollutants are trapped above the mixing layer and below the subsidence inversion. This layer

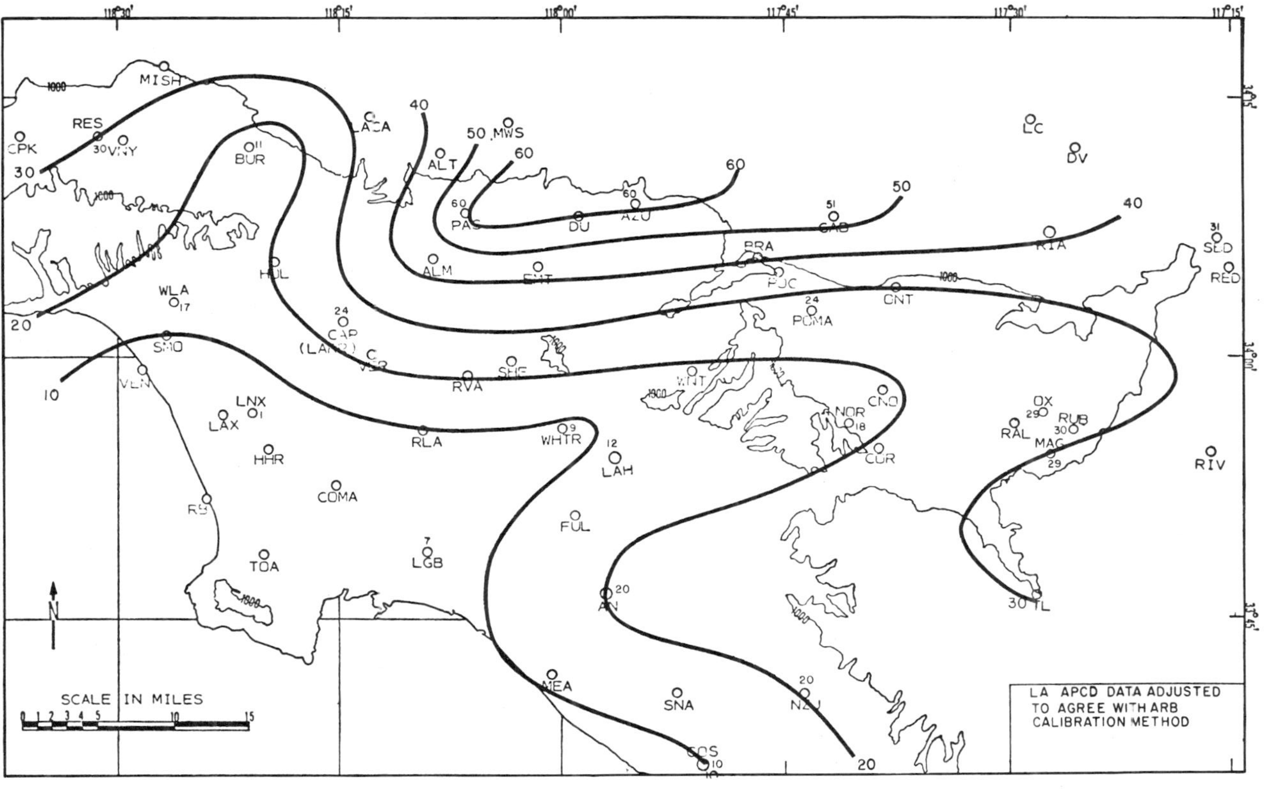

FIGURE 14. Surface ozone concentrations (pphm): 1300 PDT, July 25, 1973.

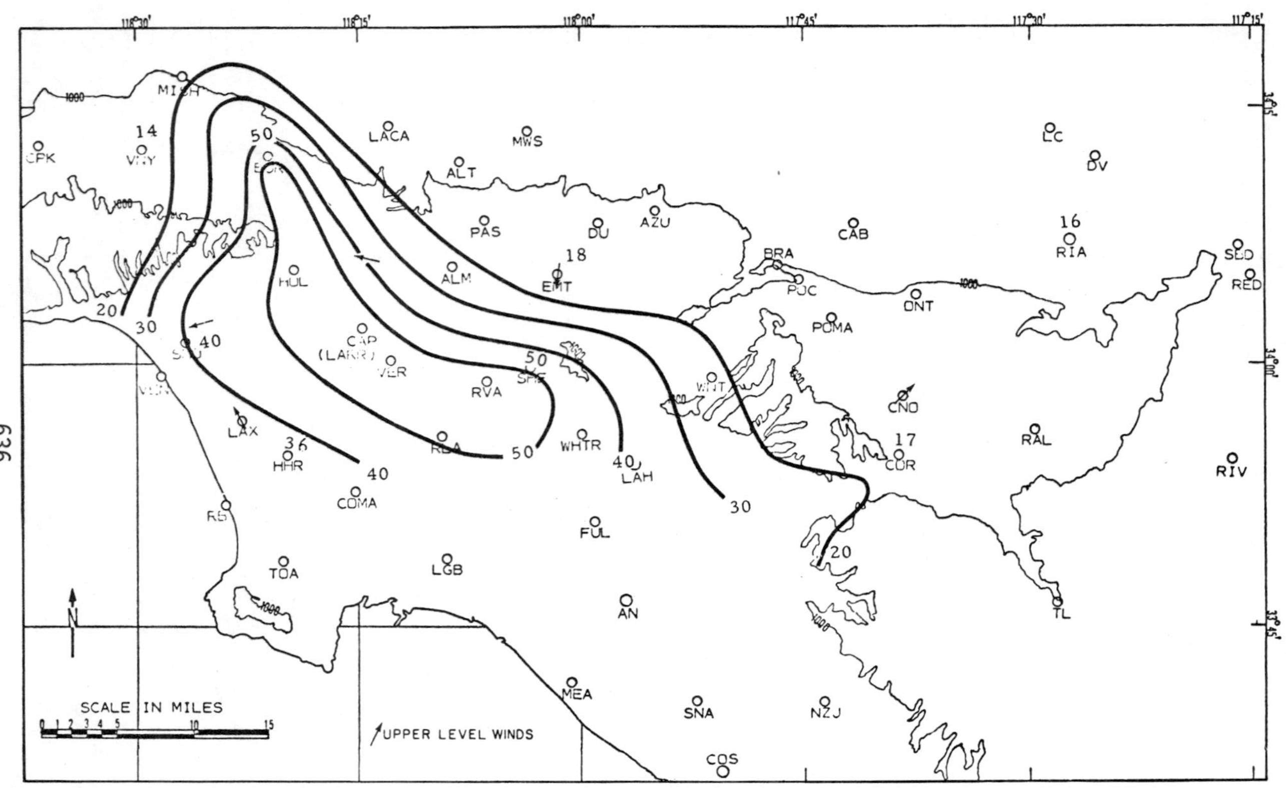

FIGURE 15. Peak O_3 above mixing layer (pphm): midday of July 25, 1973.

is also seen in the vertical cross section of Figure 9. Because of limited number of data involved, the map should be taken only as an indication and not used for detailed interpretation, but it does show that the ozone values are considerably higher aloft than at the surface in the same area.

The pollutants in this region are restrained from moving up or down by inversions and are not vented horizontally because of the light and disorganized winds. This upper layer is probably fed by ventilation of pollutants up mountain slopes and by buoyant emissions from point sources below. There is no general area source of NO emissions aloft. Thus this layer consists basically of well-aged but concentrated pollutants for which the photochemical reactions have run their course. This results in higher ozone levels aloft than in the strong primary pollutant source region immediately below.

An afternoon isopleth map of the ozone aloft is shown in Figure 16. This figure indicates that by midafternoon this layer had become quite widespread and contained ozone concentrations similar to those at Upland. It might have been possible for an ozone monitor at about the 900-m (3000-ft) level in the San Gabriel Mountains to detect this layer.

Later in the day, as shown previously in Figure 6, the area of highest surface ozone moved toward Upland. This maximum surface level ozone for the day in the basin occurred at about the time of this contour map (1600 PDT). At this time the Upland area was the only location at the surface in the basin incorporating all the features mentioned earlier. The areas to the west and south either has been partially ventilated by the sea breeze (see Figures 3 and 11) or had stronger sources of NO within the mixing layer to deplete the ozone. Since the area east of Upland is relatively source free, the far eastern portion of the basin received air that had spent less time over major source regions.

In the late afternoon, when Upland started receiving air that had come onshore during the day and thus had had a shorter residence time over land than the air received earlier, the peak ozone area shifted farther east, but by that time the solar radiation had decreased and the peak values were less. Thus, in this situation, Upland has the distinction of having the optimum combination of conditions in the basin for ozone formation, giving rise to the "Upland anomaly."

Additional information on the amount of ozone transported into the eastern basin can be obtained by doing a simple flux calculation. Figure 17 shows an estimate of the flux of ozone under the sudsidence inversion from the western basin into the eastern basin at 1700 PDT on July 25. The western face of the box drawn in the eastern basin represents the boundary across

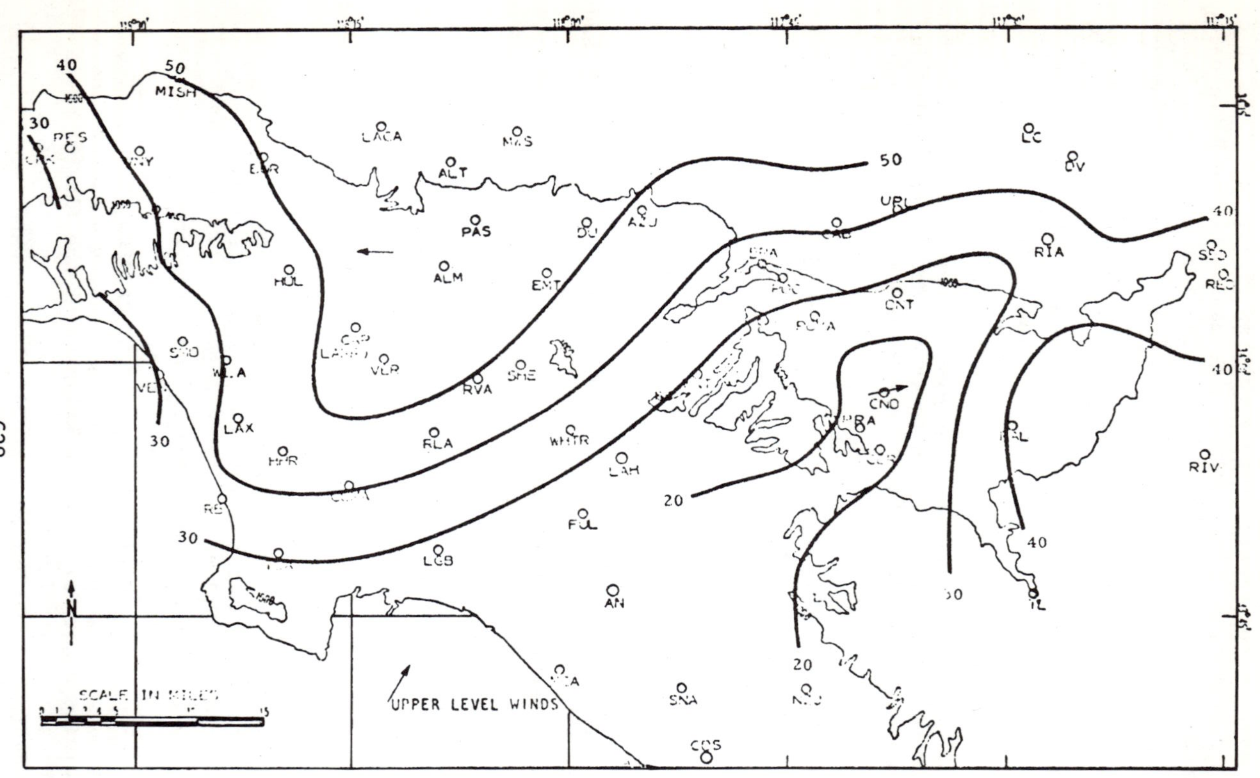

FIGURE 16. Peak O_3 above mixing layer (pphm): afternoon of July 25, 1973.

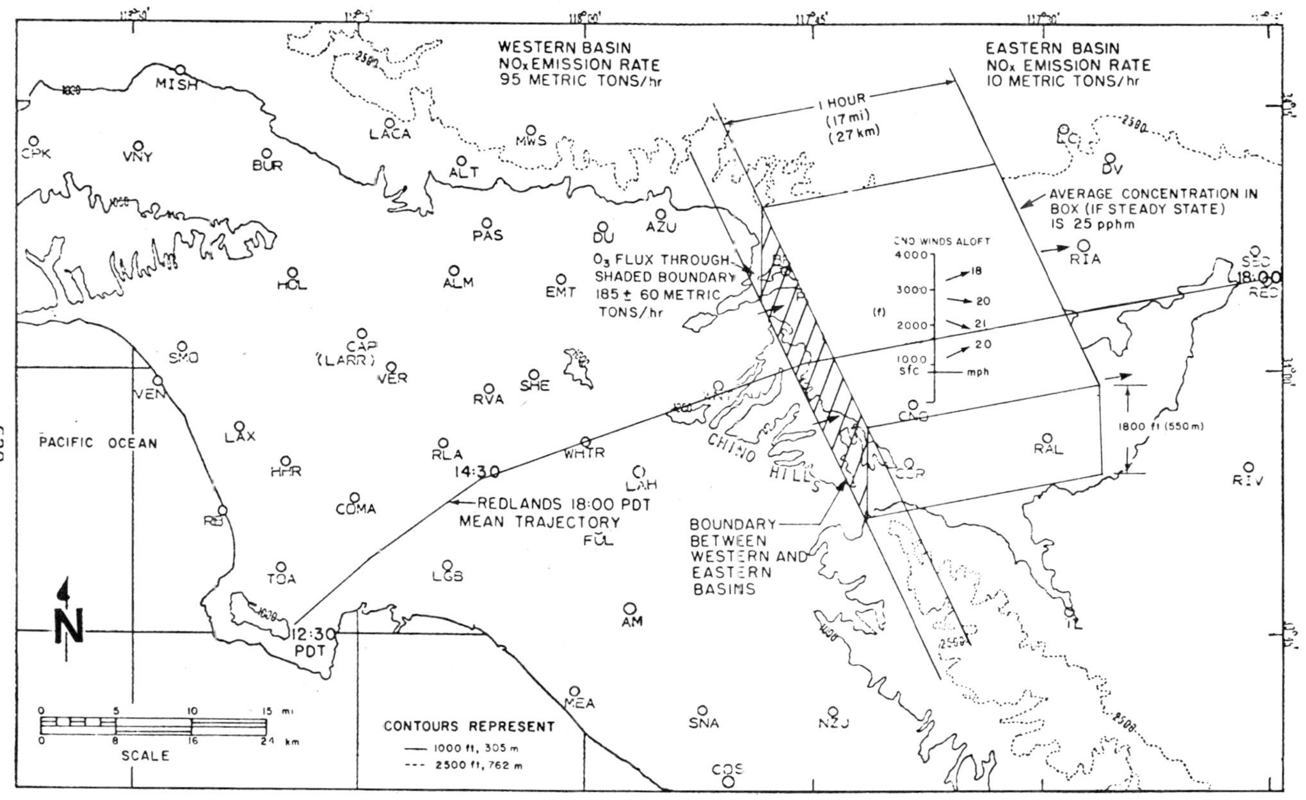

FIGURE 17. Estimated ozone flux from western to eastern basin: 1700 PDT, July 25, 1973.

which transport was calculated, and the east-west dimension of the box represents the distance covered in 1 hr by the moving air. In a steady-state situation in the absence of diffusion or diffluent winds, the box would thus contain about 185 metric tons of ozone, corresponding to a uniform concentration of about 25 pphm. For comparison the average late afternoon concentrations of ozone within the surface mixed layer at Riverside (RAL), Redlands (RED), Rialto (RIA), and Ontario (ONT) were, respectively, 36, 24, 25, and 13 pphm. (Data were taken from aircraft soundings.) The flux was calculated for about 1700 PDT since soundings were available for roughly that time, but 1700 PDT was also roughly the time of arrival of the sea breeze front. Crude flux calculations for about 2 hr earlier indicated slightly greater flux and a correspondingly higher average concentration in the box.

In the absence of photochemistry 1 ton (as NO_2) of freshly emitted NO_x (mostly NO) could scavenge about 1 ton of ozone under conditions of good mixing. The daily emissions of NO_x in the entire eastern basin are estimated to be only about 120 metric tons. Roth et al. (3) found that, in the western basin, about 16% of the total car mileage for a weekday was driven between 1600 and 1800 PDT. If this 8%/hr figure is used to scale all NO_x emissions to hourly rates (for around 1700 PDT), one obtains an estimate of 10 metric tons/hr for the NO_x emission rate in the eastern basin. This is inadequate to scavenge more than a small fraction of the advected ozone.

The estimates of the ozone flux and the corresponding average concentration in the eastern basin "box" were based on the following considerations:

1. A quasi-steady state prevailed for at least 1 hr at the boundary between the basins. This assumption was quite good at the boundary, even though this was not the case over the whole western basin, as the progress of the sea breeze in Figure 7 shows. Pollutants that had accumulated in the western basin for many hours were flushed out into the eastern basin by the sea breeze starting about midday. For this reason pollutant fluxes out of the western basin are not directly comparable to pollutant emissions within the western basin. The western basin was "emptied" more rapidly than it was "filled," giving rise to fluxes that were large in comparison with emission rates.

2. With metric units used for convenience of calculation, afternoon soundings at Brackett (BRA) and Corona (COR) showed a capping subsidence inversion at about 860 m (2800 ft) msl. Ground elevations east of the Chino Hills slope from roughly 310 m (1020 ft) msl at Brackett to 165 m (540 ft) msl at Corona; the Chino Hills themselves range from 400 to 500 m (1300 to 1600 ft) msl, and the width of the pass between the western and eastern basins is 25 to 35 km (16 to 22 miles). The area across which the flux was calculated was taken as 25 km x (860 m - 310 m) $\simeq 1.4 \times 10^7$ m^2.

3. The 1700 PDT winds aloft at Chino (CNO) were above 32 km (20 miles)/hr throughout the mixed layer, and the 1700 PDT surface winds at Chino, La Verne College (POC), and Ontario (ONT) were all 27 km (17 miles)/hr. All wind directions in this area were near westerly. The rate at which air was transported into the eastern basin was taken to be

$$1.4 \times 10^7 \text{ m}^2 \times 27 \text{ km/hr} \simeq 3.8 \times 10^{11} \text{ m}^3\text{/hr}$$

The average ozone concentrations measured between 310 and 860 m msl on the 1700 PDT Brackett and Corona soundings were, respectively, 33.5 pphm (660 $\mu g/m^3$) and 16.8 pphm (330 $\mu g/m^3$). These were slightly lower than the corresponding values for the 1500 PDT soundings. The distribution of ozone concentrations between Brackett and Corona were generally intermediate in value. The rate at which ozone was advected into the eastern basin was estimated to lie between 3.8×10^{11} m^3/hr x 600 μg/hr $\simeq$ 250 metric tons/hr and 3.8×10^{11} m^3/hr x 330 μg/hr $\simeq$ 125 metric tons/hr.

Using this crude technique, it is clear that much of the ozone loading in the eastern basin on the afternoon of July 25 can be accounted for by advection from the western basin. In light of the preceding analyses it is probable that a significant portion of the 24 pphm of ozone within the mixed layer arriving at Redlands at about 1800 PDT had its origin in the western basin. Redlands is approximately 100 km (65 miles) from the area south of Los Angeles where the trajectory originated; thus it is reasonably certain that large source areas such as Los Angeles and Orange Counties can and do export a significant amount of ozone and ozone precursors to distant surrounding areas. This is not to say that the eastern basin does not contribute to its own ozone concentrations. Ozone values exceeding the federal standard also occur occasionally in the morning before the onset of the sea breeze.

VI. END OF THE EPISODE - BASIN VENTILATION ON JULY 26

July 26 started similarly to July 25, with fog in the coastal areas and a late developing sea breeze. The high-pressure ridge, however, was moving northward and weakening. By midmorning the temperature aloft had dropped substantially (see Table 1), weakening and raising the subsidence inversion and thus allowing deeper mixing and lowering the surface concentrations of pollutants. By 1300 PDT the winds aloft had become organized and westerly at 16 km/hr (10 mph) or greater from the surface to 2100 m msl (7000 ft). Thus pollutants that had remained aloft from previous days or had been vented up the mountain slopes during the morning were finally swept away.

To show the ventilation processes in the eastern portion of the basin, the scattering coefficient data (b_{scat}) were integrated from the surface to 1500 m (5000 ft) for each time and location in the eastern basin. The result of the integration is an optical depth for each location and time. The average of the optical depth for the whole eastern basin was then found and plotted as a function of time in Figure 18. Since b_{scat} is roughly correlated with aerosol mass in the 0.1- to 1.0-m size range, this average optical depth is a crude indication of the total aerosol mass in the eastern basin. Figure 18 nicely shows the gradual buildup of mass from the morning of July 25 to the morning of July 26 and then indicates the sharp drop as the winds aloft picked up, the inversion weakened, and the basin ventilated.

Another indication of this ventilation is seen in the vertical cross-sectional plots of b_{scat} for the three daytime flights on July 26 (Figure 19). The morning cross-section and soundings show independent layers existing up to 1500 m (5000 ft). The morning soundings indicate that the subsidence had weakened, but that the atmosphere was still stable to the 1500 m (5000 ft) level. The winds were light and variable, so these upper layers were not ventilated. Most of the b_{scat}, however, was confined to layers below 750 m (2500 ft). A distinct dense layer is also

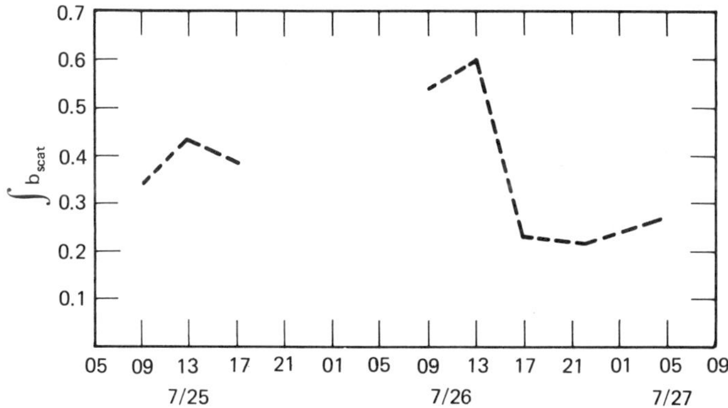

FIGURE 18. Values of the average of $\int b_{scat}\, dz$ from surface to 5000 ft over the Riverside route.

apparent between about 600 m (2000 ft) and 750 m (2500 ft) msl. (This layer will be discussed in more detail later.)

By midday, surface heating had caused mixing of the various layers from the surface to the top of the stable layer. The independent layers existing earlier had merged, and the dense layer at 600 m (2000 ft) was ventilated to the surface as well as upward. The winds aloft became organized, and the basin started to ventilate to the east.

The afternoon cross-section shows a considerably decreased particulate burden. By this time the inversions had disappeared, and mixing was limited only by some isothermal layers. In the eastern portion of the basin at Redlands, the atmosphere was neutral to over 2400 m msl (8000 ft msl) and pollutants were ventilating upward and being carried off by the winds aloft. This is the normal state on nonepisode, nonsubsidence days during the summer.

VII. CHARACTERISTICS OF THE LAYERS - DAY AND NIGHT

It was pointed out earlier that in the morning cross-section shown in Figure 19 several distinct and independent layers aloft existed. Figure 20 is the vertical profile taken at Ontario at 9:52 PDT which corresponds to the morning cross-section. This figure points out the necessity of measuring several parameters

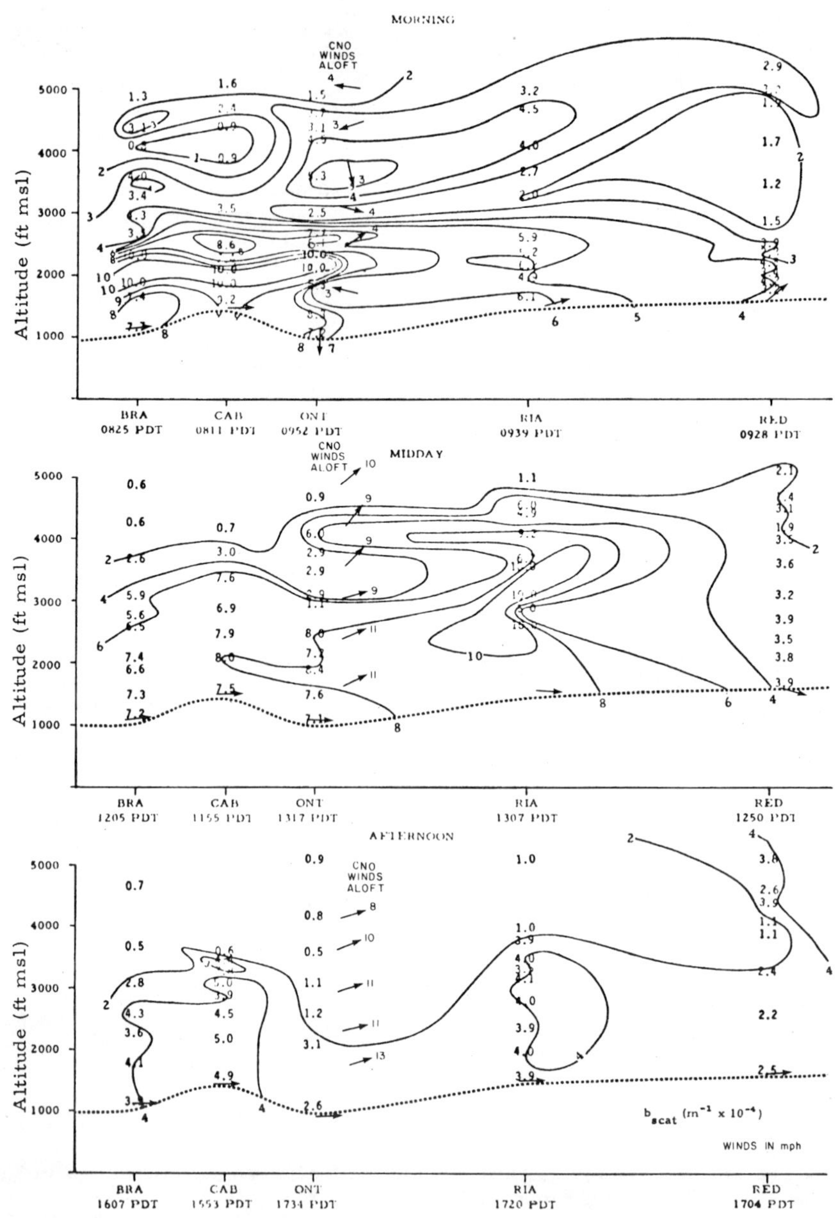

FIGURE 19. Vertical cross-sections of b_{scat}: July 26, 1973.

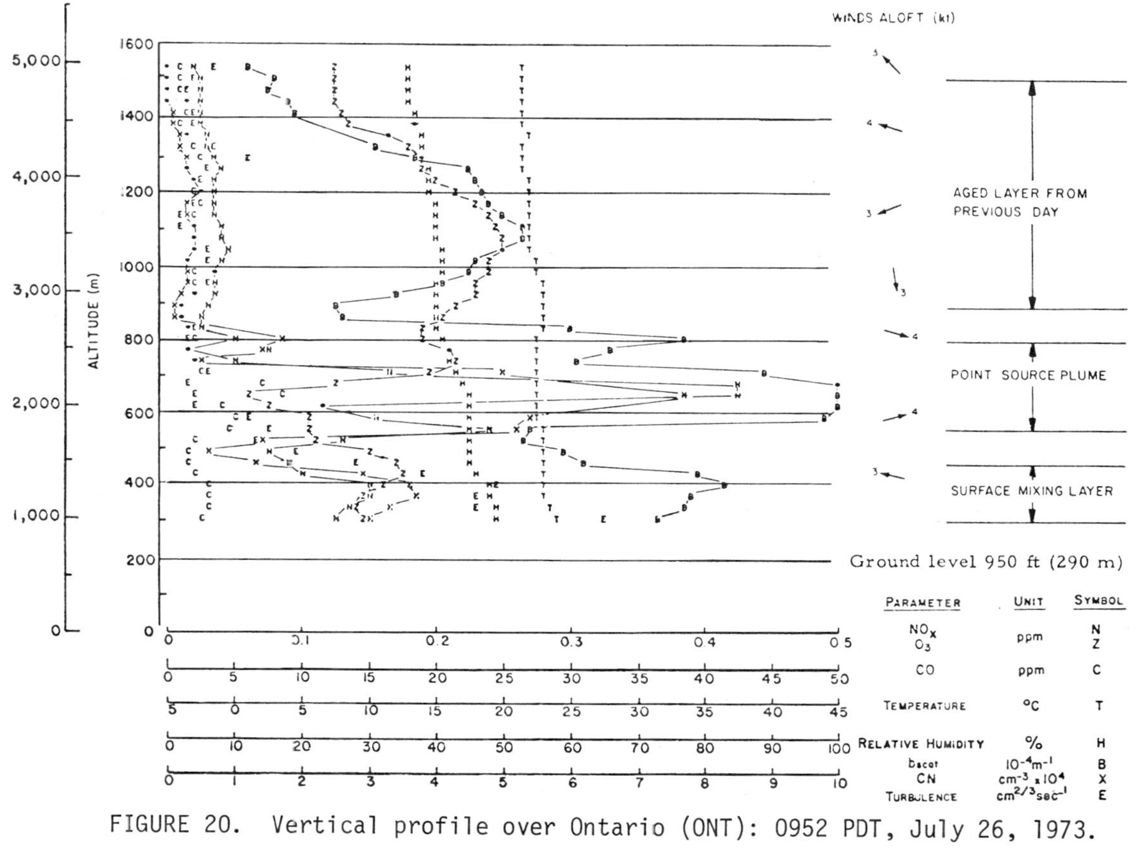

FIGURE 20. Vertical profile over Ontario (ONT): 0952 PDT, July 26, 1973.

simultaneously in order to be able to interpret the three-dimensional structure. Using temperature alone, it would be almost impossible to determine the surface mixing depth; but by using a combination of parameters, one can assign a mixing height of about 450 m (1500 ft) msl. Below that level the NO, b_{scat}, CO, and condensation nuclei increase while the ozone drops slightly. These all indicate entrainment of relatively fresh emissions. In addition, the turbulence increases dramatically below this point, indicating good mixing.

Between about 580 m (1900 ft) and 750 m (2500 ft) msl, another layer is apparent. This layer was seen covering much of the eastern basin in Figure 19. From the combination of parameters in the layer it is reasonable to conclude that this layer is the result of injection of a plume into the stable region above the surface mixing layer by a large elevated or buoyant point source. The high values of b_{scat}, NO_x, CO, and especially condensation nuclei, as well as the sharp deficit of ozone, suggest fresh emissions, yet this layer is separated from the surface and would be slow to receive pollutants from the surface by diffusion. This same layer is seen on many of the other soundings in the area. It also is evident in soundings taken on the night of July 26 and appears to originate in the area of the steel mill and power generation complex in Fontana [between Ontario and Rialto (RIA) on the cross-sections].

Another layer, between about 880 m (2900 ft) and 1370 m (4500 ft) msl, clearly consists of aged pollutants and is well separated from the ground. We have seen from our data that, in general, for a well-aged air mass, b_{scat} and ozone are usually well correlated, indicating that both are secondary products. Primary emissions such as NO_x, and especially condensation nuclei, are usually well depleted by reactions and coagulation, respectively. Thus this upper layer is well aged and has probably remained from the previous day.

In an air mass that contains both primary and aged pollutants almost any combination is possible. If fresh NO emissions are present, however, the ozone is usually depleted. Secondary aerosol-caused b_{scat}, on the other hand, is stable once formed. Although b_{scat} can also be due to primary emissions, as in the plume shown above or from field burning, these sources are becomming rarer because of more stringent emission controls.

On the night of July 26 the atmosphere again became stable to at least 1500 m (5000 ft) and the winds aloft died down. Figure 18 indicates a slow buildup of b_{scat} during the night over the eastern basin. This can be due to humidity effects or to fresh emissions; a combination of both is possible. Figure 21 shows the vertical cross-sectional plots of b_{scat} during the

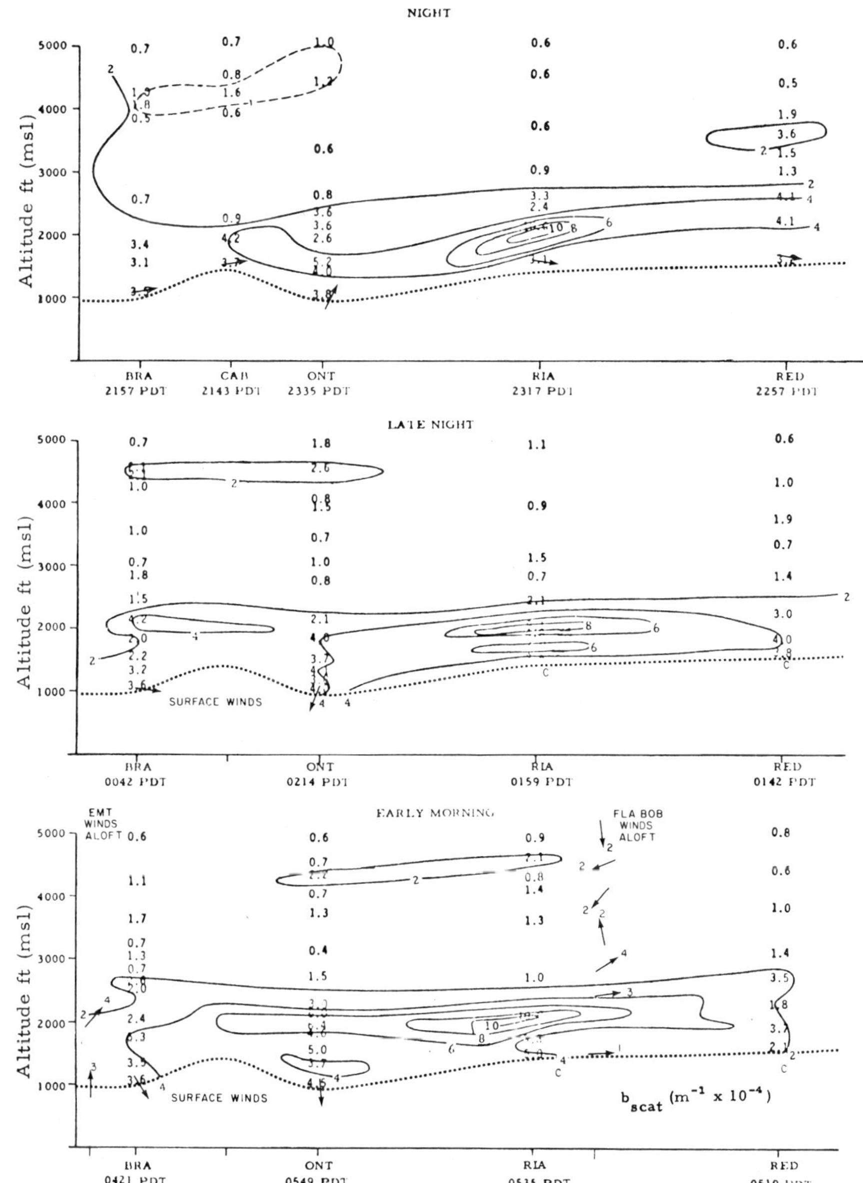

FIGURE 21. Vertical cross-section of b_{scat}: July 26-27, 1973.

night. It is clear from the plots that the eastern portion of the basin contributes to its own pollutant burden, at least during the night. Both the b_{scat} and NO_x burdens (not shown) increased over the eastern basin during the night when winds were generally light and transport was minimal. Much of the burden was in the raised layer at about 600 m (2000 ft), assumed to be due to the sources in Fontana. In the morning this upper layer should have been ventilated to the surface as the mixing layer deepened and entrained it.

Although the eastern basin is not a strong source area, the burden accumulated during the night may be photochemically reactive and contribute to the morning ozone concentration when the sun comes up. In general, later in the day, as the sea breeze picks up, the pollutant burden in the eastern basin is probably due largely to transport in from other areas.

Finally, Figure 22 is a vertical profile taken at Rialto shortly before sunrise. As in Figure 20, a surface mixing layer, a plume aloft, and aged pollutants above that are visible. This figure is important in that it points out the stability of ozone in an aged air mass. The ozone in the layers aloft existed all night at concentrations greater than 16 pphm, in the absence of fresh emissions to scavenge it. All other soundings during the night also showed these high ozone values. Within the surface mixing layer or in the plume, however, the ozone was almost totally depleted by fresh emissions, indicating that low ozone values do not necessarily imply clean air.

VIII. CONCLUSIONS

1. Air pollution in the Los Angeles basin is a regional problem, and transport of large quantities of pollutants from one side of the basin to the other is a normal occurrence.

2. In the Los Angeles basin the pollutants accumulated in source areas in the stagnant morning air can be transported downwind to areas of lesser source strength (receptor areas) with the onset of the sea breeze. The ultimate concentrations in the downwind areas depend on the mixing layer structure, as well as the duration of the stagnation period. If the mixing layer deepens, entrains clean air, and ventilates pollutants upward, concentrations will be lowered. If the mixing depth and thus the ventilation are

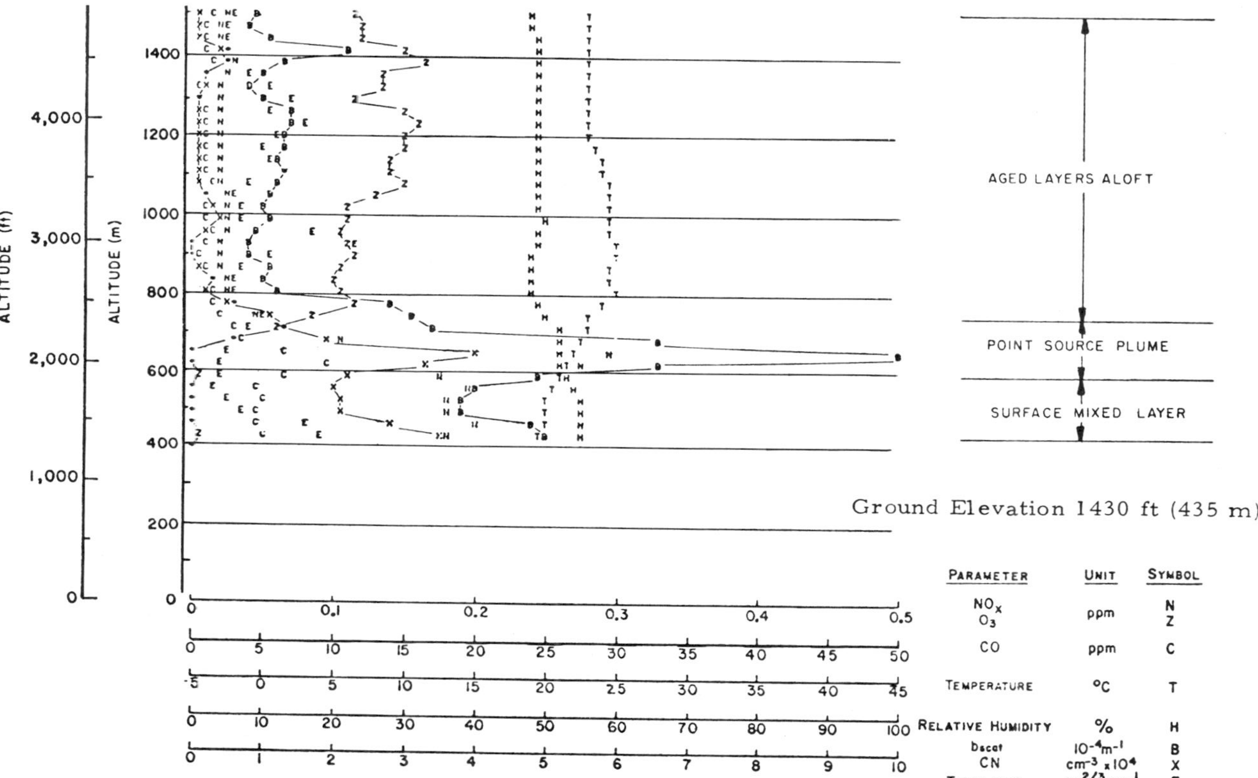

FIGURE 22. Vertical profile over Rialto (RIA): 0535 PDT, July 27, 1973.

limited by subsidence, however, pollutants will be confined to a shallow layer, with resulting high concentrations.

3. The Azusa-Upland region is located so that it often bears the maximum photochemical impact from the pollutant pulse as it is transported across the basin.

4. Pollutants, including ozone and point source emissions, can remain aloft above the surface mixing layer overnight during an episode. They can then be reentrained on subsequent days, contributing to surface concentrations.

5. Air pollution is a three-dimensional phenomenon. To understand the variations in pollutant concentration and chemistry at any point on the ground, one must have a knowledge of the history of the pollutants reaching that point.

REFERENCES

1. Blumenthal, D. L., et al. 1974. "Three-Dimensional Pollutant Gradient Study - 1972-1973 Program," Tech. Rept. No. MRI 74 FR-1262 to California Air Resources Board, Agreement Nos. ARB-631 and ARB2-1245, 333 pp.

2. De More, W. B., et al. 1975. "Comparison of Oxidant Calibration Procedures," Report of the Ad Hoc Oxidant Measurement Committee of the California Air Resources Board.

3. Roth, P. M., et al. 1974. Mathematical Modeling of Photochemical Air Pollution - II. A Model and Inventory of Pollutant Emissions, <u>Atmos. Environ.</u> <u>8</u>, 90-130.

A Lagrangian Model of the Pasadena Smog Aerosol*

WARREN H. WHITE† AND RUDOLF B. HUSAR††
W. M. Keck Laboratories of Environmental
Health Engineering
California Institute of Technology
Pasadena, California

Abstract

Smog aerosol can be related to sources, atmospheric transport, and particle growth through a Lagrangian model. Sample

---

*Revision of "A Study of Los Angeles Smog Aerosol Dynamics by Air Trajectory Analysis," Paper No. 73-111, presented at the 1973 Annual Meeting of the Air Pollution Control Association, Chicago, Illinois, June 24-28, 1973. Published in part in J. Air Poll. Control Assoc. 26, 32 (1976).

†Present address: 1180 N. Chester Avenue, Pasadena, CA 91104.

††Present address: Department of Mechanical Engineering, Washington University, St. Louis, MO 63130

calculations indicate that the midday Pasadena aerosol is dominated by material produced in the atmosphere, and that the afternoon drop in aerosol mass is due to the advection of cleaner air. Numerical experiments suggest that control of primary particulate emissions without corresponding control of reactive gases will not substantially improve visibility.

I. INTRODUCTION

In recent years a substantial amount has been learned about the physical and chemical nature of the Pasadena aerosol [Mueller et al. (1), Husar et al. (2), Ensor et al. (3), Whitby et al. (4), Grosjean and Friedlander (5), Hidy et al. (6)]. One interesting feature that has been identified is a characteristic diurnal pattern in aerosol concentration and composition. This paper describes a simple dynamic model of the aerosol which accounts for this pattern in terms of source emissions, atmospheric transport, and particle growth. Our a priori model, which calculates the characteristics of the ambient aerosol from data on emissions and meteorology, is complementary to the a posteriori models developed by Friedlander and his co-workers, which calculate the relative contributions of major sources from data on the ambient aerosol [Miller et al. (7), Friedlander (8), Heisler et al. (9), Gartrell and Friedlander (10)].

II. MODEL DESCRIPTION

The incorporation of atmospheric transport into source receptor models is most simply accomplished from the Lagrangian point of view. Our analysis is based on a conceptual model of a vertical column that maintains its identity as it is transported with the airflow. The column extends from ground level up through the mixing layer, and material is gained or lost only through the base, not through the sides. The neglect of horizontal dispersion is appropriate for an environment dominated by area sources [Eschenroeder and Martinez (11)].

The trajectory such an air column would follow can be calculated from the wind field [Petterssen (12)]. The trajectories arriving at Pasadena, calculated from surface winds on September 3, 1969, are shown in Figure 1. These trajectories reflect the characteristic shift from a leisurely southerly flow of air in the morning to the usual westerly sea breeze in the afternoon [Angell et al. (13), Neiburger and Edinger (14)].

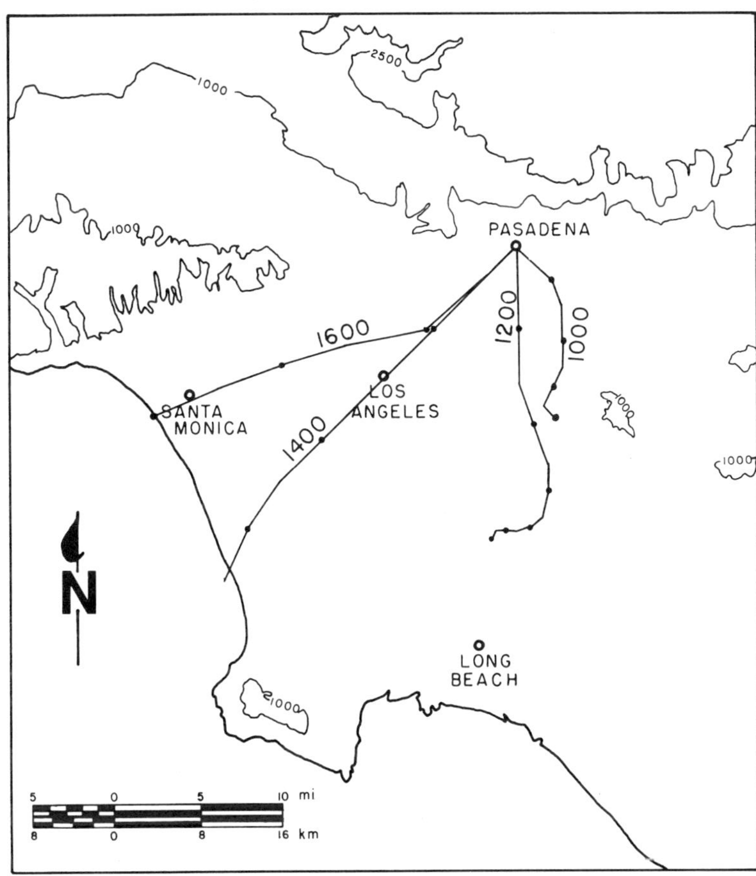

FIGURE 1. Calculated air trajectories arriving at Pasadena on September 3, 1969. Numbers give time of arrival; dots denote hour intervals.

Mixing heights are estimated from the model of Edinger (15). The mixing heights for the beginning and end of a trajectory are taken from his averages, and a uniform lifting rate is assumed. The initial aerosol for each trajectory is set equal to a marine aerosol measured by the authors at El Segundo in September 1972. The volume concentration of this aerosol was 16 $\mu m^3/cm^3$ at 45% relative humidity, and the size distribution is shown in Figure 2.

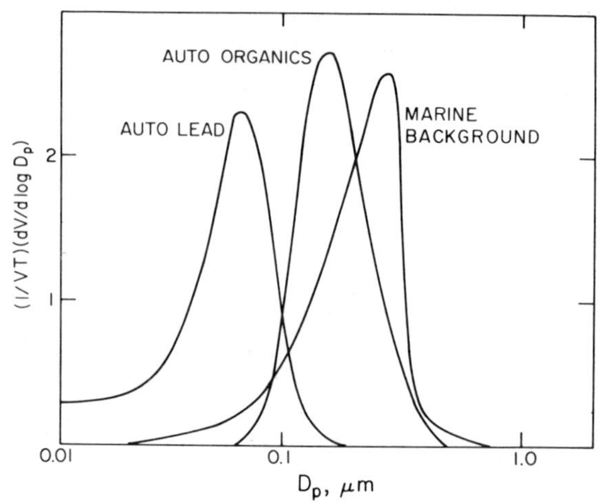

FIGURE 2. Normalized volume distributions of primary aerosols.

The relative importance of various sources and transformation mechanisms depends on the range of particle sizes under consideration. The large-particle (D_p > 1.0 μm) fraction of the ambient aerosol is dominated by dusts, and the small-particle (D_p < 0.1 μm) fraction by primary emissions from combustion sources [Heisler et al. (9)]. Because surface deposition and coagulation are effective removal mechanisms for the very large and very small particles, the quantity of material in each of these size ranges is strongly affected by transients from nearby primary sources.

In contrast, most of the material in the intermediate (0.1 μm < D_p < 1.0 μm) size range appears to be produced in the atmosphere [Husar et al. (2)]. Material tends to accumulate in this relatively stable size range, which is thus more representative of large-scale effects. Particles in the intermediate size range account for nearly all of the visibility reduction [Ensor et al. (3)] and are the ones inhaled most deeply into the lungs [Bell and Friedlander (16)]. Accordingly, sources and transformation mechanisms considered here are those most important for the size range (0.1 μm < D_p < 1.0 μm), although the entire submicron fraction must be treated in the analysis to allow for growth into this size range.

The major sources of primary aerosol identified by Friedlander (18) are soil dust, motor vehicle emissions, and industrial emissions. The mass of soil dust is concentrated in particles larger than 1 μm [Blifford (17)], above the size range of interest to us. Industrial emissions, although important in some areas of the Los Angeles Basin, are estimated to play a secondary role in the development of the Pasadena aerosol [Friedlander (8), Gartrell and Friedlander (10)]. Therefore, the only major primary source considered in the calculations is motor vehicle exhaust.

The exhaust of automobiles using leaded gasoline contains roughly equal mass concentrations of organic and lead halide particles [Colucci and Begeman (18)]. Figure 2 shows the measured size distribution of emitted lead-containing particles [Ter Haar and Stephens (19)] and the calculated size distribution of the remaining exhaust aerosol [Whitby et al. (20)], which is assumed to be organic material. Emission rates are calculated by combining data on the spatio-temporal distribution of traffic [Roberts et al. (21)] (Figure 3) with a total particulate emission of 0.3g/car-mile [Habibi (22)].

As noted earlier, much of the Pasadena smog aerosol is not emitted directly from sources but is produced in the atmosphere from reactive gases. Existing smog chamber observations [Husar and Whitby (23), Goetz and Pueschal (24), Wilson et al. (25)] are consistent with the interpretation that a pseudo-first-order gas-phase reaction is the rate-limiting step in this process. In such a system the rate at which secondary aerosol is produced is determined by the rate at which condensable species are generated and is independent of the existing aerosol. The distribution of condensable species onto the aerosol is governed by gas-phase diffusion and by the affinity of individual particles for the species.

In our calculations the rate of particulate production in the atmosphere is set proportional to the intensity of solar radiation and to the accumulated emissions of motor vehicle exhaust, with a 1-hr time lag:

$$R\ (t + 1) = bI(t) \int_0^t E(\tau)\ d\tau \qquad (1)$$

Here R is the rate at which secondary aerosol mass is produced, I is the intensity of solar radiation, and E is the rate at which exhaust is emitted. The constant b is chosen so that the calculated average concentration of secondary material in the Pasadena aerosol during the period 0900 to 2000 on September 3, 1969, is 25 $\mu m^3/cm^3$, in agreement with Friedlander's estimate (8). The significant feature of Eq. 1, as will be discussed in the next section, is the proportionality of R to the integral of E, rather

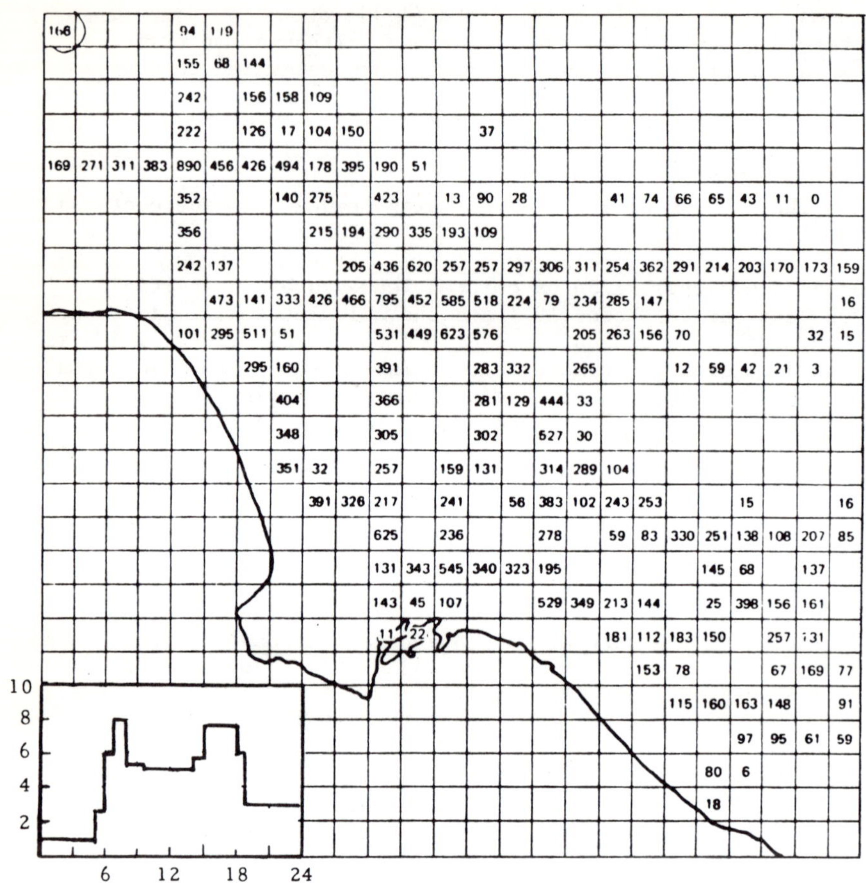

FIGURE 3. Geographical distribution of freeway traffic in the Los Angeles basin, in thousands of car-miles per day. Insert shows percent of day's mileage traveled during each hour. A similar grid gives distribution of non-freeway traffic. [From Roberts et al. (21).]

than to E itself. This choice is suggested by the evidence from the experiments cited above that a given charge of reacting gases is not "used up" quickly, but continues to produce particulate material over an extended time interval.

Chamber experiments and field measurements in Pasadena indicate that secondary particulate mass is added to the aerosol primarily through heterogeneous condensation on existing nuclei [Husar et al. (2), Husar and Whitby (23)]. Electron microscopic observations of ambient particles suggest that the chainlike aggregates of lead halide particles are not effective condensation nuclei and that most of the material is deposited on other types of nuclei, such as automobile organics and sea salt. In our calculations the secondary material is assumed to condense on existing non-lead-halide particles according to the diffusional rate law [Fuchs (26)]:

$$\frac{dD_p}{dt} = \frac{a(C_\infty - C_o)}{D_p + 2\lambda\ \ell} \tag{2}$$

In expression 2, D_p is the particle diameter, λ is the mean free path of the condensing vapor, C_∞ and C_o are the concentrations of this vapor far from the particle and at the surface, respectively, and ℓ and a are constants.

The vapor concentration gradient driving the diffusional transfer represents an equilibrium between production of condensable species in the gas phase and loss of these species to the aerosol. On the one hand, the rate at which secondary aerosol mass is produced is determined by the difference $C_\infty - C_o$, together with the existing aerosol; Eq. 2 gives

$$R = \int \frac{d}{dt}\left(\rho \frac{\pi}{6} D_p^3\right) f(D_p)\, dD_p$$

$$= \rho a \left(C_\infty - C_o\right) \int \frac{\frac{\pi}{2} D_p}{1 + 2\lambda\ell/D_p} f(D_p)\, dD_p$$

where ρ is density and f is the number distribution of the non-lead-halide aerosol. On the other hand, the rate at which secondary aerosol mass is produced is determined independently by gas-phase chemistry, as summarized in Eq. 1. The diffusional driving force appearing in Eq. 2 is therefore calculated as

$$a\left(C_\infty - C_o\right) = \frac{R}{\rho M} \tag{3}$$

where R is given by Eq. 1 and M is the diffusional transfer moment of the aerosol:

$$M = \int \frac{\frac{\pi}{2} D_p}{1 + 2\lambda \ell / D_p} f(D_p)\, dD_p$$

III. RESULTS

Results of calculations based on the above model are shown in Figures 4 and 5. These figures were produced by integrating the rate equations for aerosol growth along the hourly trajectories arriving at Pasadena, taking into account the spatio-temporal distribution of emissions. Both atmospheric transport and gas-to-particle conversion thereby influenced the characteristics of the calculated Pasadena aerosol.

As Figure 4 shows, the diurnal variation of the calculated aerosol concentration is similar to that actually observed.

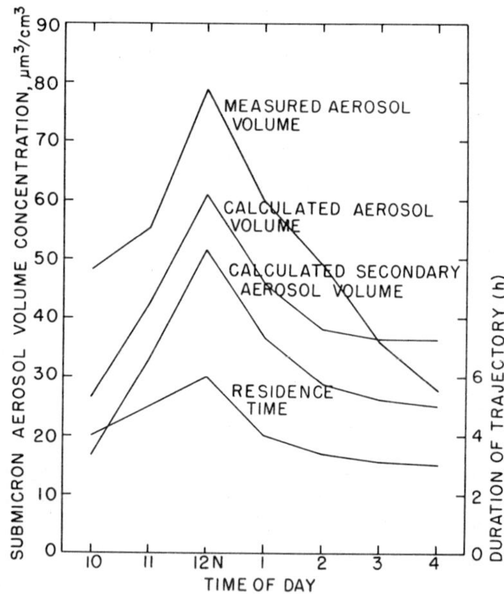

FIGURE 4. Diurnal patterns in Pasadena aerosol on September 3, 1969. Top three curves show calculated and measured submicron aerosol volume concentrations; bottom curve shows estimated time spent over land since 0600 by air mass.

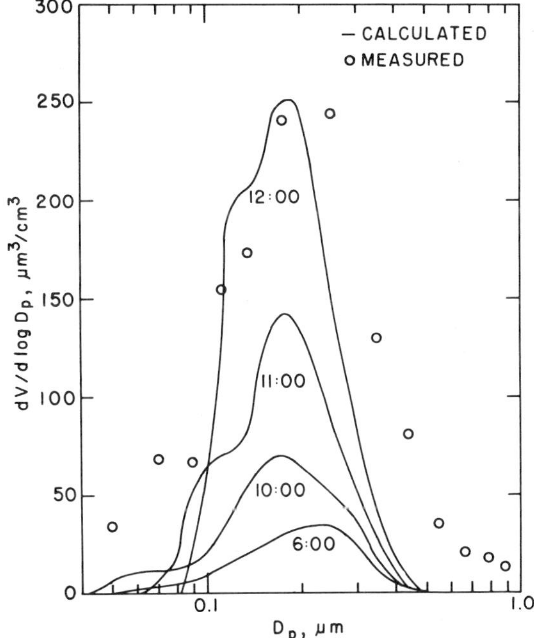

FIGURE 5. Volume distribution of midday Pasadena aerosol on September 3, 1969. Curves show development of calculated volume distribution along air trajectory arriving in Pasadena at 1200; dots show volume distribution measured in Pasadena at 1200.

Both calculated and observed aerosol volumes increase rapidly during the late morning hours, reach a peak at midday, and drop sharply with the arrival of the marine layer in the afternoon. The variation of the calculated concentration of secondary material indicates that the composition of the Pasadena aerosol undergoes marked changes during the day. Although primary emissions and natural and aged urban backgrounds are important constituents of the aerosol in the early morning and late afternoon, the midday aerosol is dominated by material recently produced in the atmosphere.

It is instructive to compare the variation of aerosol concentration with the variation of residence time, also shown in Figure 4. If the air were clean at 0600, and the mixing height and the emission rates were constant, the concentration of primary material in an air parcel would be proportional to the length of time, T, that the parcel had spent over land since 0600. If solar radiation were also constant, then, according to Eq. 1, the rate of gas-to-particle conversion would be proportional to T - 1, so that the accumulation of secondary material in the parcel would be proportional to the time integral:

$$\int_0^T (T - 1)\, dT = \frac{T^2}{2} - T$$

It is primarily due to this roughly quadratic dependence on residence time that the calculated concentration of secondary material increases so rapidly at midday.

The evolution of the calculated aerosol size distribution along the midday trajectory is shown in Figure 5. Because the nonlinear dependence of the conversion process on residence time, most of the aerosol growth occurs late in the trajectory, between 1000 and 1200. The calculated final size distribution is comparable with the size distribution actually measured, the calculated mass median particle diameter being slightly smaller than observed.

The sensitivity of the model was tested with a large number of calculations in which assumptions and data were systematically varied. A sample of the results is given in Table 1, which shows how predicted light-scattering coefficients depend on the growth law in Eq. 2 and on the rate of emission of organic particulates by motor vehicles, when all other factors are held constant. According to these figures, control of particulate emissions without corresponding control of gaseous emissions will not produce much improvement in visibility.

ACKNOWLEDGMENTS

This work was strongly influenced by Dr. S. K. Friedlander of the California Institute of Technology. It was supported by the California Air Resources Board Aerosol Characterization Study and by the National Institute of Environmental Health Sciences Training Grant No. ES 00080-06. The contents do not necessarily reflect the views and policies of the sponsoring agencies.

TABLE 1. Dependence of Calculated Light-Scattering Coefficient on Rate of Organic Particulate Emissions from Motor Vehicles and Form of Particle Growth Law

Growth law	Calculated b_{scat} (10^{-4} m^{-1}) Emission Rate g/mile			
	0.05	0.10	0.15	0.20
$\frac{dD_p}{dt} \propto \frac{1}{D_p + 2\ell\lambda}$	3.2	3.4	3.6	3.7
$\frac{dD_p}{dt} \propto 1$	3.3	3.6	3.8	4.0
$\frac{dD_p}{dt} \propto D_p$	2.6	3.1	3.5	3.8

REFERENCES

1. Mueller, P. K., Mosley, R. W., and Pierce, L. B. 1972. Chemical Composition of Pasadena Aerosol by Particle Size and Time of Day - IV. Carbonate and Non-Carbonate Carbon Content, J. Colloid Interface Sci. 39, 235-239.

2. Husar, R. B., Whitby, K. T., and Liu, B. Y. H. 1972. Physical Mechanisms Governing the Dynamics of Los Angeles Smog Aerosol, J. Colloid Interface Sci. 39, 211-224.

3. Ensor, D. S., Charlson, R. J., Ahlquist, N. C., Whitby, K. T., Husar, R. B., and Liu, B. Y. H. 1972. Multiwavelength Nephelometer Measurements in Los Angeles Smog Aerosol - I. Comparison of Calculated and Measured Light Scattering, J. Colloid Interface Sci. 39, 242-251.

4. Whitby, K. T., Husar, R. B., and Liu, B. Y. H. 1972. The Aerosol Size Distribution of Los Angeles Smog, J. Colloid Interface Sci. 39, 177-204.

5. Grosjean, D. and Friedlander, S. K. 1975. Gas-Particle Distribution Factors for Organic and Other Pollutants in the Los Angeles Atmosphere, J. Air Pollut. Control Assoc. 25, 1038-1044.

6. Hidy, G. M., et al. 1974. "Characterization of Aerosols in California," Rep. No. SC524.25FR, Rockwell International Science Center, Thousand Oaks, California.

7. Miller, M. S., Friedlander, S. K., and Hidy, G. M. 1972. A Chemical Element Balance for the Pasadena Aerosol, J. Colloid Interface Sci. 39, 165-176.

8. Friedlander, S. K. 1973. Chemical Element Balances and the Identification of Air Pollution Sources, Environ. Sci. Technol. 7, 235-240.

9. Heisler, S. L., Friedlander, S. K., and Husar, R. B. 1973. The Relationship of Smog Aerosol Size and Chemical Element Distributions to Source Characteristics, Atmos. Environ. 7, 633-649.

10. Gartrell, G., Jr. and Friedlander, S. K. 1975. Relating Particulate Pollution to Sources: the 1972 California Aerosol Characterization Study, Atmos. Environ. 9, 279-299.

11. Eschenroeder, A. Q. and Martinez, J. R. 1971. "Further Development of the Photochemical Smog Model for the Los Angeles Basin," Rep. CR-1-191, General Research Corporation, Santa Barbara, California.

12. Petterssen, S. 1956. Weather Analysis and Forecasting. McGraw-Hill, New York, p. 27.

13. Angell, J. K., Pack, D. H., Machta, L., Dickson, D. R., and Hoecher, W. H. 1972. Three-Dimensional Air Trajectories Determined from Tetroon Flights in the Planetary Boundary Layer of the Los Angeles Basin," J. Appl. Meteorol. 11, 451-471.

14. Neiburger, M. and Edinger, J. G. 1954. "Summary Report on Meteorology of the Los Angeles Basin with Particular Respect to the 'smog' Problem," Southern California Air Pollution Foundation Rep. No. 1, Los Angeles, California.

15. Edinger, J. G. 1959. Changes in the Depth of the Marine Layer over the Los Angeles Basin, J. Meteorol. 16, 219-226.

16. Bell, K. A. and Friedlander, S. K. 1973. Aerosol Deposition in Models of a Human Lung Bifurcation, Staub 33, 178-181.

17. Blifford, I. H., Jr. 1970. Tropospheric Aerosols, J. Geophys. Res. 75, 3099-3103.

18. Colucci, J. M. and Begeman, C. R. 1965. The Automotive Contribution to Air-Borne Polynuclear Aromatic Hydrocarbons in Detroit, J. Air Pollut. Control Assoc. 15, 113-122.

19. Ter Haar, G. L. and Stephens, R. E. 1971. "The Effects of Automobile Exhaust Particulates on Visibility," paper presented at the Twelfth Conference on Methods in Air Pollution and Industrial Hygiene Studies, Los Angeles, California, April.

20. Whitby, K. T., Husar, R. B., McFarland, A. R., and Tomaides, M. 1969. "Generation and Decay of Small Ions," Particle Technology Laboratory Publ. No. 137, Department of Mechanical Engineering, University of Minnesota, Minneapolis.

21. Roberts, P. J. W., Roth, P. M., and Nelson, C. L. 1971. Contaminant Emissions in the Los Angeles Basin, Appendix A of "Development of a Simulation Model for Estimating Ground Level Concentrations of Photochemical Pollutants," Rep. No. 71SAI-6, Systems Applications, Inc., Beverly Hills, California.

22. Habibi, K. 1973. Characterization of the Particulate Matter in Vehicle Exhaust, Environ. Sci. Technol. 7, 223-234.

23. Husar, R. B. and Whitby, K. T. 1973. The Growth Rate and Size Spectrum of Photochemical Aerosols, Environ. Sci. Technol. 7, 241-247.

24. Goetz, A. and Pueschel, R. F. 1965. The Effect of Nucleating Particulates on Photochemical Aerosol Formation, J. Air Pollut. Control Assoc. 15, 91-94.

25. Wilson, W. E., Jr., Miller, D. F., Levy, A., and Stone, R. K. 1973. The Effect of Fuel Composition on Atmospheric Aerosol Due to Auto Exhaust, J. Air Pollut. Control Assoc. 23, 949-956.

26. Fuchs, N. A. 1969. "Recent Progress in the Theory of Transfer Processes in Aerosols at Intermediate Values of Knudsen Number," Proceedings of the 7th International Conference on Condensation and Ice Nuclei, September 18-24, Prague and Vienna.

Relating Particulate Properties to Sources: The Results of the California Aerosol Characterization Experiment

GREGORY GARTRELL, JR., STEVEN L. HEISLER,*
AND SHELDON K. FRIEDLANDER†
California Institute of Technology
Pasadena, California

*Present address: Environmental Research and Technology, Inc., 2625 Townsgate, Suite 560, Westlake Village, California 91361.

†Present address: Department of Chemical Engineering, University of California, Los Angeles, California 90024.

Abstract

Pollution aerosol particles are composed of a mixture of natural background material, solids, and liquids introduced into the atmosphere from man-made sources and the products of certain gas-to-particle transformation processes. The relative contributions of various types of sources to the primary and secondary aerosol can be estimated from the chemical composition and emission inventories. More detailed information on size and chemical element distributions is obtained by making coagulation and growth calculations. Contributions of different sources to visibility degradation can be estimated from the changes in light scattering produced by the various components.

The methodology developed from the Aerosol Characterization Experiment (ACHEX) and earlier studies is applied to several cases selected from the California field observations in 1972 and 1973.

The mass contributions of various sources have been estimated based on chemical element balances and source inventories. These calculations were checked for consistency using data for particle size distributions.

Source contributions to visibility degradation in Pasadena and Pomona were estimated. The principal factors were ascribed to nitrates, sulfates, condensed organics, and ammonium, converted from the gas phase together with water.

I. INTRODUCTION

A major objective of the ACHEX was to develop methodologies for relating the physical and chemical properties of aerosol particles to their sources. The exclusive use of knowledge of physical properties of particulate emissions and ambient particulate samples provides only limited information on sources and ambient pollution. Considerably more can be learned about the origins of pollution aerosol by taking into account information contained in the chemical composition of collected particulate material. This approach has been illustrated in a series of papers with an application to the results of measurements of the Pasadena Smog Aerosol Study in 1969 [1, 2, 3]. Basically, the method consists of (a) estimating certain primary source contributions to the pollution at a point, using chemical elements as tracers, (b) supplementing these estimates with emission inventory data for sources for which characteristic tracers are not available, and (c) calculating the contributions of gas-to-particle conversion from the measured values of sulfates, nitrates, organics, and ammonium ion. In this way data of different types are reduced to a common basis: source contributions to the mass of particulate at a given sampling point.

The products of gas-to-particle transformation processes are then distributed with respect to size by calculations made according to certain growth laws, based on theory and/or experiment. The calculated distribution of chemical elements with respect to size can then be checked against the experimentally measured distributions. Finally, light scattering calculated from the size distribution can be compared with the experimentally measured value. In this way the contributions of different sources to the deterioration of visibility can be estimated.

The method has been summarized by Friedlander [2] in a set of fundamental equations relating the output of standard measurement systems to the input from sources. An integrated measurement system suitable for the determination of the aerosol properties used in the calculations includes (a) a total filter to collect material for total mass and chemical analysis and/or a fractionating device such as the cascade impactor to collect material according to size for chemical analysis, (b) analytical methods for determining the concentrations of certain trace elements, (c) particle size analyzers to provide a detailed breakdown on particle size distribution, and (d) an instrument that measures light scattering by the aerosol.

Such information was obtained in the ACHEX. In this section some of the data obtained in the ACHEX are analyzed by the methods described by Heisler et al. [3] and Gartrell and Friedlander [4].

The method, once perfected, will have a variety of applications. Contributions of natural background and those of primary and secondary contributors to the total particulate can be estimated; these estimates are important for air pollution control strategies. If airflow patterns and the source area distributions are known, complete particulate pollution models for urban and industrial basins can be developed as suggested by White et al. [5].

Chemical element balances were carried out for the ACHEX data collected at Pasadena and Riverside on September 20, 1972; San Jose on October 20; Fresno on September 1; and Pomona on October 24. For all of these sites 24-hr averages for the chemical data were used in the calculations. In addition, the data for two 2-hr periods were analyzed: one set from Pasadena, on September 20, 1200 to 1400, and the other from Pomona, on October 24, 1200 to 1400. The Pasadena time period corresponded to a low total mass of aerosol and high visibility, while for Pomona the total mass was high and the visibility low.

Source-mass contributions were also estimated for eight time periods, using the ACHEX data collected in 1973. The time periods and locations used were as follows: West Covina, 1200 to 1400 PST, July 25 and 2100 PST, July 24 to 2100 PST, July 25; Pomona, 1200 to 1400 PST, August 17 and 2100 PST, August 16 to 2100 PST, August 17; Rubidoux, 1200 to 1400, September 6 and 2300 PST, September 5 to 2300 PST, September 6; and Dominguez Hills, 1000 to 1200 PST, October 5 and 2100 PST, October 4 to 2100 PST, October 5.

II. CHEMICAL ELEMENT BALANCE

The concentration, x_i, of an element, i, found in the aerosol at a receptor site is given by the following relation:

$$x_i = \Sigma_j z_{ij} \alpha_{ij} m_j \tag{1}$$

where m_j is the mass of material from source j per unit mass of receptor material or unit volume of air, z_{ij} is the fraction of element i found in source j, and α_{ij} is the coefficient of fractionation between source and atmosphere. By measuring elemental concentrations in the particulate, x_i, the value of m_j can be determined if z_{ij} and α_{ij} are known. There are, however, several difficulties with this method. The α_{ij} values are difficult to determine and may be functions of local meteorology. In addition, if the aerosol chemical composition is a function of particle size, sedimentation and other deposition processes modify α_{ij}. Thus it is necessary to choose elements for which there is reason to believe that α_{ij} is near unity. A further consideration in choosing an element to be used in the determination of m_j is that all the important sources of that element must be known, along with the corresponding concentrations. For example, one source of zinc is tire dust, but there are other significant sources as well.

As in the references cited above, Na, Al, Ca, Pb, K, V, and Mg were used as the tracer elements in calculating contributions by the primary sources. For these elements automobile exhaust, soil dust, sea salt, cement dust and fly ash from fuel combustion, as explained in detail by Miller et al. (1), are believed to be the dominant sources. The coefficients z_{ij} were those given by Miller et al. (1) and Friedlander (2) and are shown in Table I. Since there are seven elements identified with five sources, the calculation was carried out by minimizing the mean-square error between measured and calculated values, weighted by the experimental error.

Magnesium was measured using neutron activation analysis, and in some cases the concentration of this element was below the limits of detection of the method. In those cases the concen-

TABLE 1. Values of z_{ij} for Different Sources, Based on Data in Friedlander (2)

	Percentages					
Constituent	Sea salt	Soil dust	Auto Exhaust	Fuel-oil fly ash	Portland cement	Tire dust
C as carbon	--[a]	--	--	U[b]	--	29.3
C compounds	--	--	40.3[c]	U	--	59[d]
Na	30.6	2.5	U	5	0.4	U
Mg	3.7	1.4	U	0.06	0.48	U
Al	--	8.2	U	0.8	2.4	U
Si	--	20	U	1	10.7	U
S	2.6	--	U	U	--	U
Cl	55.0	--	6.8	--	--	U
K	1.1	1.5	U	0.2	0.53	U
Ca	1.16	1.5	U	1.3	46.0	U
Ti	--	0.4	U	0.06	0.166	U
V	--	0.006	U	7	--	U
Cr	--	--	U	0.1	--	U
Mn	--	0.11	U	0.06	--	U
Fe	--	3.2	0.4	6	1.09	U
Co	--	0.002	U	0.2	--	U
Ni	--	0.004	U	2	--	U
Cu	--	0.008	U	0.2	--	U
Zn	--	<0.01	0.14	0.02	--	1.5
Br	0.19	--	7.9	--	--	U
I	1.4×10^{-4}	--	U	--	--	
Ba	--	0.06	U	0.1	--	U
Pb	--	0.02	40.0	0.07	--	U

[a] Negligible.

[b] Unknown.

[c] As a tarry substance assumed to be 90% C.

[d] Mostly as a copolymer of styrene and butadiene.

tration and the associated error of the measurement were taken to be one-half the value of the lower limit of detection for magnesium. This had little effect on the results since the calculations were insensitive to magnesium concentrations.

The concentrations used for the 2-hr periods were determined with a Lundgren impactor and the impactor after-filters. Potassium was not used in this set of calculations. Since wall losses in the impactor may be a significant problem, particularly for larger particles, the element contributions, as well as the source strengths for these periods, should be considered a minimum.

The values for the 24-hr periods are from high-volume filters, unless otherwise noted. The measurements of Na, Ca, Al, V, and Mg were made using neutron activation analysis; Pb and K were measured by X-ray fluorescence. In some cases, which are noted, Na was measured using wet chemistry, and Ca by X-ray fluorescence. Errors associated with the measurements are analytical errors.

The data used in the calculations were taken from the aerosol chemistry portion of the ACHEX data bank. The numerical values and the associated (estimated) errors are tabulated by Hidy et al. (6).

A. Automobile Exhaust Composition

The chemical makeup of automobile exhaust and its concentration in air play a key role as a scale factor for major primary sources. Therefore it is important to ensure that estimates of automobile emissions are as representative as possible.

The impactor and filter samples that were collected near the Harbor Freeway in Los Angeles on September 20, 1972, can be used to obtain an improved freeway primary aerosol chemical composition to be used in the analysis of the data. The resulting elemental breakdown provides values of z_{ij} for all freeway sources: automobile exhaust, diesel exhaust, and tire dust.

The wind was blowing directly from the freeway during two 2-hr periods in which samples were collected. These periods were 0500 to 0700 PST and 1700 to 1900 PST. Elemental analyses from the Lundgren impactor after-filters and total carbon analyses from silver membrane total filters were used. The after-filters collected particles with diameters less than 0.6 μm and were chosen for the calculations for two reasons. First, large particles fall out near the freeway; although particles larger than 0.6 μm will remain airborne, no data on mass concentrations were available for the impactor stages. Second, during both time periods 85% of the lead in particles smaller than 1.5 μm (sum of after-filter and last impactor stage) was on the after-filter, indicating that most of the freeway aerosol mass is below 0.6 μm. Data on carbon concentration were not available for the after-filters. It was assumed that the carbon on the silver membrane total filters was associated with particles smaller than 0.6 μm and was solely from the freeway. The resulting carbon concentration is therefore an upper limit. The results of the chemical analyses are shown in Table 2.

TABLE 2. Chemical Element Concentrations from Harbor Freeway, September 20, 1972

Species	After-filter concentration ($\mu g/m^3$) 0500–0700 PST	1700–1900 PST
Al (N)[a]	0.0334 ± 0.0099	0.0473 ± 0.0092
Br (N)	6.32 ± 0.14	4.40 ± 0.10
Ca (x)[b]	0.097 ± 0.024	0.054 ± 0.022
Cl (N)	3.19 ± 0.10	1.57 ± 0.07
	(1.75 ± 0.27)	(1.10 ± 0.13)[c]
Fe (x)	0.112 ± 0.005	0.126 ± 0.005
Na (N)	0.136 ± 0.032	0.158 ± 0.030
Pb (x)	13.3 ± 0.5	8.91 ± 0.36
V (N)	0.00712 ± 0.00039	0.00454 ± 0.00027
Zn (x)	0.094 ± 0.004	0.053 ± 0.002
S+ (E)[d]	1.20 ± 0.3	0.5 ± 0.5[e]
S- (E)	1.50 ± 0.3	0.5 ± 0.5[e]
Mass	68.6 ± 7.6	50.3 ± 5.5
C[f]	U	14.9 ± 2.2

[a]Neutron activation analysis.

[b]X-ray fluorescence analysis.

[c]From charge balance with Pb and Br used in concentration calculations.

[d]ESCA analysis.

[e]Not detected. Value is one-half of detection limit.

[f]As carbon. From silver membrane total filter.

[g]Unknown.

The concentrations of Na, Al, Ca, and V were used to estimate the contributions of sea salt, soil dust, dement dust, and fuel-oil fly ash to the material on the after-filters. The oxidized S concentrations (S+) were scaled up to give sulfate concentrations. These estimates are shown in Table 3 and are less than 7% of the total mass, the sulfate dominating. The concentrations of Pb, Br, Cl, Zn, C, and reduced S (S-) in the remaining masses were then calculated; the results are shown in Table 4. Since some of the chloride in the samples was contributed by sea salt, the freeway chloride was calculated by an electrical charge balance on Pb, Br, and Cl, the Pb expected to be present as lead halides [Hirschler et al. (7)]. These species, along with some H and O associated with the C, are expected to comprise the freeway aerosol (1). It is seen in Table 4 that when the C compounds are assumed to be 15% H and O the sum is only 78.5 ± 7.8% of the mass. This indicates that sources other than those accounted for, probably industrial, for which no tracers are available, contributed to the samples. The concentrations were therefore normalized to 100%; the results are shown in Table 4.

TABLE 3. Contributions of Non-freeway Sources to Harbor Freeway Filter Samples, September 20, 1972

Source	Contribution ($\mu g/m^3$)	
	0500–0700 PST	1700–1900 PST
Sea salt	0.415 ± 0.108	0.475 ± 0.030
Soil dust	0.346 ± 0.122	0.503 ± 0.116
Cement dust	0.211 ± 0.052	0.117 ± 0.047
Fuel-oil fly ash	0.102 ± 0.006	0.065 ± 0.0039
SO_4^{2-} [a]	3.60 ± 0.9	1.5 ± 1.5[b]
Total	4.67 ± 0.9	2.66 ± 1.5
Remaining mass	63.9 ± 7.7	47.6 ± 5.7

[a] Scaled up from S+ measurements by ESCA using scale factor of 1.5.

[b] Not detected. Value is one-half of lower limit of detection.

TABLE 4. Elemental Concentrations in Primary Freeway Aerosol

Element	Percent 0500-0700 PST	Percent 1700-1900 PST	Normalized	Friedlander (2)
Pb	20.8 ± 2.5[a]	18.7 ± 2.4	28.3 ± 3.4	29.2
Br	9.90 ± 1.2[a]	0.24 ± 1.12	13.5 ± 1.6	5.77
Cl	2.74 ± 0.54[a]	2.30 ± 0.39	3.72 ± 0.73	4.96
Zn	0.147 ± 0.019[a]	0.111 ± 0.014	0.200 ± 0.026	0.264
S-	2.3 ± 0.5[a]	1.05 ± 1.05	3.13 ± 0.68	-
C	U	31.3 ± 6.0[a]	42.6 ± 8.2	52.0
		(37.6 ± 7.2)[b]	(51.2 ± 9.8)[b]	(59.4)

Total freeway 78.5 ± 7.8[c]

[a]Used in calculation of freeway composition. Morning sample had higher mass loadings and was used with afternoon carbon.

[b]As carbon compounds, assumed to be 85% C.

[c]Used for normalization and includes carbon as carbon compounds.

Also shown in Table 4 is the freeway chemical composition used by Friedlander (2). The estimates for the lead concentration are essentially the same, while the new carbon estimate is lower. The lower C/Pb ratio leads to the calculation of lower primary freeway carbon in carbon balances than accounted for previously [e.g., Gartrell and Friedlander (4)]. The total primary freeway mass contributions from chemical element balances will be the same with the composition in Table 4 as in older estimates.

For the purpose of detailed source breakdown it was necessary to estimate the individual contributions of automobile exhaust, diesel exhaust, and tire dust. Following Gartrell and Friedlander (4), emission inventories were used to estimate their relative contributions.

B. Emission Inventory Scale-Up

Emission inventory data were used to estimate contributions of industrial, aircraft, and agricultural sources for which characteristic tracers were not available. The total mass inputs of primary particulate pollutants from various sources into the air basins of California have been estimated (8). The inputs, relative to the automobile exhaust, are listed in Table 5 for various sources and locations. If the air basins into which these particulates are emitted are well mixed, the contributions to the aerosol can be approximated by scaling to the automobile exhaust contribution found in the chemical element balance. This assumption cannot be applied, however, to a site where there is a large local source of particulate pollution. For example, at the Harbor Freeway, high ambient concentrations of lead were measured; other sources cannot be scaled to the automobile contribution to the aerosol because of the proximity of the freeway to the sampling site. The errors associated with these mass input estimates are not known; since these inputs are annual averages and do not reflect short-term fluctuations, the errors could reasonably be assumed to be as large as $\pm$100%. For most of the cases considered, the contributions of the sources estimated in this way are not large and the effect of error on the overall balance is not great.

C. Treatment of Secondary Constituents

The secondary constituents of aerosol particles consist of product materials from chemical reactions in the atmosphere and constitute a major fraction of the mass concentration. They include sulfate, nitrate, liquid water, and a portion of the

TABLE 5. Primary Particulate Inputs Relative to Automobile (8)

Input	Los Angeles County	San Francisco Basin	Riverside County	Fresno County
Automobile	1.0	1.0	1.0	1.0
Diesel	0.27	0.26	0.24	0.26
Industrial (less fuel combustion)	0.92	2.8	4.3	8.7
Aircraft	0.25	0.59	1.9	0.41
Agriculture	--	0.51	0.95	8.9
Automobile[a] (tons/day)	44.5	28.1	2.1	2.7
Total[b] (tons/day)	129	174	18.8	54.1

[a] Errors in these estimates are unknown.

[b] Includes fuel combustion.

noncarbonate carbon component. The origins of the precursor gases must be traced to the emissions of SO_2, NO_x, and certain organic vapors.

The data were treated without considering indirect means of accounting for the compositions of sulfate and nitrate salts. Direct measurements of these anions were used in the material balances. To estimate the equilibrium water content associated with sulfate and nitrate salts, it was assumed that compounds in the particles consisted of $NaHSO_4$, NH_4NO_3, NH_4HSO_4, and $(NH_4)_2SO_4$ based on stoichiometry. Replacement of chloride by sulfate in sea salt was included after Miller et al. (1). Water was estimated assuming that the sulfate and nitrate salts were present as saturated solutions. Solubilities of salts were obtained from Hodgman (16).

Table 6 shows the results of the assignment of the nonorganic secondary sources, along with associated water based on assumption of saturated salt solutions. The calculated water concentrations ranged from 8.4 to 14.3% for the midday 1973 samples at West Covina, Pomona, and Rubidoux, while the 2-hr concentration for Dominguez Hills was 24.5%. The 24-hr concentrations were on the order of 30% except at Dominguez Hills, where the volume was 44%. The values measured in 1972 were 10 to 20% using the waterometer and filter results [Hidy et al. (6)], indicating that the present estimates are reasonable. The

TABLE 6. Assignment of Nonorganic Secondary Species to Chemical Compounds and Associated Water[a]

	Total[b]	SO_4^{2-}	NO_3^-	NH_4^+	H_2O
West Covina, 1200-1400, 7/24/73					
$NaHSO_4$	7.15 ± 0.63	5.72 ± 0.50	0	0	25.0 ± 2.2
NH_4NO_3	15.5 ± 1.5	0	12.0 ± 1.2	3.48 ± 0.30	6.41 ± 0.62
NH_4HSO_4	44.0 ± 5.2	36.78 ± 4.33	0	6.89 ± 0.81	0
Excess		0	0	2.84 ± 1.58	
Total[c]		42.5 ± 4.3	12.0 ± 1.2	13.2 ± 1.33[d]	
Pomona, 1200-1400, 8/17/73					
$NaHSO_4$	2.44 ± 0.59	1.95 ± 0.47	0	0	8.54 ± 2.07
NH_4NO_3	18.8 ± 1.9	0	14.6 ± 1.5	4.24 ± 0.42	7.80 ± 0.78
NH_4HSO_4	40.0 ± 4.2	33.4 ± 3.6	0	6.25 ± 0.67	
Excess		0	0	0	
Total[c]		35.3 ± 3.5	14.6 ± 1.5	10.5 ± 0.8[d]	
Rubidoux, 1200-1400, 9/6/73					
$NaHSO_4$	0	0	0	0	0
NH_4NO_3	55.1 ± 5.5	0	42.7 ± 4.3	12.4 ± 1.3	22.8 ± 2.3

(continued)

TABLE 6 (cont.)

	Total[b]	SO_4^{2-}	NO_3^-	NH_4^+	H_2O
NH_4HSO_4	0	0	0	0	0
Excess		13.1 + 1.3	27.3 + 8.2	0	
Total[c]		13.1 ± 1.3	70.0 ± 7.0	12.4 ± 1.3[d]	
Dominguez Hills, 1000-1200, 10/5/73					
$NaHSO_4$	10.1 ± 0.6	8.07 ± 4.8	0	0	35.3 ± 2.1
NH_4SO_3	6.6 ± 0.7	0	5.1 ± 0.51	1.48 ± 0.15	2.72 ± 0.27
NH_4HSO_4	29.3 ± 4.1	24.5 ± 3.4	0	4.59 ± 0.64	
Excess		17.7 ± 6.1	0	-	
Total[c]		50.3 ± 5.0	5.1 ± 0.51	6.07 ± 0.66[d]	
West Covina, 24-hr, 7/24/73					
$NaHSO_4$	17.6 ± 0.6	14.1 ± 0.5	0	0	61.5 ± 2.2
NH_4NO_3	8.76 ± 0.88	0	6.79 ± 0.68	1.97 ± 0.20	3.63 ± 0.36
$(NH_4)_2SO_4$	24.8 ± 3.5	18.0 ± 3.3	0	6.76 ± 1.22	
Excess		0	0	1.67 ± 1.24	
Total[c]		32.1 ± 3.2	6.79 ± 0.68	10.4 ± 1.0	
Pomona, 24-hr, 8/17/73					
$NaHSO_4$	9.71 ± 0.41	7.77 ± 0.33	0	0	34.0 ± 1.4

(continued)

TABLE 6 (cont.)

	Total[b]	SO_4^{2-}	NO_3^-	NH_4^+	H_2O
NH_4NO_3	6.21 ± 0.62	0	4.81 ± 0.48	1.40 ± 0.14	2.57 ± 0.26
$(NH_4)_2SO_4$	14.1 ± 2.5	10.2 ± 1.8	0	3.83 ± 0.69	
Excess		0	0	2.09 ± 1.01	
Total[c]		18.0 ± 1.8	4.81 ± 0.48	7.32 ± 0.73	
Rubidoux, 24-hr, 9/6/73					
$NaHSO_4$	12.0 ± 0.3	9.61 ± 0.38	0	0	42.0 ± 1.7
NH_4NO_3	54.7 ± 5.5	0	42.4 ± 4.2	12.3 ± 1.2	22.7 ± 2.3
NH_4HSO_4	0	0	0	0	
Excess		0	0	0	
Total		9.61 ± 0.38	42.4 ± 4.2	12.3 ± 1.2	
Dominquez Hills, 24-hr, 10/5/73					
$NaHSO_4$	19.6 ± 0.7	15.7 ± 0.5	0	0	
NH_4NO_3	1.21 ± 0.12	0	0.936 ± 0.094	0.272 ± 0.027	68.7 ± 2.3
$(NH_4)_2SO_4$	21.6 ± 4.4	15.7 ± 3.2	0	5.88 ± 1.19	0.501 ± 0.050
Excess		0	0	0.548 ± 1.4	
Total		31.4 ± 3.3	0.936 ± 0.094	6.70 ± 0.67	

[a] Values in $\mu g/m^3$.

[b] Excludes water.

[c] Measured.

[d] Mass fraction assumed equal to 24-hr sample.

assignment of the nonorganic secondary components to chemical compounds was generally satisfactory, the exceptions being the 2-hr samples at Rubidoux and Dominguez Hills, where exceptionally large nitrate and sulfate loadings were measured.

D. Carbon Balance

The assignment of carbon to primary and secondary sources can be made by using a carbon balance, first discussed by Friedlander (2).

All of the diesel exhaust, aircraft, and agricultural aerosols were assumed to be carbon in treating the 1972 data. Frey and Corn (9) found that 85 to 93% of diesel exhaust aerosol was comprised of inorganic and organic carbon. For automobile exhaust, carbon was assumed to comprise 90% of the tar in the particulates. The carbon concentration in the motor vehicle sources (auto exhaust, diesel exhaust, and tire dust) was assumed to be that in Table 5 in treating the 1973 data. In the absence of other information we assumed that the fraction of carbon in industrial emissions was the same as the fraction measured in the total particulate, about 25% for the daily average for the 1972 samples and 6 to 18% for the 1973 data. As shown below, the carbon balance is not sensitive to this assumption. In this way, estimates were made of the amounts of carbon from primary sources in the aerosol. The difference between the measured carbon and the primary carbon was assigned to organic compounds resulting from gases converted by atmospheric reactions. (As shown in Table 7, most of the aerosol carbon in the 1972 samples came from secondary conversion with only small contributions from the individual primary sources.) The results of the carbon balance for the 1973 period are shown in Table 8. The large primary carbon contributions at Rubidoux were due primarily to the large scaling for estimating the contributions of aircraft from those of automobiles in Riverside County and the large carbon content assumed for aircraft emissions.

Grosjean's measurements (10) indicate that about two-thirds of the organic fraction of the Pasadena aerosol is carbon on smoggy days when the aerosol carbon is dominated by atmospheric conversion. This is in good agreement with the results reported by Mader et al. (11). Thus the organic fraction of the aerosol that results from gas-to-particle transformation was assumed to be 1.5 of the excess carbon found from the carbon balance.

E. Results of Source Breadkown Calculations

Tables 9 and 10 show the results of the source breakdown calculations for various locations, based on 24-hr and 2-hr

TABLE 7. Carbon Balance Based on 1972 Data[a]

Carbon Source	Pasadena 9/20 24-hr	Pomona 10/24 24-hr	Riverside 9/20 24-hr	Pasadena 9/20 1200-1400	Pomona 10/24 1200-1400
Automobile exhaust[b]	2.1	2.9	1.6	2.4	2.6
Diesel exhaust[c]	1.4	1.9	0.9	1.6	1.3
Industrial emissions[d]	1.2	1.7	4.2	1.4	1.6
Aircraft exhaust[c]	1.3	1.8	7.4	1.5	1.7
Tire dust[e]	0.4	0.6	0.3	0.5	0.6
Agriculture[c]	-	-	3.7	-	-
Organic conversion as C (by difference)	19.7	19.5	16.5	22.3	28.0

[a]Values in $\mu g/m^3$ are taken to 0.1 $\mu g/m^3$ for bookkeeping purposes only. Since diesel, industrial, aircraft, and agriculture emissions are scaled to auto exhaust using Table 5, it is not possible to give error values for these estimates.

[b]Carbon assumed to comprise 90% of tar in particulates.

[c]Assumed to be 100% carbon.

[d]Assumed to be equal to fraction of carbon found in aerosol, ∿25%.

[e]87.3% carbon, after Friedlander (2).

TABLE 8. Carbon Balance[a]

Source	West Covina 7/24/73 1200–1400	Pomona 8/17/73 1200–1400	Rubidoux 9/6/73 1200–1400	Dominguez Hills 10/5/73 1200–1400
Auto exhaust + Tire dust + Diesel exhaust	5.83 ± 0.24	4.54 ± 0.17	2.24 ± 0.09	1.92 ± 0.087
Aircraft[b]	2.51 ± 0.10	1.95 ± 0.07	7.45 ± 0.28	0.82 ± 0.03
Industrial[c]	1.64 ± 0.07	1.15 ± 0.06	1.34 ± 0.12	0.44 ± 0.02
Agricultural[b]	0	0	5.37 ± 0.22	0
Sum of primary carbon	9.99 ± 0.41	7.64 ± 0.30	16.4 ± 0.7	3.18 ± 0.13
Total carbon	47.6 ± 2.4[d]	39.8 ± 2.0[d]	20.66 ± 1.52[e]	19.58 ± 1.35[e]
Secondary carbon	37.6 ± 2.4	32.2 ± 2.0	4.26 ± 1.7	16.4 ± 1.4
Organics-secondary	56.4 ± 3.6	48.3 ± 3.0	6.39 ± 2.55	24.6 ± 2.1
	24-hr	24-hr	24-hr	24-hr
Auto exhaust + Tire dust + Diesel exhaust	5.62 ± 0.23	3.02 ± 0.12	2.17 ± 0.09	1.63 ± 0.06
Aircraft[b]	2.42 ± 0.10	1.29 ± 0.06	7.23 ± 0.31	0.703 ± 0.28
Industrial[c]	1.02 ± 0.04	0.434 ± 0.03	0.947 ± 0.042	0.185 ± 0.008
Agricultural[b]	0	0	3.62 ± 0.15	0
Sum of primary carbon	9.06 ± 0.37	4.74 ± 0.21	14.0 ± 0.6	2.52 ± 0.10
Total carbon	24.1 ± 1.2[f]	16.4 ± 0.8[f]	15.1 ± 0.8[f]	10.6 ± 0.5[f]
Secondary carbon	15.0 ± 1.3	11.7 ± 0.8	1.10 ± 1.00	8.08 ± 0.51
Secondary-organics	22.5 ± 2.0	17.6 ± 1.2	1.65 ± 1.50	12.1 ± 0.8

(continued)

TABLE 8 - footnotes

[a]Values in $\mu g/m^3$.

[b]Assumed to be 100% carbon.

[c]Assumed to be 19% carbon.

[d]Total carbon by conversion to CO_2.

[e]Scaled from FID analysis using $y = 1.75 + 1.26x$, where y is value shown and x is measurement by FID. Equation is least-squares best fit from data where both types of measurements were available.

[f]In particles smaller than 3-5 μm diameter. Measured by conversion to CO_2.

Source	Pasadena 9/20/72	Pomona 10/24/72	Riverside 9/20/72	Fresno 9/1/72	San Jose 10/20/72
Sea salt	0.7 ± 0.06	5.7 ± 0.6	1.3 ± 0.1	$0^{b} \pm 0.4$	19.4 ± 0.5
Soil dust	19.8 ± 0.1	15.1 ± 0.5	28.5 ± 0.9	51.1 ± 2.8	29.6 ± 1.1
Auto exhaust	5.1 ± 0.15	7.2 ± 0.3	3.9 ± 0.15	2.2 ± 0.1	8.3 ± 0.33
Cement dust	1.4 ± 0.15	3.3 ± 0.6	2.3 ± 0.15	0.5 ± 0.14	4.5 ± 1.3
Fly ash	0.1 ± 0.01	0.2 ± 0.01	0.1 ± 0.01	$<0.1 \pm 0.01$	0.1 ± 0.01
Diesel exhaust[c]	1.4	1.9	0.9	0.6	2.2
Tire dust[d]	0.5	0.7	0.4	0.2	0.8
Industrial and agricultural[c]	4.7	6.6	20.5	37.8	27.6
Aircraft[c]	1.3	1.8	7.4	0.9	4.9
SO_4^{2-} [e]	2.9 ± 0.7	19 ± 5	5.9 ± 1.5	4.2 ± 1.0	16.6 ± 3.3
NO_3^-	4.9 ± 0.4	36.4 ± 2.7	12.9 ± 1.0	7.9 ± 0.6	12.3 ± 0.9
NH_4^+	2.3 ± 0.1	16.3 ± 0.8	5.7 ± 0.3	3.1 ± 0.15	7.2 ± 0.72
Organics[f]	29.6	29.3	24.8	U[g]	U
Water[e]	12 ± 6	18 ± 9	U	U	U
Total mass	86.7	161.5	114.6	108.5	133.5
Measured mass	64 ± 7	180 ± 20	125 ± 14	207 ± 23	189 ± 21

[a]Values in $\mu g/m^3$. Errors associated with sea salt, soil dust, auto exhaust, cement dust, and fly ash are standard errors from the least-squares fit for the chemical element balance; errors associated with SO_4, NO_3, NH_4, water, and measured total mass concentrations are analytical errors.

[b]Values actually found to be slightly negative.

[c]Scaled to auto exhaust based on relative inputs (Table 5).

[d]Assumed to be 10% of auto exhaust component.

[e]Measured values.

[f]Based on carbon balance (Table 8).

[g]Unknown.

TABLE 10. Two-hour Source Breakdown Based on Chemical Element Balance[a]

Source	Pasadena 9/20/72 1200-1400	Pomona 10/24/72 1200-1400
Sea salt	1.1 ± 0.4	2.9 ± 0.4
with $SO_4^{2=}$ replacement[b]	(1.3)	(3.5)
Soil dust	7.5 ± 0.5	10.6 ± 0.44
Auto exhaust	6.0 ± 0.26	6.8 ± 0.4
Cement dust	1.4 ± 1.3	0.6 ± 0.6
Fly ash	0.1 ± 0.01	0.1 ± 0.01
Diesel exhaust[c]	1.6	1.8
Tire dust	0.6	0.7
Industrial and agricultural[c]	5.5	6.2
Aircraft[c]	1.5	1.7
SO_4^{2-} [d]	3.0 ± 0.6	23 ± 4.5
Less replacement[e]	(2.3)	(20.1)
NO_3^- [f]	4.0	46.0
NH_4^+ [d]	U[g]	18 ± 3.7
Organics[h]	33.5	42.0
Water[d]	13 ± 7	U
Total mass	78.8	160.4
Measured mass	65 ± 7	227 ± 20

[a]Values in $\mu g/m^3$; errors as in Table 9.

[b]SO_4^{2-} has been assumed to replace Cl^- in sea salt, and these values reflect the mass increase.

[c]Tire dust assumed to be 10% of auto exhaust component; diesel, aircraft, industrial, and agricultural emissions scaled to auto exhaust using Table 5.

[d]Measured values.

[e]SO_4^{2-} left after Cl^- replacement.

[f]Fraction of nitrate assumed equal to 24-hr average.

[g]Unknown.

[h]Based on carbon balance.

sampling periods, respectively, in 1972. The sea salt, soil dust, automobile exhaust, cement dust, and fly ash contributions were based on the chemical element balance, while diesel exhaust, tire dust, agricultural and industrial, and aircraft aerosols were scaled to the automobile through an emission inventory. Secondary conversion products--sulfate, nitrate, ammonia, and water--were measured directly, while the organics were corrected for primary contributions by the carbon balance. The measurements for water are from water filters or the waterometer experiments at the sites.

The results for Pasadena, Pomona, and Riverside show quite acceptable agreement between the sum of the calculated source contributions and the measured total mass. As in previous studies, the products of the conversion of organic vapors, nitrogen oxides, and sulfur dioxide represent a significant fraction of the total aerosol. The data for Pomona show much higher concentrations of sulfate, nitrate, and ammonia than those for Pasadena, although organic concentrations are about the same. In both cases there is about enough ammonium ion on a molar basis to account for ammonium salts of nitrate and sulfate.

Table 11 compares measured concentrations of the elements with values calculated from the concentrations of soil dust, sea salt, automobile exhaust, fly ash, and cement dust obtained from the chemical element balance. The species on which the calculation was based (Al, Na, Pb, Ca, V, Mg, and K) are listed first in the table and show generally good agreement between measured and calculated values. The results are not normalized to specific species, so the good agreement adds confidence that the sources identified are the major ones for this set of elements.

A chemical element breakdown was not carried out for the industrial sources. Hence the calculated concentrations for elements in the industrial sources would be expected to be smaller than the measured values. This probably explains the deviations between measured and calculated concentrations for Cu, Cr, and Zn in Table 11. The discrepancy would not appear in the material balance (Tables 9 and 10) since these species would be included, on a mass basis, in the industrial contribution. Except at Fresno, chlorine was present in smaller concentrations than calculated because it is lost from the aerosol by chemical reactions [Miller et al. (1), Friedlander (2)].

The results for the mass contribution calculations for 1973 data are shown in Table 12. The agreement is acceptable in all cases. It is interesting that the sea salt contribution is larger for the 24-hr samples than for the 2-hr midday samples, indicating that the peak in the marine contribution does not occur at midday. West Covina, Pomona, and Dominguez Hills are

TABLE 11. Measured and Calculated Element Concentrations[a]

Element	Pasadena, 9/20/72 Calculated	Pasadena, 9/20/72 Measured	Pomona, 10/24/72 Calculated	Pomona, 10/24/72 Measured	Riverside, 9/20/72 Calculated	Riverside, 9/20/72 Measured
Na[b]	0.716	0.716 $\pm$ 0.014	2.13	2.06 $\pm$ 0.20	1.13	1.13 $\pm$ 0.019
Al[b]	1.66	1.03 $\pm$ 0.33	1.32	1.30 $\pm$ 0.42	2.35	2.35 $\pm$ 0.07
Ca[b]	0.94	0.94 $\pm$ 0.10	1.83	0.8 $\pm$ 0.28	1.49	1.49 $\pm$ 0.07
V[b]	4.8×10^{-3}	$4.8\times10^{-3} \pm 2.6\times10^{-4}$	0.014	$0.014 \pm 5.2\times10^{-4}$	8.8×10^{-3}	8.8×10^{-3} $\pm$ 0.000
Pb[b]	2.03	2.03 $\pm$ 0.077	2.89	2.89 $\pm$ 0.12	1.55	1.55 $\pm$ 0.006
Mg[b]	0.34	<0.4	0.50	<0.4	0.50	0.97 $\pm$ 0.38
K[b]	0.31	0.36 $\pm$ 0.036	0.31	0.44 $\pm$ 0.044	0.45	0.68 $\pm$ 0.068
Mn	0.022	0.023 $\pm$ 0.002	0.017	0.037 $\pm$ 0.001	0.031	0.040 $\pm$ 0.001
Cu	1.7×10^{-3}	0.043 $\pm$ 0.002	1.6×10^{-3}	0.025 $\pm$ 0.002	2.4×10^{-3}	0.29 $\pm$ 0.01
Cl	0.73	0.44 $\pm$ 0.22	3.6	0.93 $\pm$ 0.035	1.0	0.76 $\pm$ 0.026
Br	0.40	0.50 $\pm$ 0.011	0.58	0.81 $\pm$ 0.018	0.31	0.39 $\pm$ 0.009
Cr	5.2×10^{-5}	0.006 $\pm$ 0.002	1.9×10^{-4}	0.025 $\pm$ 0.004	1.0×10^{-4}	0.013 $\pm$ 0.002
Ni	1.8×10^{-3}	0.006 $\pm$ 0.001	4.4×10^{-3}	0.024 $\pm$ 0.002	3.2×10^{-3}	0.014 $\pm$ 0.001
I	9.7×10^{-7}	0.0023 $\pm$ 0.0005	7.9×10^{-6}	$0.0042 \pm 7.4\times10^{-4}$	1.9×10^{-6}	$6.4\times10^{-3} \pm 7\times10^{-4}$
Fe	0.67	1.15 $\pm$ 0.044	0.56	1.55 $\pm$ 0.06	0.94	2.31 $\pm$ 0.09
Zn	0.009	0.097 $\pm$ 0.004	0.012	0.23 $\pm$ 0.009	0.008	0.13 $\pm$ 0.005

[a]Values in $\mu g/m^3$; errors represent analytical errors. Calculated values are from estimates of soil dust, sea salt, auto exhaust, cement dust, and fly ash from the chemical element balances.

[b]Used in making the chemical element balance, hence close agreement is expected.

(continued)

TABLE 11 (continued)

Element	Fresno, 9/1/72		San Jose, 10/20/72	
	Calculated	Measured	Calculated	Measured
Na[b]	0.84	0.838 ± 0.084	6.70	6.71 ± 0.15
Al[b]	4.20	4.97 ± 0.75	2.54	2.56 ± 0.082
Ca[b]	0.98	0.98 ± 0.049	2.75	2.72 ± 0.58
V[b]	0.005	<0.02	0.011	$0.011 \pm 7\times10^{-4}$
Pb[b]	0.87	0.87 ± 0.04	3.34	3.34 ± 0.13
Mg[b]	0.67	0.68 ± 0.035	1.24	3.5 ± 0.91
K[b]	0.75	0.70 ± 0.070	0.68	0.62 ± 0.062
Mn	0.056	0.051 ± 0.003	0.033	0.047 ± 0.003
Cu	4.1×10^{-3}	0.012 ± 0.001	2.6×10^{-3}	1.5 ± 0.006
Cl	0.15	0.38 ± 0.06	11.2	7.9 ± 0.20
Br	0.17	0.204 ± 0.008	0.69	1.19 ± 0.027
Cr	2.8×10^{-5}	0.012 ± 0.003	1.4×10^{-4}	0.032 ± 0.005
Ni	2.6×10^{-3}	0.011 ± 0.001	3.9×10^{-3}	0.025 ± 0.004
I	0	U[c]	2.7×10^{-5}	0.011 ± 0.002
Fe	1.65	2.13 ± 0.08	1.0	1.95 ± 0.08
Zn	0.003	0.115 ± 0.005	0.015	0.16 ± 0.006

[a] Values in $\mu g/m^3$; errors represent analytical errors. Calculated values are from estimates of soil dust, sea salt, auto exhaust, cement dust, and fly ash from the chemical element balances.

[b] Used in making the chemical element balance, hence close agreement is expected.

[c] Unknown.

TABLE 12. Estimated Source Contributions - 2-hr Filters[a]

Source	West Covina 7/24/73 1200-1400	Pomona 8/17/74 1200-1400	Rubidoux 9/6/73 1200-1400	Dominguez Hills 10/5/73 1000-1200
Sea salt[b]	8.15 ± 0.7	2.78 ± 6.7	0.0	11.5 ± 0.69
	(4.48 ± 0.38)	(1.53 ± 0.37)	(0.0)	(6.32 ± 0.38)
Soil dust	31.5 ± 1.2	43.9 ± 1.7	40.4 ± 1.6	18.0 ± 0.8
Auto exhaust	10.0 ± 0.4	7.80 ± 0.28	3.93 ± 0.16	3.29 ± 0.13
Diesel exhaust	2.70 ± 0.10	2.10 ± 0.08	0.947 ± 0.035	0.891 ± 0.035
Tire dust	1.00 ± 0.04	0.780 ± 0.028	0.393 ± 0.016	0.329
Aircraft	2.51 ± 0.10	1.95 ± 0.07	7.10 ± 0.28	0.821 ± 0.013
Industrial	9.19 ± 0.42	7.17 ± 0.28	16.9 ± 0.7	3.03 ± 0.12
Fuel oil	0.647 ± 0.028	0.910 ± 0.038	0.190 ± 0.014	0.831 ± 0.026
Cement dust	3.93 ± 0.18	3.09 ± 0.18	5.82 ± 0.29	0.737 ± 0.085
Agricultural	0	0	3.74 ± 0.15	0
Total primary	69.1 ± 1.4	70.5 ± 1.9	79.4 ± 2.1	39.4 ± 1.1
Organics	56.4 ± 3.6	48.3 ± 3.0	6.39 ± 2.55	24.6 ± 2.1
SO_4^{2-} [c,d]	36.0 ± 4.3	33.4 ± 3.6	13.1 ± 1.3	42.2 ± 5.0
NO_3^- [d]	12.0 ± 1.2	14.6 ± 1.5	70.0 ± 7.0	5.1 ± 0.5
NH_4^+	13.2 ± 1.3	10.5 ± 0.8	12.4 ± 1.3	6.07 ± 0.66
Total secondary	118 ± 7	107 ± 6	102 ± 7	78.0 ± 5.4
Water	31.4 ± 2.3	16.3 ± 2.2	22.8 ± 2.3	38.0 ± 2.1
Total mass	219 ± 8	194 ± 7	204 ± 8	155 ± 6
Measured mass	268 ± 2	249 ± 2	260 ± 2	134 ± 2

(continued)

TABLE 12 (cont.)

Source	West Covina 7/24/73	Pomona 8/17/73	Rubidoux 9/6/73	Dominguez Hills 10/5/73
Sea salt[b]	20.0 $\pm$ 0.7	11.1 $\pm$ 0.5	13.7 $\pm$ 0.5	22.4 $\pm$ 0.7
	(11.0 $\pm$ 0.4)	(6.09 $\pm$ 0.26)	(7.5 $\pm$ 0.30)	(12.3 $\pm$ 0.4)
Soil dust	30.8 $\pm$ 0.2	33.0 $\pm$ 0.2	31.3 $\pm$ 0.2	20.5 $\pm$ 0.2
Auto exhaust	9.68 $\pm$ 0.42	5.19 $\pm$ 0.2	3.81 $\pm$ 0.16	2.80 $\pm$ 0.10
Diesel exhaust	2.60 $\pm$ 0.10	1.40 $\pm$ 0.06	0.912 $\pm$ 0.035	0.758 $\pm$ 0.028
Tire dust	1.31 $\pm$ 0.06	0.519 $\pm$ 0.021	0.381 $\pm$ 0.016	0.280 $\pm$ 0.010
Aircraft	2.42 $\pm$ 0.10	1.29 $\pm$ 0.06	7.24 $\pm$ 0.28	0.703 $\pm$ 0.028
Industrial	8.91 $\pm$ 0.35	4.77 $\pm$ 0.19	16.4 $\pm$ 0.7	2.58 $\pm$ 0.10
Fuel oil	0.453 $\pm$ 0.009	0.543 $\pm$ 0.009	0.302 $\pm$ 0.008	0.952 $\pm$ 0.010
Cement dust	3.19 $\pm$ 0.22	2.87 $\pm$ 0.22	7.01 $\pm$ 0.42	1.78 $\pm$ 0.13
Agricultural	0	0	3.62 $\pm$ 0.15	0
Total primary	79.0 $\pm$ 1.2	60.7 $\pm$ 0.81	84.7 $\pm$ 1.4	52.8 $\pm$ 0.8
Organics	22.5 $\pm$ 2.0	17.6 $\pm$ 1.2	1.65 $\pm$ 1.50	12.1 $\pm$ 0.8
SO_4^{2-} c,d	18.0 $\pm$ 3.3	10.2 $\pm$ 1.8	0	15.7 $\pm$ 3.2
NO_3^- d	6.79 $\pm$ 0.68	4.81 $\pm$ 0.48	42.4 $\pm$ 4.2	0.936 $\pm$ 0.094
NH_4^+ d	10.4 $\pm$ 1.0	7.32 $\pm$ 0.73	12.3 $\pm$ 1.2	6.70 $\pm$ 0.67
Total secondary	57.5 $\pm$ 4.0	39.9 $\pm$ 2.4	56.4 $\pm$ 4.6	35.4 $\pm$ 3.4
Water	65.1 $\pm$ 2.2	36.6 $\pm$ 1.5	64.7 $\pm$ 2.9	69.2 $\pm$ 2.3
Total mass	203 $\pm$ 5	137 $\pm$ 3	206 $\pm$ 6	157 $\pm$ 4
Measured mass	211 $\pm$ 2	180 $\pm$ 2	262 $\pm$ 2	148 $\pm$ 2

(continued)

TABLE 12 (cont.)

[a]Contributions in $\mu g/m^3$. Errors reflect analytical errors in chemical analyses only.

[b]First number is with Cl^- replaced by HSO_4^-. Second number is without replacement.

[c]Does not include HSO_4^- used in Cl^- replacement in sea salt.

[d]Measured.

[e]Mass fraction assumed equal to 24-hr filter.

dominated by a mixture of secondary organics, ammonium, and sulfate contributions, while Rubidoux contains a small secondary organic contribution but is dominated by nitrate and sulfate.

Table 13 compares measured concentrations of several elements with those predicted from the results of the chemical element balance. The values for Na, Al, Pb, Ca, and V agree quite well; Mg and K show poorer agreement, although they were also used in the chemical element balances. The measured concentrations of the magnesium and potassium were considerably higher than those found in 1972, when good agreement between measured and calculated values was obtained. Elements for which the measured concentrations are higher than the calculated values may come from industrial sources not accounted for in the inventory. Elements for which the concentrations are lower, such as bromine and chlorine, may be lost from the aerosol as a result of atmospheric chemical reactions.

III. SIZE DISTRIBUTIONS OF AEROSOLS

A. Primary Sources

The method of estimating source contributions can be checked in part by comparing predicted and measured size distributions after accounting for particle growth. For the primary sources, Heisler et al. (3) used the size distributions of automobile exhaust, sea salt, and soil dust reported by Whitby et al. (12), Woodcock (13), and Blifford (14), respectively. More recent data have since become available for the size distributions of these sources.

Woodcock (13) made measurements of sea salt over the Pacific Ocean near Hawaii for a range of particle diameters from approximately 1.25 to 10 μm. More recently, Junge (15) has reported the size distribution of marine aerosol collected over the Atlantic Ocean for periods with and without contributions of particles from the Sahara. His measurements include particles smaller than 1 μm in diameter. Junge's data for particles with diameters ranging from 0.25 to 12 μm (without the contribution of dust from the Sahara Desert) were used as the characteristic sea salt size distribution in the calculations. The sea salt size distribution used is shown in Figure 1.

The size distributions for soil dust are based on recent measurements by the Air Resources Board mobile van in Goldstone, California, a desert region well away from other sources. The volume distributions for several time periods (normalized to the total volume measured in the sampling period) are shown in

TABLE 13. Measured and Calculated Element Concentrations[a]

Element	West Covina 7/24/73, 1200-1400		Pomona 8/17/73, 1200-1400	
	Calculated	Measured	Calculated	Measured
Na (N)[b]	2.20	2.12 ± 0.11	1.62	1.61 ± 0.11
Al (N)	2.67	2.39 ± 0.10	3.68	3.58 ± 0.14
Ca (x)	2.09	2.09 ± 0.08	2.11	2.11 ± 0.08
U (x)	0.0472	0.0472 ± 0.0020	0.0663	0.0663 ± 0.0027
Pb (x)	3.89	3.89 ± 0.15	3.02	3.02 ± 0.12
Mg (N)	0.689	1.71 ± 0.86	0.746	1.13 ± 1.13[c]
K (x)	0.542	1.69 ± 0.08	0.693	1.15 ± 0.12
Mn (x)	0.0350	0.093 ± 0.007	0.0488	0.084 ± 0.007
Cu (x)	0.00381	0.061 ± 0.004	0.00533	0.028 ± 0.004
Cl (N)	2.46	0.537 ± 0.063	0.841	0.444 ± 0.061
Br (N)	1.85	0.670 ± 0.019	1.85	0.456 ± 0.013
Cr (x)	0.000647	0.035 ± 0.008	0.00091	0.012 ± 0.012[c]
Ni (x)	0.0142	0.053 ± 0.004	0.0200	0.067 ± 0.004
I (N)	6.27×10^{-6}	0.0105 ± 0.0011	2.14×10^{-6}	0.00880 ± 0.00117
Fe (x)	1.09	3.58 ± 0.14	1.49	3.42 ± 0.14
Zn (x)	0.0275	0.456 ± 0.018	0.0213	0.193 ± 0.008

Element	West Covina 7/24/73, 24-hr		Pomona 8/17/73, 24-hr	
	Calculated	Measured	Calculated	Measured
Na (N)[b]	4.16	4.13 ± 0.12	2.73	2.72 ± 0.08
Al (N)	2.61	2.60 ± 0.02	2.78	2.77 ± 0.02
Ca (x)	2.06	2.06 ± 0.10	1.89	1.89 ± 0.10
V (N)	0.0336	0.0336 ± 0.007	0.0400	0.0400 ± 0.0006

(continued)

TABLE 13 (cont.)

Element	West Covina 7/24/73, 24-hr		Pomona 8/17/73, 24-hr	
	Calculated	Measured	Calculated	Measured
Pb (x)	3.75	3.75 $\pm$ 0.15	2.01	2.01 $\pm$ 0.08
Mg (N)	0.913	1.39 $\pm$ 0.59	0.756	0.839 $\pm$ 0.466
K (x)	0.600	0.957 $\pm$ 0.096	0.578	0.933 $\pm$ 0.093
Mn (x)	0.0341	0.076 $\pm$ 0.005	0.0366	0.071 $\pm$ 0.004
Cu (x)	0.00337	0.041 $\pm$ 0.002	0.00373	0.026 $\pm$ 0.002
Cl (N)	6.05	0.425 $\pm$ 0.024	3.35	0.343 $\pm$ 0.018
Br (N)	1.80	0.555 $\pm$ 0.016	0.965	0.433 $\pm$ 0.016
Cr (x)	0.000453	0.015 $\pm$ 0.005	0.000543	0.018 $\pm$ 0.004
Ni (x)	0.0103	0.050 $\pm$ 0.002	0.0122	0.042 $\pm$ 0.002
I (N)	1.54×10^{-5}	0.0116 $\pm$ 0.0012	8.53×10^{-6}	0.00767 $\pm$ 0.00073
Fe (x)	1.05	2.92 $\pm$ 0.12	1.12	2.77 $\pm$ 0.11
Zn (x)	0.0265	0.462 $\pm$ 0.018	0.0144	0.169 $\pm$ 0.007

[a]Concentrations in $\mu g/m^3$. Errors in measured values reflect uncertainties in chemical analysis. Elements above line (Na-K) used in chemical element balance.

[b]Method of analysis: N = neutron activation analysis; x = x-ray fluorescence analysis.

[c]Not detected. Assumed equal to one-half of lower limit of detection with 100% uncertainty.

TABLE 13 (cont.)

Element	Rubidoux 9/6/73, 1200-1400 Calculated	Rubidoux 9/6/73, 1200-1400 Measured	Dominguez Hills 10/5/73, 1000-1200 Calculated	Dominguez Hills 10/5/73, 1000-1200 Measured
Na (N)	0.983	0.969 ± 0.094	2.43	2.43 ± 0.11
Al (N)	3.45	3.34 ± 0.13	1.50	1.50 ± 0.06
Ca (x)	3.28	3.28 ± 0.13	0.693	0.693 ± 0.004
V (N)	0.0157	0.0157 ± 0.0010	0.0592	0.0592 ± 0.0018
Pb (x)	1.50	1.50 ± 0.06	1.28	1.28 ± 0.05
Mg (N)	0.698	2.35 ± 0.92	0.504	0.552 ± 0.622
K (x)	0.635	1.37 ± 0.14	0.345	0.355 ± 0.063
Mn (x)	0.0446	0.069 ± 0.006	0.0203	0.033 ± 0.006
Cu (x)	0.00361	0.042 ± 0.003	0.00310	0.021 ± 0.003
Cl (N)	0.0	0.560 ± 0.061	3.45	0.156 ± 0.53
Br (N)	0.708	0.266 ± 0.008	0.618	0.154 ± 0.004
Cr (x)	0.00019	0.011 ± 0.011[c]	0.000831	0.0105 ± 0.015[c]
Ni (x)	0.00545	0.010 ± 0.003	0.0173	0.047 ± 0.003
I (N)	0	0.00629 ± 0.00091	8.85×10^{-6}	0.0142 ± 0.0010
Fe (x)	1.37	3.15 ± 0.13	0.634	1.01 ± 0.4
Zn (x)	0.0106	0.142 ± 0.006	0.00918	0.142 ± 0.006

Element	Rubidoux 9/6/73, 24-hr Calculated	Rubidoux 9/6/73, 24-hr Measured	Dominguez Hills 10/5/73, 24-hr Calculated	Dominguez Hills 10/5/73, 24-hr Measured
Na (N)	3.13	3.12 ± 0.09	4.34	4.31 ± 0.12
Al (N)	2.74	2.74 ± 0.02	1.73	1.73 ± 0.12
Ca (x)	3.79	3.78 ± 0.02	1.28	1.28 ± 0.06
V (N)	0.0230	0.0230 ± 0.0005	0.0678	0.0678 ± 0.0007

(continued)

TABLE 13 (cont.)

Element	Rubidoux 9/6/73, 24-hr		Dominguez Hills 10/5/73, 24-hr	
	Calculated	Measured	Calculated	Measured
Pb (x)	1.45	1.45 $\pm$ 0.06	1.09	1.09 $\pm$ 0.04
Mg (N)	0.886	1.49 $\pm$ 0.50	0.786	1.57 $\pm$ 0.46
K (x)	0.591	1.28 $\pm$ 0.13	0.454	0.568 $\pm$ 0.57
Mn (x)	0.0346	0.092 $\pm$ 0.004	0.0231	0.042 $\pm$ 0.003
Cu (x)	0.00311	0.027 $\pm$ 0.002	0.00355	0.020 $\pm$ 0.001
Cl (N)	4.14	0.844 $\pm$ 0.0343	6.77	
Br (N)	0.700	0.217 $\pm$ 0.006	0.539	0.203 $\pm$ 0.006
Cr (x)	0.000302	0.024 $\pm$ 0.004	0.000952	0.020 $\pm$ 0.003
Ni (x)	0.00732	0.034 $\pm$ 0.002	0.0199	0.065 $\pm$ 0.003
I (N)	1.05×10^{-5}	0.00576 $\pm$ 0.0089	1.72×10^{-5}	0.0104 $\pm$ 0.0008
Fe (x)	1.10	3.52 $\pm$ 0.14	0.733	1.54 $\pm$ 0.06
Zn (x)	0.0103	0.172 $\pm$ 0.007	0.00785	0.353 $\pm$ 0.014

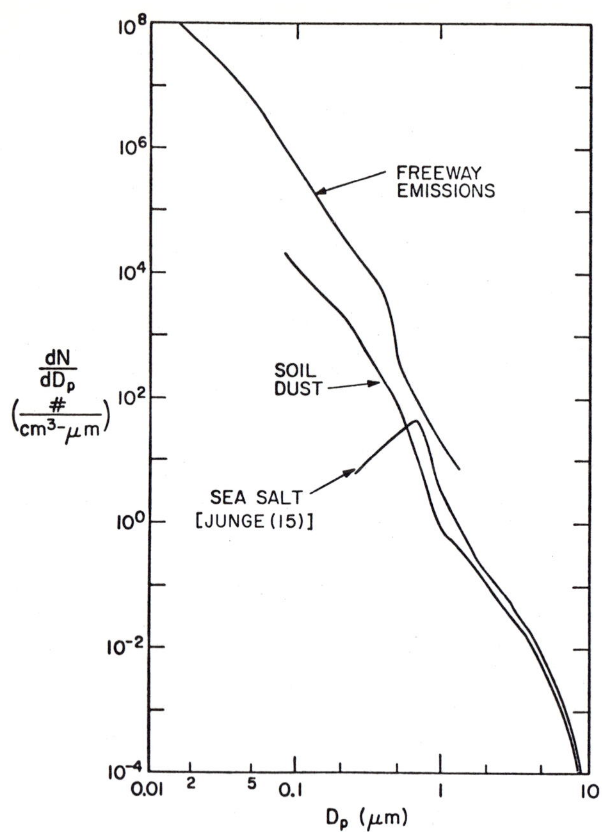

FIGURE 1. Size distributions of sea salt, soil dust, and freeway aerosols. Total volumes are 7.5, 6.7, and 80.6 $\mu m^3/cm^3$, respectively. [Data of Whitby et al. (6).]

Figure 2. These volume distributions are bimodal; but, as discussed below, the lower mode is probably only in part soil dust. Included in Figure 2 is a volume distribution of aerosol that was sampled over several hours during which the wind had been blowing across a dry lake. The solid line in Figure 2 represents the volume distribution used for soil dust in the calculations; the number distribution appears in Figure 1.

Size distributions of aerosol from the Harbor Freeway in Los Angeles were obtained by the ACHEX mobile van in September

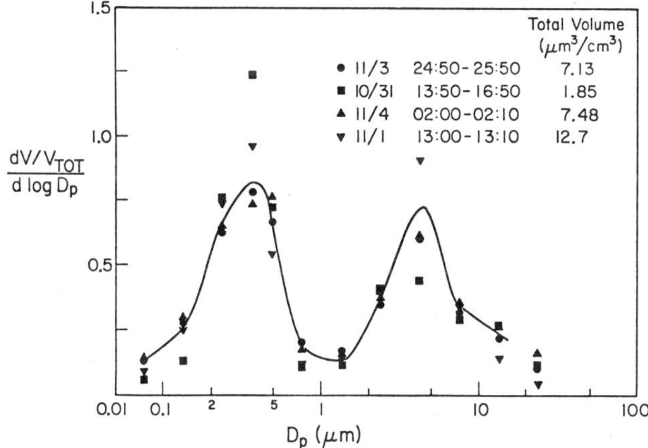

FIGURE 2. Volume distribution of Goldstone aerosol, primarily soil dust in origin, normalized to total volume. Curve indicates distribution used for soil dust in calculations. [Data of Whitby et al. (12).]

1972. During two sampling periods on September 20, the wind came directly from the freeway to the mobile van. During the first of these two periods (0500 to 0700), the ambient lead concentration measured at the sampling site was 23 $\mu g/m^3$, while during the second period (1700 to 1900) it was 11.5 $\mu g/m^3$. The measured volume distributions of aerosol for these two 2-hr periods were averaged and normalized to their respective lead concentrations. The resulting normalized distributions are shown in Figure 3.* For particles with a diameter in the range of 0.1 to 1.0 μm the normalized distributions are quite similar. Below 0.1 μm it is believed that small particles from other local sources cause the deviation between the two distributions. Contributions from sea salt and soil dust probably result in the deviations above 1.0 μm. Since Figure 3 is normalized to the measured lead concentration, particles not associated with lead (i.e., particles from sources other than freeway emissions) would not be expected to have similar size distributions on this type of graph.

*It should be noted that the approach of Whitby et al. (6) gives a somewhat different aerosol distribution for motor vehicle exhaust than the Gartrell and Friedlander (4) method described here. This difference has not been resolved as yet.

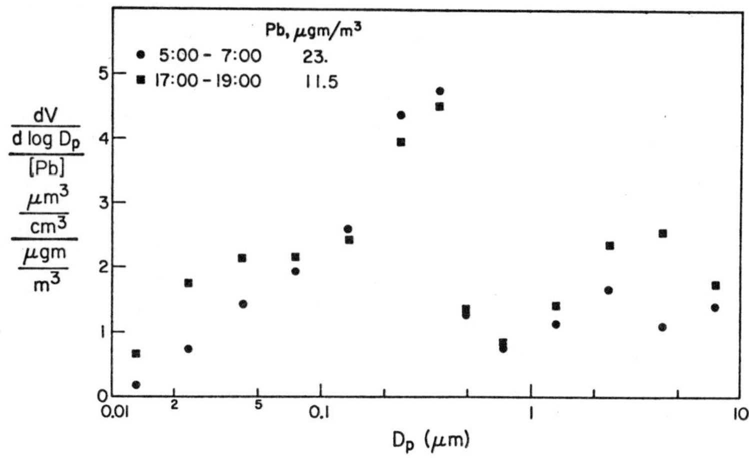

FIGURE 3. Freeway aerosol volume distribution normalized to lead. Size distribution for 0500-0700 was used in the calculations. [Data of Whitby et al. (12).]

Data from the Lundgren impactor and the associated after-filters used at the freeway indicate that most of the lead is associated with particles smaller than 0.8 μm. For this reason, and because of the fact that much of the volume above 1.0 μm came from sources other than the automobile, the portion of the size distribution for particles with diameters greater than 1.0 μm was truncated. Figure 1 shows the size distribution used for automobile exhaust in the calculations. This distribution for particles in the range 0.01 to 1 μm was averaged over the period from 0500 to 0700 (chosen because of the high lead concentration reported in that period).

There are few data on the size distributions of aerosol from other primary sources. The freeway size distributions include diesel exhaust, as well as automobile emissions. In accordance with the assumptions made by Heisler et al. (3), aircraft exhaust was assumed to be similar to the freeway aerosol, and cement and tire dusts to soil dust. The industrial emissions were assumed to be similar to the sum of all the other scaled distributions. The chemical element balances indicated a very small fuel-oil fly ash contribution, which was omitted in the calculations.

B. Coagulation of the Primary Distribution

The automobile is the major source of atmospheric lead pollution in Los Angeles and the major source of primary particulate pollution in the submicron size range as well. Most of the lead fallout occurs near the roadways and is in the large-particle size range (17). Hence the ratio of total number concentration to lead concentration is a measure of the extent to which coagulation has occurred in the automobile emissions and is an indication of the age of the aerosol. As shown in Table 14, this ratio was highest at the freeway and smaller elsewhere. For this reason it was necessary to take coagulation into account in estimating the change in the freeway emissions away from their source.

TABLE 14. Ratio of Number Concentration to Lead Concentration (number/cm^3)/(μg/m^3)

Pasadena, 9/20, 1200-1400 PST[a]	5.8×10^4
Pomona, 10/24, 1200-1400 PST[b]	2.2×10^4
Harbor Freeway, 9/20, 0500-0700 PST[b]	8.4×10^4

[a]Based on condensation nuclei counter.

[b]Based on Whitby analyzer.

In the calculation of coagulation rates a "Monte Carlo" method was employed. The technique, as well as the computer program used, are described in detail by Husar (18). The technique amounts to determining the probability, b_{ij}, that two particles of sizes D_{pi} and D_{pj} will collide. A number between 0 and 1 is chosen at random, and this number is used to determine the particles involved in the collision. After each collision the b_{ij}'s are re-evaluated (since the number distribution has changed), and the next random number is chosen.

The probability, b_{ij}, that a particle of size D_{pi} will collide with a particle of size D_{pj} is given by

$$b_{ij} = \frac{c_{ij} N_i N_j}{\sum_{i=1} \sum_{j=i} c_{ij} N_i N_j} \tag{2}$$

where c_{ij} is a rate constant of coagulation and is a function of particle size, and N_i is the number of particles with diameter D_{pi}. For particles in the transition regime (i.e., with a diameter such that 0.01 μm < D_p < 0.1 μm for air) c_{ij} is given by Fuchs (19) as

$$c_{ij} = 2\pi\beta_{ij}(\underline{D}_i + \underline{D}_j)(D_{pi} + D_{pj}) \qquad (3)$$

where $\underline{D}_i$ and $\underline{D}_j$ are the diffusion coefficients and β_{ij} is a number depending on particle size and rms velocity. For the calculations the particles were grouped into size ranges of equal-log intervals of D_p, with the log mean diameter of the interval used for D_{pi}.

Because the number of particles with D_p > 0.1 μm is small compared to the total number, the coagulatin process was carried out only for particles with D_p < 0.1 μm. To reduce the total computation time, a unit volume was chosen for which the total number of particles was about 50,000. If the number of particles in a given size range became too small, random fluctuations in the process caused excessive errors for the range. Therefore the size distributions resulting from the coagulation process were cut off above the smallest size range for which the final number of particles was less than 200. Points between the cutoff size and the original size distribution (which was used for particles greater than 0.1 μm in diameter) were interpolated as shown in Figure 4. The two size distributions that resulted from the coagulation calculation were used as the freeway size distributions in calculations for the 2-hr periods at Pomona and Pasadena.

The average time between collision of two particles is given by

$$\Delta T = \left(\sum_{i=1}^{\infty} \sum_{j=1}^{\infty} c_{ij} N_i N_j\right)^{-1} \qquad (4)$$

Calculations were made, using Eq. 4, to determine the average time for the coagulation process. The interval ΔT is a function of the particle concentration and must be re-evaluated after each collision. The total time for the coagulation is

$$T = \sum_{k=1}^{N_t - N_f} \Delta T_k \qquad (5)$$

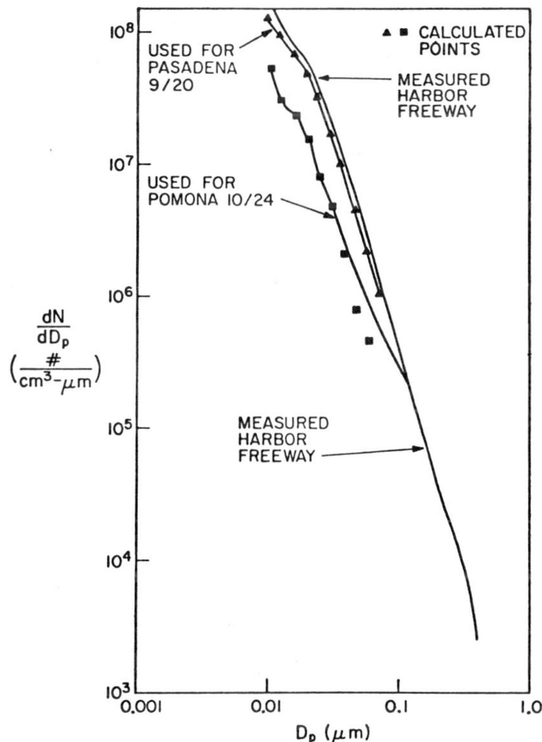

FIGURE 4. Results of coagulation calculations. Measured freeway distribution is shown. Points indicate calculated values; lines connecting points are interpolated values used in growth calculations.

where ΔT_k is the average time between the $(k - 1)^{th}$ and the kth collisions. The results of this calculation showed that for $N_t/N_f = 4$, that is, total number of particles reduced one-fourth the initial number, the time scale of the process was about 400 sec. This time is of the same order as the time found by Husar (18) in experiments on the coagulation of particles of similar concentration and size range. In the atmosphere the time would actually be greater because of the mixing processes.

C. Growth Calculations

Heisler et al. (3) considered three processes for the conversion of organic gases and sulfur dioxide to the particulate phase and found that a diffusion-controlled growth law which accounted for the influence of droplet radius of curvature on vapor pressure (Kelvin effect) agreed best with the experimental data. One difficulty with this approach is that it requires a somewhat arbitrary cutoff of particle diameter, below which no condensation of organic material is allowed. Other growth processes considered by Heisler et al. were a diffusion-controlled process without a Kelvin effect, which showed poor agreement with the experimental data, and a homogeneous, irreversible chemical reaction in the particulate phase, which follows the volume growth law:

$$\frac{dv}{dt} = K(t)D_p^{\gamma} \tag{6}$$

where v is the particle volume, K(t) is a rate coefficient, and $\gamma = 3$. In this paper the power law growth relation, Eq. 6, has been adopted with values of γ adjusted to fit the experimental results.

To evaluate the number and volume distributions resulting from the growth calculations, the following method was used. Integrating Eq. 6 with $v = \pi D_p^3/6$, one finds that for $\gamma \neq 3$

$$\frac{1}{(3-\gamma)} (D_p^{3-\gamma} - D_{po}^{3-\gamma}) = \int_0^t \frac{2K(t)}{\pi} dt \tag{7}$$

and for $\gamma = 3$

$$\frac{D_p}{D_{po}} = \exp\left[\int_0^t \frac{2K(t)}{\pi}\, dt\right] \tag{8}$$

where D_{po} is the initial particle diameter and D_p is the diameter after growth. The right-hand sides of these equations are functions of time only, and are chosen so that the total volume after the growth is equal to the sum of the total volume before growth and the volume of material added to the aerosol.

For the new number distributions the result is

$$\frac{dN}{dD_p} = \frac{dN}{dD_{po}}\,\frac{dD_{po}}{dD_p} = \frac{dN}{dD_{po}}\left(\frac{D_p}{D_{po}}\right)^{2-\gamma} \tag{9}$$

and for the new volume distributions

$$\frac{dV}{d\log D_p} = \frac{dV_o}{d\log D_{po}}\left(\frac{D_p}{D_{po}}\right)^4 \frac{4dD_{po}}{dD_p} = \frac{dV_o}{d\log D_{po}}\left(\frac{D_p}{D_{po}}\right)^{6-\gamma} \tag{10}$$

These equations hold for $\gamma = 3$ as well as $\gamma < 3$.

The composition of the aerosol as determined from chemical element balances and from the experimental data is given in micrograms per cubic meter. Since the conversion processes are concerned with aerosol volumes and not masses, densities were required for the various constituents of the aerosol. The densities used in the calculations were those estimated by Heisler et al. (3). The value for nitrate was assumed to be 1.3. For the growth calculations it was assumed that nitrate was present in the aerosol as saturated ammonium nitrate. In this way, some, if not all, of the reported ammonium found in the aerosol could be accounted for in the growth calculations.

D. Results of the Coagulation and Growth Calculations

Coagulation and secondary conversion calculations were carried out for the 2-hr periods at Pomona and Pasadena in 1972. Based on the chemical element balance for Pomona on October 24 and Pasadena on September 20 (both covering the period 1200 to 1400 PST), the sea salt, soil dust, and coagulated freeway distributions were scaled so that the total volume of a given source was equal to the mass of the source divided by the appropriate density. The sea salt component was assumed to be saturated, and water (3.7 μg/m^3 for Pasadena and 6.0 μg/m^3 for Pomona) was included in the sea salt contribution before scaling. The primary size and volume distributions calculated in this way are shown in Figures 5 through 8. Following Heisler et al. (3), it was assumed that the sulfate and nitrate conversion processes followed a cubic growth law ($\gamma = 3$).

Several values of γ were used for the organic conversion process, and it was found that, as γ decreased, growth occurred increasingly in the smaller particles. Best agreement with the data for Pomona was found with $\gamma = 2.7$ for the organic conversion. Figures 5 and 6 show the results of the secondary conversion calculations for Pomona, along with the experimental data. Agreement is best in the size range $0.1\ \mu m < D_p < 1.0\ \mu m$. There is a deviation between the calculated and measured distributions in the upper and lower portions of the size spectrum. The spike in the volume distribution at about 1.1 μm is caused by a discontinuity in the primary size distribution, and presumably would be smoothed out by coagulation for a well-aged aerosol. No attempt was made, however, to "age" the primary aerosol with the Monte Carlo method of coagulation except for the lower end of the freeway spectrum.

For the Pasadena data, shown in Figures 7 and 8, a value of $\gamma = 2.3$ for the organic conversion gave the best agreement between the calculated and experimental distributions. However, no measurements were made for particles with diameters less than 0.3 μm; thus, it is not known whether the lower end of the size distribution is correct. It is possible that for the smaller particles the Kelvin effect plays a role in the distribution of organics according to size, and that the chemical composition of the particles is an important factor in the growth processes.

IV. SIZE DISTRIBUTION OF CHEMICAL ELEMENTS

Chemical elements in the atmospheric aerosol are distributed with respect to size in a way that depends on the sources of particulates and on atmospheric conversion processes. Air quality standards may eventually take this into account by considering separately particulates above and below 1 or 2 μm in size.

Calculations of the distributions of elements with respect to size and comparison with experiment permit a more severe test of the source breakdown calculations made in the preceding sections. We have carried out such calculations, assuming that

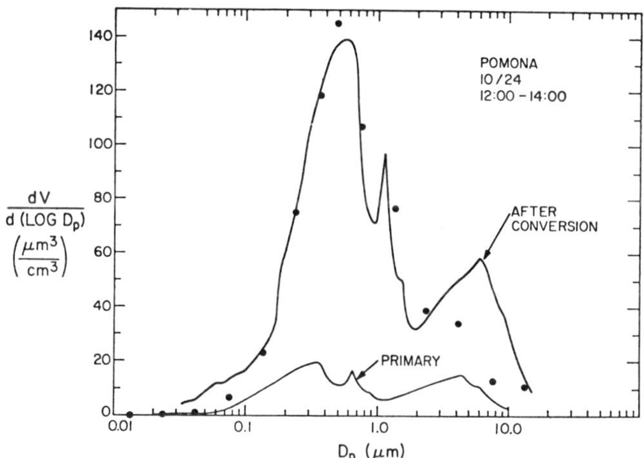

FIGURE 5. Primary volume distribution based on chemical element balance, and final volume distribution after secondary conversion calculations for Pomona. Points are experimental data.

the mass fraction of an element in each primary source is independent of particle size. Table 15 shows the calculated fraction of certain chemical species in particles smaller than 1.5 μm, together with measured values when available. Measured values are based on Lundgren impactor data calibrated for unit density spherical particles; the data have not been corrected for density effects.

Calculated values for sodium and lead are in good agreement, while agreement for aluminum is poor. If the volume distribution measured at Goldstone (Figure 2) is all soil dust, the calculations predict more aluminum in the smaller sizes than was actually found. If, on the other hand, only particles larger than 1 μm are soil dust, the calculations predict too little aluminum in the smaller sizes. Most of the particles greater than 1 μm found at Goldstone are probably of soil dust origin, while only a fraction of the smaller particles are from soil dust.

Table 15 also includes calculated values for secondary conversion products; the calculations indicate that most of these species exist on particles smaller than 1.5 μm. This is true also for lead, and the calculations are confirmed by the

TABLE 15. Fraction of Chemical Species on Particles Smaller than 1.5 μm

	Pasadena 9/20/72, 1200-1400		Pomona 10/24/72, 1200-1400	
Species	Calculated	Measured	Calculated	Measured
Na	0.41	0.36 ± 0.07	0.29	0.20 ± 0.03
Al	0.56, 0.06[a]	0.38 ± 0.02	0.52, 0.01[a]	0.10 ± 0.01
Pb	0.99	U[b]	0.99	0.95 ± 0.01
SO_4^{2-}	0.89	U	0.72	U
NO_3^{2-}	0.89	U	0.72	U
Organics	0.98	U	0.80	U

[a]Higher value assumes that Figure 2 represents all soil dust; lower value assumes only particles greater than 1 μm in Figure 2 are soil dust.

[b]Unknown.

experiments. Sodium, on the other hand, comes primarily from soil dust and sea salt and is associated with the larger particles.

Calculations of the detailed distributions with respect to size, not shown, are in only fair agreement with experiments. One source of inaccuracy in the experiments is the bouncing of large particles that are carried to lower stages of the impactor. Needed, also, are better data for the distribution of chemical elements with respect to size for the primary sources, as well as more information on atmospheric growth laws.

V. LIGHT SCATTERING AND SECONDARY CONVERSION

Measurements of the light-scattering coefficient, b_s, were made with an integrating nephelometer while filter samples of aerosol were being collected during the intensive sampling periods. At Pasadena, on September 20, 1972, from 1200 to 1400 PST, the average value for b_s was 2.3×10^{-4} m^{-1}, while the relative humidity was approximately 20% over the period. At Pomona, on October 24, from 1200 to 1400, the average value for b_s was 8.3×10^{-4} m^{-1} with a relative humidity near 50%.

The extinction coefficient can be calculated from the size distribution using the following equation:

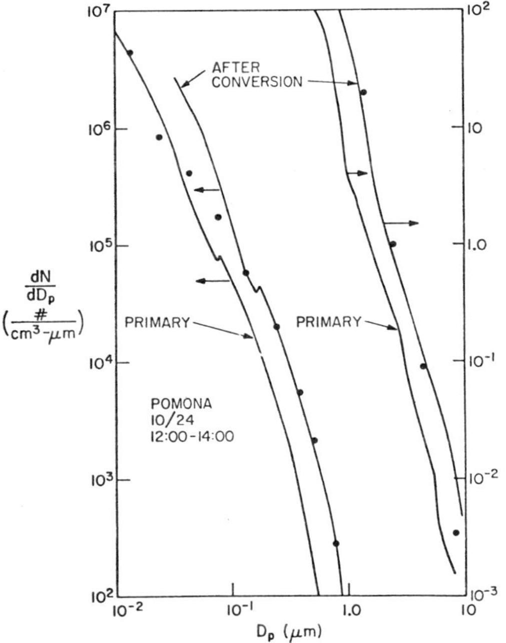

FIGURE 6. Primary number distribution based on chemical element balance, and final number distribution after secondary conversion calculations for Pomona. Points are experimental data.

$$b_s = \int_{\lambda_1}^{\lambda_2} \int_{D_{p_1}}^{D_{p_2}} g(\lambda) K \left(\frac{\pi D_p}{\lambda}, m \right) \frac{\pi D_p^2}{4} \frac{dN}{dD_p} \, dD_p \, d\lambda \qquad (11)$$

where

$$\int_{\lambda_1}^{\lambda_2} g(\lambda) \, d\lambda = 1$$

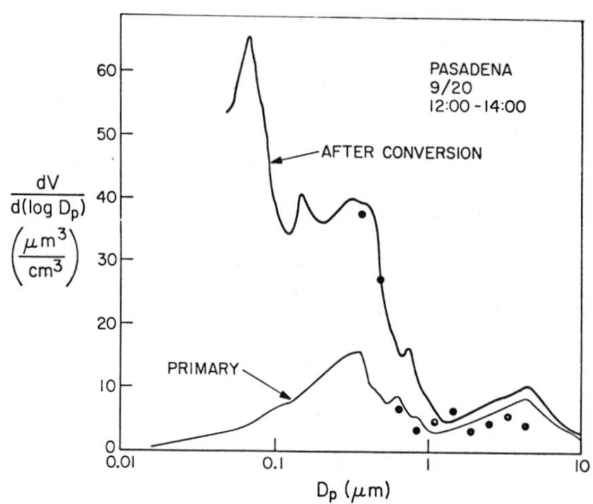

FIGURE 7. Primary volume distribution based on chemical element balance, and final volume distribution after secondary conversion calculations for Pasadena. Points are experimental data.

Here λ is the wavelength of the incident light, g(λ) is the spectral distribution of sunlight at the surface of the earth, and K, the light-scattering efficiency factor, is a function of the index of refraction of the particle, m. The index of refraction for smog aerosol was assumed to be 1.5, a value that Ensor et al. (20) found to give good agreement between measured and calculated values for b_s, using the 1969 data for the Pasadena aerosol. Values for K were calculated using an IBM subroutine, DAMIE. The double integral was evaluated numerically, using the size distributions obtained from the results of the coagulation and growth calculations. The limits of the integration were 0.390 to 0.700 µm for the wavelengths of light and 0.05 to 4.0 µm for the particle diameters. These calculations yield values for b_s of $1.6 \times 10^{-4}\ m^{-1}$ for Pasadena and $6.5 \times 10^{-4}\ m^{-1}$ for Pomona. For the primary size distributions (Figure 5) before growth, b_s was calculated to be $0.6 \times 10^{-4}\ m^{-1}$ and $0.8 \times 10^{-4}\ m^{-1}$ for Pasadena and Pomona, respectively. One can easily see the effects of the formation of secondary particulate matter.

There are several possible reasons why calculated scattering coefficients are low compared with measured values. In

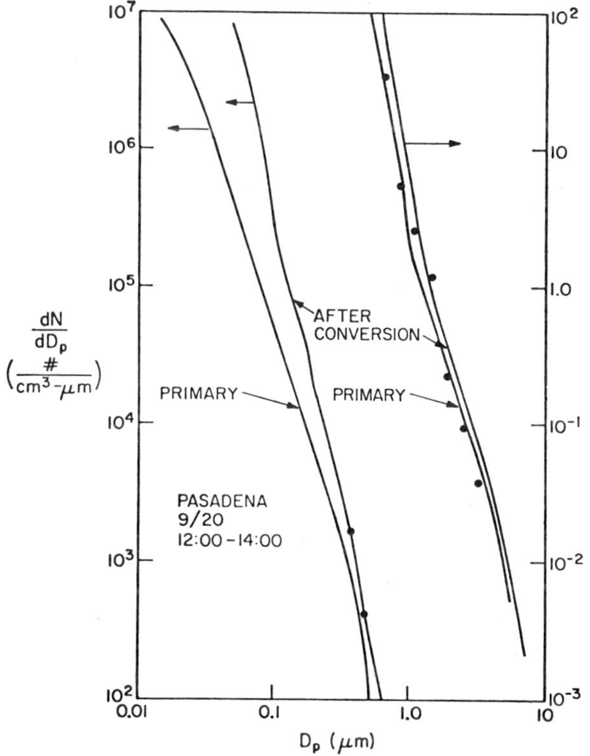

FIGURE 8. Primary number distribution based on chemical element balance, and final number distribution after secondary conversion calculations for Pasadena. Points are experimental data.

the secondary conversion process, water other than that associated with the sea salt and the ammonium nitrate conversion was not added to the aerosol. The water content for Pomona was not known. Better agreement would have been obtained for Pasadena if the additional 8 $\mu g/m^3$ of water had been included, but a suitable growth law is not yet available. In addition, the flashtube used in the nephelometer has a wavelength distribution function which does not correspond exactly to that of sunlight at the earth's surface.

Calculations were made to determine the functional relationship between b_s and the total volume of the secondary conversion products. A growth law with $\gamma = 2.8$ was applied to the calculated primary particle size distribution for the September 20 time period in Pasadena. The value of 2.8 was chosen as a rough average between the lower values found in experiments on organic conversion and the value assumed for sulfate and nitrate conversion. As shown in Figure 9, the calculated values for b_s are nearly a linear function of the volume of secondary material. The function is given approximately by the expression

$$\text{Converted volume}\left(\frac{\mu m^3}{cm^3}\right) = 21.2\left[b_s \cdot \left(\frac{10^4}{m^{-1}}\right) - 0.6\right]$$

Assuming a density of 1.4 for the aerosol, and including the primary mass, one finds:

$$\text{Total mass}\left(\frac{\mu g}{m^3}\right) \approx 33\left[b_s \cdot \left(\frac{10^4}{m^{-1}}\right) + 0.4\right]$$

which compares well with the relation given by Charlson et al. (21):

$$\text{Mass}\left(\frac{\mu g}{m^3}\right) = 38 b_s \cdot \left(\frac{10^4}{m^{-1}}\right)$$

Although b_s may often correlate well with total aerosol mass, this may not always be the case. Measurements at the Harbor Freeway showed that, while the total mass may be quite high, b_s may be rather small because the aerosol is made up of very small particles that do not scatter light efficiently. In addition, relative humidity plays an important role in the light-scattering properties of an aerosol, and since water that has been associated with the aerosol may be lost between the time the aerosol was sampled and the time it is weighed, further discrepancies may occur in relating mass to the light-scattering coefficient.

VI. CONCLUSIONS

The results of the analysis of the 1972 ACHEX data support, in a general way, the method developed in previous publications

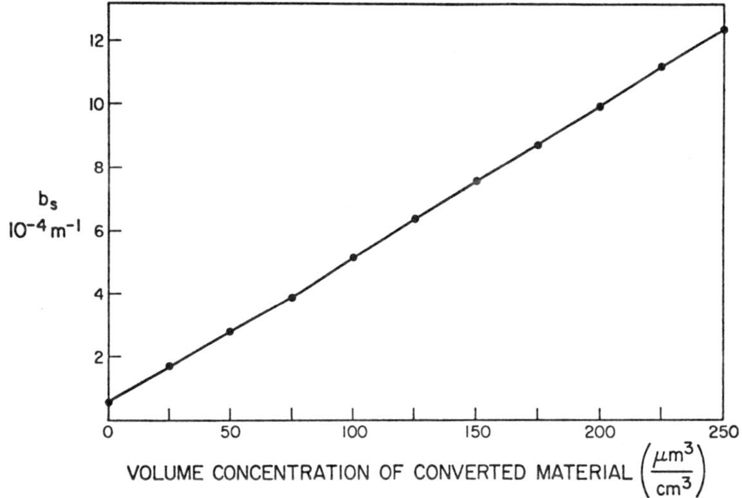

FIGURE 9. Calculated values of b_S as a function of volume of converted material: power law growth equation with $\gamma = 2.8$. Points represent calculations, and the straight line an approximate best fit.

for relating the characteristics of pollution aerosol to its sources. The calculation of the relative contributions of the various sources is based on a chemical element balance and is independent of the wind field. Hence it is particularly useful in dealing with long-term average data such as those given in the air quality summaries of the regulatory agencies. Both coagulation and growth processes, as well as sedimentation, depend on the time history of the aerosol. The growth calculations are empirically based and can be improved by taking into account the geographical distribution of sources and the wind field. Such calculations should be verified by comparing predicted size and chemical element distributions with measured values.

The visibility depends approximately on the total volume of material condensed from the gas phase. This includes sulfates, nitrates, organics, and ammonia, together with associated water. Only two 2-hr sampling periods were analyzed in detail, but the results indicate that all of these species contributed significantly. Improving visibility on days when pollution (and not sea fogs) play the major role will require a major reduction in aerosol precursor gases.

The results of the calculations on the 1973 ACHEX observations tend to support the general method and are basically consistent with the 1972 ACHEX data analysis. The estimated chemical composition of the primary freeway aerosol is similar to the composition estimated previously on the basis of data from various sources for exhaust from individual automobiles.

REFERENCES

1. Miller, M. S., Friedlander, S. K., and Hidy, G. M. 1972. J. Colloid Interface Sci. 39, 165-179.

2. Friedlander, S. K. 1973. Environ. Sci. Technol. 7, 235-240.

3. Heisler, S. L., Friedlander, S. K., and Husar, R. B. 1973. Atmos. Environ. 1, 633-649.

4. Gartrell, G., Jr. and Friedlander, S. K. 1975. Atmos. Environ. 9, 279-299.

5. White, W., Husar, R. B., and Friedlander, S. K. 1973. "A Study of Los Angeles Smog Aerosol Dynamics by Air Trajectory Analyses," presented at the June 1973 Annual APCA Meeting, Chicago, Illinois.

6. Hidy, G. M., et al. 1975. "Characterization of Aerosols in California (ACHEX)," Vols. II, III, IV, Final report to California Air Resources Board, Rockwell Science Center, Rept. SC524.25FR, Thousand Oaks, California.

7. Hirschler, D. A. 1975. Ind. Eng. Chem. 49, 1131.

8. California Air Resources Board. 1972. "The Study of California Implementation Plan for Achieving and Maintaining the National Ambient Air Quality Standards," Sacramento, California.

9. Frey, J. H. and Corn, M. 1967. Amer. Ind. Hyg. Assoc. J. 28, 468-478.

10. Grosjean, D. 1977. "Aerosols," Chap. 3 in Ozone and Other Photochemical Oxidants, National Academy of Sciences, Washington, D. C.

11. Mader, P. P., MacPhu, R. D., Latberg, R. T., and Larson, C. P. 1952. Ind. Eng. Chem. 44, 1352-1355.

12. Whitby, K. T., Husar, R. B., McFarland, A. R., and Tomaides, M. 1969. "Generation and Decay of Small Ions," Particle Technology Laboratory Publ. No. 137, Dept. of Mechanical Engineering, University of Minnesota, Minneapolis.

13. Woodcock, A. H. 1953. J. Meteorol. 10, 362-371.

14. Blifford, I. H., Jr. 1970. J. Geophys. Res. 75, 3099-3101.

15. Junge, C. E. 1972. J. Geophys. Res. 77, 5183-5200.

16. Hodgman, C. D. (Ed.). 1961. Handbook of Chemistry and Physics. Chemical Rubber Publishing Co., Cleveland, Ohio, pp. 661-1701.

17. Huntzicker, J. J. and Friedlander, S. K. 1973. "The Flow of Automobile Emitted Lead through the Los Angeles Basin," presented at the 166th National American Chemical Society Meeting.

18. Husar, R. B. 1971. "Coagulation of Knudsen Aerosol," Ph.D. thesis, University of Minnesota, Minneapolis.

19. Fuchs, W. A. 1964. The Mechanics of Aerosols. MacMillan Co., New York, p. 294.

20. Ensor, D. S., Porch, W. M., Pilat, M. J., and Charlson, R. J. 1971. J. Appl. Meteorol. 10, 1303-1306.

21. Charlson, R. J., Ahlquist, N. C., Selridge, H., and MacCready, P. B., Jr. 1969. J. Air Pollut. Control Assoc. 19, 937-942.

On the Nature and Origins of Visibility-Reducing Aerosols in the Los Angeles Air Basin*

WARREN H. WHITE† AND PAUL T. ROBERTS††
W. M. Keck Laboratories
California Institute of Technology
Pasadena, California

---

*Published in part in Atmospheric Environment, Vol. 11, pp. 803-812, 1977. Reprinted by permission of Pergamon Press. Presented in part as Paper No. 75-28.6 at the 68th Annual Meeting of the Air Pollution Control Association, Boston, June 15-20, 1975.

†Present address: 1180 N. Chester Avenue, Pasadena, CA 91104.

††Present address: Chevron Research Co., P. O. Box 1627, Richmond, CA 94802.

Abstract

Simultaneous measurements of the light-scattering coefficient and chemical composition of ambient aerosols were made during selected smog episodes in the Los Angeles air basin. These data are statistically analyzed to determine the effective scattering efficiencies of the major secondary aerosol species. The individual scattering efficiencies are then used to estimate the contributions of major sources of reactive gases to the reduction of visibility in Los Angeles.

Sulfate and nitrate compounds appear to have scattered more light at a given mass concentration than did other chemical fractions of the aerosol. The observed relationship of SO_2 and NO_x concentrations to the concentrations of tracers for major source types was consistent with existing inventories of SO_2 and NO_x emissions in the basin. Because of the high scattering efficiency of sulfates, the estimated contribution of large stationary sources of SO_2 to the reduction of visibility was comparable with that of the automobile.

I. INTRODUCTION

"This day they saw on land great signal smokes. It is a good land in appearance, and there are great valleys, and in the interior there are high ridges."

The Los Angeles air basin as seen clearly from offshore by the Cabrillo expedition [Ferrel (5)] in 1542.

"Objects on the plain below us dwindled till it began to look like an airman's map; but though the bungalows became white specks among their tufted orchards, they seemed to retain their sharpness of outline, and every tree stood like a toy on a little platform of darker green shadow. One began to realize a little more clearly what the lucid air of California might mean to the astronomer."

The Los Angeles air basin as seen from the road to Mount Wilson in 1917 by the English poet Alfred Noyes [Noyes (17)].

Modern inhabitants of the Los Angeles air basin seldom enjoy the magnificent views that once charmed explorers, poets, and astronomers. The deterioration of the visual environment characteristic of smoggy days is due primarily to the scattering of light by a complex atmospheric suspension of fine solid and liquid particles. This aerosol is a mixture of natural components such as fog and dust, anthropogenic primary particulates from mobile and stationary sources, and secondary material produced in the atmosphere from gases. To develop a control strategy which will improve visibility, it is first necessary to relate the optics of the atmosphere to the emissions of aerosols and their gas-phase precursors.

The most important optical parameter of the urban atmosphere is the total scattering coefficient, b_{scat}, which determines the rate (fraction per distance) at which a beam of light is scattered in all directions. Light scattered into an observer's line-of-sight by the ambient aerosol is perceived as a visible haze obscuring distant objects. While some features of this haze depend in an essential way on the angular distribution of the scattered light [Husar and White (14)] or on its polarization [White (23)], prevailing visibility within the haze is described adequately by the expression

$$\text{Visibility} = \frac{2.9}{b_{scat}}$$

[Samuels et al. (18)].

An aerosol is most often characterized for control purposes by its mass concentration, MASS, so that it is convenient to relate visibility to the aerosol through the aerosol's <u>scattering efficiency</u>, which is just the ratio b_{scat}/MASS. Because of the variability of this ratio for different aerosols, however, a model which simply scales the light-scattering coefficient to the ambient mass concentration is not an adequate base for a control strategy [Samuels et al. (18)]. If the ambient aerosol were simply a mixture of directly emitted primary aerosols, then the scattering efficiency of each component aerosol could be measured at the source, and the optical effects of specific control measures could readily be calculated. Unfortunately, much of the Los Angeles smog aerosol is formed in the atmosphere, and modeling studies [Gartrell and Friedlander (9), White and Husar (22)] have identified this secondary material as a major contributor to reduced visibility; its importance can be judged from the fact that the scattering efficiencies of Los Angeles smog aerosol are much higher than the scattering efficiencies generally measured for primary aerosols [Findley and Rossano (6), Elder et al. (4)].

Most of the secondary aerosol is added to existing particles [Husar et al. (15)] and cannot be isolated for direct <u>in situ</u> measurement. Moreover, the light-scattering efficiency of a particle depends strongly on its size [van de Hulst (20)], so that the optical contributions of individual aerosol species cannot be reconstructed accurately from the chemical analysis of a filter sample. In view of these experimental difficulties, it is natural to look for a statistical relationship between the scattering efficiency of the ambient aerosol and its chemical composition. In this paper we derive effective scattering efficiencies for the major secondary aerosol species, through the statistical analysis of data from the 1973 Aerosol Characterization Experiment (ACHEX) in Los Angeles [Hidy et al. (12)]. We then use these individual scattering efficiencies to estimate the contributions of major sources of reactive gases to the reduction of visibility in Los Angeles.

II. DATA BASE AND STATISTICAL PROCEDURES

During summer and early fall of 1973, an instrumented mobile van was used to monitor aerosol composition, scattering coefficients, and gas concentrations during selected smog episodes in the Los Angeles air basin. The data base for this paper is summarized in Appendix A; a more detailed description is given in Hidy et al. (12).

Our conclusions are based primarily on empirical relationships obtained by stepwise multiple linear regressions performed with a standard computer routine. In general, regression of a variable Y on variables $X_1, \ldots, X_n$ determines coefficients $c_1, \ldots, c_n$ which minimize the variance ε^2 of the quantity $Y - \Sigma_i c_i X_i$. The result is an approximate linear relationship

$$Y = c_o + \Sigma_i c_i X_i \pm \varepsilon$$

with a standard error of ε. If the linear model is physically appropriate, then the coefficients c_i can be determined with arbitrary accuracy by taking enough samples, even though the uncertainty in the relationship may be large for individual samples. In most of the regressions reported here, the coefficients are estimated accurate to within 20%; coefficients whose calculated standard error exceeds 20% are marked with an asterisk (*). Coefficients are set to zero unless the F-test shows the corresponding variables to be statistically significant at the 95% confidence level in reducing ε^2. In some of the regression models, the constant term is constrained to be zero on physical grounds.

The notation $(n, \bar{y}, r)$ at the end of a regression relationship means that regression was performed on n observations for which the mean value of the dependent variable was $\bar{y}$, and produced a multiple correlation coefficient of r. The number of observations can fluctuate slightly with changes in the set of variables, due to missing data. The square of the correlation coefficient is the proportion of the variance of the dependent variable which is explained by the regression relationship:

$$r^2 = \frac{\sigma^2 - \varepsilon^2}{\sigma^2} .$$

Here σ^2 is the original variance of the dependent variable and ε^2 is the variance remaining after regression.

III. AEROSOL COMPOSITION

In terms of mass concentration, the three most important elemental and ionic species found in the aerosols sampled were NO_3^-, SO_4^{2-}, and C. In this section we discuss the probable nature of the nitrate, sulfate, and carbonaceous compounds these represented in the aerosol, and describe their distribution with respect to particle size.

Likely nitrate and sulfate compounds in the aerosol include ammonium salts formed by absorption of ammonia from the gas phase, and sodium salts formed by replacement of chloride in sea salt. A mole balance on the 24-hr hi-vol samples indicates that most of the nitrate and sulfate ion can be accounted for in this way:

$$\frac{NH_4^+}{18} = 0.98\,\frac{NO_3^-}{62} + 1.74\,\frac{(SO_4^{2-} - (96/46)\ Na_{sea\ salt})}{96} \pm 0.05$$

$$(n = 11,\ \bar{y} = 0.43,\ r = 0.98) \qquad (1)$$

This relationship was determined by regression of NH_4^+ on NO_3^- and $(SO_4^{2-} - (96/46)\ Na_{sea\ salt})$, where $Na_{sea\ salt}$ is calculated from sodium and aluminum concentrations as in Miller et al [16]. The quantity $(96/46)\ Na_{sea\ salt}$ in the sulfate term accounts for the conversion of all marine sodium chloride to sodium sulfate. Other possible models for regression, in which sodium chloride is converted to sodium bisulfate or sodium nitrate, produce slightly poorer fits.

In Eq. 1 the regression relationship is expressed with ionic concentrations divided by ionic weights, so that the empirical coefficients of the nitrate and sulfate terms have direct stoichiometric significance. These coefficients, and the small degree of scatter in the relationship, suggest that nearly all of the nitrate is ammonium nitrate, and that most of the sulfate not associated with sea salt is ammonium sulfate. The molecular weight of ammonium nitrate is approximately 1.3 times the weight of the nitrate ion, and we will assume in what follows that

$$\text{NITRATES}/NO_3^- = 1.3$$

The same compound-to-ion ratio (C/I) is representative of the likely atmospheric mix of $(NH_4)_2SO_4$ (C/I = 1.38), $(NH_4)HSO_4$ (C/I = 1.20), Na_2SO_4 (C/I = 1.48), $NaHSO_4$ (C/I = 1.25), and H_2SO_4 (C/I = 1.02); we will also assume that

$$\text{SULFATES}/SO_4^{2-} = 1.3$$

The bulk of the carbon in the aerosol was contained in organic compounds [Appel et al. (1)]. In analyses of hi-vol filter samples taken at Pasadena during the period of ACHEX, Grosjean and Friedlander (10) found that the carbon content of these compounds ranged from 67% for the polar fractions to 85% for the nonpolar fractions, with an average for their data of 73%. The reciprocal of 0.73 is approximately 1.4, and we will assume that

$$\text{ORGANICS}/C = 1.4$$

Table 1 gives the average mass concentration of nitrates, sulfates, and organics in the ACHEX 2-hr samples, based on the scaling factors estimated above. These three components account for over half of the average aerosol mass. Judging from the eleven 2-hr samples analyzed for aluminum, soil dust accounts for 35 to 40% of the remainder, leaving an average of about 60 $\mu g/m^3$ to be attributed to water and primary anthropogenic aerosols. The figures in Table 1 summarize data taken at four locations over a limited experimental period, and do not represent true basinwide or yearly averages.

The light-scattering effectiveness of an aerosol depends not so much its chemical composition, as on its distribution of particle sizes as well. Table 2 shows the average mass fraction of each aerosol component which was found in particles less than 0.5 μm in diameter.

It is evident from Table 2 that the chemical composition of the aerosol varies with particle size. The smaller particles were found to be enriched with carbon, while nitrate and aluminum were concentrated in the larger particles. Although its

TABLE 1. Estimated Breakdown of Aerosol Mass Concentration for ACHEX 2-hr Aerosol Samples

	Average mass concentration (μg/m^3) total of 199 μg/m^3 derived from:		
	All sources	Gasoline	Fuel and crude oil
Nitrate compounds	41	32	9
Sulfate compounds	33	5	28
Organic compounds	30	30	0
Sum	104	67	37

TABLE 2. Ratio of Mass on After-Filter (Particle Diameter Less than 0.5 μm) to Mass on Total Filter (Particle Diameter Less than 20 μm) for Individual Aerosol Components (Average and Standard Deviation of ACHEX 2-hr Aerosol Samples)

Aerosol component	AF/TF Average Overall	RH < 60%	RH > 60%
Total aerosol	0.37 ± 0.10	0.43 ± 0.06	0.30 ± 0.09
Nitrate	0.33 ± 0.16	0.39 ± 0.10	0.28 ± 0.19
Sulfate	0.49 + 0.20	0.64 + 0.12	0.34 + 0.14
Carbon	0.72 ± 0.13		
Aluminum	0.08 ± 0.09		

effect cannot be quantified from our data, ambient relative humidity appears as an important determinant of particle size [cf. Sinclair et al. (19)]. Nitrate and sulfate size distributions are particularly affected by relative humidity; this is consistent with the known hygroscopicity of the probable nitrate and sulfate compounds in the aerosol [e.g., Charlson et al. (3)], but may also reflect humidity-dependent formation mechanisms for these compounds.

IV. RELATIONSHIP OF THE LIGHT-SCATTERING COEFFICIENT TO AEROSOL COMPOSITION

The scattering due to a given mass concentration of aerosol depends on the size, shape, and composition of the aerosol particles. If the particles are uniform, homogeneous spheres of diameter D_p, density ρ, and refractive index m, then the ratio of aerosol scattering coefficient to aerosol mass concentration is given by

$$b_{scat}/\text{MASS} = G(D_p) = \frac{3}{2\rho D_p} \int K(\pi D_p/\lambda, m) f(\lambda)\, d\lambda, \qquad (2)$$

where K is the Mie scattering function and $f(\lambda)$ is the normalized wavelength distribution of solar radiation. The function $G(D_p)$, plotted in Figure 1, peaks sharply when particle diameters are near the wavelengths of visible light.

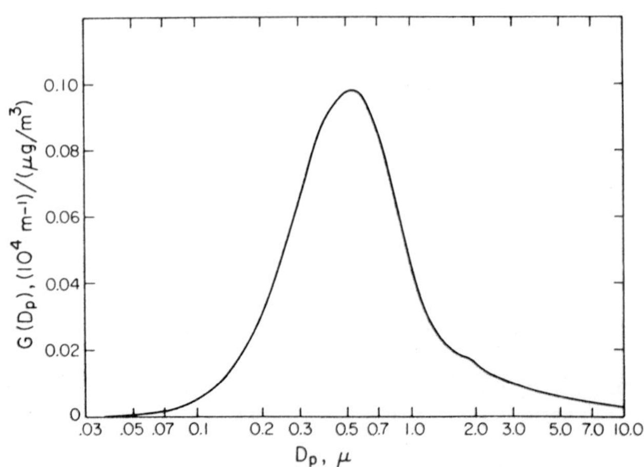

FIGURE 1. Ratio of light-scattering coefficient to mass concentration for uniform spherical particles of unit density, refractive index 1.5, and diameter D_p.

Ambient aerosols are distributed with respect to particle size, and b_{scat}/MASS for such an aerosol is obtained by averaging:

$$b_{scat}/\text{MASS} = \int G(D_p) \text{ mass } (D_p) \, dD_p, \qquad (3)$$

where $G(D_p)$ is the ratio for a monodisperse aerosol, given in Eq. 2, and mass (D_p) is the normalized aerosol mass distribution.

Previous studies by Charlson et al. (2) and Samuels et al. (18) have shown that this ratio has a fairly narrow distribution for ambient Los Angeles aerosols. Figure 2 illustrates the approximate proportionality between b_{scat} and MASS in the ACHEX 2-hr samples; the average b_{scat}/MASS for these samples was 0.032 with a standard deviation of 0.009:

$$b_{scat}/\text{MASS} = 0.032 \pm 0.009, \qquad (4)$$

which is consistent with the earlier measurements.

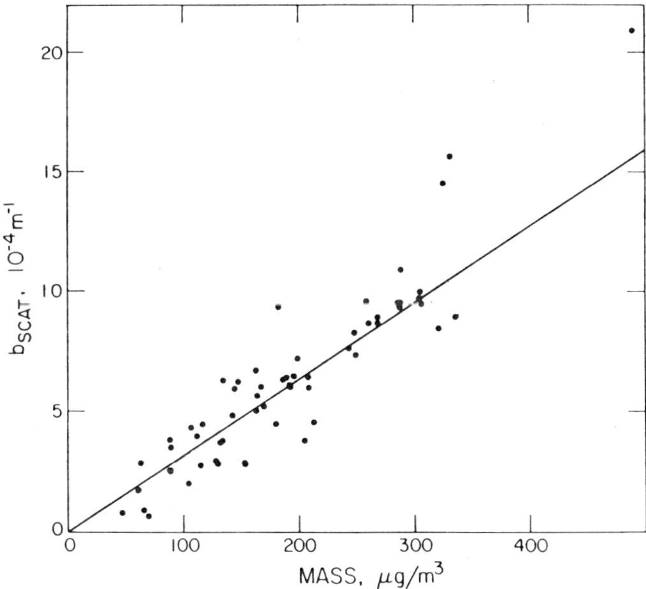

FIGURE 2. Correlation of light-scattering coefficient with aerosol mass concentration in ACHEX 2-hr aerosol samples. Solid line shows average ratio b_{scat}/MASS.

The variability in the ratio $b_{scat}/MASS$ is of course due to the variable composition of the ambient aerosol. Regression shows that much of the deviation in Eq. 4 was statistically associated with the sulfate and nitrate fractions of the aerosol:

$$b_{scat}/MASS = 0.020 + 0.056 \frac{SO_4{}^{2-}}{MASS} + 0.064\,\mu^2 \frac{NO_3{}^{-}}{MASS} \pm 0.006 \qquad (5)$$

$$(n = 58,\ \bar{y} = 0.032,\ r = .72)$$

Here μ denotes the ambient relative humidity: $\mu = \%\ RH/100$. As shown in Figure 3, the ratio $b_{scat}/MASS$ tended to be higher than average for aerosols rich in sulfates and nitrates, and lower than average for aerosols poor in these compounds.

We can rewrite Eq. 5 in a suggestive form if we use the conclusions of the last section that $NITRATES/NO_3{}^{-} = SULFATES/SO_4{}^{2-} = 1.3$. Simple algebraic manipulation of Eq. 5 then gives

$$\begin{aligned} b_{scat}/MASS = {} & 0.062 \frac{SULFATES}{MASS} \\ & + (0.020 + 0.049\,\mu^2) \frac{NITRATES}{MASS} \qquad (6) \\ & + 0.020 \frac{MASS - SULFATES - NITRATES}{MASS} \pm 0.006 \end{aligned}$$

$$(n = 58,\ \bar{y} = 0.032,\ r = .72)$$

In Eq. 6, $b_{scat}/MASS$ is presented as the average of three different values, weighted according to the composition of the aerosol.

The form of regression 6 can be derived theoretically from the observation that the (unnormalized) aerosol mass distribution is the sum of the (unnormalized) component mass distributions:

$$\text{mass}\ (D_p)\ MASS = \sum_i \text{mass}_i\ (D_p)\ MASS_i\ , \qquad (7)$$

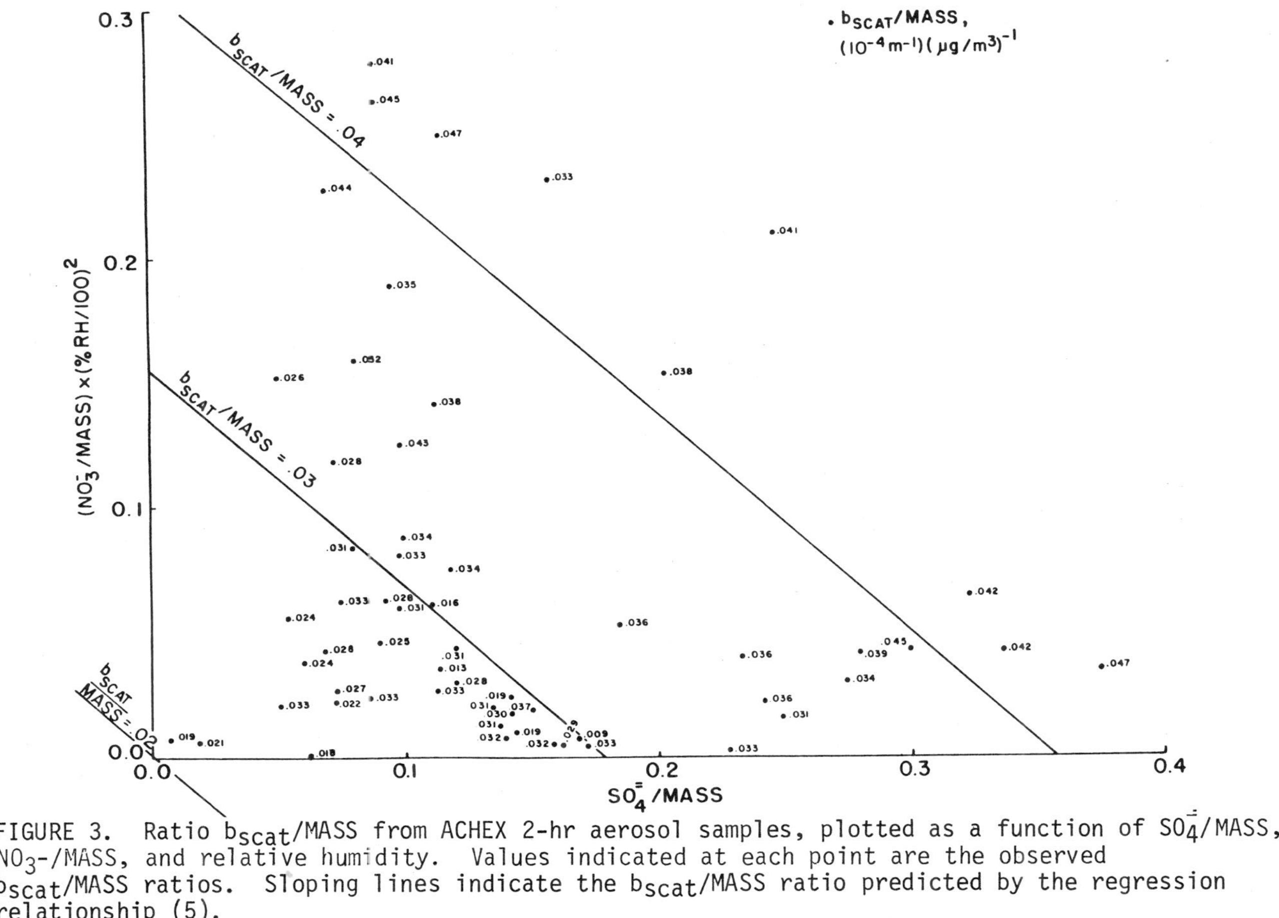

FIGURE 3. Ratio b_{scat}/MASS from ACHEX 2-hr aerosol samples, plotted as a function of $SO_4^=$/MASS, NO_3^-/MASS, and relative humidity. Values indicated at each point are the observed b_{scat}/MASS ratios. Sloping lines indicate the b_{scat}/MASS ratio predicted by the regression relationship (5).

where $MASS_i$ is the mass concentration of the ith component and $mass_i\ (D_p)$ is the normalized component mass distribution. Using Eq. 7, we can rewrite Eq. 3 to resemble regression 6:

$$b_{scat}/MASS = \sum_i C_i \frac{MASS_i}{MASS} \tag{8a}$$

where

$$C_i = \int G(D_p) mass_i\ (D_p)\ dD_p \tag{8b}$$

Comparison of Eqs. 8b and 3 shows that the coefficients C_i of Eq. 8b are analogues of the ratio $b_{scat}/MASS$ for individual fractions of the aerosol. It is natural, then, to identify C_i as the scattering efficiency of the ith aerosol component. If each component of the aerosol possessed a characteristic normalized mass distribution which depended only on the relative humidity, so that the component scattering efficiencies $C_i = C_i(\mu)$ were independent of the component mass concentrations $MASS_i$, then $C_i MASS_i$ would be the contribution of the ith component to the light-scattering coefficient. The success of regression 6 indicates that this is to some extent the case. The functional relationship between C_i and relative humidity is in general difficult to predict on theoretical grounds, and we do not attribute any physical significance to the specific form of the nitrate coefficient beyond the observation that it is nonlinear, as might be expected for a deliquescent material.

The coefficients of regression 6 are empirical estimators for the C_i, so the contribution of the ith aerosol component to the light-scattering coefficient can be judged from these coefficients and the measured component mass concentrations. On this basis, sulfates and nitrates together account for more than half of the light-scattering observed during the ACHEX 2-hr sampling periods, as shown in Table 3. The figures in Table 3 summarize data taken at four locations over a limited experimental period, and do not represent true basinwide or yearly averages.

V. DISCUSSIONS OF THE AEROSOL-SCATTERING RELATIONSHIP

There is some scatter in the b_{scat} values observed at a given level of SULFATES, NITRATES, and MASS. The coefficients of Eq. 6 are chosen to minimize this scatter relative to aerosol mass concentration; the coefficients which minimize the absolute scatter are only slightly different:

TABLE 3. Estimated Breakdown of Light-Scattering Coefficient for ACHEX 2-hr Aerosol Samples

	Average contribution to b_{scat} (10^{-4} m^{-1}) total of 6.4 x 10^{-4} m^{-1} derived from:		
	All sources	Gasoline	Fuel and crude oil
Nitrate compounds	1.7	1.4	0.3
Sulfate compounds	2.0	0.3	1.7
Organic compounds	0.6	0.6	0.0
Sum	4.3	2.3	2.0

$$b_{scat} = 0.062 \text{ SULFATES} + (0.022 + 0.050\,\mu^2) \text{ NITRATES} + 0.022 \text{ (MASS - SULFATES - NITRATES)} \pm 1.0 \quad (9)$$

$$(n = 58,\ \bar{y} = 6.4,\ r = .96)$$

Because the strong correlation of b_{scat} with MASS has not been factored out of Eq. 9, the correlation coefficient for this regression is much more impressive than the correlation coefficient for Eq. 6. On the other hand, precisely because the correlation of b_{scat} with MASS has been factored out of Eq. 6, the evidence from this regression for the optical importance of sulfates and nitrates is the more convincing.

The functional relationships in Eqs. 6 and 9 are but two of scores which were tested in regressions. Although carbon was a major constituent of the aerosol, in its light-scattering effectiveness it could not be distinguished statistically from the rest of the nonsulfate, nonnitrate fraction. The inclusion of nickel and lead as tracers for primary aerosols did not significantly improve the relationships. Of the functional forms tested for the relative-humidity-dependent coefficients $C_i(\mu)$, quadratics generally gave the best fit to the data. No significant nonlinear relationships were found between light scattering and any of the aerosol components. Regression of MASS on aerosol composition indicated that the sulfate and nitrate terms in Eqs. 6 and 9 do represent sulfate and nitrate compounds, and do not include significant contributions from unrelated material statistically associated with SO_4^{2-} and NO_3^- [White (24)].

An important chemical factor contributing to the superior scattering efficiency of sulfates and nitrates is the affinity of these compounds for water. Much of the scattering statistically associated with sulfate and nitrate may have been due to unbound water which was not measured gravimetrically due to losses in sampling and equilibration. In the case of nitrate compounds, this hypothesis is supported by the relative humidity dependence of the empirical scattering efficiency. No such dependence could be identified for sulfate compounds; even the deletion of all data taken at relative humidities greater than 70% does not significantly change the coefficient of the sulfate term in regression 9. This is surprising since the distribution of sulfates with respect to particle size does depend on relative humidity, as Table 2 shows.

The importance of sulfate compounds for visibility at all relative humidities probably stems chiefly from their distribution with respect to particle size. The size resolution of the impactor was not sufficient for a direct calculation of the efficiencies C_i, but the after-filter/total-filter ratios in Table 2 show that the mass median particle diameter of sulfates averaged a little over 0.5 μm, right in the middle of the particle size range producing the most scattering for a given mass concentration. For comparison, the mass median diameter of carbon was well below 0.5 μm; many of the carbonaceous particles were too small to scatter effectively. The difference between the complex refractive index of much of the carbonaceous material and the essentially real refractive index of the sulfates may have augmented the difference in their effective scattering efficiencies.

The aerosol-scattering relationship presented here was fitted to a limited set of data from a limited geographic area. Pending further measurements to test its generality, it should be applied to other situations only with considerably caution. It is worth noting, however, that the reciprocal of the sulfate coefficient in Eqs. 6 and 9 falls within the range of sulfate/b_{scat} ratios reported by Waggoner et al. (21) for a variety of sulfate-dominated aerosols. The distribution of sulfate compounds with respect to particle size is governed largely by the mechanics of the SO_2-SO_4^{2-} conversion process, and may be self-preserving [Friedlander (8)]. In this case one might expect the scattering efficiency of sulfates to be more or less universal.

VI. SOURCES OF AEROSOL PRECURSORS

Mass balances based on known emissions of primary aerosol account for only a small fraction of the nitrates, sulfates, and organics found in the Los Angeles smog aerosol, and the bulk of each of these species is thought to result from gas-to-particle conversion in the atmosphere [Heisler et al. (11), Gartrell and Friedlander (9)]. The mechanisms involved are poorly understood, but the presumed precursors of these species in the gas phase are the nitrogen and sulfur oxides and certain hydrocarbons, particularly the C_6 and larger olefins and some aromatics. Table 4 presents an abbreviated inventory of the sources of these gases.

Table 4 shows that two broadly defined groups of sources, with mutually distinct geographical distributions and characteristic emissions, are responsible for nearly all of the production of reactive gases in Los Angeles. The gasoline sources are distributed fairly uniformly over a large area of the basin, and produce nearly all of the reactive hydrocarbons. The fuel and crude oil sources are clustered along the coast and produce most of the sulfur oxides. In this section we identify and validate chemical tracers for the emissions of these two source groups.

TABLE 4. Inventory of Estimated Reactive Gas Emissions (Tonnes per Day) in the Los Angeles Basin, Compiled from Most Recent Inventories by Local Air Pollution Control Districts: Los Angeles (1973), Orange County (1972), San Bernardino (1972), and Riverside (1970)

Source type	NO_x (T/d)	SO_x (T/d)	RHC[a] (T/d)
1. Gasoline	1040	40	1070
Motor vehicle exhaust and blow-by[b]	1040	40	770
Evaporation			300
2. Fuel oil and crude oil	430	345	10
Power plants	150	190	
Industry and commerce	160	20	
Ships, railroads, jets, diesel vehicles	90	20	5
Petroleum refining and sulfur recovery	30	115	5
3. Other	10	55	70

[a]Reactive hydrocarbons as defined by the APCD's include some species, such as the light olefins, which are not converted to aerosol in significant amounts.

[b]Based on the seven-mode cycle.

Lead aerosol and carbon monoxide are important constituents of the exhaust from gasoline-powered motor vehicles. This single source is estimated to account for 97 to 98% of all emissions of these species in the Los Angeles air basin, about 9 tons/day of airborne lead [estimated from Huntzicker et al. (13)] and 10,000 tons/day of carbon monoxide (from APCD figures). The reliability of lead and carbon monoxide as tracers can be judged from their high correlation in the ACHEX 2-hr samples:

$$\begin{aligned} &\mathrm{CO} = 874 + 954\ \mathrm{Pb} \pm 555 \\ &(n = 53,\ \bar{y} = 3775,\ r = .96) \end{aligned} \tag{10}$$

The lead coefficient in Eq. 10 is close to the estimated mass ratio of carbon monoxide emissions to lead emissions, 10,000/9 = 1111.

Nickel and vanadium are, like sulfur, trace constituents of the heavier fractions of petroleum, and are exhausted to the atmosphere in aerosols during the regeneration of refinery catalyst and the combustion of fuel oil. It is estimated that, nationwide, more vanadium is released to the atmosphere in this manner than is consumed in all metallurgical applications combined [Fischer (7)]. Nickel correlates with vanadium in the 11 2-hr samples analyzed for vanadium and in the 24-hr hi-vol samples:

$$\begin{aligned} &\mathrm{V} = 0.011 + 0.72^{*}\mathrm{Ni} \pm 0.014 \\ &(n = 11,\ \bar{y} = 0.051,\ r = .83)\ \text{(2-hr)} \end{aligned} \tag{11a}$$

$$\begin{aligned} &\mathrm{V} = 0.006 + 0.91\ \mathrm{Ni} \pm 0.008 \\ &(n = 11,\ \bar{y} = 0.030,\ r = .91)\ \text{(24-hr)} \end{aligned} \tag{11b}$$

The ratio of Ni to V in fuel and crude oils fluctuates greatly from batch to batch, and we are unable to estimate the ratio of the emissions of these elements for comparison with the Ni coefficients of Eqs. 11.

Of the four source tracers discussed above, only lead and nickel were monitored for all 2-hr samples taken at the mobile van and satellite stations. Figure 4 shows the geographical distribution of the nickel/lead ratio calculated from the average values of lead and nickel during the sea breeze regime. The limited data suggest a nickel-rich plume originating in the area between El Segundo and Alamitos Bay and following wind streamlines to the north and east.

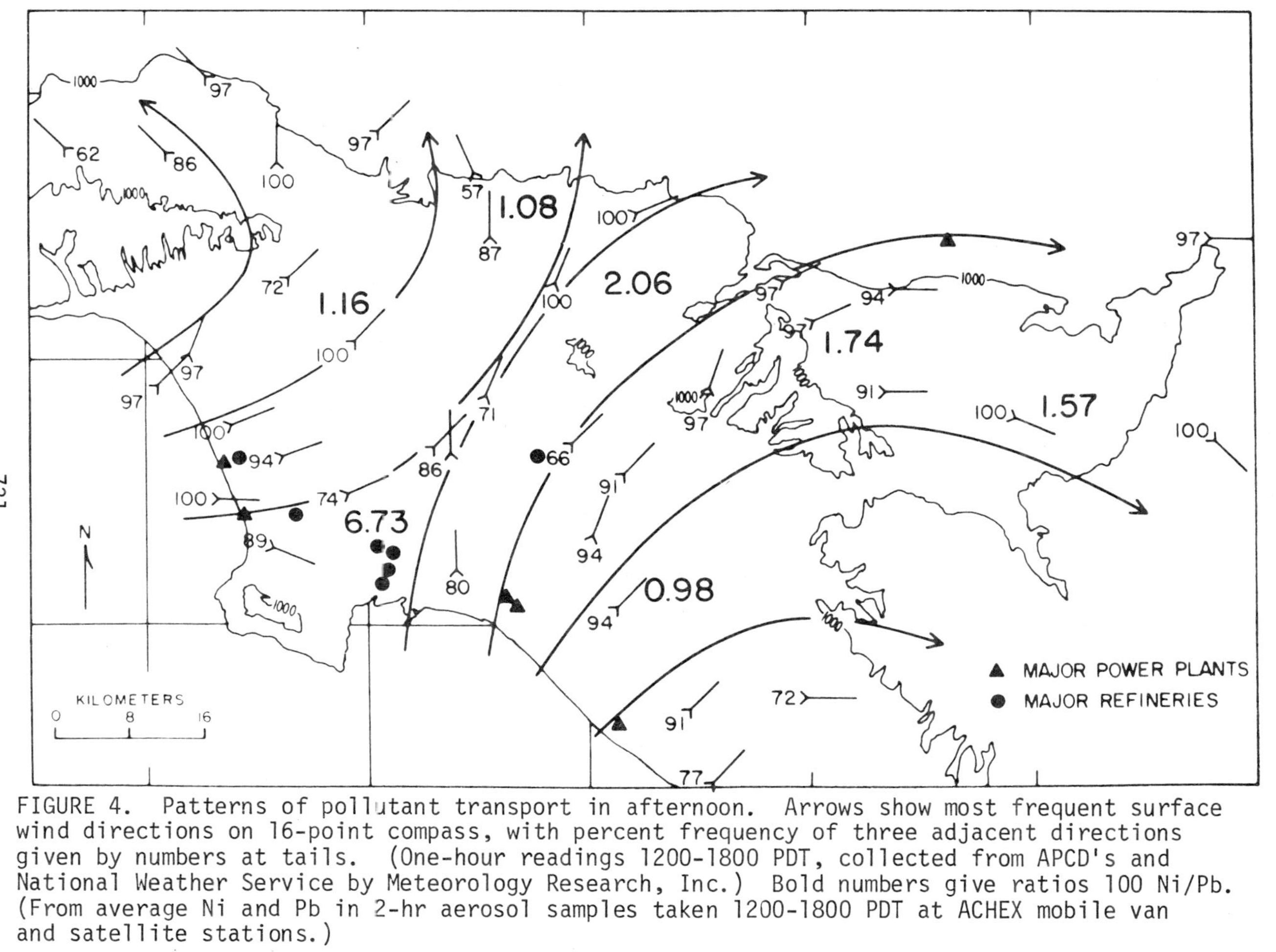

FIGURE 4. Patterns of pollutant transport in afternoon. Arrows show most frequent surface wind directions on 16-point compass, with percent frequency of three adjacent directions given by numbers at tails. (One-hour readings 1200-1800 PDT, collected from APCD's and National Weather Service by Meteorology Research, Inc.) Bold numbers give ratios 100 Ni/Pb. (From average Ni and Pb in 2-hr aerosol samples taken 1200-1800 PDT at ACHEX mobile van and satellite stations.)

Oxides of nitrogen and sulfur were measured during the 2-hr sampling periods at the mobile van. If we set $N_T = N_{NO_x} + N_{NO_3^-}$ and $S_T = S_{SO_2} + S_{SO_4^=}$, and regress N_T and S_T on Pb and Ni, we find:

$$N_T = -11 + 34\text{Pb} \pm 32 \quad (n = 17,\ \bar{y} = 145,\ r = .95) \quad \text{before 10:00 PST} \tag{12a}$$

$$N_T = 7 + 17\text{Pb} + 199^{*}\text{Ni} \pm 10 \quad (n = 43,\ \bar{y} = 57,\ r = .88) \quad \text{after 10:00 PST} \tag{12b}$$

$$S_T = 4 + 746\text{Ni} \pm 11 \quad (n = 14,\ \bar{y} = 45,\ r = .95) \quad \text{before 10:00 PST} \tag{13a}$$

$$S_T = 25 + 430^{*}\text{Ni} \pm 16 \quad (n = 30,\ \bar{y} = 49,\ r = .56) \quad \text{after 10:00 PST} \tag{13b}$$

These relationships are in qualitative agreement with the inventory, linking nitrogen primarily to gasoline sources and sulfur to fuel and crude oil sources.

The scatter in both morning relationships is small, and we can use the coefficient of Eq. 13a, together with known sulfur emissions, to estimate emissions of nickel in the Los Angeles air basin: $[(32/64)\ 385\ \text{tons/day}]/746 \simeq 1/4$ ton/day. The coefficient of lead in Eq. 12a is close to the estimated ratio of nitrogen emissions to lead emissions for motor vehicles, $[(14/46)\ 1040]/9 = 35$. There is a good deal of scatter in the afternoon relationships, and the coefficients of Pb and Ni are lower than they are in the morning. These lower coefficients may reflect the loss of nitrogen and sulfur oxides to atmospheric and surface sinks.

VII. SOURCE CONTRIBUTIONS TO AEROSOL

In this section we develop rough estimates for the contributions of the two source groups identified in the preceding section to ambient levels of particulate nitrates, sulfates, and organics. To simplify notation, let X_1, X_2, and X_3 represent secondary nitrates, sulfates, and organics, and let $[X_j]_i$ be the concentration, in a given air parcel, of X_j produced from the emissions of source group i. If source group i is considered as a single emitter, then $[X_j]_i$ is given formally by

$$[X_j]_i = f_{ij} \frac{\phi_i}{V} e_{ij}$$

where e_{ij} (tons/day) is the emission rate of X_j precursor by source group i, ϕ_i (days) is the fraction of source group i's daily emissions which is exhausted into the air parcel, V (m^3) is the volume of the air parcel, and f_{ij} (μg/ton) is the mass of X_j produced from a ton of precursor emitted by source group i.

We have no way of measuring f_{ij}, and will assume for simplicity that $f_{1j} = f_{2j}$; for each species, the fraction of the gas-phase precursor converted to particulate is independent of the precursor source. If we assume also that the two source groups of Table 4 account for all emissions of X_j precursor, then the fraction of X_j attributable to each source group is:

$$\frac{[X_j]_1}{[X_j]} = \frac{e_{1j}}{(e_{1j} + we_{2j})}$$

$$\frac{[X_j]_2}{[X_j]} = \frac{we_{2j}}{(e_{1j} + we_{2j})} \qquad (14)$$

where $w = \phi_2/\phi_1$. The quantities e_{ij} are given in Table 4 and w can be determined from the source tracers identified in the preceding section:

$$w = \frac{Ni/e_{2,Ni}}{Pb/e_{1,Pb}} = \frac{Ni/(1/4)}{Pb/9} = 36Ni/Pb \qquad (15)$$

The quantity $[X_j]$ is just the measured concentration of particulate nitrates, sulfates, or organics, if we neglect the contributions of primary aerosols.

Table 1 gives average values for $[X_j]_i$, calculated from Eqs. 14 and 15 using concentrations of compounds and tracers sampled during ACHEX. Table 3 presents corresponding breakdowns of the aerosol light-scattering coefficient. The calculations are not very sensitive to the crude assumptions on which they are based, because most nitrogen oxides and reactive hydrocarbons come from gasoline sources, while most sulfur oxides come from fuel and crude oil sources. Indeed, the estimated total contributions of the two source groups are not very different if we simply attribute all nitrates and organics to gasoline sources and all sulfates to fuel and crude oil sources.

It should be noted that, while the sources itemized in Table 4 are involved in producing a large fraction of the ambient aerosol, they are not the sole agents of this production. For example, sulfur oxides are essential to the formation of sulfate aerosol, but the process is facilitated by the presence of ammonia and reactive hydrocarbons. The major sources of sulfur oxides are not major sources of ammonia and reactive hydrocarbons, so that it is somewhat misleading to attribute all effects of sulfates to the sources of sulfur oxides.

VIII. GEOGRAPHICAL VARIATIONS IN AEROSOL COMPOSITION

The ACHEX revealed substantial geographical variations in aerosol concentration and composition. The relative contributions of individual aerosol components and precursor source groups depended strongly on location, as shown in Figures 5 to 8. Generally, sulfates and oil-based sources were more important in the western portion of the basin, and nitrates were more important in the eastern portion. The sampling periods at the different sites were overlapping but not coincident, so that some of the observed differences in average concentration and composition may be due to temporal, rather than spatial, variations.

Figures 5 to 8 include data from the satellite stations. Since relative humidity was not monitored at these stations, the breakdown of b_{scat} in Figures 6 and 8 is based on the following regression, rather than regression 6:

$$b_{scat}/\text{MASS} = 0.071\,\frac{\text{SULFATES}}{\text{MASS}} + 0.050\,\frac{\text{NITRATES}}{\text{MASS}}$$

$$+ 0.015\,\frac{\text{MASS} - \text{SULFATES} - \text{NITRATES}}{\text{MASS}} \pm 0.007$$

$$(n = 58,\ \bar{y} = 0.032,\ r = .60)$$

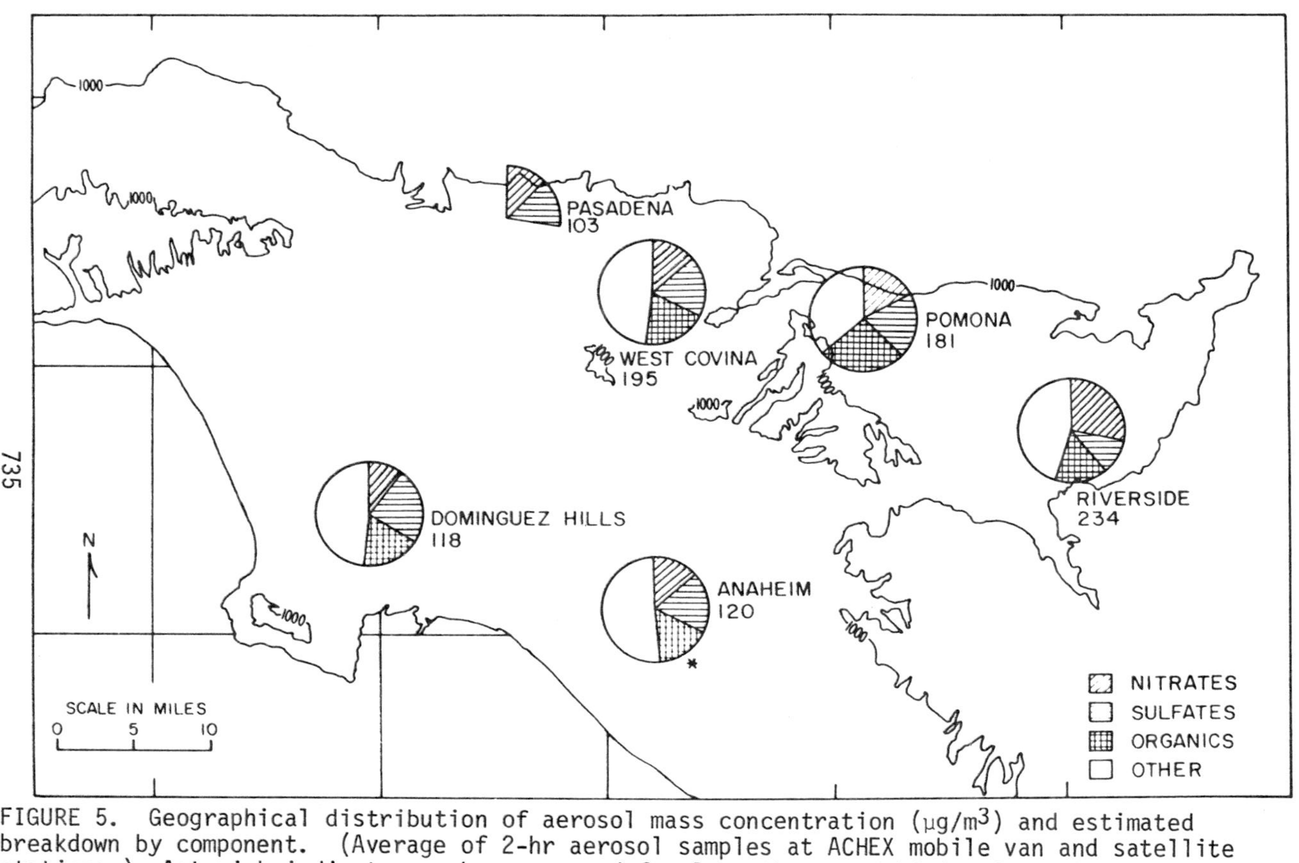

FIGURE 5. Geographical distribution of aerosol mass concentration ($\mu g/m^3$) and estimated breakdown by component. (Average of 2-hr aerosol samples at ACHEX mobile van and satellite stations.) Asterisk indicates carbon measured for less than two-thirds of samples at that location.

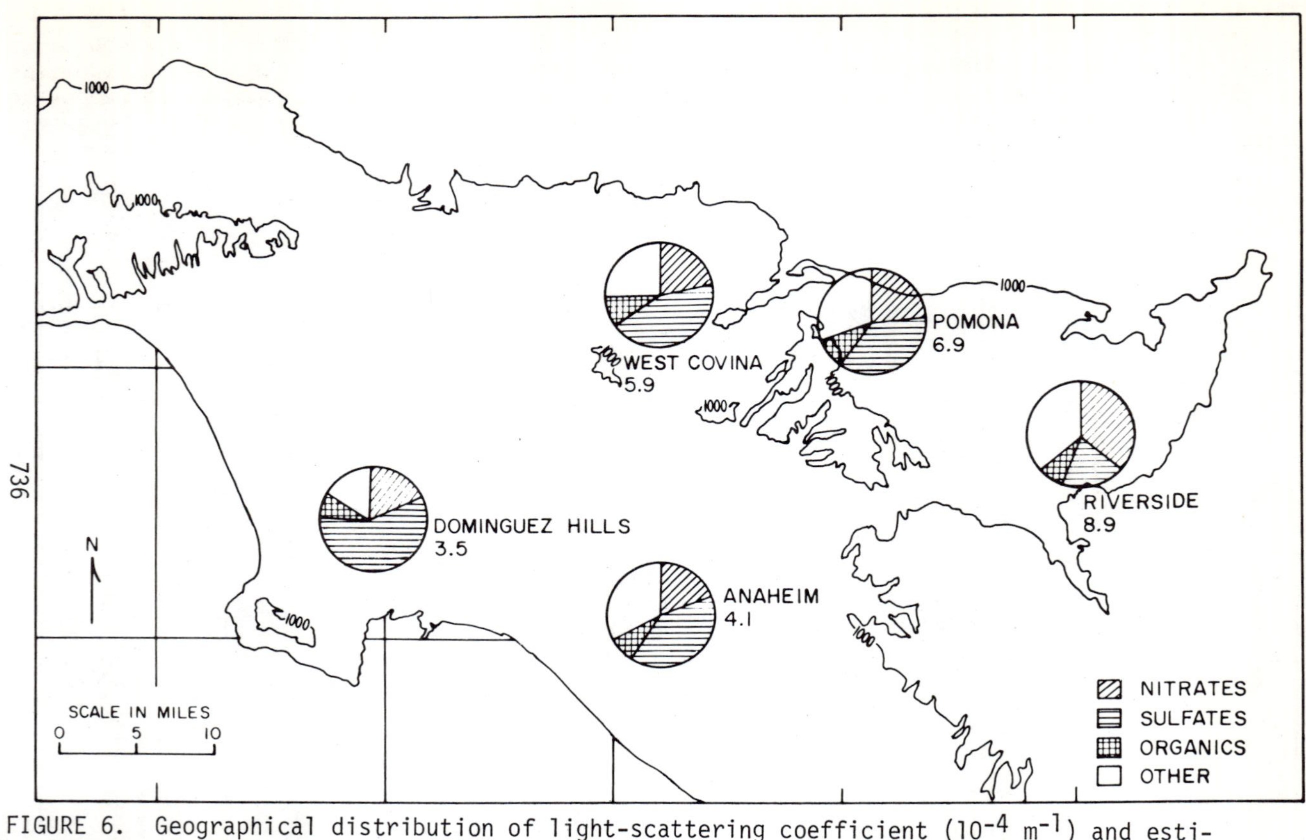

FIGURE 6. Geographical distribution of light-scattering coefficient (10^{-4} m^{-1}) and estimated breakdown by component. (Average of 2-hr aerosol samples at ACHEX mobile van and satellite stations.)

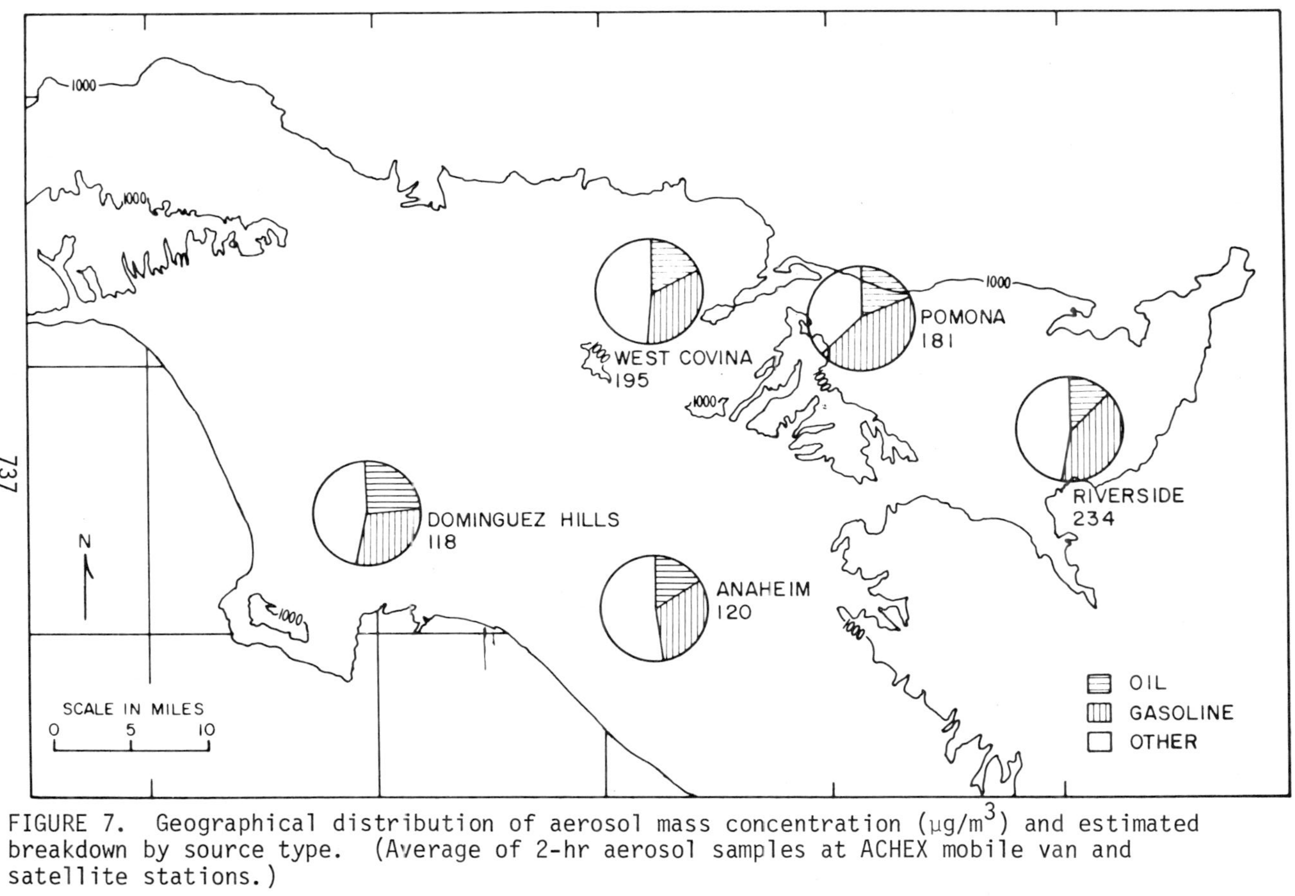

FIGURE 7. Geographical distribution of aerosol mass concentration ($\mu g/m^3$) and estimated breakdown by source type. (Average of 2-hr aerosol samples at ACHEX mobile van and satellite stations.)

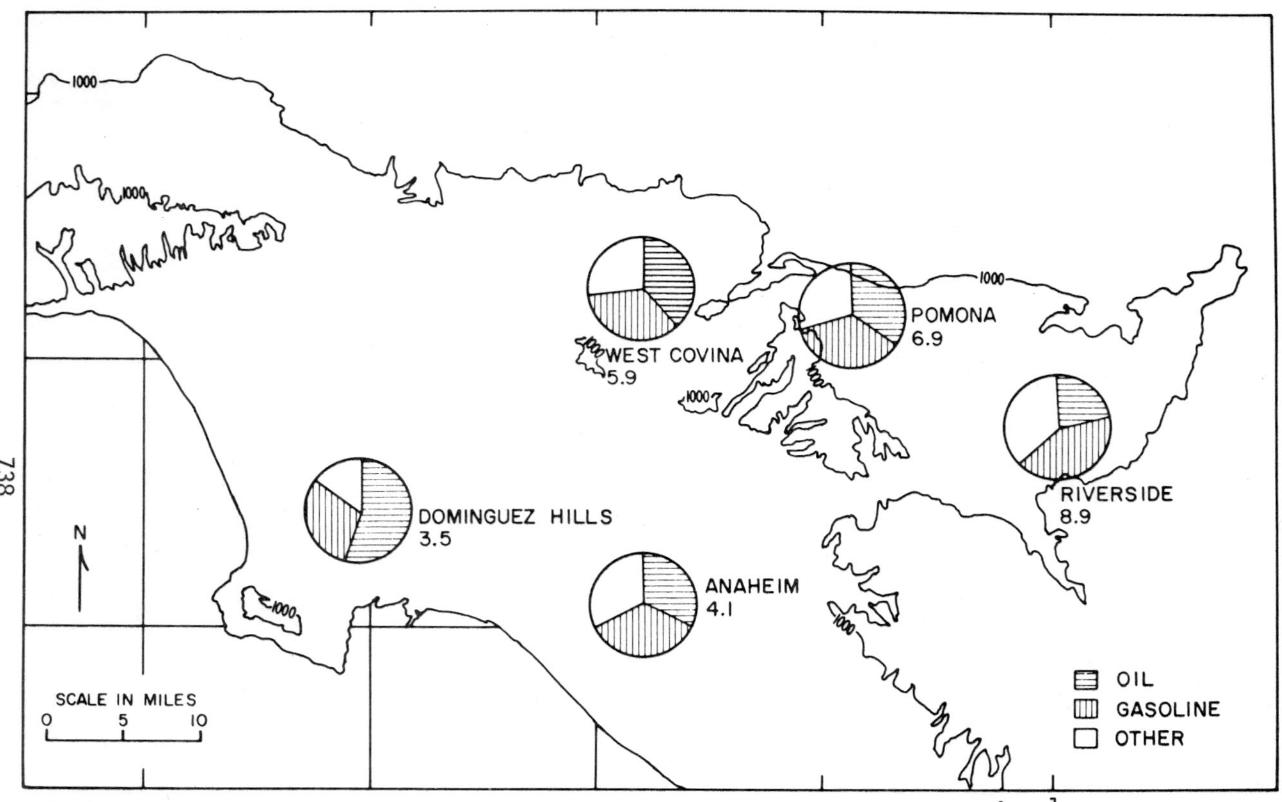

FIGURE 8. Geographical distribution of light-scattering coefficient (10^{-4} m^{-1}) and estimated breakdown by source type. (Average of 2-hr aerosol samples at ACHEX mobile van and satellite stations.)

This regression, based on mobile van data, gives a relationship which is "averaged" over the distribution of relative humidities experienced at the van.

ACKNOWLEDGMENTS

The California Aerosol Characterization Study was sponsored by the State of California Air Resources Board. The analysis presented in this paper was supported by National Institutes of Environmental Health Sciences Training Grants 3T01 ES0004-13S1 and 3T01 ES00080-0751 and by Meteorology Research, Inc. L. Hashimoto and S. L. Heisler assisted in the preparation of figures and tables. The authors are particularly grateful to Professor S. K. Friedlander for helpful advice which strongly influenced the finished work.

REFERENCES

1. Appel, B. R., Colodny, P., and Weslowski, J. J. 1976. Analysis of Carbonaceous Materials in Southern California Atmospheric Aerosols, Environ. Sci. Technol. 10, 359-363.

2. Charlson, R. J., Ahlquist, N. C., Selvidge, H., and MacCready, P. B., Jr. 1969. Monitoring of Atmospheric Aerosol Parameters with the Integrating Nephelometer, J. Air Pollut. Control Assoc. 19, 937-942.

3. Charlson, R. J., Vanderpol, A. H., Covert, D. S., Waggoner, A. P., and Ahlquist, N. C. 1974. $H_2SO_4/(NH_4)_2SO_4$ Background Aerosol: Optical Detection in St. Louis Region, Atmos. Environ. 8, 1257-1267.

4. Elder, J. C., Ettinger, H. J., and Nelson, R. Y. 1974. Chamber Studies of Visibility-Reducing Aerosols, Atmos. Environ. 8, 1035-1048.

5. Ferrel. 1542. Translation from the Spanish of the account by the pilot Ferrel of the voyage of Cabrillo along the west coast of North America in 1542. Appendix to Part 1, Vol. 7, "Report upon United States Geological Surveys West of the One Hundredth Meridian," Engineer Department, U. S. Army, 1879.

6. Findley, C. E. and Rossano, A. T., Jr. 1972. "Continuous Monitoring of Particulate Matter in Automobile Exhaust," presented at the 65th Annual Meeting of the Air Pollution Control Association, Miami Beach, Florida, June 18-22, 1972.

7. Fischer, R. P. 1973. "Can American Oil Refineries Yield Vanadium?" presented before the Division of Petroleum Chemistry of the American Petroleum Society, Chicago, Illinois, August 26-31, 1973.

8. Friedlander, S. K. 1977. Smoke, Dust, and Haze: Fundamentals of Aerosol Behavior. Wiley-Interscience, New York.

9. Gartrell, G., Jr. and Friedlander, S. K. 1975. Relating Particulate Pollution to Sources: the 1972 California Aerosol Characterization Study, Atmos. Environ. 9, 279-299.

10. Grosjean, D. and Friedlander, S. K. 1975. Gas-Particle Distribution Factors for Organic and Other Pollutants in the Los Angeles Atmosphere, J. Air Pollut. Control Assoc. 25, 1038-1044.

11. Heisler, S. L., Friedlander, S. K., and Husar, R. B. 1973. The Relationship of Smog Aerosol Size and Chemical Element Distributions to Source Characteristics, Atmos. Environ. 7, 633-649.

12. Hidy, G. M., et al. 1974. "Characterization of Aerosols in California (ACHEX)," Final Report, Air Resources Board, State of California.

13. Huntzicker, J. J., Friedlander, S. K., and Davidson, C. I. 1975. A Material Balance for Automobile Emitted Lead in the Los Angeles Basin, Environ. Sci. Technol. 9, 448-457.

14. Husar, R. B. and White, W. H. 1976. On the Color of the Los Angeles Smog," Atmos. Environ. 10, 199-204.

15. Husar, R. B., White, W. H., and Blumenthal, D. L. 1976. Direct Evidence of Heterogeneous Aerosol Formation in Los Angeles Smog, Environ. Sci. Technol. 10, 490-491.

16. Miller, M. S., Friedlander, S. K., and Hidy, G. M. 1972. A Chemical Element Balance for the Pasadena Aerosol, J. Colloid and Interface Sci. 39, 165-176.

17. Noyes, A. 1927. New Essays and American Impressions. Henry Holt and Co.

18. Samuels, H. J., Twiss, S., and Wong, E. W. 1973. "Visibility, Light Scattering, and Mass Concentration of Particulate Matter," Report of the California Tri-City Aerosol Sampling Project, California Air Resources Board.

19. Sinclair, D., Countess, R. J., and Hoppes, G. S. 1974. Effect of Relative Humidity on the Size of Atmospheric Aerosol Particles, Atmos. Environ. 8, 1111-1117.

20. van de Hulst, H. C. 1957. Light Scattering by Small Particles. John Wiley.

21. Waggoner, A. P., Vanderpol, A. J., Charlson, R. J., Larsen, S., Granat, L., and Tragardh, C. 1976. Sulphate-Light Scattering Ratio as an Index of the Role of Sulphur in Tropospheric Optics, Nature 261, 120-122.

22. White, W. H. and Husar, R. B. 1976. A Lagrangian Model of the Los Angeles Smog Aerosol, J. Air Pollut. Control Assoc. 26, 32-35.

23. White, W. H. 1975. Estimating the Size Range of Smog Aerosol Particles with a Pair of Sunglasses, Atmos. Environ. 9, 1036-1037.

24. White, W. H. 1976. Reduction of Visibility by Sulfates in Photochemical Smog, Nature 264, 735-736.

APPENDIX A. NOTATION, UNITS, AND ANALYTICAL TECHNIQUE

b_{scat} (10^{-4} m^{-1})	Aerosol light-scattering coefficient, nephelometer
MASS ($\mu g/m^3$)	Aerosol mass concentration, gravimetric
NO_3^- ($\mu g/m^3$)	Particulate nitrate concentration, wet chemistry
SO_4^{2-} ($\mu g/m^3$)	Particulate sulfate concentration, wet chemistry
C ($\mu g/m^3$)	Particulate carbon concentration, flame ionization detector and ashing
NITRATES ($\mu g/m^3$)	Estimated concentration of particulate nitrate compounds, NITRATES = 1.3 NO_3^-
SULFATES ($\mu g/m^3$)	Estimated concentration of particulate sulfate compounds, SULFATES = 1.3 SO_4^{2-}
ORGANICS ($\mu g/m^3$)	Estimated concentration of particulate organic compounds, ORGANICS = 1.4 C
NH_4^+ ($\mu g/m^3$)	Particulate ammonium ion concentration, wet chemistry
Na ($\mu g/m^3$)	Particulate sodium concentration, neutron activation
Al ($\mu g/m^3$)	Particulate aluminum concentration, neutron activation
$Na_{sea\ salt}$ ($\mu g/m^3$)	Estimated concentration of particulate sodium due to sea salt, $Na_{sea\ salt}$ = Na - (0.025/0.082) Al.
Pb ($\mu g/m^3$)	Particulate lead concentration, X-ray fluorescence
Ni ($\mu g/m^3$)	Particulate nickel concentration, X-ray fluorescence
V ($\mu g/m^3$)	Particulate vanadium concentration, neutron activation

CO ($\mu g/m^3$)	Carbon monoxide concentration, gas chromatography followed by flame ionization detector
NO_x (ppm)	Nitrogen oxides concentration, chemiluminescence detector
SO_2 (ppm)	Sulfur dioxide concentration, gas chromatography followed by flame photometric detector
N_T ($\mu g/m^3$)	Nitrogen concentration, $N_T = 573 \times NO_x + (14/46) \times NO_3^-$
S_T ($\mu g/m^3$)	Sulfur concentration, $S_T = 1309 \times SO_2 + (32/96) \times SO_4^{2-}$

Except for those presented in Figures 4-8, all data are from the ACHEX mobile van, which was operated at Dominguez Hills, West Covina, Pomona, and Riverside. Gravimetric and chemical analyses were performed on three types of filters. Unless otherwise noted, the data used in this paper are taken from 2-hr samples on a total filter with an upper cutoff at about D_p = 20 μm. This filter was operated in parallel with a Lundgren impactor having an afterfilter designed to collect particles with $D_p \le 0.5$ μm. Sampling intervals are summarized in Table A. Certain analyses, such as that for ammonium, were performed only on 24-hr hi-vol filter samples.

TABLE A1. "Two-Hour" Samples Used in Regression Analysis and Mobile Van Averages. All Samples Were Weighted Equally, Regardless of Length. Particulates Were Measured on Total Filter Except Where Noted in Text. SO_2 Was Not Measured at Rubidoux

Mobile van site	Date	Pacific Standard Time																				
		3	4	5	6	7	8	9	10	11	12	13	14	15	16	17	18	19	20	21	22	23
West Covina	7-24-73				—	—	—	—	—	—	—	—	—	—	—	—	—	—	—	—		
West Covina	7-25-73			—	—	—	—	—	—	—	—	—	—	—	—							
West Covina	7-26-73				—	—	—	—	—	—	—	—	—	—	—	—	—					
West Covina	8-9-73				—	—	—	—	—	—	—	—	—	—	—	—	—					
Pomona	8-17-73				—	—	—	—	—	—	—	—	—	—	—	—	—	—	—	—		
Rubidoux	9-6-73	—	—	—	—	—	—	—	—	—	—	—	—	—	—	—	—	—	—	—	—	—
Rubidoux	9-19-73	—	—	—	—	—	—	—	—	—	—	—	—	—	—	—	—	—	—	—	—	—
Dominguez Hills	10-5-73				—	—	—	—	—	—	—	—	—	—	—	—	—	—	—	—		
Dominguez Hills	10-11-73				—	—	—	—	—	—	—	—	—	—	—	—	—	—	—	—		

ADDENDUM

In our analysis for ACHEX (White et al., 1974), we set out to construct a scattering budget for the Los Angeles aerosol, in analogy with the more familiar mass budgets developed by S. K. Friedlander and his coworkers. A trivializing approximation would have been to set each component's scattering proportional to its mass, via the Charlson proportionality coefficient relating total scattering to total mass. We found, however, that we could do better than this; through multiple regression we could derive distinct proportionality coefficients for individual components, in effect using chemical composition as a key to particle size and hygroscopicity. Our methodology has subsequently been employed to advantage by a number of investigators, notably G. R. Cass and J. Trijonis. Cass (1976) showed how to extend the analysis to routine monitoring data, greatly broadening its utility. Trijonis and Yuan (1978a) applied Cass's approach to the Southwestern United States, where visibility is currently a critical issue. These and other results are abstracted in Tables A1 and A2.

Sulfates. It is by now well established that high sulfate concentrations tend to correlate with reduced visibilities. The seeming ubiquity of this correlation [e.g., England ($r = .71$, $n = 27$), Eggleton, 1969; Sweden ($r = .73$, $n = 77$), Waggoner et al., 1976; New York ($r = .89$, $n = 43$), Leaderer et al., 1978] is persuasive evidence for the widespread control of visibility by sulfates. That it is not by itself conclusive evidence is nicely illustrated by the Riverside data of Lundgren (1970). As indicated in Table A1, Lundgren found a degree of correlation between sulfates and scattering comparable to those reported by Eggleton and Waggoner et al. Had Lundgren not chosen to measure nitrates as well, we might be tempted to add Riverside to the list of locations where sulfates are thought to be the dominant scatterers.

What the ACHEX study and subsequent work has shown is that sulfates correlate with reduced visibilities even when the remainder of the aerosol is taken into account. The multivariate analyses summarized in Table A2 all discount for the simultaneous influences of other components of the aerosol. That sulfates consistently emerge as statistically significant correlates of reduced visibilities, in competition with other potential tracers (nitrates and organics) for the submicron aerosol, is strong evidence of an underlying physical relationship.

To date, multivariate analyses have succeeded in resolving at most three components of the light-scattering aerosol: sulfates, nitrates, and "everything else." This is a rather

TABLE A1. Multivariate Aerosol/Visibility Studies

	Lundgren, 1970 Riverside	White, Roberts, 1977 Los Angeles Basin	Grosjean, Friedlander, 1975 Pasadena	Cass, 1976 Downtown Los Angeles	Grosjean et al., 1976 Riverside	Trijonis, Yuan, 1978 Phoenix (Maricopa)	Trijonis, Yuan, 1978 Salt Lake City	Leaderer et al., 1979 Santa Monica - LAX
Data Description								
Number of observations	88	58	8	413	159	87	130	293
Measurement type								
Aerosol (a)	L	L	H	H	H	H	H	H
Optical (b)	S	S	S	E	S	E	E	E
Averaging time, hours								
Aerosol	4	2	1	24	24	24	24	24
Optical	4	2	1	9	24	9	9	24
Correlation Coefficients								
B(b) - Mass	.94	.91*	.82*	.40	.66	.31	.54	
B - Sulfates	.72	.58*	.00*	.62		.81	.50	
B - Nitrates	.94	.78*	.56*	.09		.71	.59	
B - Organics		.59*	.19*			.31		
B - Lead		.09*						
B - Nickel		.08*						
Mass - Sulfates	.72	.46*	.16*	.32*		.44	.45	
Mass - Nitrates	.92	.79*	.69*	.43*		.34	.61	
Mass - Organics		.73*	.41*			.51		
Mass - Lead		.23*						
Mass - Nickel		.05*						
Sulfates - Nitrates	.74	.08*	.07*	-.03		.56	.59	
Sulfates - Organics		.46*	.75*			.45		
Nitrates - Organics		.25*	.30*			.43		
Lead - Nickel		-.04*						

Footnotes for Table A1

Asterisks (*) denote data presented here for the first time (White and Roberts, 1977), or derived here from data presented elsewhere (Grosjean and Friedlander, 1976; Cass, 1976).

[a] L, H denote low-volume, high-volume sampling.

[b] S, E denote scattering, extinction. Scattering coefficients (b_{scat}) were measured directly, by nephelometry. Extinction coefficients (b_{ext}) were calculated from visual range, VR, via the Koschmieder formula $b_{ext} = -(\log \varepsilon)/VR$ (Middleton, 1952). The distinction between scattering and extinction should be unimportant for sulfates and nitrates, which do not absorb significantly. Both parameters are designated by B in the correlation table.

TABLE A2. Component Scattering and Extinction Efficiencies*

	White, Roberts, 1977 Equation 9	White, 1976 Double Regression	Cass, 1976 Table III, Entry 3	Grosjean et al., 1976 Page 5	Trijonis, Yuan, 1978 Phoenix - Regression	Trijonis, Yuan, 1978 Phoenix - Strike	Trijonis, Yuan, 1978 Salt Lake City	Leaderer et al., 1979 Santa Monica - LAX
Efficiencies, 10^{-4} m^2/μg			b		b	b	b	b
Sulfates								
30% RH[a]			.101		.033	.039	.031	
50% RH			.126				.043	
70% RH[a]	.062	.085	.177	.165			.071	.14
Nitrates								
30% RH[a]	.027	.033	.045		.045		.072	
50% RH	.035	.037	.057				.101	
70% RH[a]	.047	.042	.080	.028			.169	.03
Other[c]								
All RH	.022	.022	<.01				.003	
Fraction of variance accounted for (r^2)	.93	--	.64	.59	.76	--	.66	.83

*Efficiencies represent statistical increment in total scattering or extinction coefficient (10^{-4} m^{-1}) per increment in individual compound concentration (μg/m^3). The data of Lundgren (excessive colinearity), and of Grosjean and Friedlander (insufficient observations), do not support distinct component efficiencies.

[a]Efficiencies not differentiated by RH are entered nearest the average RH for the data set.

[b]Extinction efficiencies (cf. note to Table A1) are based on Middleton's experimental value ε = .031 for the contrast threshold (Middleton, 1952, p. 220), rather than the traditional but arbitrary value ε = .020 used by Cass, Trijonis, and Yuan, and Leaderer et al.

coarse breakdown, which conceivably overlooks minor nonsulfate, nonnitrate subcomponents which are efficient scatterers. To the extent that these unrepresented subcomponents are statistically associated with sulfates and nitrates, their scattering may show up in the guise of inflated estimates for the sulfate and nitrate scattering efficiencies. Two different approaches have been used to verify the "purity" of estimates for the sulfate scattering efficiency.

White (1976) tested the ACHEX data for the presence of statistical sulfate mimics by regressing MASS on aerosol composition:

$$\text{MASS} = 20 + .9\ \text{SULFATES} + 1.5\ \text{NITRATES} + 2.9\ \text{ORGANICS} \pm 22$$

$$(n = 58,\ \bar{y} = 199,\ r = .98)$$

The point of this regression is that the physically correct coefficients are known; to within our uncertainties over molecular identities, they should all be unity. The empirical SULFATES coefficient is in fact within one standard error of unity, demonstrating that the sulfate fraction is sufficiently decoupled from the remainder of the Los Angeles aerosol that even its contribution to mass, which in relative terms is less than its contribution to scattering, can be distinguished statistically. (The NITRATES coefficient is 50% high; regressions involving aluminum suggest that the excess may represent soil dust whose geographical distribution in the Basin is similar to that of nitrates. The ORGANICS coefficient confirms that carbon cannot be distinguished statistically from the rest of the nonsulfate, nonnitrate aerosol.) The analogous regression for b_{scat} gives:

$$b_{scat} = -.5 + .076\ \text{SULFATES} + (.049 + .034\mu^2)\ \text{NITRATES}$$
$$+ .062\ \text{ORGANICS} \pm 1.0$$

$$(n = 58,\ \bar{y} = 6.4,\ r = .97)$$

This regression attributes much scattering to organics because there is no other term representing the nonsulfate, nonnitrate aerosol. Dividing the coefficients of the b_{scat} regression by the corresponding coefficients of the MASS regression yields alternative estimates for the component scattering efficiencies, which are compatible with those obtained by direct regression.

Trijonis and Yuan (1978a) tested their Southwest analyses, in the classical manner of experimental science, by introducing a well-defined perturbation of the atmosphere. The "experiment" consisted of a strike which shut down the copper industry over a nine-month period in 1967-1968. In the southwest at this time, copper production accounted for over 90% of the SO_x emissions, less than 1% of the NO_x emissions, and less than 10% of the conventional particulate emissions (Marians and Trijonis, 1978), and should, therefore, have impacted visibility primarily through its contribution to sulfate loadings. Comparing measurements during the strike with those of the surrounding four or six years, Trijonis and Yuan found a large decrease in Phoenix sulfate loadings, accompanied by a substantial improvement in visibility. Attributing the improvement in visibility wholly to the drop in sulfate levels yields an estimated extinction efficiency only 18% higher than that obtained by regression, which utilized a disjoint subset (1973-1974) of the data base.

The various estimates of White/Roberts and Trijonia/Yuan for sulfate scattering efficiencies--in the Los Angeles Basin, Phoenix, and Salt Lake City--are in good agreement with one another. (They also agree with later estimates by Trijonis and Yuan (1978b) for Newark, New Jersey, Cleveland, Ohio, and Lexington, Kentucky.) The remaining estimates in Table A2 are significantly higher. Cass (1976) noted that his value should be high; his regression model relates visibilities measured during nine daylight hours, when Los Angeles sulfate concentrations can display strong diurnal maxima (Appel et al., 1978), to aerosol loadings measured over a full 24-hour day. This, of course, detracts little from its utility as a site-specific visibility predictor. The explanation for the similarly high values obtained by Grosjean et al. (1976) and Leaderer et al. (1979) is not obvious, although it may lie in the nonlinear influence of relative humidity, which undergoes wide swings over a typical 24-hour period.

Nitrates. There has been some tendency to discount the atmospheric importance of particulate nitrates in the wake of recent evidence (Spicer and Schumacher, 1977) that perhaps 90 to 95% of the nitrate collected on glass fiber filters in Upland is artifact. This seems premature. Whatever the sampling problems associated with various filter media, they do not affect *in situ* measurements of scattering and visual range, which in Los Angeles and the southwest are consistently found to correlate with "particulate nitrate."* Whatever is limiting visibility here, it is not artifact nitrate. (Specifically, it is not HNO_3 in the gas phase, thought to be the principle source of artifact; it is not PAN,

*It is not surprising that this correlation should be harder to discern in the northeast United States (Trijonis and Yuan, 1978b), given that sulfate/nitrate ratios there are several times those typical of the west.

which Lundgren (1970) found to correlate strongly with scattering ($r = .94$, $n = 88$); it is not NO_2 at ambient concentrations, for which Cass (1976) found an empirical extinction efficiency much higher than physically expected.)

The substantial reductions in visibility associated with "particulate nitrate" must result from physical scattering by ambient aerosol--either genuine particulate nitrates, or non-nitrate particulate statistically associated with nitrates. The first alternative implies that the nitrate scattering efficiencies in Table A2 should be multiplied by the ratio (filter particulate nitrate)/(ambient particulate nitrate) to correct for sampling error. If this ratio is large, the resulting values would be difficult to justify physically, save by invoking large volumes of unbound water (cf. Figure 1 of our original paper). The second alternative implies that we are overlooking an aerosol component which, in order to account for the scattering attributed to nitrates, must be present in substantial concentration and also scatter effectively. There is no evidence that the postulated component is organics (see below), and it is difficult to see what other candidate fraction would so closely mimic nitrates. Under either alternative, the effect of a pronounced increase in artifact production at high humidities should be to decrease the apparent scattering efficiency. No inverse relationship between scattering efficiencies and relative humidity is observed, however.

Other. Much of the nonsulfate, nonnitrate mass is contributed by large-particle dusts, which are very inefficient scatterers. The relatively low upper particle-size cutoff of the ACHEX total filters is reflected in relatively high estimates for the scattering efficiency of the nonsulfate, nonnitrate aerosol.

In Los Angeles smog, much of the nonsulfate, nonnitrate mass is contributed by organics. Perhaps the least-expected result of our ACHEX analysis was the finding that this submicron fraction is relatively unimportant in the scattering budget. This hesitantly-voiced conclusion has so far received consistent support in the work of Grosjean and Friedlander (1976), Grosjean et al. (1976), and Trijonis and Yuan (1978a). The limited data of Grosjean and Friedlander are particularly interesting, as they record a classic smog episode at peak intensity (Blumenthal et al., 1978); organics loadings of up to 141 $\mu g/m^3$ had no apparent effect on the scattering coefficient.

Water can in principle be regarded as a separate aerosol component, on an equal footing with sulfates and nitrates. From the regulatory point of view, however, it makes sense to lump its effects together with those of the hygroscopic materials responsible for its presence in the particulate phase, as we have done. The strong humidity dependences evident in Table A2 suggest that much of the scattering attributed to both sulfates and nitrates is in fact due to their associated waters.

REFERENCES

1. Appel, B. R., Kothny, E. L., Hoffer, E. M., Hidy, G. M., and Wesolowski, J. J. 1978. Environ. Sci. Technol. 12, 418.

2. Blumenthal, D. L., White, W. H., and Smith, T. B. 1978. Atmos. Environ. 12, 893.

3. Cass, G. R. 1976. EQL Memorandum No. 18, California Institute of Technology, Pasadena, California.

4. Eggleton, A. E. J. 1969. Atmos. Environ. 3, 355.

5. Grosjean, D., Doyle, G. J., Mischke, T. M., Poe, M. P., Fitz, D. R., Smith, J. P., and Pitts, J. N., Jr. 1976. Annual Meeting Paper No. 76-20.3, Air Pollution Control Association.

6. Grosjean, D. and Friedlander, S. K. 1975. J. Air Pollut. Control Assoc. 25, 1038.

7. Leaderer, B. P., et al. 1978. J. Air Pollut. Control Assoc. 28, 321.

8. Leaderer, B. P., Holford, T. R., and Stolwijk, J. A. J. 1979. J. Air Pollut. Control Assoc. 29, 154. Regression presented at New York Academy of Sciences, 1979.

9. Lundgren, D. A. 1970. J. Air Pollut. Control Assoc. 20, 603.

10. Marians, M. and Trijonis, J. 1979. Report to USEPA OAQPS, Technology Service Corporation.

11. Middleton, W. E. K. 1952. Vision Through the Atmosphere. University of Toronto Press, Canada.

12. Spicer, C. W. and Schumacher, P. M. 1977. Atmos. Environ. 11, 873.

13. Trijonis, J. and Yuan, K. 1978a. EPA-600/3-78-039.

14. Trijonis, J. and Yuan, K. 1978b. EPA-600/3-78-075.

15. Waggoner, A. P., Vanderpol, A. J., Charlson, R. J., Larsen, S., Granat, L., and Tragardh, C. 1976. Nature 261, 120.

16. White, W. H. 1976. Nature 264, 735.

17. White, W. H. and Roberts, P. T. 1977. Atmos. Environ. 11, 803.

18. White, W. H., Roberts, P. T., and Friedlander, S. K. 1974. Keck Laboratories Report, California Institute of Technology.

Epilogue

GEORGE M. HIDY AND PETER K. MUELLER
Environmental Research and Technology, Inc.
Westlake Village, California 91361

This volume is being published approximately four years after completion of the Aerosol Characterization Experiment (ACHEX) analysis and contract reporting. Over this duration the technology of aerosol measurement and characterization in the atmosphere has progressed well beyond that used in the experiment in some areas, but not in others. It would be presumptuous for us to evaluate the impact of the ACHEX on the science of atmospheric aerosols since completing the program. However, it is of historical interest to trace some known derivatives of the study.

The experimental design of the ACHEX relied heavily on short-term intensive case studies. This method has been used subsequently by the investigators in formulating parts of the U.S. Environmental Protection Agency's (EPA) Regional Air Pollution Study in St. Louis (RAPS) [e.g., Burton ahd Hidy (6)] and the aircraft and ground studies associated with the EPA's Midwest Interstate Sulfur Transport and Transformation (MISTT) experiments [e.g., Wilson et al. (30)]. More recently, the experience of the ACHEX and related studies has been used as a part of the basis for the design of the Sulfate Regional Experiment (SURE) [Hidy et al. (14)].

The ACHEX and related studies, and the RAPS, basically represented projects that attempted to characterize aspects of air pollution in a large metropolitan area over horizontal distances of about 100 km. The time scale of investigation for pollution episode behavior for such phenomena was hypothesized to be 1 to 3 days. In the RAPS the spatial resolution for phenomena was estimated to be about 10 km. In the ACHEX the urban scale processes were speculated to be resolved over similar distances even though the experimental design tacitly assumed a resolution more like 50 km. In contrast, the SURE is a major field program which began operations in mid-1977 covering the greater northeastern United States. The aim of this study is to elucidate the origins and evolution of aerosols containing sulfate* on a regional scale. The regional scale has a spatial extent of about 1000 km or greater with a resolution of 50 to 100 km. It has a temporal resolution of 1 to 5 days. The interest in tracing the behavior of aerosols over widespread regions is a natural extension of the concern for defining a zone of influence of sources of pollution on receptors located over a range of distance from the source.

The ACHEX concept evolved into later studies that were aimed at characterizing directly the evolution of aerosols from (gaseous and particulate) source emissions. In California an attempt was made in 1974 to study sulfate formation from power plant plumes mixing into nonpolluted and smoggy air over distances up to 50 km [e.g., Appel et al. (1), Richards et al. (23), Smith et al. (24)]. At the same time the California Air Resources Board (ARB) attempted an experiment in Los Angeles to estimate the amount of sulfate associated with catalyst-equipped automobiles. This work was complemented by the EPA's continuing automobile sulfate studies in Los Angeles. the MISTT plume experiments initiated in 1974 may be identified with the analogous source-receptor investigations attempted in California. The California experience, in combination with recent reports of MISTT investigators [e.g., Wilson et al. (30), Husar et al. (17)], has yielded significant new information on plume chemistry. It is particularly interesting that the MISTT investigations suggest differences between the day and night oxidation rates of sulfur dioxide. However, neither the California plume experiments nor the MISTT experience indicates maximum oxidation rates above a few percent per hour. This is lower than the upper limit rates of 10 to 15%/hr for oxidation at ground level in urban photochemical smog deduced from the ACHEX and related observations in 1973. New plume studies are contemplated by the Electric Power Research Institute, the U. S. Department of Energy, and the EPA to elucidate further the sulfur dioxide oxidation process.

A major constraint in the design of the ACHEX was its inability to provide long-term monitoring observations to check the representativeness for "pollution climatology" of the episode studies. This later proved to be a severe limitation, which

---

*Although the focus of the SURE is on sulfur oxides, it was recognized that particulate sulfate must be considered a part of a complex mixture of material including nitrate, organics, and trace metal compounds. This mixture has been referred to by some workers as the Sulfur Oxide Particulate Complex (SPC)

made it very difficult to apply the ACHEX results to control strategies requiring the projection of long-term averaged conditions. This problem has been avoided in later experiments like the RAPS and the SURE, which have incorporated basic long-term monitoring networks for air quality data along with intensive episode studies.

The ACHEX experience in operating instrument systems with on-line computing and data acquisition was transferred and improved in two major EPA monitoring networks, one in St. Louis supporting the RAPS, and the other, the Community Health Air Monitoring Program. As a part of RAPS and MISTT, a second aerosol trailer was built for the EPA. This transportable research facility is similar to that used in the ACHEX and has been operated by University of Minnesota workers for episode studies in St. Louis and other locations from 1973 to 1975. The aircraft instrumentation system of Meteorology Research, Inc. has been improved and is serving the upper level measurement requirements of both the MISTT and the SURE. The ACHEX mobile laboratory has continued to be used intermittently for episodes studied in southern California by the Statewide Air Pollution Research Center [e.g., Pitts et al. (22)]. The results obtained are generally similar to those reported in the ACHEX.

Further development of the ACHEX techniques for physical and chemical characterization of aerosol particles has supported the evolution of a number of research programs, and provided many of the data to bring about perspective in regards to the usefulness and limitations of new methods, including X-ray fluorescence [Stevens et al. (26), Flocchini et al. (12)], photoelectron spectroscopy, high-resolution mass spectroscopy [Crittendon (10)], and selective organic solvent extraction [Appel et al. (3)]. Basic wet chemical methods for sulfate were used which subsequently have been tested extensively by the Air and Industrial Hygiene Laboratory for accuracy of measurement [Appel et al. (2)]. Coupled to the chemical methods has been an extensive effort to improve the reliability of sampling the total mass and submicron mass concentrations of particulate material [Stevens et al. (26)] to evaluate the collection surfaces of cascade impactors [Wesolowski et al. (28)], and to understand the interferences of acid gas absorption on filter substrate [e.g., Mueller (20)]. The work emerging from the ACHEX and other experiments has provided convincing evidence for the multimodal mass distribution of particulate material [e.g., Whitby et al. p. 519 of this book). This recognition, as well as the acknowledged requirement for sampling characterizing the submicron fraction of the atmospheric aerosol particles, probably will influence heavily consideration of an air quality standard by particle size range.

A limited monitoring network for sampling aerosols by size fraction and X-ray fluorescence analysis was implemented by the ARB after completion of the ACHEX [e.g., Flocchini et al. (12)]. The ACHEX and later work have led to a substantial improvement in substrate selection for sulfate analysis, which should be extrapolatable to nitrate [Mueller (20), Appel et al. (4), Stevens et al. (26)].

The results of Spicer et al. (25) call attention to the potentially severe errors in the estimation of apparent particulate nitrate in urban air if glass fiber or cellulose filters are used. In contrast to gas absorption interferences of $\pm 20\%$ for sulfate collection, existing nitrate data may be in error (too high) by as much as a factor of 2 to 5. These results are not consistent with sampling done by impactor and filter at Pasadena, which have shown levels similar to those reported in the ACHEX [Moskowitz (19)]. Yet the results of a new 1977 aerosol characterization study in the Los Angeles area have shown lower nitrate concentrations and less diurnal variation as compared with the ACHEX work. The substrate used in 1977 was Fluoropore, while cellulose fiber and cellulose ester were used in the ACHEX [Heisler et al. (13)]. In a side-by-side field sampling study about twice as much nitrate was collected on cellulose and cellulose acetate filters as on Fluoropore. The results were consistent with the relative amounts of nitric acid collected by these filters in laboratory exposures [Appel et al. (4)]. Because of the interest in stability of aerosol samples after filter collection in smog, the 1977 Los Angeles study has attempted to look at (water-soluble) particulate sulfur oxide (as sulfate), nitrate, and chloride chilled to $-4°C$ and analyzed immediately after collection versus after aging at room temperature. The results suggest that sulfate and nitrate are stable on Fluoropore filters for considerable lengths of time. Such analytical results are important for the interpretation of the ACHEX and subsequent data in aerosol characteristics.

The work of the ACHEX provided support for a variety of subsequent investigations relating aerosols in polluted air to visibility degradation. The research of Charlson and colleagues in this area in St. Louis and other locations has shown that humidified nephelometry can be used to distinguish between a sulfate species in air [e.g., Charlson et al. (19)]. Waggoner et al. (27) obtained a surprising result from examination of simultaneous light-scattering and sulfate observations from several locations. They found that the ratio

$$[SO_4^{\ 2-}]/b_{scat} \approx 0.2 \begin{smallmatrix} +0.49 \\ -0.06 \end{smallmatrix} \ (g/m^2)$$

for a number of different locations, including the Los Angeles area, based in part on the ACHEX.

Even after 4 years the ACHEX data constitute the only accessible set of its kind which has reported detailed trace gas observations with measurements of physical and chemical properties of airborne particles. Additional data of this kind have been taken as part of the RAPS, but are not available extensively in the literature as yet. As indicated in this volume, the orientation of the ACHEX data analysis focused heavily on the phenomenological aspects of chemical components associated with chemical processes in photochemical smog. The data do indeed provide a very detailed picture of the behavior of particulate matter at a fixed location as smog evolves through the day. An important limitation in the interpretation of such data was found to be the lack of ability to account for chemical processes going on in the same air mass traveling across the air basin with the winds. This cannot be done using a single station fixed in space. The aerosol behavior characterized by the ACHEX data set requires quantitative analysis taking into account air mass movement with coupled chemical kinetics. Within the ACHEX a preliminary attempt was made to do this, as reported in the paper by White et al. (p. 519 in this book). Because of limitations in funding, further efforts to improve such questions for interpretation of the ACHEX results have been restricted. Only recently have attempts to use the ACHEX data with air quality modeling been reported [e.g., Burton et al. (5), Hidy et al. (15)]. The former study showed that an air transport model coupled with a photochemical aerosol chemistry mechanism could explain surprisingly well daytime data for sulfate, nitrate, and organics taken in July 1973, at West Covina. The latter test case was confined to analysis of 24-hr averaged observations for sulfate at West Covina and other sites on July 25, 1973. An air transport model using simple oxidation chemistry, first order in sulfur dioxide, simulated semiquantitatively the sulfur dioxide and the sulfate distribution on this day satisfactorily. These recent calculations depend strongly on specification of the spatial distribution of pollutant emissions in the Los Angeles area. Detailed information of this kind was not readily available at the time of the ACHEX, and in retrospect was another limitation of the study. Reliance was placed on aggregated emissions, using the chemical element balance approach as described in papers in this volume.

Accounting for the sources of particulate matter in urban air by using the chemical element balance has been attempted, employing data from cities other than Los Angeles. Heisler* has

* Personal communication.

applied the technique to observations taken in St. Louis, Miami, and Pittsburgh. The results showed a source distribution which differed somewhat from that of the Los Angeles area, with poorer material balance than was achieved in the ACHEX data. Differences consisted partly of lower contributions to the secondary nitrogen and organic components, and in the ratios of primary sources associated with transportation and industry as identified by a tracer material.

Recently Novakov* has criticized the findings reported in the ACHEX as being unduly biased toward explaining the behavior of the secondary particle components in terms of photochemical processes. In his opinion, alternate explanations using heterogeneous processes involving carbon or metal oxides and salts should be tested to clarify the role of photochemical processes in the summer aerosol, as contrasted with other mechanisms. The wealth of unanalyzed metal data and the carbon-organics data from the ACHEX could certainly be examined for this alternative and others as more fundamental knowledge of aerosol-forming mechanisms becomes available.

Although the ACHEX experience may have found its way mostly into scientific programs in atmospheric chemistry, it has had significant practical application in the development of regulatory policy regarding aerosols in California. The results provided supporting data indicating elevated ambient concentrations of particulate sulfate in southern California, despite the fact that the air over Los Angeles rarely contains sulfur dioxide concentrations exceeding the National Ambient Air quality Standards or the California standards. The ACHEX analysis gave circumstantial evidence that the inordinately high concentrations of sulfates and light-scattering aerosols in the Los Angeles basin corresponded to the changes in intensity of oxidizing species in photochemical smog. Thus control of sulfate, for example, would require reduction in oxidant, as well as in the gaseous precursor, SO_2. Holmes (16) extended the approach to reduction of total suspended particulate levels, using a simple rollback model† that accounted for direct reduction in sulfur dioxide emissions, and for nonmethane hydrocarbons, an oxidant precursor.

In 1976 the ARB promulgated a state ambient quality standard for airborne sulfates based on <u>suspected</u> health effects of the material. The application of this standard for very stringent emissions control of SO_2 has led to a detailed reexamination of the simple rollback concept, as well as the early interpretation

---

*Personal communication.

†For a description of a rollback approach see, for example, National Academy of Sciences (21).

of the ACHEX and other data in the South Coast Air Basin [e.g., testimony of California Public Utilities Commission (7)]. Analysis presented during this series of hearings suggested that the rollback approach was too simplistic to provide a satisfactory basis for a Los Angeles basin control strategy. As a result, new investigations of sulfur oxide behavior have been initiated in California to engineer improved methods to relate sulfur oxide emissions to ambient air quality [e.g., Cass (8), Joint Project (18), White et al. (29), Hidy et al. (15), Burton et al. (5)].

The issue of the validity of suspected health effects of particulate sulfate has created a continuing controversy about the basis for the California sulfate standard. It remains to be proven that sulfates as a broad class of compounds, including ammonium salts, sulfuric acid, and metal salts, and organic sulfur compounds, are deleterious to health. The studies to date fail to support such a contention at concentrations observed in the atmosphere. Therefore new investigations have been initiated to examine the physical and chemical nature of the "sulfur oxide particulate complex" to better identify components that actually may be important toxicants at atmospheric levels. Such studies will extend to speciation of condensed sulfur compounds as well as nitrogen and organic materials. This work will extend significantly the knowledge base developed by the ACHEX. The new experiments include the investigations of Moskowitz (19), Heisler et al. (13), and the Statewide Air Pollution Research Center [e.g., Pitts et al. (22)].

Like the earlier Pasadena aerosol experiment in 1969, the ACHEX has proven to be only a milestone in continuing research on atmospheric aerosols in California and in other geographical areas. Despite the major effort of the ACHEX, the details of the relationship between human activities and the intense haze over southern California remain incompletely understood. This continues to be especially true of the processes involved in the evolution of nitrogen oxides and condensable organic compounds. Because of the highly complex chemical nature, and the close connection with gas-phase chemical reactions, one can expect productive research to continue for many years to come. It is to be hoped that the practical need to make judgments for early regulation of particulate sulfate in California will not provide a public disincentive to pursue the more concise and accurate supporting technology needed for intelligent control measures of specific contaminants.

REFERENCES

1. Appel, B. et al. 1976. "Sulfate Formation from Stationary SO_2 Sources in Clean and Polluted Air," Final Report, Interagency Agreement ARB-948, Air and Industrial Hygiene Laboratory, California State Department of Health, Berkeley, California.

2. Appel, B. et al. 1977. "A Comparative Study of Wet Chemical and Instrumental Methods for Sulfate Determination," American Chemical Society, New Orleans, LA, Preprints, Div. Environ. Chem., p. 117.

3. Appel, B. et al. 1977. "Characterization of Organic Particulate Matter," Final Report to California Air Resources Board under Contract No. 5-682.

4. Appel, B. et al. 1977. "Effect of Environmental Variables and Sampling Media on the Collection of Atmospheric Sulfate and Nitrate," Contract No. ARB 5-1032 (report in preparation).

5. Burton, C. S., Jerskey, T. N., and Reynolds, S. D. 1976. "A Preliminary Investigation of Expected Visibility Improvements in the Los Angeles Basin from Oxidant Precursor Gases and Particulate Emission Control," presented at International Symposium on Oxidants, Raleigh-Durham, NC, September.

6. Burton, C. S. and Hidy, G. M. 1975. "Regional Air Pollution Study Program Objectives and Plans," Environmental Protection Agency Rep. EPA-650/3-75-009 (NTIS PB-247-769), Rockwell International Science Center, Thousand Oaks, CA.

7. California Public Utilities Commission (CPUC). 1976. "Hearings on CARB Petition for Natural Gas Reallocation from Northern California to Southern California," CPUC Case No. 9642, Sacramento, CA.

8. Cass, G. 1975. "Dimensions of the Los Angeles SO_2/Sulfate Problem," No. 15, Environmental Quality Laboratory, California Institute of Technology, Pasadena, CA.

9. Charlson, R. J. et al. 1974. Atmos. Environ. 8, 1257.

10. Crittendon, A. L. 1976. "Analysis of Atmospheric Organic Aerosols by Mass Spectroscopy," U. S. Environmental Protection Agency Rep. EPA-600/3-76-093, Environmental Science Research Laboratory, Research Triangle, NC.

11. Cunningham, P. and Holt, B. D. 1975. "Chemical Engineering, Division of Environmental Chemistry," Annual Report 1974-1975, Argonne National Laboratory Rep. ANL-75-51, Argonne, ILL.

12. Flocchini, R. et al. 1976. Environ. Sci. Technol. 10, 76-82.

13. Heisler, S. L. et al. 1978. "Investigation of Aerosol Chemistry in the Los Angeles Area," Rep. P-5618 to Southern California Edison Co., Environmental Research and Technology, Inc., Westlake Village, CA.

14. Hidy, G. M. et al. 1976. "Design of the Sulfate Regional Experiment (SURE)," Vol. I, Electric Power Research Inst. Rep. EC-125, Palo Alto, CA.

15. Hidy, G. M. et al. 1977. "Preliminary Air Quality Analysis of Urban and Regional Sulfate Distributions," Report No. 4295, American Petroleum Institute, Washington, D. C.

16. Holmes, J. 1975. "An Assessment of the Aerosol-Visibility Problem in the South Coast Air Basin," California Air Resources Board Staff Rep. 75-02-03, Sacramento, CA.

17. Husar, R. B. et al. 1978. Sulfur Budget of a Power Plant Plume, Atmos. Environ. 12, 549-568.

18. Joint Project. 1977. "Interim Report, Joint Project for Relating Emissions of Sulfur Dioxide to Air Quality," prepared by Environmental Research and Technology, Inc., for the South Coast Air Quality Management District, Rosemead, CA.

19. Moskowitz, A. H. 1976. "The Distribution of Aerosol Nitrate Compounds with Respect to Particle Size Vaporization Analysis with the Low Pressure Impactor," Master's Thesis, California Institute of Technology, Pasadena, CA.

20. Mueller, P. K. 1976. Sulfate Measurement Technology, Appendix D in "Design of the Sulfate Regional Experiment (SURE)," Vol. III, Electric Power Research Institute Rep. EC-125, Palo Alto, CA, p. D1 et seq.

21. National Academy of Sciences. 1974. "Air Quality and Automobile Emission Control," Vol. III: "Relationship of Emissions to Ambient Air Quality," Report by the Coordinating Committee on Air Quality Studies for the Committee on Public Works, U. S. Senate, Serial No. 93-24, U. S. Government Printing Office, Washington, D. C., 137 pp.

22. Pitts, J. N., Jr. et al. 1976. "Detailed Characterization of Gaseous and Site-Resolved Particulate Pollutants at a South Coast Air Basin Smog Receptor Site: Levels and Modes of Formation of Sulfate, Nitrate and Organic Particulates and Implications for Control Strategies," Report, Contract No. ARB 5-387, Statewide Air Pollution Research Center, University of California, Riverside, CA.

23. Richards, L. W., Avol, E., and Harker, A. B. 1976. "The Chemistry, Dispersion, and Transport of Air Pollutants Emitted from Fossil Fuel Power Plants in California: Ground Level Pollutant Measurement and Analysis," Final Report, Contract No. ARB 3-916, Rockwell International, Air Monitoring Center, Newbury Park, CA.

24. Smith, T. B. et al. 1976. "The Chemistry, Dispersion, and Transport of Air Pollutants Emitted from Fossil Fuel Power Plants in California: Airborne Pollutant Measurement and Analysis," Final Report, Contract No. ARB 3-929, Meteorology Research, Inc., Altadena, CA.

25. Spicer, C. et al. 1976. "The Fate of Nitrogen Oxides Gases in the Atmosphere," Report to Coordinating Research Council, Battelle Memorial Institute, Columbus, OH.

26. Stevens, R. K. et al. 1978. Sampling and Analysis of Atmospheric Sulfates and Related Species, Atmos. Environ. 12, 55-68.

27. Waggoner, A. et al. 1976. Nature 261, 120-122.

28. Wesolowski, J. J. et al. 1977. Collection Surfaces of Cascade Impactors, in X-Ray Fluorescence Analysis of Environmental Samples, (T. G. Dzubay, Ed.), Ann Arbor, Science, Ann Arbor, MI.

29. White, W. et al. 1978. The Same Day Impact of Power Plant Emissions on Sulfate Levels in the Los Angeles Air Basin, Atmos. Environ. 12, 779-784.

30. Wilson, W. E., Jr. et al. 1976. "Sulfate in the Atmosphere," presented at the 69th Annual Meeting of the Air Pollution Control Association, Portland, OR.

Index